OPTIC FLOW AND BEYOND

SYNTHESE LIBRARY

STUDIES IN EPISTEMOLOGY,

LOGIC, METHODOLOGY, AND PHILOSOPHY OF SCIENCE

VOLUME 324

OPTIC FLOW AND BEYOND

Edited by

LUCIA M. VAINA
Boston University and Harvard University,
U.S.A.

SCOTT A. BEARDSLEY
Boston University,
U.S.A.

and

SIMON K. RUSHTON
Cardiff University,
U.K.

SPRINGER SCIENCE+BUSINESS MEDIA, B.V.

A C.I.P. Catalogue record for this book is available from the Library of Congress.

ISBN 978-90-481-6589-6 ISBN 978-1-4020-2092-6 (eBook)
DOI 10.1007/978-1-4020-2092-6

Printed on acid-free paper

Contents

Contributors

José F. Barraza Department of Biomedical Engineering, University of Southern California, University Park, Olin Hall of Engineering 500, 3650 S. McClintock Avenue, Los Angeles, CA 90089, USA

Andrew C. Beall Department of Psychology, University of California, Santa Barbara, CA 93106, USA

Scott A. Beardsley Brain and Vison Research Laboratory, Department of Biomedical Engineering, Boston University, 44 Cummington Street, Boston, MA 02215, USA

Jaap A. Beintema Functionele Neurobiologie, Universiteit Utrecht, Hugo R. Kruijtgebouw, Padualaan 8, 3584 CH Utrecht, The Netherlands

C. Busettini Department of Physiological Optics, University of Alabama at Birmingham, Birminham, AL 35294, USA

James E. Cutting Department of Psychology, Cornell University, Ithaca, NY 14853, USA

Charles J. Duffy Departments of Neurology, Neurobiology and Anatomy, Ophthalmology, Brain and Cognitive Sciences, and The Center for Visual Science, University of Rochester Medical Center, 601 Elmwood Ave, Box 673 Rochester, NY 14642, USA

Brett R. Fajen Department of Cognitive Science, Rensselaer Polytechnic Institute, Carnegie 108, 110 8th Street, Troy, NY 12180, USA

Martin H. Fischer Department of Psychology, University of Dundee, Dundee, DD1 4HN, Scotland, UK

Martin A. Giese ARL, Department of Cognitive Neurology, University Clinic Tübingen, Max Planck Institute for Biological Cybernetics, Spemannstr. 34, D-72076, Tübingen, Germany

Norberto M. Grzywacz Department of Biomedical Engineering, University of Southern California, University Park, Olin Hall of Engineering 500, 3650 S. McClintock Avenue, Los Angeles, CA, USA

Julie M. Harris School of Biology (Psychology), King George VI building, University of Newcastle-upon-Tyne, Newcastle-upon-Tyne NE1 7RU, UK

Heiko Hecht Psychologisches Institut, Johannes-Gutenberg Universität, D-55099 Mainz, Germany

Marta Iordanova Department of Psychology, Concordia University, Drummond Science Building, Rm. 413, 7141 Sherbrooke Street West, Montréal, Québec H4B 1R6, Canada

Michael S. Langer School of Computer Science, McGill University, 3480 University Street, Room 410, Montréal, Québec H3A 2A7, Canada

Markus Lappe Allgemeine Zoölogie & Neurobiologie, Ruhr-Universität Bochum, D-44780 Bochum, Germany

M. Anthony Lewis Iguana Robotics, Inc., P.O. Box 625, Urbana, IL 61803, USA

Jack M. Loomis Department of Psychology, University of California, Santa Barbara, CA 93160-9660, USA

Richard Mann School of Computer Science, University of Waterloo, 200 University Avenue West, Waterloo, Ontario N2L 3G1, Canada

Guillaume S. Masson Institut de Neurosciences Physiologiques et Cognitivesm, C.N.R.S., Marseilles 13402 cedex 20, France

Fred A. Miles Laboratory of Sensorimotor Research, National Eye Institute, Building 49, 49 Convent Drive, Bethesda, MD 20892-4435, USA

William K. Page Department of Neurology, and The Center for Visual Science, University of Rochester Medical Center, Rochester, NY 14642, USA

Aftab E. Patla Gait & Posture Lab, Department of Kinesiology, University of Waterloo, 200 University Ave. W., Waterloo, Ontario N2L 3G1, Canada

Milena Raffi Center for Molecular and Behavioral Neuroscience, Rutgers University, 197 University Avenue, Newark, NJ 07102, USA

Constance S. Royden Department of Mathematics and Computer Science, College of the Holy Cross, 1 College Street, Worcester, MA 01610, USA

Simon K. Rushton Centre for Vision Research & Department of Psychology, York University, 4700 Keele Street, Toronto, Ontario M3J 1P3, Canada

Ralph M. Siegel Center for Molecular and Behavioral Neuroscience, Rutgers University, 197 University Avenue, Newark, NJ 07102, USA

Sergei Soloviev Brain and Vision Research Laboratory, Department of Biomedical Engineering, Boston University, 44 Cummington Street, Boston, MA 02215, USA

Venkataraman Sundareswaran Rockwell Scientific Company, 1049 Camino Dos Rios, Thousand Oaks, CA 91360, USA

Lucia M. Vaina Brain and Vision Research Laboratory, Department of Biomedical Engineering, Boston University, 44 Cummington Street, Boston, MA 02215, USA

Albert V. van den Berg Functionele Neurobiologie, Universiteit Utrecht, Hugo R. Kruijtgebouw, Padualaan 8, 3584 CH Utrecht, The Netherlands

Michael W. von Grünau Department of Psychology, Concordia University, Drummond Science Building, Rm. 413, 7141 Sherbrooke Street West, Montréal, Québec, H4B 1R6, Canada

Ranxiao F. Wang Department of Psychology, University of Illinois, 603 E. Daniel Street, Champaign, IL 61820, USA

John P. Wann School of Psychology, University of Reading, Harry Pitt Building, Earley Gate, Whiteknights Road, Reading, Berkshire RG6 6AL, UK

William H. Warren Department of Cognitive & Linguistic Sciences, Brown University, Box 1978, Providence, RI 02912, USA

Richard M. Wilkie School of Psychology, University of Reading, Harry Pitt Building, Earley Gate, Whiteknights Road, Reading, Berkshire RG6 6AL, UK

D.S. Yang Department of Ophthalmology, Columbus Children's Hospital, Columbus, OH 43205, USA

Preface

The idea that locomotion is guided by optic flow is very compelling. It is based upon the recognition that movement of an observation point produces characteristic patterns of visual motion sampled at the observation point. For example a radial expansion pattern indicates the observation point is moving forward, and position of the centre of the radial pattern, relative to images of scene objects, is informative about the trajectory through the environment. It doesn't matter if the optical device that samples the visual motion is an eye or a camera, whether the locomotion is by foot or by wheel, which direction the eye or camera is pointing. A simple pattern of motion carries all the information needed to determine direction of locomotor direction and guide target interception.

As with most ideas in vision, there are historical antecedents for the optic flow theory. Ptolemy describes the illusion of self-motion that occurs when a boat is moored in a flowing river (see p.124, Smith, 1996). Indeed it is argued that this example can be traced even further back (footnote 146, p.124, Smith, 1996).

The current period of interest in optic flow begins in the early 1940s with Grindley. In a report to the British Air Ministry he concluded that "a pilot could land an aeroplane safely *on the basis of velocity cues alone*" and provided a schematic flow diagram (see Mollon, 1997). This observation underpins all the work that followed. In 1950 Gibson introduced the idea to the academic world and contemporaneously Calvert (e.g., Calvert, 1950) was publishing in engineering journals similar ideas due to his interest in runways and the landing of planes. Calvert's work has slipped from memory, but maybe someday his work will be revisited and the importance of his contribution quantified. Today however it is indisputably Gibson's name that is associated with optic flow theory, his books and 1958 paper in particular having brought the idea to the scientific community.

Empirical studies on the use of optic flow have been commonplace for a number of decades now and a nice summary of the early work was provided by Cutting (1986). Looking back with hindsight, notable amongst them was Llewellyn's (1971) paper. Llewellyn concluded that observers do not use optic flow and reasoned that it was more likely that they used drift (the change in egocentric direction of an object they were travelling towards). Llewellyn's empirical findings can be accounted for today (his choice of stimuli was a vertical place), but in retrospect his target drift idea can be said to prefigure a more recent suggestion that locomotion is based upon perceived egocentric direction.

Koenderink & van Doorn (1975) provided probably the first mathematical treatment of optic flow and that original work along with a paper by Longuet-Higgins & Prazdny (1980) has formed the basis for most models today.

In 1982 Regan & Beverley published an important paper that prompted a consideration of the importance of eye-movements. They pointed out that the idealised or abstract optic flow pattern that provided the inspiration for the work on optic flow was not actually the flow pattern that was found at the back of the eye. The reason for this is because humans have mobile gaze systems, i.e. eyes and a head that can be rotated. Gaze rotation introduces additional flow that disrupts the use of simple features such as the focus of expansion. This paper should be seen as the progenitor of later work on extra-retinal flow decomposition that was the focus of much of the psychophysical research in the 90's.

In 1988, Warren & Hannon published data that showed that an observer could judge their direction of heading from optic flow with a high degree of accuracy, they also claimed that they could do so solely from optic flow when the flow pattern was 'contaminated' by an eye movement. Following Warren's paper the field took off and since that time there have been a vast number of papers on optic flow – computational modelling, neurophysiology and psychophysics. We will not even attempt to summarise them, many are discussed in chapters of this book and a number of reviews are available elsewhere (e.g., Lappe et al, 1999).

Today we find the field at a very interesting stage, a considerable amount of work has been published on psychophysics, neurophysiology and modelling and the literature is now very comprehensive, also the optic flow and locomotion story is very well known, finding a prominent place in many a textbook. Some lines of research have converged, for example today the importance of relative object motion is broadly recognised after having been championed in comparative isolation by Cutting (e.g., Cutting et al, 1992) for many years, also debates about the role of extra-retinal signals in the interpretation of flow now feel old. However at the same time some of the basic assumptions of the flow theory are being questioned and alternative accounts of the visual guidance of locomotion are being proposed and evaluated. Thus this volume is timely, providing an opportunity at this critical juncture to review what has been learnt and look forward to where the field is going.

This book resulted from a workshop on optic flow held at Boston University in 2001. The workshop was intended to bring together researchers in the Boston area. However, the meeting ended up with some researchers from well beyond the greater Boston area attending. Most of the chapters were contributed by participants at the meeting; others were invited to broaden and round the book.

REFERENCES

Calvert, E.S. (1950). Visual aids for landing in bad visibility, with particular reference to the transition from instrument to visual flight. *Trans. Illum. Eng. Soc. Lond.*, *15,* 183-219.

Cutting, J.E. (1986). *Perception with an eye for motion.* Cambridge: MIT Press.

Cutting, J.E., Springer, K., Baren, P.A., & Johnson, S.H. (1992). Wayfinding on foot from information in retinal, not optical, flow. *J. Exp. Psychol. Hum. Percept. Perform.*, *121,*41-72.

Gibson, J.J. (1958). Visually controlled locomotion and visual orientation in animals. *Br. J. Psychol., 49,* 182-194.

Koenderink, J.J., & van Doorn, A.J., (1975). Invariant features of the motion parallax field due to motion of rigid bodies relative to the observer. *Optica Acta, 22,* 773-791.

Lappe, M., Bremmer, F., & van den Berg, A.V. (1999). Perception of self-motion from visual flow. *Trends Cogn. Sci., 3,* 329-336.

Longuet-Higgins, H.C., & Prazdny, K. (1980). The interpretation of a moving retinal image. *Proc. R. Soc., 208,* 385-397.

Llewellyn, K.R. (1971). Visual guidance of locomotion. *J. Exp. Psychol., 91,* 245-261.

Mollon, J. (1997). "....on the basis of velocity cues alone": some perceptual themes. *Q. J. Exp. Psychol., 50A,* 859-878.

Regan, D., & Beverley, K.I. (1982). How do we avoid confounding the direction we are looking and the direction we are moving. *Science, 215,* 194-196.

Smith, A.M. (1996). *Ptolemy's theory of visual perception: an english translation of the optics with introduction and commentary.* Philadephia: The American Philosophical Society.

Warren, W.H., & Hannon, D.J. (1988). Direction of self-motion is perceived from optical-flow. *Nature, 336,* 162-163.

Section 1

Optic Flow - Neurophysiology & Psychophysics

1. Multiple Cortical Representations of Optic Flow Processing

Milena Raffi and Ralph M. Siegel

Center for Molecular and Behavioral Neuroscience
Rutgers University, Newark, NJ, USA

1 INTRODUCTION

When an observer moves through the environment, moving images form on his or her retinae. The visual perception of self-motion is provided by expanding or contracting visual fields projected on the retina. Gibson (1950) called this particular motion, originated by the observer's own navigation, "Optic Flow". He noted that when an observer moves forward, fixating his or her final destination, the expanding visual field seems to originate from a specific point and he named this point "Focus of Expansion" (FOE). In everyday life, self-motion perception is more complicated, because eye and vestibular movements almost always occur together with the optic flow. For example, during locomotion, we experience retinal flow, composed of the translational and rotational components of eye, head and body movements, and optic flow (Lappe et al., 1999). Although all the self-motion perception mechanisms are not clear yet, it seems that the visual system analyzes the visual component, i.e. the optic flow, first and then it combines the optic flow with the other retinal and extraretinal signals in order to construct a dynamic map of extrapersonal space suitable for self-motion guidance (Regan & Beverley, 1982; Warren & Hannon, 1990; Lappe et al., 1999).

A related issue to self-motion perception is the role of stationary and/or moving objects. Self-motion perception can be affected by these cues (Cutting, 1996; Cutting et al., 1999; Wang & Cutting, 1999; Vaina & Rushton, 2000). Moving objects are parts of the retinal flow and by analyzing speed, location in space and relation with both other objects and the observer, the system can derive detailed information for heading discrimination (Cutting, 1996; Cutting et al., 1999; Wang & Cutting, 1999).

L.M. Vaina, S.A. Beardsley and S.K. Rushton (eds.), Optic Flow and Beyond, 3–22.

Gibson (1950, 1966) initiated a long line of research, on how optic flow can be used for guiding locomotion. Psychophysical studies on both monkeys and humans and neurophysiological studies on birds, cats and monkeys are the two main lines of research followed for answering the question (see Frost et al., 1994, and Sherk & Fowler, 2001 for review). In this chapter we will focus on neurophysiological studies of optic flow on monkeys.

2 FUNCTIONAL CHARACTERISTICS OF OPTIC FLOW CORTICAL NEURONS IN THE MACAQUE

Neuronal activity correlated with the perception of self-motion has been demonstrated in several cortical areas of the monkey and human brain, but the neural mechanisms of this perception are unknown. Psychophysical stimuli simulating self-motion have been used extensively to study the neuronal activity and understand the neural substrate of the optic flow processing.

In monkeys, many areas show optic flow responsiveness. The tuning characteristics of neurons in each area can vary. There has been a substantial consideration of the progressive refinement of feature selectivity in the visual processing stream culminating in optic flow responses in the medial superior temporal (MST) area together with its computational aspect (see Duffy, 2000 and Lappe, 2000 for review). By contrast, there has been almost no discussion of the signal processing performed by visual association areas and beyond. We now know there are neurons in many cortical areas that report different aspects of optic flow in the context of other sensory-motor modalities. Besides MST, these areas (Figure 1) are the ventral intraparietal (VIP) area (Schaafsma & Duysens, 1996; Schaafsma et al., 1997; Bremmer et al., 1997b, 2002a), area 7a (Siegel & Read, 1997a; Read & Siegel, 1997; Merchant et al., 2001), the anterior superior temporal polysensory (STPa) area (Anderson & Siegel, 1999), the caudal portion of area PE (PEc) (Raffi et al., 2002) and the motor cortex (Merchant et al., 2001).

Optic flow is an important visual feature that enables us to perceive and interact with our environment. The manner in which these cortical areas contribute to actual perception remains an open question. In order to describe the perceptual role of the optic flow, particular attention has to be paid to the specific cortico-cortical connections of each area and to the activity of target neurons.

The aim of this chapter is to propose a novel concept of the neuronal processing of optic flow: That no hierarchy is necessary in the utilization of optic flow and that each cortical area uses the processing of optic flow in order to form a stable visuo-spatial representation necessary for the functions particular to that region.

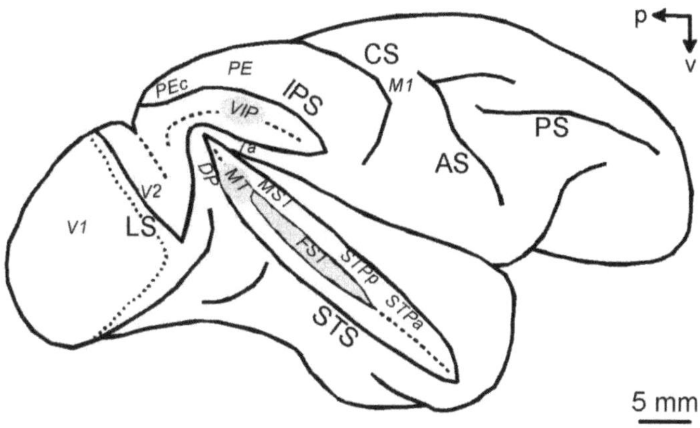

Figure 1. Anatomic location of visual and optic flow areas. Lateral view of a macaque's right hemisphere. The intraparietal sulcus (IPS), superior temporal sulcus (STS) and lunate sulcus (LS) are open in order to show areas lying on their walls. AS: arcuate sulcus; PS: principal sulcus; CS: central sulcus; V1: primary visual area; V2: visual area V2; DP: dorsal prelunate area; 7a: area 7a; PE: parietal area PE (cfr. Von Bonin & Bayley, 1947); PEc: caudal portion of area PE; VIP: ventral intraparietal area; STPa: anterior superior temporal polysensory area; STPp: posterior superior temporal polysensory area; FST: fundus of the superior temporal sulcus; MT: middle temporal area: MST: medial superior temporal area; M1: motor cortex; p: posterior; v: ventral [Modified from Raffi, 2001; used by permission].

The functional characteristics of cortical neurons sensitive to optic flow in these different areas are summarized in Table 1. Areas are discussed in historical precedence. It can be seen that while there are differences in representation, there does not appear to be a strict functional hierarchy.

2.1 The Dorsal Superior Temporal Sulcus: Area MST

Area MST is found in the depths of the superior temporal sulcus (STS) and it is considered part of the parietal stream (Maunsell & Van Essen, 1983). In area MST, a neuronal selectivity for optic flow stimuli was initially described by Saito and co-workers (Saito et al., 1986; Tanaka et al., 1989; Tanaka & Saito, 1989). They suggested that MST neurons perform an optic flow analysis by linearly combining motion information from the middle temporal (MT) neurons.

Duffy and Wurtz (1991a,b, 1995, 1997) performed a series of detailed studies of the functional features of MST cells. MST neurons best respond to one type of optic flow (i.e., rotation, radial or spiral). Moreover, the population shows a *continuum* of responses to different optic flow stimuli and not a clear classification in clusters (Duffy & Wurtz, 1991a). MST neurons

Table 1. Comparison of different functional features in different areas. Y: presence of the effect; N: lack of the effect; RF: receptive field; FLO-G: flow general neuron; FLO-P: flow particular neuron (for FLO-G and FLO-P see Section 2.1.2). **AS99**: Anderson & Siegel, 1999; **B81**: Bruce et al., 1981; **B97**: Bremmer et al., 1997a; **B99**: Bremmer et al., 1999; **B02a**: Bremmer et al., 2002a; **B02b**: Bremmer et al., 2002b; **C93**: Colby et al., 1993; **DS79**: Desimone & Gross, 1979; **DW91a**: Duffy & Wurtz, 1991a; **DW91b**: Duffy & Wurtz, 1991b; **DW95**: Duffy & Wurtz, 1995; **DW97**: Duffy & Wurtz, 1997; **D98**: Duhamel et al., 1998; **G95**: Graziano et al., 1995; **J00**: Jellema et al., 2000; **M01**: Merchant et al., 2001; **PD99**: Page & Duffy, 1999; **PS00**: Phinney & Siegel, 2000; **RS97**: Read & Siegel, 1997; **R02**: Raffi et al, 2002; **RS02**: Raffi & Siegel, 2002; **S96**: Schaafsma & Duysen, 1996; **S96**: Schaafsma et al., 1997; **SR97**: Siegel & Read, 1997a; **S99**: Shenoy et al., 1999; **S02**: Shenoy et al., 2002.

	MST	**VIP**	**7a**	**STPa**	**PEc**	**M1**
Speed	Y–DW95	Expected (C93)	Y–PS00	Unknown	Unknown	Unknown
Heading perception	Y–DW97	Y - B02a	Y–SR97a	Y–AS99	Y–R02	Unknown
Effect of eye position	Expected (B97a)	Expected (B99)	Y–RS97	Y–J00	Expected	Unknown
Effect of eye movement	Y–PD99 Y–S02	Unlikely	Unlikely	Unknown	Unknown	Unknown
Cortical topography	Unknown	Unknown	Y–RS02	Y–DS79	Unknown	Y–M01 Motor Space
FLO-P	Y–DW91a Y–G95	Y–S96, 97 Y–B02a	Y–SR97	Y–AS99	Y–R02	Y–M01
FLO-G	N–DW91a	Unknown	Y–R97	Y–AS99	Expected	Unknown
Tactile	Unknown	Y–D98	Unlikely	Y–DS79 Y–B81	Expected	Unknown
Vestibular	Y–S99	Y–B02b	Unknown	Unknown	Expected	Unknown
Independence from RF features	Y–DW91b	Y–B02	Y–SR97	Y–AS99	Y–R02	Absence of visual RF

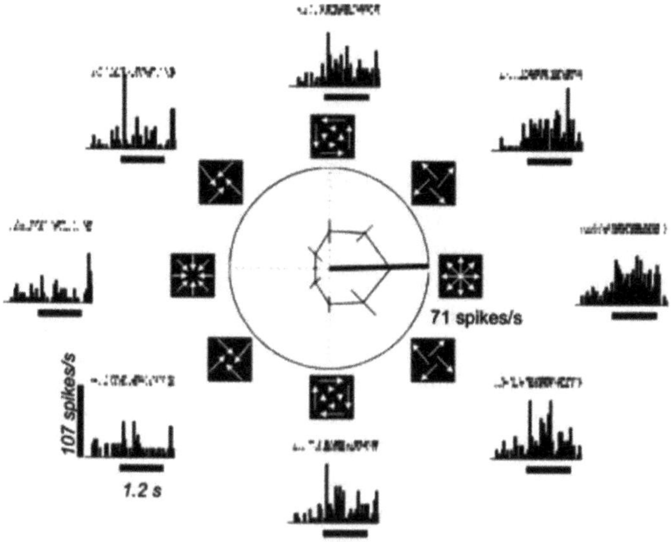

Figure. 2. Response of a single neuron to eight spiral space optic-flow stimuli. Stimulus icons and associated neural responses are positioned around a circle according to their angle in spiral space (e.g., expansion at 0°, counterclockwise rotation at 90°). Spike rasters and peri-stimulus time histograms (PSTHs) indicate that this neuron responds vigorously to the expansion pattern but has little or no response to the contraction pattern. The horizontal bar indicates when the stimulus is visible (1.2 s) and the vertical bar indicates the response scale (107 spikes/s) for all PSTHs (50-ms bins, average of 3 trials). [Modified from Shenoy et al., 2002; used by permission].

are selective for optic flow stimuli, because the responses to such stimuli do not derive from the functional features of the receptive field (Duffy & Wurtz, 1991b). MST cells are selective for the position of the FOE (Duffy & Wurtz, 1995) and tuned for different speeds (Duffy & Wurtz, 1997). This suggests that the MST neuronal population encodes heading during different types of self-motion.

Graziano and co-workers (1994) studied the functional features of MSTd (dorsal MST) neurons using complex optic flow stimuli (i.e., spiral motions). In this study they were interested in determining if the self-motion perception is realized by decomposing the complex scene into each single component. Using spiral motion stimuli they found that many cells in MSTd are preferentially selective for spiral motions and expansion optic flow. This suggests that area MSTd could process self-motion information by using population encoding. An example of spiral motion tuning is illustrated in Figure 2.

Page and Duffy (1999) studied the influence of pursuit eye movement during optic flow stimuli presentation. MSTd neurons change their FOE selective responses during pursuit eye movements, suggesting a role of MSTd neurons in heading stabilization during extra-retinal input. This result has been confirmed in a later paper (Shenoy et al., 2002). Shenoy and co-workers also studied the influence of speed on pursuit eye movements, showing that MSTd neurons change their FOE tuning depending on the pursuit speed. MST neurons are also modulated by vestibular input (Shenoy et al., 1999), and this is not surprising, because a vestibular influence in extrastriate visual areas (i.e., V2 and V3/V3a) has also been reported (Sauvan & Peterhans, 1999). To test the vestibular influence on MST neurons, the monkey was stationary relative to the stimulus, but the whole body was moved while the eyes fixated the target. It seems that area MSTd could use vestibular signals in order to create a compensated heading representation. This is important because heading compensation is perceptually accessible only when additional cues related to active head movements are present.

All of the reported studies have been performed while the monkeys were stationary and psychophysical stimuli simulating self-motion were presented on a screen. Indeed, the monkeys' behavior did not reflect a perception of motion. Recently, Froehler and Duffy (2002) recorded the activity of single neurons while the monkey was moving on a motorized sled looking at a stationary luminous background. Using this system they showed that MST neurons could encode true heading direction by combining vestibular signals with optic flow. However, a population of neurons is activated only when the heading is part of a specific path. This result suggests that MST neurons encode visuo-spatial orientation; the role of attentional and intentional modulation of MST remains unexplored.

2.2 The Intraparietal Sulcus: Area VIP

Area VIP lies on the deep lateral bank of the intraparietal sulcus (IPS) and receives a direct projection from area MST (Boussaoud et al., 1990; Baizer et al., 1991). VIP neurons show directional and speed selectivity to moving visual stimuli (Colby et al., 1993) and they seem to be more selective for expansion optic flow, (Figure 3), (Schaafsma & Duysen, 1996; Bremmer et al., 2002a). In addition they are multimodal, being also activated by tactile (Duhamel et al., 1998) and vestibular stimulation (Bremmer et al., 1997b, 2002b). An eye position tuning for VIP visual neurons has been also described (Bremmer et al., 1999).

Given the presence of tactile receptive fields, VIP neurons could use the optic flow information for movement guidance in the near extrapersonal space. These cells could be involved in the avoidance of collision with near objects in motion. This is a difference between the optic flow analysis

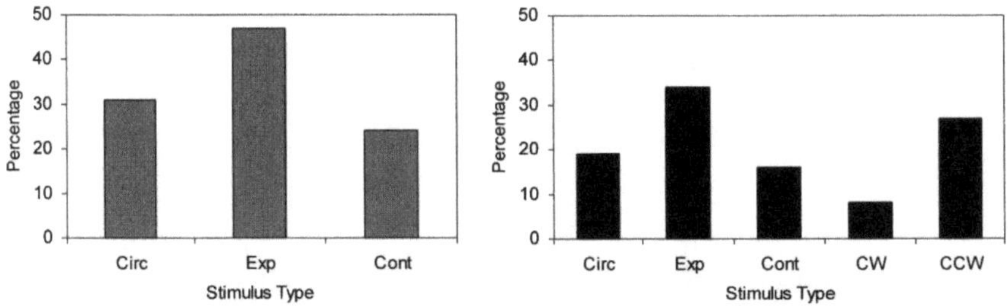

Figure 3. Optic flow tuning of VIP neurons. Distribution of optic flow preferences when cells were tested with three (left panel) or five (right panel) optic flow stimuli. Most neurons had a maximum discharge for expansion stimuli. Circ: circular motion; Exp: expansion; Cont: contraction; CW: rotation clockwise; CCW: rotation counterclockwise. [Redrawn from Bremmer et al., 2002a; used by permission].

performed in MST and VIP, because MST seems to be involved in the analysis of heading perception in the far extrapersonal space while VIP is analyzing space closer to the animal (Bremmer et al., 1997b; Schaafsma et al., 1997).

2.3 Inferior Parietal Lobule: Area 7a

A number of studies have demonstrated that the visual analysis of self-motion has substantial expression in the area 7a (see Siegel & Read, 1997b for review). Area 7a neurons show several functional features important for spatial analysis. Firstly, 7a neurons show gain fields, i.e. the activity of the neuron is enhanced when the gaze is in a particular zone of the visual field (Andersen et al., 1985). Motter and Mountcastle (1981) described neurons that respond to inward or outward moving stimuli with respect to the fovea; these neurons have been called opponent vector neurons. In a following paper Steinmetz and co-workers (1987) postulated that the radial organization of the receptive fields in the opponent vector neurons could be important for the optic flow responsiveness.

Subsequent studies have shown that area 7a neurons, while selective to optic flow, do not derive these properties from opponent vector organization (Siegel & Read, 1997a). Studying optic flow, two new properties appear in area 7a. Firstly, two new classes of motion selectivity termed “Flow General Neurons” (FLO-G) and “Flow Particular Neurons” (FLO-P) emerge. The FLO-G neurons are able to distinguish between particular types of optic flow (i.e., radial vs. rotation), while the FLO-P are tuned for a particular direction of optic flow (i.e., expansion vs. contraction), (Siegel & Read, 1997).

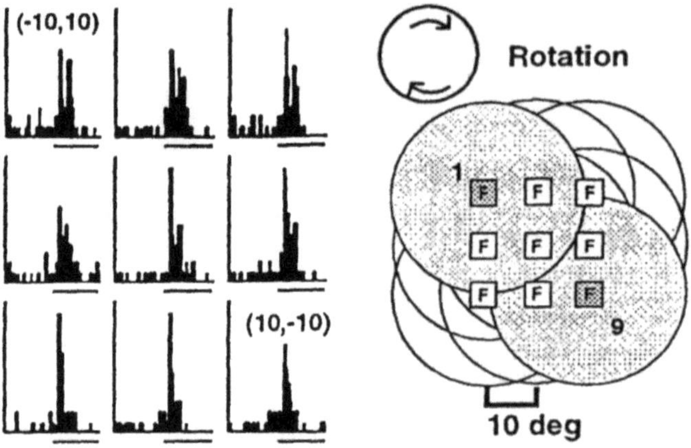

Figure 4. On the left, neuronal response from a single neuron to a clockwise rotation optic flow with a modified angle-of-gaze. The horizontal bar indicates 1 sec. The vertical bar indicates 90 Hz. On the right, a schematic of the fixation point and optic flow stimulus is shown. The fixation point is illustrated by a square (F) while the structured optic flow is a circle. The displacement step for fixation and display is given in degrees of arc and variation of the fixation point occurs together with the concentric stimulus. [Modified from Read & Siegel, 1997; used by permission].

Presumably FLO-G cells are formed through a superimposition of FLO-P cells existing in either MST or 7a. Optic flow responses are modulated by the retinotopic position of the stimulus (Read & Siegel, 1997) and the neurons' responses are not explained by the functional feature of the receptive field (Siegel & Read, 1997a). The second new property is that 7a optic flow neurons show angle of gaze tuning (Read & Siegel, 1997), (Figure 4). Like MST, 7a neurons are tuned for speed (Phinney & Siegel, 2000). Interestingly, this speed selectivity is dynamic. Area 7a neurons can change their speed tuning during a trial, and speed dependence and optic flow selectivity interact (i.e., a neuron could respond to one type of optic flow at a slow speed and a different type of optic flow at a fast speed).

All of the described properties of 7a neurons indicate that they may play a role in the speed representation of multiple objects. Hence, area 7a neurons do not seem to be involved in the analysis of the optic flow direction. They seem to utilize the analysis of optic flow in order to make a spatial representation of extra-personal space.

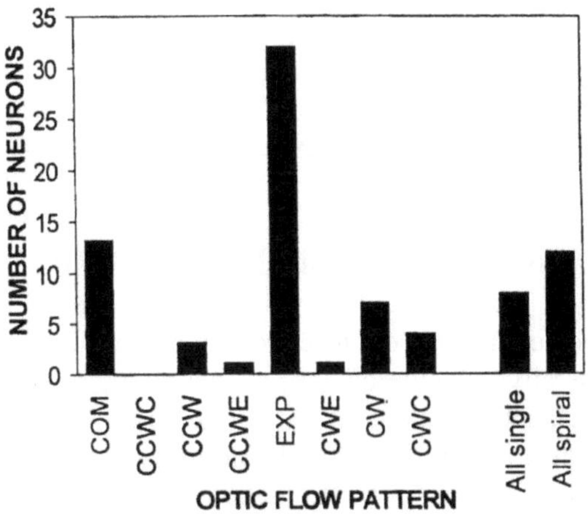

Figure 5. Distribution of neurons responding maximally to the onset of each optic flow display. Neurons were tested with eight different displays. COM: compression; CCWC: counterclockwise compression; CCW: counterclockwise; CCWE: counterclockwise expansion; EXP: expansion; CWE: clockwise expansion; CW: clockwise; CWC: clockwise compression. All single and all spiral are cells that had significantly different responses in firing rate to the single-component compared to the spiral display, but did not show differences for a specific single component or spiral display. [Modified from Anderson & Siegel, 1999; used by permission].

2.4 The Temporal Cortex: STPa

More ventral to area MST in the STS lie the superior temporal polysensory (STP) area (Bruce et al., 1981), and the fundus superior temporal (FST) area (Desimone & Ungerleider, 1986). Bruce and co-workers (1981) described STP neurons activated by visual moving stimuli. On the basis of different cortico-cortical connections (Seltzer & Pandya, 1989a,b; Boussaoud et al., 1990), STP was divided into an anterior (STPa) and a posterior (STPp) portion, or upper and lower banks. A substantial sub-parcellation of the STS based upon cytoatchitectonics, myeloarchitectonics and connectional criteria has been also done by Cusick and co-workers (1995).

Area STPa receives projections from area MST and 7a (Andersen et al., 1990; Boussaoud et al., 1990; Baizer et al., 1991; Cusick et al., 1995) and its neurons respond during object- and self-motion (Hietanen & Perrett, 1996). STPa cells fire when an object is moving in a particular direction during self-motion in the opposite direction, but they do not fire when self- and object-motion occur at the same speed and direction (so to remove observer-relative motion cues). This suggests that STPa neurons process self-motion perception (Hietanen & Perrett, 1996). STPa neurons also respond to different optic flow stimuli (Anderson & Siegel, 1999). Although these neurons have been tested with several stimuli, including complex motion (i.e. spiral stimuli), they seem to prefer radial expansion (Figure 5). This could mean that STPa, unlike MST, does not analyze the direction of motion, but the coding of specific signals are used to control forward locomotion (Anderson & Siegel, 1999).

2.5 The Superior Parietal Lobule: Area PEc

Area PEc, located in the dorso-caudal portion of the superior parietal lobule, was classically considered as a division of somato-sensory cortex (Duffy & Burchfiel, 1971; Sakata et al., 1973; Mountcastle et al., 1975), but recent physiological findings show that its neurons are activated by optic flow stimuli (Raffi et al., 2002). Area PEc sends a direct input to the premotor cortex (Matelli et al., 1998). PEc neurons seem to be involved in the integration of visuo-motor signals because their neuronal activity is related to visual stimuli and hand-movements (Squatrito et al., 2001; Ferraina et al., 2001). The optic flow processing contributes to the occipito-frontal cortical stream, linking visual input to motor output. The optic flow responsiveness is not entirely explained by the small field sensitivity in PEc cells (Figure 6), suggesting that optic flow and object motion responses might serve different mechanisms in the integration of visuo-motor signals to prepare body movements.

PEc neurons represent forward and backward body movements (Figure 7). Like MST and VIP, the selectivity for the FOE position with respect to the fovea could provide the system with detailed heading information.

2.6 The Motor Cortex

The motor cortex (M1) is a recent and wholly unexpected addition to the pantheon of optic flow areas. It is well known that M1 contains a specific movement representation and that each motor region plays a specific role in motor control based on the specific cortico-cortical connections (see Rizzolatti et al., 1998 for review). It was recently determined that M1 neurons also respond to optic flow (Merchant et al., 2001). M1 optic flow neurons show a bias for expansion optic flow, although they do not possess a clear visual receptive field; therefore they are optic flow specific. A large number of M1 neurons respond to only one optic flow stimulus (Figure 8). Like other posterior areas, M1 neurons seem to have a small bias for radial expansion, although this effect is not as strong as in VIP and STPa. In comparison with area 7a (Figure 8), a difference in processing appears. Area 7a is more visual

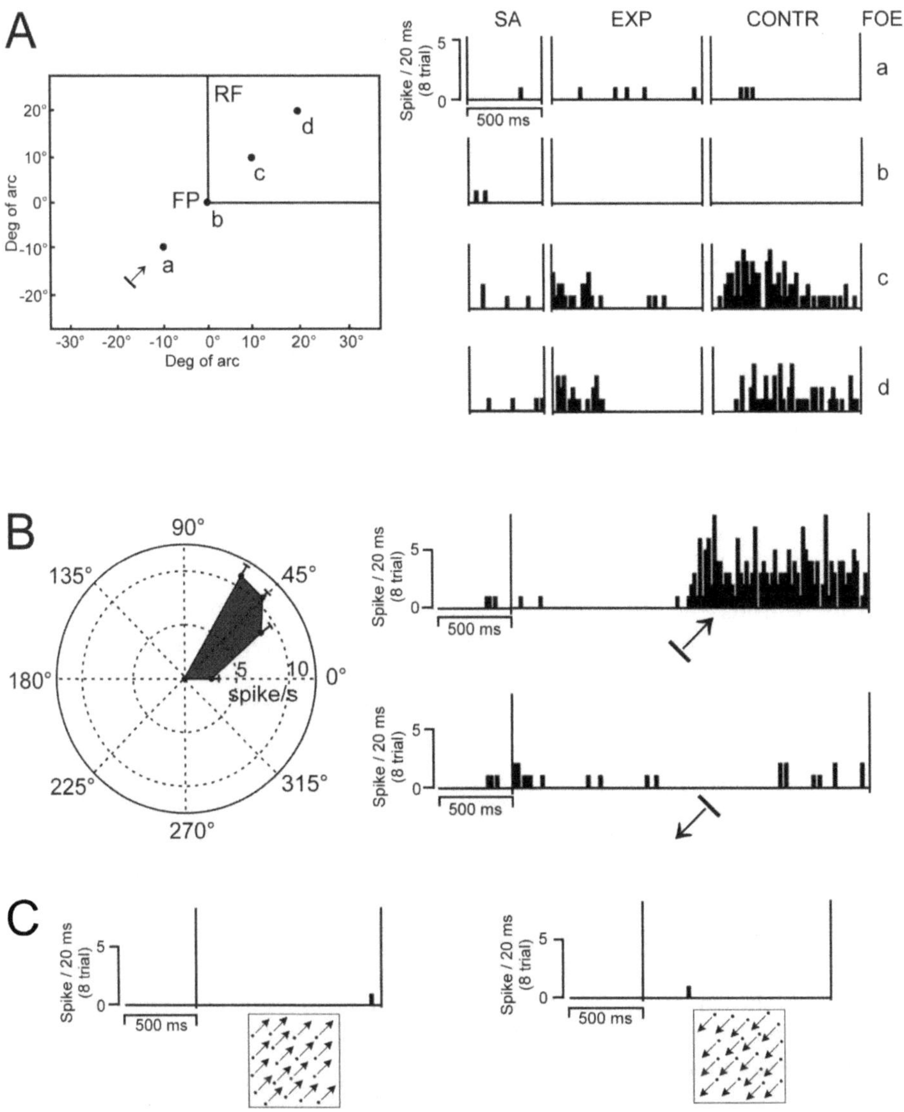

Figure 6. Neuronal response to optic flow, planar motion and a moving bar. A. Radial optic flow stimulation. On the left of the figure, the receptive field (RF) and four different positions of the focus of expansion (FOE) are indicated with a, b, c and d. In b, the FOE and the fixation point (FP) are in the same location. On the right side, responses to expansion (EXP) and contraction (CONTR), compared to the spontaneous activity (SA), are shown by Peri-Stimulus Time Histograms (PSTH). B. Moving bar stimulation. The polar-plot on the left shows the directional tuning. The PSTH on the right illustrates the neuron's response in the 45° and 225° directions. C. Planar motion stimulation. The void PSTH illustrates the lack of neuron's response to random-dot backgrounds moving at 45° (left) and 225° (right) directions. [From Raffi et al., 2002; used by permission].

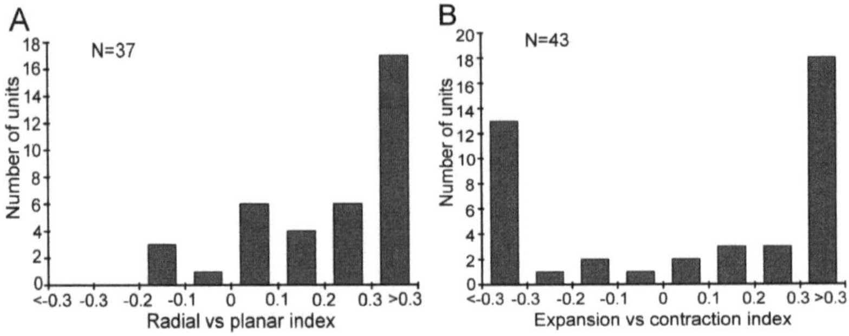

Figure 7. Comparison of firing rates for radial optic flow and planar motion A. Frequency distribution histogram of activity indices, PD= $(St1_{max} - St2_{max})/(St1_{max} + St2_{max})$, computed for the best direction (expansion or contraction) of radial optic flow versus preferred direction of planar motion. Positive values on the abscissa indicate a higher firing rate with radial than with planar motion; negative values the opposite. B. Histogram of indices computed for expansion versus contraction of radial optic flow. Positive values on the abscissa indicate greater firing rates during expansion than during contraction. Negative values the opposite. [Redrawn from Raffi et al., 2002; used by permission].

and its neurons encode different types of visual motion. Area M1 is clearly motor and its neurons are tuned to detect a particular type of motion in order to guide an appropriate movement.

It is well known that each area of the parietal cortex is connected with a distinctive set of frontal areas involved in processing different sensory information (Cavada & Goldman-Rakic, 1989). Therefore, the parieto-frontal circuit is formed by a series of segregated anatomical circuits involved in specific sensorimotor transformations. These circuits are believed responsible for the transformation from sensory input into action. This has been confirmed by analyzing the latency at the onset of the optic flow stimulus. The shortest latencies are shown by MT and MST neurons (Lagae et al., 1994; Duffy & Wurtz, 1997). Longer latencies are found in 7a neurons, while the longest is shown by M1 neurons (Merchant et al., 2001). The result of a comparison of latency between area 7a and M1 shows that M1 neurons have a longer latency of about 40 ms (Figure 9).

3 HOW THE BRAIN COULD USE OPTIC FLOW

It is clear that the representation of optic flow is very common in cortex. However, the tuning of each region differs. Area MST seems to be involved in the first analysis for self-motion perception. Area STPa seems to be tuned to forward locomotion and could be integrating this with other temporally coded signals (e.g., form). Area 7a seems to use optic flow processing in order

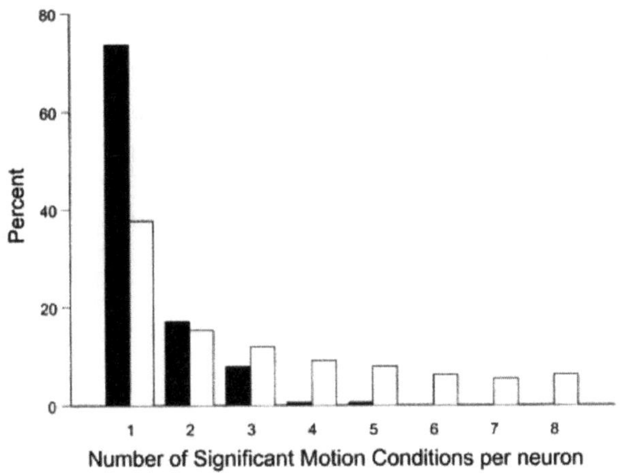

Figure 8. Percentages of times that a cell showed significant responses in the Poisson train analysis to different numbers of stimulus motion conditions in the motor cortex (black square, n=152) and area 7a (white square, n=353). [From Merchant et al., 2001; used by permission].

to make a spatial representation of extra-personal space. Area VIP seems to integrate somatosensory cues and optic flow signals in order to guide movements in near extrapersonal space. Area PEc could play a role in planning and control of navigation, sending signals related to a particular aspect of optic flow to premotor neurons.

Based on a hierarchical model of visual processing, much of the representation beyond area MST does not seem right. Why should STPa, which receives inputs from area 7a, lack FLO-G cells? The tuning of near-extrapersonal space representations in area VIP appears lost in area 7a. Area STP is deep in the temporal lobe, presumably segregated for form analysis, yet it responds well to motion. Why does area M1 represent flow? It is not even visual in a classical sense (i.e. absence of visual receptive fields). About the only results that consistently seems to fit, is the over representation of expansion flow. Many areas seem to be more selective for expansion optic flow. This seems to be consistent with respect to the daily behavior. In both monkey and human, forward locomotion is predominant. These functional properties of optic flow areas certainly suggest that optic flow processing is used in action and it evolves in some way across cortical areas. Various other somatosensory, visual, proprioceptive and motor eye position signals are incorporated as needed. However a strict hierarchical model is not supported by the existing data.

The final location for the convergence of these signals into a coherent representation is unknown and may indeed never occur. There may be successive integration and modulation ultimately producing a final common

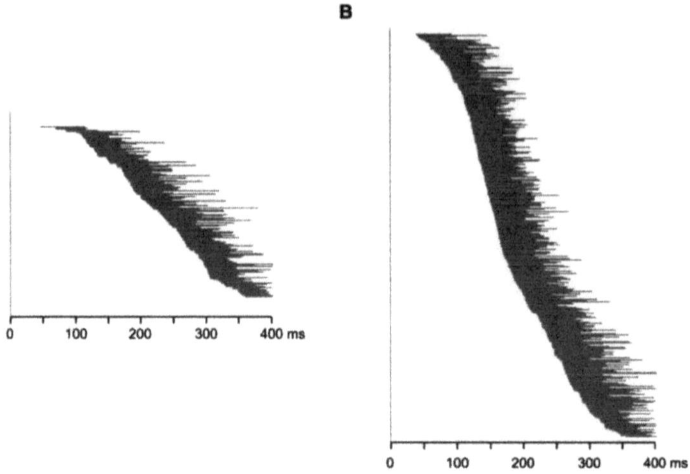

Figure 9. Response onset latency and duration for neurons with a significant activation effect in the Poisson train analysis. A. Motor cortex (n =150); B. Area 7a (n= 353). Each line represents a cell, and its beginning and end are the average times of onset and offset of the response. [Redrawn from Merchant et al., 2001; used by permission].

motor output. It is likely that modification in the optic flow representation arises from substantial mediation by feed-forward and feedback cortico-cortico circuits. As well as sensory modalities, these cortico-cortico signals could support cognitive processing such as attention, intention and memory. Perhaps the optic flow signals found in M1 are a final, highly extracted and abstracted, version and are not visual in a classical sense.

Ultimately, the processing and utilization of optic flow – in each area – involves the unique combination of cortical signals. These combinations require a specific connectivity, not just between pairs of cortical areas, or pairs of neurons, but between specific parts of each cortical representation. In both the early visual system and late motor system, the topography of representation is of the utmost importance. It provides a substrate for efficient connectivity between areas as well as the substrate, which permits the self-organizing principles of cortical development and plasticity to exist (Linsker, 1986, 1990). The topography of the representation of optic flow in each cortical area remains our next challenge in unveiling the cortical analysis of optic flow.

ACKNOWLEDGMENTS

We wish to thank Dr. Barbara Heider and Dr. Kurt F. Ahrens for their comments on this manuscript. Supported by NIH EY09223 and the Whitehall Foundation.

LIST OF ABBREVIATIONS

7a: Area 7a
AS: Arcuate sulcus
CCW: Counterclockwise
CCWC: Counterclockwise compression
CCWE: Counterclockwise expansion
COM: Compression (same as contraction)
CONTR: Contraction
CS: Central sulcus
CW: Clockwise
CWC: Clockwise compression
CWE: Clockwise expansion
DP: Dorsal prelunate area
EXP: Expansion
FOE: Focus of expansion
FLO-G: Flow general neurons
FLO-P: Flow particular neurons
FP: Fixation point
FST: Fundus of the superior temporal sulcus
IPS: Intraparietal sulcus
LS: Lunate sulcus
M1: Motor cortex
MST: Medial superior temporal area
MT: Middle temporal area
p: Posterior
PD: Preferred direction
PE: Parietal area PE
PEc: Caudal portion of area PE
PS: Principal sulcus
PSTH: Peri-stimulus time histogram
RF: Receptive field
SA: Spontaneous activity
STPa: Anterior superior temporal polysensory area

STPp: Posterior superior temporal polysensory area
STS: Superior temporal sulcus
v: Ventral
***V1*:** Primary visual area
V2: Visual area V2
VIP: Ventral intraparietal area

REFERENCES

Andersen, R. A., Asanuma, C., Essick, G. K., & Siegel, R. M. (1990). Cortico-cortical connections of anatomically and physiologically defined subdivisions within inferior parietal lobule. *J. Comp. Neurol., 296*, 65-113.

Andersen, R. A., Essick, G. K., & Siegel, R. M. (1985). Encoding of spatial location by posterior parietal neurons. *Science, 230*, 456-458.

Anderson, K. C., & Siegel, R. M. (1999). Optic flow selectivity in the anterior superior temporal polysensory area, STPa, of the behaving monkey. *J. Neurosci., 19*, 2681-2692.

Baizer, J. S., Ungerleider, L. G., & Desimone, R. (1994). Organization of visual inputs to the inferior temporal and and posterior parietal cortex in macaques. *J. Neurosci., 11*, 168-190.

Bremmer, F., Ilg, U. J., Thiele, A., Distler, C., & Hoffmann, K. P. (1997a). Eye position effects in monkey cortex. I. Visual and pursuit-related activity in extrastriate areas MT and MST. *J. Neurophysiol., 77*, 944-961.

Bremmer, F., Duhamel, J.-R., Ben Hamed, S., & Graf, W. (1997b). The representation of movement in near extrapersonal space in the macaque ventral intraparietal area (VIP). In: P. Thier, & H.O. Karnath (Eds.), *Parietal lobe contributions to orientation in 3-D space.* Exp. Brain Res. 25 (pp. 619-630), Series Springer, Berlin.

Bremmer, F., Graf, W., Ben Hamed, S., & Duhamel, J.-R. (1999). Eye position encoding in the macaque ventral intraparietal area (VIP). *Neuroreport, 10*, 873-878.

Bremmer, F., Duhamel, J.-R., Ben Hamed, S., & Graf, W. (2002a). Heading encoding in the macaque ventral intraparietal area (VIP). *Eur. J. Neurosci., 16*, 1554-1568.

Bremmer, F., Klam, F., Duhamel, J.-R., Ben Hamed, S., & Graf, W. (2002b). Visual vestibular interactive responses in the macaque ventral intraparietal area (VIP). *Eur. J. Neurosci., 16*, 1569-1586.

Bruce, C., Desimone, R., & Gross, C.G. (1981). Visual properties of neurons in a polysensory area in the superior temporal sulcus of the macaque. *J. Neurophysiol., 46,* 369-384.

Boussaoud, D., Ungerleider, L. G., & Desimone, R. (1990). Pathways for motion analysis: cortical connections of the medial superior temporal and fundus of superior temporal visual areas in the macaque. *J. Comp. Neurol., 296,* 462-495.

Cavada, C., & Goldman-Rakic P. S. (1989). Posterior parietal cortex in rhesus monkey: II. Evidence for segregated corticocortical networks linking sensory and limbic areas with the frontal lobe. *J. Comp. Neurol., 287*, 422-445.

Colby, C. L., Duhamel, J-R., & Goldberg, M. E. (1993). Ventral intraparietal area of the macaque: anatomic location and visual response properties. *J. Neurophysiol., 69*, 902-914.

Cusick, C. G., Seltzer, B., Cola, M., & Griggs, E. (1995). Chemoarchitectonics and corticocortical terminations within the superior temporal sulcus of the rhesus monkey: evidence for subdivisions of superior temporal polysensory cortex. *J. Comp. Neurol., 360*, 513-535.

Cutting, J. E. (1996). Wayfinding from multiple sources of local information in retinal flow. *Percept. Perform., 22*, 1299-1313.

Cutting, J. E., Wang, R.F., Fluckiger, M., & Baumberger, B. (1999). Human heading judgements and object-based motion information. *Vision Res., 39*, 1079-1105.

Desimone, R., & Gross, C. G. (1979). Visual areas in the temporal cortex of the macaque. *Brain Res., 178,* 363-380.

Desimone, R., & Ungerleider, L. G. (1986). Multiple visual areas in the caudal superior temporal sulcus of the macaque. *J. Comp. Neurol., 248,* 164-190.

Duffy, C. J. (2000). Optic flow analysis for self-movement perception. In: M. Lappe (Ed.), *Neuronal processing of optic flow.* Int. Rev. Neurobiol. 44 (pp. 199-218). Academic Press.

Duffy, C. J., & Wurtz, R. H. (1991a). Sensitivity of MST neurons to optic flow stimuli. I. A continuum of response selectivity to large field stimuli. *J. Neurophisiol., 65,* 1329-1345.

Duffy, C. J., & Wurtz, R. H. (1991b). Sensitivity of MST neurons to optic flow stimuli. II. Mechanism of response selectivity revealed by small-field stimuli. *J. Neurophisiol., 65,* 1346-1359.

Duffy, C. J., & Wurtz, R. H. (1995). Response of monkey MST neurons to optic flow stimuli with shifted centers of motion. *J. Neurosci., 15*, 5192-5208.

Duffy, C. J., & Wurtz, R. H. (1997). Medial superior temporal area neurons respond to speed patterns in optic flow. *J. Neurosci., 17,* 2839-2851.

Duffy, F. H., & Burchfiel, J. L. (1971). Somatosensory system: organizational hierarchy from single unit in monkey area 5. *Science, 172*, 273-275.

Duhamel, J-R., Colby, C. L., & Goldberg, M. E. (1998). Ventral intraparietal area of the macaque: congruent visual and somatic response properties. *J. Neurophysiol., 79*, 126-136.

Ferraina, S., Battaglia-Mayer, A., Genovesio, A., Marconi, B., Onorati, P., & Caminiti, R. (2001). Early coding of visuomanual coordination during reaching in parietal area PEc. *J. Neurophysiol., 85,* 462-467.

Froehler, M. T., & Duffy, C. J. (2002). Cortical neurons encoding path and place: where you go is where you are. *Science, 295*, 2462-2465.

Frost, B. J., Wylie, D. R., & Wang, Y. C. (1994). The analysis of motion in the visual system of birds. In: P. Green, & M. Davies (Eds.), *Perception and motor control in birds* (pp. 249-266). Springer-Verlag, Berlin.

Gibson, J. J. (1950). *The perception of the visual world.* Boston, Houghton Mifflin.

Gibson, J. J. (1966). *The sense considered as perceptual system.* Boston, Houghton Mifflin.

Graziano, M. S., Andersen, R. A., & Snowden, R. J. (1994). Tuning of MST neurons to spiral motions. *J. Neurosci., 14,* 54-67.

Hietanen, J. K., & Perrett, D. I. (1996). A comparison of visual responses to object- and ego-motion in the macaque superior temporal polysensory area. *Exp. Brain Res., 108*, 341-345.

Jellema, T., Baker, C. I., Wicker, B., & Perrett, D. I. (2000). Neural representation for the perception of the intentionality of actions. *Brain Cogn., 44*, 280-302.

Lagae, L., Maes, H., Raiguel, S., Xiao, D. K., & Orban, G. A., (1994). Responses of macaque STS neurons to optic flow components: a comparison of areas MT and MST, *J. Neurophysiol., 71,* 1597-1626.

Lappe, M., Bremmer, F., & van den Berg, A. V. (1999). Perception of self-motion from visual flow. *Trends Cogn. Science,* 3, 329-336.

Lappe, M. (2000). Computational mechanisms for optic flow analysis in primate cortex. In: M. Lappe (Ed.), *Neuronal processing of optic flow.* Int. Rev. Neurobiol. 44 (pp. 235-268). Academic Press.

Linsker, R. (1986). From basic network principles to neural architecture: emergence of spatial-opponent cells. *Proc. Natl. Acad. Sci. USA, 83,* 7508-7512.

Linsker, R. (1990). Perceptual neural organization: some approaches based on network models and information theory. *Annu. Rev. Neurosci., 13*, 257-281.

Matelli, M., Govoni, P., Galletti, C., Kutz, D. F., & Luppino, G. (1998). Superior area 6 afferents from the superior parietal lobule in the macaque monkey. *J Comp. Neurol., 402*, 327-352.

Maunsell, J. H. R., & Van Essen, D. C. (1983). The connections of the middle temporal visual area (MT) and their relationship to a cortical hierarchy in the macaque monkey. *J. Neurophysiol., 3*, 2563-2586.

Merchant, H., Battaglia-Mayer, A., & Georgopoulos, A. P. (2001). Effects of optic flow in motor cortex and area 7a. *J. Neurophysiol., 86*, 1937–1954.

Motter, B. C., & Mountcastle,. V. B. (1981). The functional properties of the light-sensitive neurons of the posterior parietal cortex studied in waking monkeys: foveal sparing and opponent vector organization. *J. Neurosci., 1*, 3-26.

Mountcastle, V. B., Lynch, J. C., Georgopoulos, A., Sakata, H., & Acuna, C., (1975). Posterior parietal association cortex of the monkey: command functions for operations within extrapersonal space. *J. Neurophysiol., 38*, 871-908.

Page, W. K., & Duffy, C. J. (1999). MST neuronal responses to heading direction during pursuit eye movements. *J. Neurophysiol., 81,* 596-610.

Phinney, R. E., & Siegel, R. M. (2000). Speed selectivity for optic flow in area 7a of the behaving macaque. *Cereb. Cortex., 10*, 413-421.

Raffi, M., (2001). Neuronal responsiveness to classical visual stimuli and optic flow in the superior parietal lobule of the behaving monkey. *Doctoral Thesis.*

Raffi, M., Squatrito, S., & Maioli, M. G. (2002). Neuronal responses to optic flow in the monkey parietal area PEc. *Cereb. Cortex, 12,* 639-646.

Raffi, M., & Siegel, R. M. (2002). A functional architetcture of optic flow in the inferior parietal cortex of the behaving monkey investigated with intrinsic optical imaging. *Soc.*

Neurosci. Abstr. 28 n. 56.9 (32th Annual Meeting of Society for Neuroscience, Orlando, FL, November 02-07, 2002).

Read, H. L., & Siegel, R. M. (1997). Modulation of responses to optic flow in area 7a by retinotopic and oculomotor cues in monkeys. *Cereb. Cortex, 7*, 647-661.

Regan, D., & Beverley, K. I. (1982). How do we avoid confounding the direction we are looking and the direction we are moving? *Science, 215*, 194-196.

Rizzolatti, G., Luppino, G., & Matelli, M. (1998). The organization of the cortical motor system: new concepts. *Electroencephalogy. Clin. Neurophysiol., 106,* 283-96.

Saito, H., Yukie, M., Tanaka, K., Hikosaka, K., Fukada, Y., & Iwai, E. (1986). Integration of direction signals of image motion in the superior temporal sulcus of the macaque monkey. *J. Neurosci., 6,* 145-157.

Sakata, H., Takaoka, Y., Kawarasaki, A., & Shibutani, H. (1973). Somatosensory properties of neurons in the superior parietal cortex (area 5) of the rhesus monkey. *Brain Res., 64*, 85-102.

Sauvan, X. M., & Peterhans, E. (1999). Orientation constancy in neurons of monkey visual cortex. *Visual Cognition, 6*, 43-54.

Schaafsma, S. J., & Duysens, J. (1996). Neurons in the ventral intraparietal area of awake macaque monkey closely resemble neurons in the dorsal part of the medial superior temporal area in their responses to optic flow patterns. *J. Neurophysiol., 76*, 4056-4068.

Schaafsma, S. J., Duysens, J., & Gielen, C. C. (1997). Responses in ventral intraparietal area of awake macaque monkey to optic flow patterns corresponding to rotation of planes in depth can be explained by translation and expansion effects. *Vis. Neurosci., 14* , 633-646.

Seltzer, B., & Pandya, D. N. (1989a). Frontal lobe connections of the superior temporal sulcus in the rhesus monkey. *J. Comp. Neurol., 281*, 97-113.

Seltzer, B., & Pandya, D. N. (1989b). Intrinsic connections and architectonics of the superior temporal sulcus in the rhesus monkey. *J. Comp. Neurol., 290*, 451-471.

Shenoy, K. V., Bradley, D. C., & Andersen, R.A. (1999). Influence of gaze rotation on the visual response of primate MSTd neurons. *J. Neurophysiol., 81*, 2764-2786.

Shenoy, K. V., Crowell, J. A., & Andersen, R. A. (2002). Pursuit speed compensation in cortical area MSTd. *J. Neurophysiol., 88*, 2630-2647.

Sherk, H., & Fowler, G. A. (2001). Neural analysis of visual information during locomotion. *Prog. Brain. Res., 134*, 247-64.

Siegel, R. M., & Read, H. L. (1997a). Analysis of optic flow in the monkey parietal area 7a. *Cereb. Cortex, 7*, 327-346.

Siegel, R. M., & Read, H. L. (1997b). Construction and representation of visual space in the inferior parietal lobule. In: J. Kaas, K. Rockland, & A. Peters (Eds.), *Cereb. Cortex, 12* (pp. 499-525). New York, Plenum Press.

Squatrito, S., Raffi, M., Maioli, M. G., & Battaglia-Mayer, A. (2001). Visual motion responses of neurons in the caudal area PE of the macaque monkeys. *J Neurosci., 21,* RC130 (1-5).

Steinmetz, M. A., Motter, B. C., Duffy, C. J., & Mountcastle, V. B. (1987). Functional properties of parietal visual neurons: radial organization of directionalities within the visual field. *J. Neurosci., 7*, 177-191.

Tanaka, K., Fukada, Y., & Saito, H. (1989). Underlying mechanisms of the response specificity of expansion/contraction and rotation cells in the dorsal part of the medial superior temporal area of the macaque monkey. *J. Neurophysiol., 62*, 642-656.

Tanaka, K., & Saito, H., (1989). Analysis of motion of the visual field by direction, expansion/contraction, and rotation cells clustered in the dorsal part of the medial superior temporal area of the macaque monkey. *J. Neurophysiol., 62*, 626-641.

Vaina, L. M., & Rushton, S. K. (2000). What neurological patients tell us about the use of optic flow. In: M. Lappe, (Ed.), *Neuronal processing of optic flow.* Int. Rev. Neurobiol. 44 (pp. 293-313). Academic Press.

von Bonin, G., & Bailey, P. (1947). *The neocortex of Macaca mulatta.* Urbana, Illinois: Univ. of Illinois Press.

Wang, R. F., & Cutting, J. E. (1999). Where we go with a little good information. *Psychol. Sci., 10*, 71-75.

Warren, W. H. J., & Hannon, D. J. (1990). Eye movements and optical flow. *J. Opt. Soc. Am. Ser. A., 7,* 160-169.

2. Optic Flow and Vestibular Self-Movement Cues: Multi-Sensory Interactions in Cortical Area MST

Charles J. Duffy and William K. Page

Departments of Neurology, Neurobiology and Anatomy, Ophthalmology, Brain and Cognitive Sciences, and The Center for Visual Science
University of Rochester Medical Center, Rochester, NY, USA

1 INTRODUCTION

An efficient observer must optimize the use of sensory signals about self-movement. Optimization includes adaptation to the varying availability of cues in the range of environments that are encountered. Because vision dominates sensation in primates, the critical distinction in self-movement environments is that between self-movement in darkness and in light.

Optic flow, the global pattern of visual motion that results from observer self-movement, is a rich source of information about self-movement. During observer self-movement through a lighted environment optic flow can signal heading direction, environmental layout, and postural stability. As such, optic flow is re-afferent sensory input; it occurs because of the observer's movements and it informs the observer about those movements (Gibson, 1950).

Observer self-movement is also accompanied by vestibular signals about rotation and translation. The vestibular organs provide an acceleration signal that may undergo sub-cortical integration to create mixed acceleration and speed cues. These cues provide a signal about self-movement that endures in darkness, as well as an alternative signal about self-movement in light.

The differing response dynamics of the visual and vestibular systems create the potential for specific circumstances that are best suited to the use of one or the other cue. They also create the potential utility of combining visual and vestibular signals to verify the validity of each and to extend the observer's operational range by using a composite self-movement signal.

Our view is that cerebral cortex serves the task of processing sensory cues to optimize their utility in diverse circumstances that present different cue

L.M. Vaina, S.A. Beardsley and S.K. Rushton (eds.), Optic Flow and Beyond, 23–44.

availabilities and different behavioral demands. In doing so, cortex might flexibly allocate processing capacity to the specific sensory modality that is best suited to the current situation (Duffy & Wurtz, 1994).

This implies the existence of resources that access more than one modality, so that they might support the time-sharing of their processing capacity. This arrangement might also create the opportunity for the dynamic weighting of inputs so that their relative influence on local processing is quantitatively tailored to their suitability in current circumstances. Finally, this implies that such areas in association cortex are best thought of as being devoted to some processing task, rather than to a modality or behavior.

Our work is based-on the view that the medial superior temporal area (MST) is devoted to the processing of self-movement cues for the representation of self-movement heading in support of navigation and orientation. This does not preclude other areas sharing in a distributed representation of self-movement, nor does it preclude MST's participation in distributed systems serving other functions, such as pursuit control and object motion analysis. It does imply that MST accesses multiple cues about self-movement and can exchange or combine those cues depending on circumstances. The following studies continue our efforts to examine these issues. These results have been described previously (Page & Duffy, 2003).

2 METHODS

2.1 Neurophysiology

Most methods applied in these studies are standard in cortical, single-neuron neurophysiology. These methods are described briefly here and are detailed in our other published works (Duffy, 1998; Page & Duffy, 1999) as well as the related works of other authors. Surgical preparations for single-neuron recordings include the implantation of scleral search coils to monitor gaze (Robinson, 1963; Judge et al., 1980) along with a head holder and bilateral recording cylinders placed over trephine holes above area MST (AP –2mm, ML+/-15mm, angle 0). All protocols were approved by the University of Rochester Committee on Animal Research and complied with Public Health Service and Society for Neuroscience policy on laboratory animals.

The monkeys were trained to sit in a primate chair and perform a visual fixation task. Trials began with a stationary, red fixation point centered on a rear projection screen or on the facing wall. The monkey maintained fixation (+/-3°) throughout stimulus presentation to receive a liquid reward. During movement trials, the fixation target moved to remain directly in front of the monkey. During optic flow alone trials, the fixation target remained centered on the rear-projection screen.

Single neurons were recorded using tungsten microelectrodes (FHC, Inc. and Microprobe, Inc.) that were passed through a transdural guide tube in the recording cylinder. The stereotaxic positioning of the recording chambers and the depths of microelectrode penetrations directed neuron recordings into cortical area MST. MST neurons were identified by their physiologic characteristics: large receptive fields ($>20^{\circ 2}$) which included the fixation point, a preference for large moving patterns rather than moving bars or spots, and direction selective responses (Komatsu & Wurtz, 1988b; Duffy & Wurtz, 1991, 1995). Histological analysis confirmed that the neurons studied were located in the anterior bank of the superior temporal sulcus that is included in MSTd (Komatsu & Wurtz, 1988a).

Neuronal activity was averaged over six stimulus presentations for each stimulus. Responses to each stimulus were tested for statistically significant differences from control activity that was recorded in darkness without movement, (Student's t-test, $p \leq 0.05$). Response profiles were tested for significant directional effects using one-way ANOVA ($p \leq 0.05$) with stimulus direction (8 directions) as the between class variable and repeated trials (6 trials) within classes. Significant effects across directions and stimulus conditions were identified using two-way ANOVA ($p \leq 0.05$) with direction and condition as main effects and a direction-by-condition interaction term.

2.2 Stimuli

2.2.1 Translational Movement

The monkey chair, eye coil, and video display systems were mounted on a 1 x 2 m platform on a double-rail drive apparatus (Acutronics, Figure 1). Platform movements were controlled with position feedback from the drive motors, sampled at 125 Hz, and stored as an analog record. Stimulus presentation and experimental conditions were controlled by the pc-based real-time experimental system (Hays et al., 1982).

At the start of each movement trial, the platform was moved 60 cm from the center of the room to one of eight starting positions. The platform then moved on a straight path through the center to stop after a total excursion of 120 cm. During these movements, the platform accelerated at 30 cm/s^2 for one second, maintained a constant velocity of 30 cm/s for three seconds, and then decelerated at 30 cm/s^2 for one second (Figure 1c). Such movements are

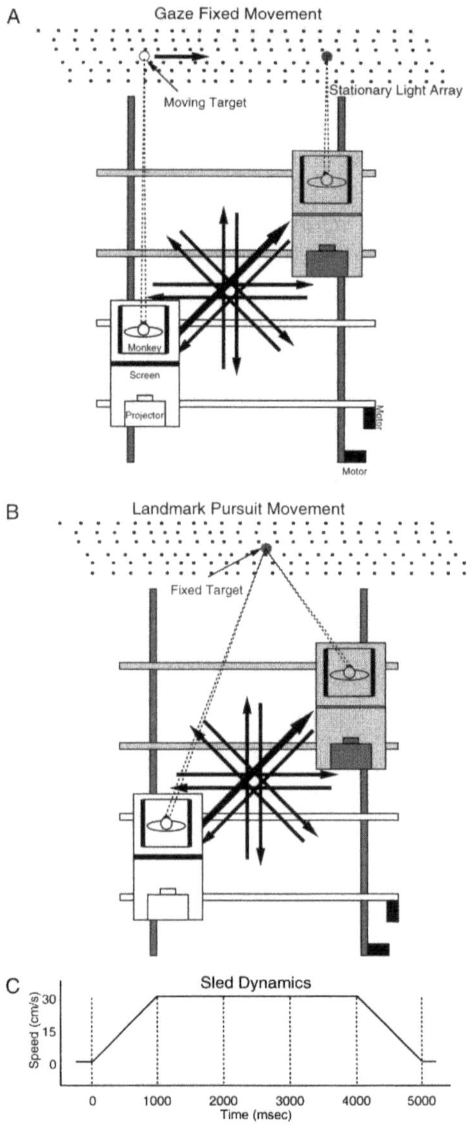

Figure 1. MST neurons were recorded during naturalistic self-movement with gaze fixed straight ahead. A. The two-axis monkey sled moved across the room in eight directions; right-forward movement is illustrated. The monkey continuously viewed the far wall that was covered by 600 small white lights while maintaining neutral gaze throughout the movement by fixating a target that moved to remain directly in front on the animal. B. The optic flow seen by the observer during gaze fixed movement in light. Each box represents the retinal flow field containing visual motion (thin arrows) during self-movement in the indicated direction (thick arrows) with the retinal flow field reflecting the heading of self-movement in the ground plane. C. The speed profile of sled movement for all eight directions included one second of acceleration, three seconds of steady-speed movement, and one second of deceleration.

well above human vestibular thresholds (Benson et al., 1986) and allow stable neuron recordings.

2.2.2 Movement in Light

During movement in light trials, a wall-mounted light array was illuminated. It contained 600 small, white lights uniformly distributed across a 322 cm x 168 cm wall that was 220 cm from the monkey's centered position. These lights were stationary so any visual motion was the result of observer movement. All of the lights were always in the monkey's field of view, although the array subtended different horizontal angles depending on the distance of the sled from the wall.

2.3 Optic Flow

In optic flow only trials, the monkey viewed a rear-projection (Electrohome ECP4100) tangent screen (Figure 1a). Eight optic flow stimuli simulated the visual motion seen by an observer during the eight directions of ground plane linear translation presented in movement trials. The movement platform and fixation target were stationary during these one-second stimuli.

The optic flow stimuli consisted of 500 white dots (0.19° at 2.61 cd/m^2) on a black background (0.18 cd/m^2) stimulating the central 90° X 90° of the visual field. All dots were replaced by lifetime expiration (33 to 1000 ms) or by a smoothing algorithm that maintained a uniform and consistent dot density across the stimulus in all frames. Dots for these radial patterns accelerated as a sine X cosine function of their distance from the focus of expansion maintaining an average speed of 40°/s.

2.4 Data Analysis

The neuronal responses were displayed as polar plots to illustrate the strength and directionality of the responses. In the polar plots, the eight thin radial lines represent responses to each of the eight directions of real or simulated self-movement (Figure 2). The length of the eight radial lines is proportionate to the neuronal firing rate during the corresponding movement direction. The thick radial line in each polar plot indicates the vector sum of the eight individual response vectors. Circular statistical analyses for data sampled at 45° intervals around 360° were used (Batschelet, 1981), including

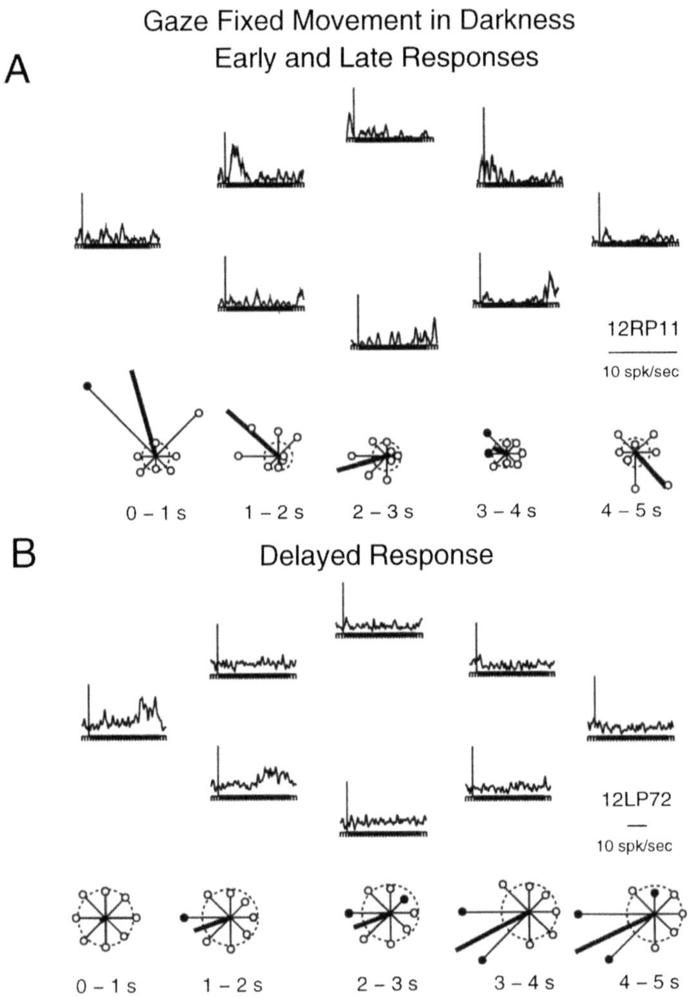

Figure 2. The time-course of MSTd neuronal responses to gaze-fixed movement in darkness with directional responses shown as spike density histograms and polar plots of activity during the five second stimulus. A. A neuron with transient responses to acceleration (left-forward) and deceleration (right-backward). B. A neuron with sustained responses and slowly increasing left-backward heading selectivity. These responses persist during movement without the fixation point (not shown) and are consistent with time-integrated vestibular effects. Significant responses to gaze fixed movement in darkness were obtained from most MST neurons (68/109). The largest number responded around the one second acceleration period. The interval histograms (top) for each neuron (rows) mark all 500 ms intervals (columns) with significant directional responses ($p<0.05$ of Z-value from circular distribution).

the Raleigh Z statistic, to test for significant directionality in a circular response profile ($p \leq 0.05$).

To represent MST's population responses to a given stimulus, we combined the net vectors from the response profiles of each neuron. We applied a vector summation algorithm that includes each neuron's contribution using its preferred heading and the amplitude of its response to the stimulus under consideration (Georgopoulos et al. 1986). Population vectors were displayed in polar plots with thin polar limbs for each neuron's vector and a thick polar limb for the population. The direction of the vector matches the angle at which its limb emanates from the origin. The length of each limb indicates the strength of the response.

3 RESULTS

We studied 130 neurons in three Rhesus monkeys. We used three stimulus conditions to test heading selectivity: movement in darkness, movement in light, and simulated optic flow without movement. The latter two conditions present optic flow with and without vestibular movement signals respectively. In both cases, the monkey sees a focus-of-expansion (FOE) during forward movement and a focus-of-contraction (FOC) during backward movement (Figure 1).

3.1 Movement in Darkness

We assessed the influence of vestibular cues on heading responses by recording MSTd neuronal activity during movement in darkness. Figure 2a shows the directional responses of a neuron with transient activation during left-forward acceleration (0-1 s) that reversed to a right-backward preference during deceleration (4-5 s). This reversal suggests an underlying vestibular mechanism.

In contrast, Figure 2b shows the directional responses of a neuron with more delayed activation during left-backward movement throughout the five second stimulus. This response shows a build-up of activity and an accumulation of direction selectivity across the five second stimulus period. This is consistent with the temporal integration of a direction selective vestibular signal in this neuron.

The latency and duration of the movement in darkness responses varied across neurons. Most neurons (63%, 68/109) showed significant responses to movement in darkness in at least one of ten 500 ms stimulus intervals. These responses usually (59%, 40/68) included one of the first three response intervals corresponding to the 1500 ms period during, and immediately after, translational acceleration (Figure 2c).

The varying time-course of these responses, and their reversal of preferred direction during deceleration, supports their vestibular origin. These responses are not attributable to the vestibulo-ocular reflex (VOR) because fixation control suppressed the VOR. These responses are not attributable to the effects of VOR cancellation because they remained when the fixation point was shut-off during movement. This placed the monkey in complete darkness with no available fixation target but the continued requirement that it maintain steady, centered gaze. These findings confirm the vestibular origin of the movement in darkness responses.

3.2 Movement in Light

Most MST neurons respond more strongly to movement in light than to movement in darkness. Figures 3a and 3b show the movement in light responses of the neurons that were used to illustrate the movement in darkness responses in Figure 2. The transient response neuron shown in Figure 2a clearly maintained its transient activation in spite of the sustained nature of the visual stimulus that accompanied movement in light (Figure 3a).

This neuron also continued to show a reversal of its heading preferences during the deceleration phase of the movement, even though the optic flow continued to reflect the same movement direction during this period. This neuron's directional responses and its direction selectivity were both clearly stronger in light, although the preferred direction was much the same as it was in darkness across all five of the one-second stimulus intervals (Figure 3a).

The sustained response neuron shown in Figure 2b showed a more substantial increase in response amplitude and duration during movement in light (Figure 3b). In this case, the response showed the same direction selectivity it had shown in darkness, but the selectivity was apparent almost immediately after stimulus onset. This direction selectivity increased throughout the acceleration and steady-speed intervals.

These responses also show opposite direction activation very late during the stimulus. This might reflect a reverse direction response during the terminal segment of deceleration. Perhaps more likely is the possibility that they reflect off-responses associated with the end of non-activating visual stimuli, that is, the anti-preferred optic flow stimuli.

The sample of neurons studied with movement in light reflected the effects shown by these examples. There was a substantial increase in the number of responsive neurons from 59% in darkness to 87% (95/109) in light (Figure 3c). There were far more sustained responses in light and almost all of

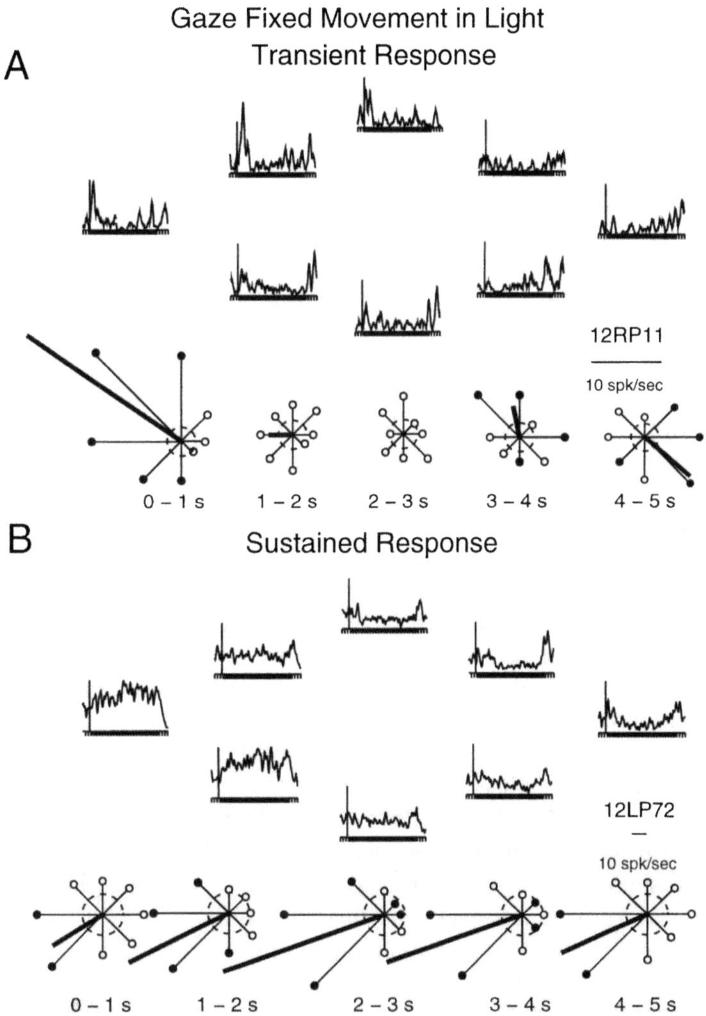

Figure 3. The time-course of MST neuronal responses to gaze fixed movement in light (neuron and format as in Figure 2). A. The same pattern of transient responses to acceleration and deceleration were seen with movement in light and darkness but light evoked much stronger activity. B. The same patterns of sustained responses were seen with movement in light and darkness but light evoked much stronger activity, especially in the acceleration interval. Significant responses to gaze fixed movement in light were obtained from almost all MST neurons (95/109), some showed transient responses like those seen with movement in darkness but many others (35/95) showed sustained responses. Thus, movement in light evoked more acceleration responses than movement in darkness.

the responses began at stimulus onset, during the acceleration period. Thus, MST neurons are more responsive during movement in light than during movement in darkness.

3.3 Optic Flow Effects

MSTd neuronal responses to movement in light were often attributable to the visual effects of optic flow. In many neurons, movement in darkness did not evoke significant responses, whereas optic flow without translational movement evoked robust responses. These directional responses to optic flow were commonly reflected in the responses to movement in light.

Individual neurons showed a large variety of different combinations of vestibular and visual effects across the five one second stimulus periods. The neuron shown in Figure 4a had a left-forward heading direction preference during movement in darkness. During optic flow stimulation without translational movement, this neuron showed a clear preference for left-backward movement.

This neuron's responses to movement in light combined the movement in darkness and optic flow effects in a manner that reflected the relative amplitude, directionality, and time-course of these responses. In light, it initially showed strong left-forward movement preferences. This preference gradually faded to be replaced by the left-backward heading preference much like that seen with optic flow alone.

The neuron illustrated in Figure 4b showed a very different pattern of responses that nonetheless were attributable to interactions between movement in darkness and optic flow effects. During movement in darkness this neuron showed strong left-forward heading selectivity building-up over the five one-second stimulus intervals. In response to optic flow alone, this neuron showed a strong left-backward heading preference.

This neuron's responses to movement in light were seemingly dominated by the optic flow response early in the stimulus with clear left-backward heading responses. Later in the movement in light stimuli, the neuron developed the left-forward heading preference like that seen during movement in darkness. These examples suggest that directional responses to movement in light reflect differences in the time-course and preferred headings of the vestibular and visual responses as quantitative response variation, not as winner-take-all effects.

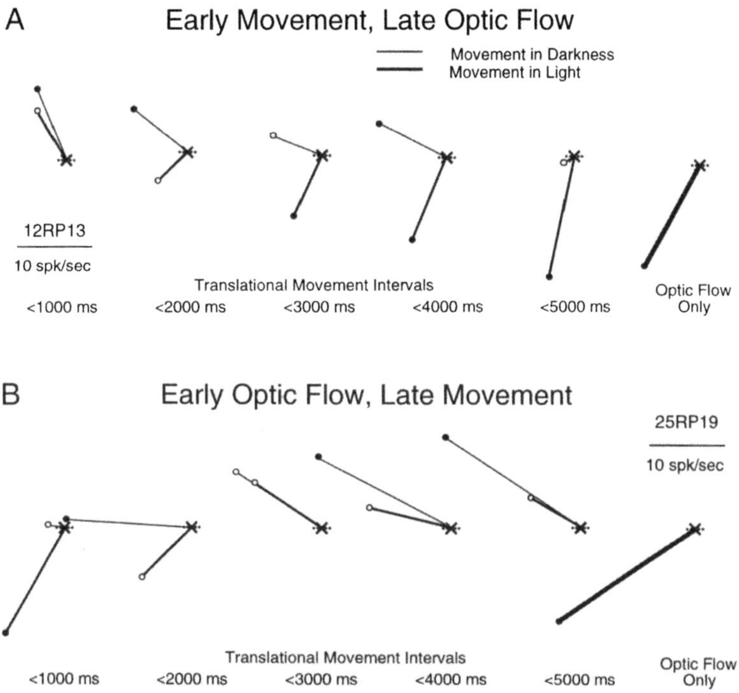

Figure 4. Movement in light responses combine optic flow and movement in darkness responses. A. A neuron with directional net vectors showing early left-forward movement in light heading preferences (thicker lines) like those evoked by movement in darkness (thin lines) and later left-backward heading preferences like those evoked by optic flow (bold line). B. A neuron with directional net vectors showing early left-backward movement in light heading preferences (thicker lines) like those evoked by optic flow (bold line) and later left-forward heading preferences like those evoked by movement in darkness (thin lines).

3.4 Visual and Vestibular Comparisons

We compared the strength of vestibular and visual responses in 66 neurons tested with movement in darkness and optic flow without translational movement. We divided the movement stimulus into five one-second intervals to match the one second duration of the optic flow stimuli (Figure 5a). Most neurons (71%, 47/66), showed significant directional responses (Z of net vector, $p \leq 0.05$), to at least one of the five one-second movement intervals or to optic flow alone. Most of those neurons (62%, 29/47), showed stronger responses to optic flow (vector length contrast >0.2), but a substantial minority (17%, 8/47) responded more strongly during at least

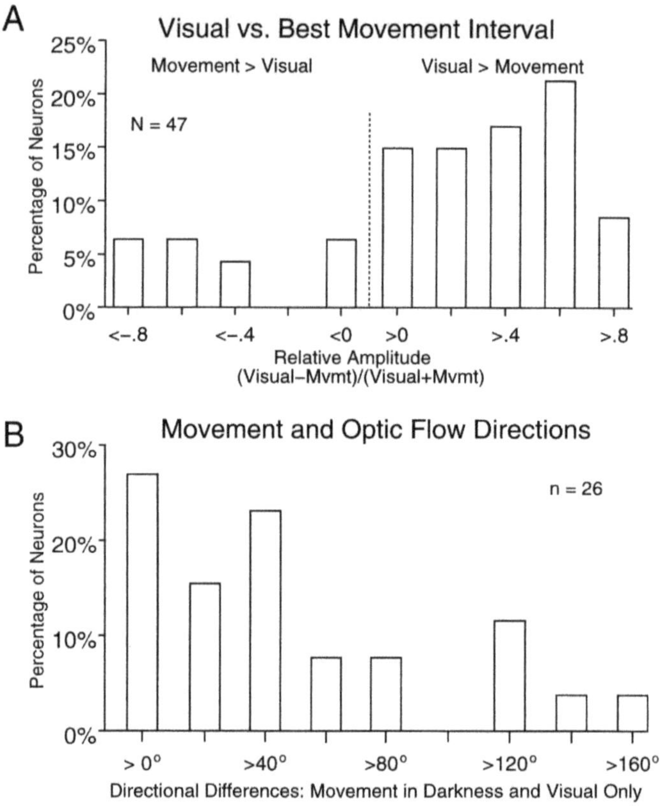

Figure 5. Comparison of movement in darkness and optic flow responses. A. Relative amplitude of optic flow and movement in darkness responses showing that 62% (29/47) of the neurons had larger optic flow responses (>0.2) and 17% (8/47) had larger (<-0.2) movement in darkness responses. B. Comparison of the best, significant preferred directions to movement in darkness and optic flow. Half of the neurons (53%, 16/30) showed similar preferred directions to both types of stimuli (<60° difference) but others showed larger differences.

one of the movement intervals (contrast <-0.2). Thus, MST neurons typically show stronger responses to optic flow than to movement in darkness.

We compared heading preferences in response to movement in darkness and optic flow. Responses to the first four seconds of movement, excluding deceleration's directional reversals, and the one second of optic flow yielded 26 significant heading comparisons from 14 neurons (Z of resultant vectors with $p \leq 0.05$ to both stimuli). Most responses (62%, 16/26), showed similar heading preferences (<60° difference), while the remainder showed larger differences in their movement in darkness and optic flow heading preferences, many (23%, 6/26) with nearly opposite preferred headings (>120° difference), (Figure 5b).

Differences in heading preferences with movement in darkness and optic flow are the likely origin of the time-dependent variation in heading preferences seen in the movement in light responses. When movement in darkness and optic flow heading preferences are nearly opposite, these effects could cancel each other such that no significant heading response was apparent with movement in light.

3.5 Population Responses

Population vector analysis was used to assess MST's potential contribution to heading estimation during movement in darkness and movement in light. Each neuron's preferred direction was derived by vector summing its responses to all eight headings. Population vectors for each heading direction were then derived by weighting each single neuron's heading vector by the amplitude of its response to that heading direction. These single neuron vectors were then summed to create a population net vector (PNV) for each heading direction stimulus.

We first compared responses from the middle one-second interval of the five second movement stimuli. The accuracy of the population vector's indication of the stimulus heading varied across directions and conditions. Overall, the population vector indicated the approximate direction of the self-movement stimuli. This was more clearly the case with movement in light responses because the population vectors from movement in darkness were consistently smaller than those obtained during movement in light (Figure 6).

We compared the population vector's representation of heading for the eight stimulus directions, in the five one-second stimulus intervals, for the six self-movement conditions. During movement in darkness, the smaller population response amplitudes yielded several non-significant responses (Figure 7). Regression lines relating stimulus heading to population vector heading showed stronger associations for movement in light than for movement in darkness. Thus, MST's population response indicates the heading of self-movement more accurately in light than in darkness.

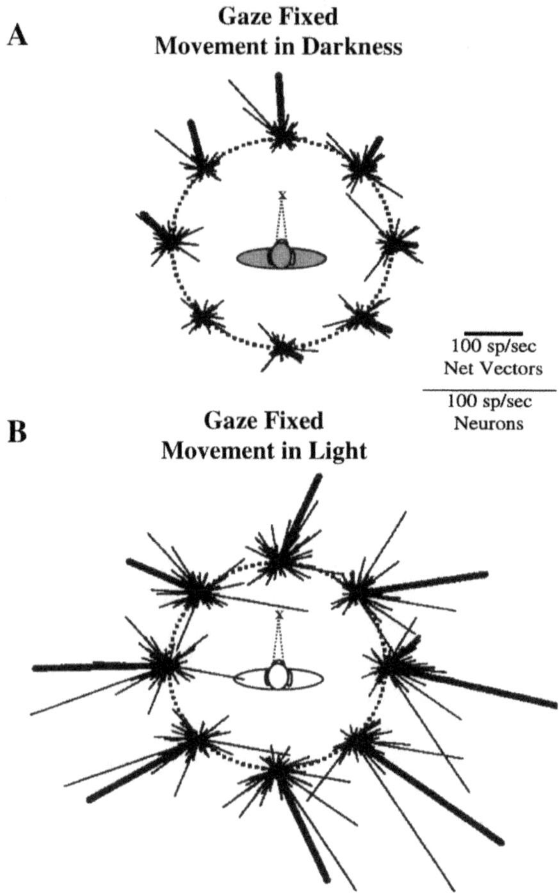

Figure 6. Population responses to movement in darkness (A) and to movement in light (B). Each cluster corresponds to one of the eight stimulus heading directions and contains a vector for each neuron (thin lines) having that neuron's preferred direction from the sum of its responses to the middle one-second of the eight directions. The length of each neuron's vector is determined by the amplitude of its response to the direction illustrated by each cluster. The population vector for each direction (bold lines) is the sum of the vectors for individual neurons. The population vectors more closely approximate the stimulus direction during movement in light.

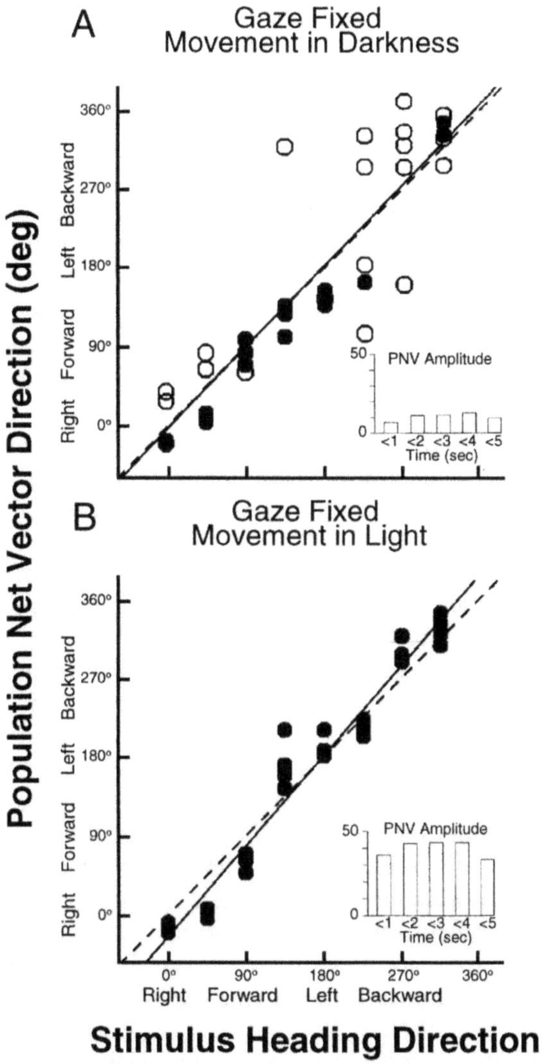

Figure 7. Accuracy of population vectors to movement in darkness (A) and movement in light (B). Each scatter plot shows the population vectors for the five, one-second movement intervals (ordinate) in relation to the direction in that movement stimulus (abscissa). A least squares linear regression was used to fit these points (solid line). Filled points represent significant net vectors (Z statistic $p<0.05$). The bar graphs show the amplitude of net vectors for each interval. A. The population vectors for movement in darkness show a clear relationship between the population vector direction and the movement direction ($r^2 = 0.79$, slope = 1.03) but relatively small population vector amplitudes. B. Population vectors for movement in light show a stronger relationship between population vector direction and movement directions ($r^2 = 0.96$, slope = 1.14) with much larger population vector amplitudes. Thus, MST provides veridical heading estimates in darkness or light.

4 DISCUSSION

4.1 Vestibular Responses in MST

These studies confirm that MST neurons receive vestibular, as well as visual, signals about self-movement. Our previous studies of optic flow and translational self-movement responses in MST suggested that these inputs combine to shape MST neuronal responses (Duffy, 1998). Those studies combined self-movement with the presentation of simulated optic flow stimuli. The current studies extend those results by demonstrating that MST neurons respond to combined translational self-movement and naturalistic optic flow; that is, optic flow that accompanies self-movement through a textured visual environment. These observations are consistent with vestibular projections to dorsal extrastriate visual cortex (Faugier-Grimaud & Ventre, 1989; Akbarian et al., 1994) that might mediate responses during movement in darkness.

MST neuronal responses to translational self-movement in darkness showed several characteristics that suggested their vestibular origin. First, these responses showed clear direction selectivity such that they responded more strongly to movement in one direction in the ground plane than to movement in opposite, or nearly opposite, directions. Second, the time-course of many of these responses showed a clear relationship to the acceleration phase of movement. This suggests a link to the accelerometer function of the vestibular otoliths rather than a link to visual or oculomotor events that may accompany translational movement. Third, the direction selective responses of these neurons often reversed their preferred direction in the deceleration phase of a movement stimulus. This suggests that these responses reflect the direction of the movement force, which reverses in deceleration such that the force experienced during acceleration in a given direction is the same as the force experienced during deceleration in the opposite direction.

Three further observations about these responses are relevant. First, responses to translational movement in darkness were recorded during visual fixation of a target that moved with the monkey. Thus, these responses could not be attributed to vestibular ocular reflex (VOR) movements triggered by translational self-movement. Second, these responses persisted when the fixation point was not present during the motion stimulus even though the monkey continued to maintain fixed gaze. Thus, these responses could not be attributed to VOR cancellation by the presence of a fixation target during translational self-movement.

Finally, the time-course of many responses to translational self-movement in darkness extended throughout the five second stimulus period. Thus, the transient responses of some neurons that were limited to the acceleration period demonstrated vestibular otolith input; the building-up or sustained responses of other neurons suggested some temporal integration of otolith input. We cannot conclude that this temporal integration occurs in MST, it may well be generated in subcortical structures. We can only conclude that it is reflected in the responses of some MST neurons.

4.2 Combined Visual and Vestibular Responses

Most of the neurons recorded in these studies showed stronger directional responses to translation self-movement in light than to translational self-movement in darkness (Figure 3). This difference was greatly attributable to optic flow responses that showed sustained activation with clear direction selectivity for movement in the ground plane.

This effect was most clearly demonstrated by comparing the responses to translational self-movement in light to those evoked by translational self-movement in darkness and by simulated optic flow presented without translational self-movement. Such comparisons commonly suggested that vestibular responses evoked by translational self-movement in darkness combined with visual responses to optic flow to generate the responses to translational self-movement in light.

The best examples of such combinations of visual and vestibular responses were seen in neurons that showed different preferred directions of visual and vestibular effects and different time-courses of visual and vestibular effects (Figure 4). In these neurons, the combination of visual and vestibular effects could be seen in changes in the preferred direction of responses to translational self-movement in light across the five second period of those stimuli.

MST neurons maintain their identity as key elements in the dorsal extrastriate visual pathway (Saito et al., 1986; Duffy & Wurtz, 1991; Orban et al., 1992; Graziano et al., 1994) by being activated selectively by optic flow stimuli simulating self-movement along a particular heading (Tanaka & Saito, 1989; Saito et al., 1986; Duffy & Wurtz 1991, 1995; Orban et al., 1992; Graziano et al., 1994) even in complex environments (Upadhyay et al., 2000). Nevertheless, they also respond to both rotation (Thier & Erickson, 1992) and translation (Duffy, 1998) signals about self-movement, likely created by vestibular canalicular and otolithic transduction.

In these studies we found that most MST neurons show stronger visual responses than vestibular responses, but some of these neurons showed stronger vestibular responses (Figure 5a). The latter group of neurons might play a particularly important role in supporting heading representation during self-movement in darkness, although all neurons with vestibular responses could contribute under such circumstances.

Neurons that showed similar preferred directions in their visual and vestibular responses might mediate an enhancement of MST's heading representation by combining multi-sensory signals about self-movement. This would seem most relevant to the naturalistic circumstance of moving with eyes open in a lighted environment. In contrast, the contribution of neurons with different preferred directions in their visual and vestibular responses is less obvious (Figure 5b).

One potential function of neurons with different visual and vestibular heading preferences is that they could enhance heading representation when the head and eyes are not aligned. For example, an observer moving with its head rotated, and the eyes counter-rotated to direct gaze toward the heading, would have a different correspondence of visual and vestibular axes than that experienced by an observer moving with neutral head and eye position. Under such circumstances, neurons with different visual and vestibular heading preferences might be optimally stimulated.

Another potential function of neurons with different visual and vestibular heading preferences is that they might undergo a sharpening of their heading tuning with combined stimulation. Such an effect would rely on the common occurrence of inhibitory effects during the presentation of non-preferred direction of one or the other sensory modality. The combination of a broadly tuned excitatory response in one modality and inhibition in the other modality could narrow the width of directional tuning in the composite response.

A special case is presented by neurons with nearly opposite visual and vestibular heading preferences. Such neurons might be optimally stimulated when translational self-movement is combined with pursuit eye movements during fixation on a target that is closer to the observer than a textured background. Under such conditions, eye rotation from pursuit stabilizes the fixation target while creating a retinal image of the optic flow that corresponds to the opposite direction of self-movement. This would combine opposite directions of visual and vestibular stimulation.

Thus, these studies reveal that MST neurons show a diverse set of self-movement responses. Neurons with stronger visual than vestibular inputs tend to show sustained heading responses throughout a movement in the preferred direction. Neurons with stronger vestibular than visual inputs tend to show transient responses that are most sensitive to acceleration, fade during steady-speed movement, and often reverse their preferred direction during deceleration. The combination of balanced visual and vestibular inputs might enhance heading direction responses in some neurons. Other neurons, that combine visual and vestibular responses might be suited to specific conditions of self-movement (Figure 8A).

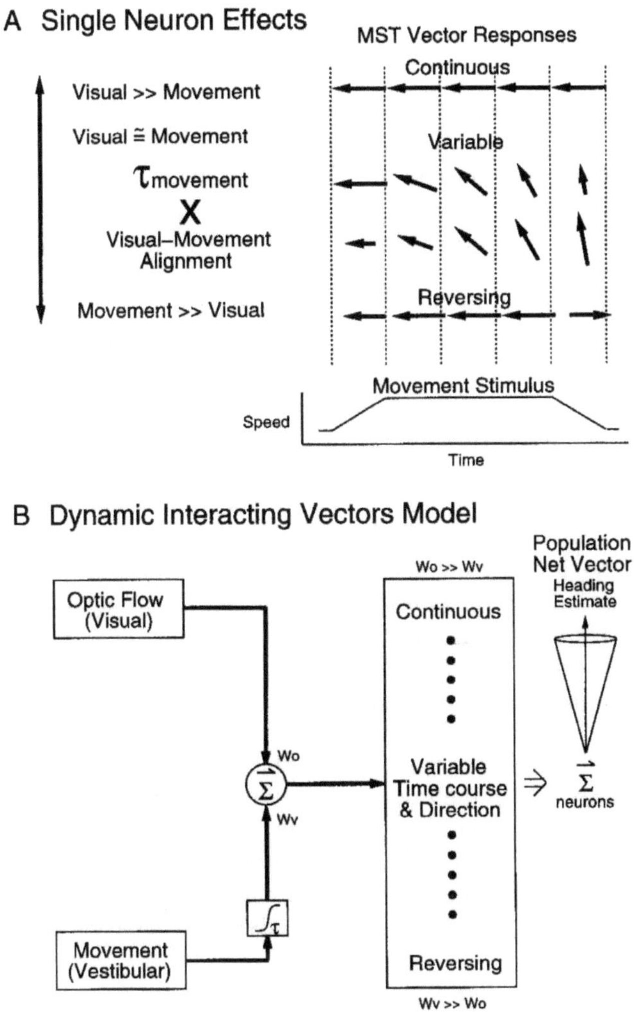

Figure 8. Summary of findings and a model of heading estimation in MST. A. Single neuron responses to observer movement in light can be classified into three groups: 1) Continuous responses that are dominated by visual effects from the optic flow maintaining the same pattern throughout the movement. 2) Variable responses combining visual and vestibular effects depending on the time-course, amplitude and relative directions of those effects. 3) Reversing responses that are dominated by vestibular effects reflecting the acceleration or speed profiles of the movement. B. The visual and vestibular responses interact in a manner that is influenced by the time-course of the vestibular effects and its directional alignment with the visual effects. The summation is governed by non-additive response dynamics that can be characterized as a sigmoidal activation-response curve that yields sub-additive effects, additive effects, and super-additive effects. These signals vector sum to provide a population representation of heading in darkness and light.

4.3 Population-Based Heading Representation

The diverse responses of single neurons in MST might contribute to the population-based representation of the heading of self-movement. The vector summation of directional responses of MST neurons provides one approach to illustrating the population coding of heading. This method illustrates that vestibular responses support veridical heading representation, even with the small numbers of neurons recorded in this study (Figures 6a and 7a).

MST's population response is substantially more robust during translational self-movement in light. In this case the combination of vestibular and visual responses creates larger population responses with greater accuracy in the population net vector's representation of the direction of movement presented in the stimulus (Figures 6b and 7b).

Naturalistic self-movement occurs in a great many more conditions than simple darkness or light. As light levels vary across conditions, the relative influence of visual and vestibular signals might also vary. Similarly, parametric changes in the quality of visual and vestibular signals might also influence the population response. For example, environments with mainly distant visual features might yield less salient optic flow patterns so that vestibular cues become more important. In contrast, sustained self-movement without accelerations or directional changes might yield less salient otolith responses so that visual cues become more important.

The variety of single neuron response characteristics seen in MST (Figure 8a) might not only reflect their relative sensitivity to different stimuli but also their relative influence on the population response under different self-movement conditions (Figure 8b). MST's population response might reflect more dynamic interactions between input signals under intermediate environmental conditions, matching the dynamic effects seen in single neurons.

REFERENCES

Akbarian, S., Grusser, O. J., & Guldin, W. O. (1994). Corticofugal connections between the cerebral cortex and brainstem vestibular nuclei in the macaque monkey. *J. Comp. Neurol., 339*, 421-437.

Batschelet, E. (1981). *Circular Statistics in Biology*. New York, Academic Press, 1-371.

Benson, A. J., Spencer, M. B., & Stott, J. R. R. (1986). Thresholds for the detection of the direction of whole-body, linear movement in the horizontal plane. *Aviat. Space Environ. Med.*, 1088-1096.

Duffy, C. J. (1998). MST neurons respond to optic flow and translational movement. *J. Neurophysiol., 80*, 1816-1827.

Duffy, C. J. & Wurtz, R. H. (1991). Sensitivity of MST neurons to optic flow stimuli. I. A continuum of response selectivity to large-field stimuli. *J. Neurophysiol., 65*, 1329-1345.

Duffy, C. J. & Wurtz, R. H. (1994). Optic flow responses of MST neurons during pursuit eye movements. *Society for Neurosci. Absts., 20*, 1279.

Duffy, C. J. & Wurtz, R. H. (1995). Response of monkey MST neurons to optic flow stimuli with shifted centers of motion. *J. Neurosci., 15*, 5192-5208.

Faugier-Grimaud, S. & Ventre, J. (1989). Anatomic connections of inferior parietal cortex (area 7) with subcortical structures related to vestibulo-ocular function in a monkey (Macaca fascicularis). *J. Comp. Neurol., 280*, 1-14.

Georgopoulos, A. P., Schwartz, A. B., & Kettner, R. E. (1986). Neuronal population coding of movement direction. *Science, 233*, 1416-1419.

Gibson, J. J. (1950). *The Perception of the Visual World.* Boston, Houghton Mifflin.

Graziano, M. S. A., Andersen, R. A., & Snowden, R. J. (1994). Tuning of MST neurons to spiral motion. *J. Neurosci., 14*, 54-67.

Hays, A. V., Richmond, B. J, & Optican, L. M. (1982). A UNIX-based multiple process system for real-time data acquisition and control. *WESCON Conf. Proc., 2*, 1-10.

Judge, S. J., Richmond, B. J, & Chu, F. C. (1980). Implantation of magnetic search coils for measurement of eye position: an improved method. *Vision Res., 20*, 535-538.

Komatsu, H. & Wurtz, R. H. (1988a). Relation of cortical areas MT and MST to pursuit eye movements. I. Localization and visual properties of neurons. *J. Neurophysiol., 60*, 580-603.

Komatsu, H. & Wurtz, R. H. (1988b). Relation of cortical areas MT and MST to pursuit eye movements. III. Interaction with full-field visual stimulation. *J. Neurophysiol., 60*, 621-644.

Orban, G. A., Lagae, L., Verri, A., Raiguel, S., Xiao, D., Maes, H., & Torre, V. (1992). First-order analysis of optical flow in monkey brain. *Proc. Nat. Acad. Sci. USA,* 89, 2595-2599.

Page, W. K. & Duffy, C. J. (1999). MST neuronal responses to heading direction during pursuit eye movements. *J. Neurophysiol., 81*, 596-610.

Page, W. K. & Duffy, C. J. (2003). Heading representation in MST: sensory interactions and population encoding. *J. Neurophysiol., 89*, 1994-2013.

Robinson, D. A. (1963). A method of measuring eye movement using a scleral search coil in a magnetic field. *IEEE Trans.Bio.-Med. Eng., 10*, 137-145.

Saito, H., Yukie, M., Tanaka, K., Hikosaka, K., Fukada, Y., & Iwai, E. (1986). Integration of direction signals of image motion in the superior temporal sulcus of the macaque monkey. *J. Neurosci., 6*, 145-157.

Tanaka, K. & Saito, H. (1989). Analysis of motion of the visual field by direction, expansion/contraction, and rotation cells clustered in the dorsal part of the medial superior temporal area of the macaque monkey. *J. Neurophysiol., 62*, 626-641.

Thier, P. & Erickson, R. G. (1992). Vestibular input to visual-tracking neurons in area MST of awake rhesus monkeys. *Annals New York Acad. Sci., 656*, 960-963.

Upadhyay, U. D., Page, W. K., & Duffy, C. J.(2000). MST Responses to pursuit across optic flow with motion parallax. *J. Neurophysiol., 84*, 818-826.

3. A Visual Mechanism for Extraction of Heading Information in Complex Flow Fields

Michael W. von Grünau and Marta Iordanova

Department of Psychology
Concordia University, Montreal, Québec, Canada

1 INTRODUCTION

When organisms move forward or backward through their environment, they produce particular patterns of optic flow on their retina, i.e., the images of objects move in characteristic ways away from or towards a focus of expansion or contraction (Gibson, 1950, 1954). For straight motion without eye or head movements, the flow pattern is radial with a speed gradient and complete symmetry (Gibson, 1950). It is then straightforward to compute the heading direction, calculate the time to collision with objects or, indeed, take action to avoid collisions (van den Berg, 2000; Warren, 1998; Warren et al., 1988, 1991). The situation becomes more difficult when eye movements (e.g., pursuit movements) are made simultaneously, which is the rule and not the exception. Fixating an object that moves from left to right would create a lateral flow on the retina, giving a rightward motion to the images of all other objects. We call this a parallel flow to express the fact that all motion vectors would be parallel. This motion would be added to the appropriate radial motion for each image point, resulting in a complex flow pattern. This complex flow still contains the original radial flow that is needed for the extraction of heading direction. But the complex pattern would have to be decomposed to yield this information.

Decomposition could occur in several ways. The brain could use feedback information about eye position and subtract this from the complex flow pattern, which would leave the radial pattern. The brain could also use information about the original output signals that created the eye movements (efference copy). Again, subtraction would result in the original radial flow pattern. Both solutions are characterized by the use of extra-retinal information to accomplish the decomposition. There is evidence that these

L.M. Vaina, S.A. Beardsley and S.K. Rushton (eds.), Optic Flow and Beyond, 45–59.

non-visual solutions are indeed employed by the brain (van den Berg, 2000), especially under conditions that are more difficult (fast speeds of eye rotation) or ambiguous (no spatial layout, no depth cues; Roydon et al., 1992, 1994; Banks et al., 1996). Another solution to the decomposition problem could be purely visual, i.e., not involving extra-retinal information. Evidence shows that visual decomposition is possible under circumstances when eye rotation speeds are slow (< 1.5 °/sec) and environments are rich in depth cues (Warren & Hannon, 1988, 1990). While visual decomposition does occur, the underlying mechanism is unknown. Again, this visual decomposition would imply some sort of subtraction process, isolating the radial flow contribution by de-emphasizing the contribution of the parallel flow. We report here some experiments, which explored the interaction between two transparently superimposed flow fields, one of which was radial, the other parallel.

In such transparent displays, which were introduced by Duffy and Wurtz (1993), the two flow components are already defined separately (Figure 1). They also appear perceptually as two distinct motions. Yet they are not independent, since the presence of the parallel flow affects the perceived location of the focus of expansion (or contraction) of the radial flow. This illusory shift of the focus in the direction of the parallel flow shows the workings of a compensatory mechanism, which may be based on induced motion (Meese et al., 1995), relative speed (Pack & Mingolla, 1998) and/or relative depth (Grigo & Lappe, 1998). This work supports the contention that the processes involved have a physiological basis that can support a level of analysis that operates on a large scale (over much of the visual field) and globally (integrating across space and time). More specifically, this would suggest the involvement of higher areas in the motion stream, such as the medial superior temporal area (MSTd in primate extrastriate cortex) and beyond (Duffy & Wurtz, 1991, 1995a,b; Ilg, 1997; Kawano et. al., 2000). At this level, information about eye movements and binocular depth is also available and can be integrated. Receptive field properties of single neurons at this level are well suited for such a global analysis (Geesaman & Anderson, 1996; Duffy & Wurtz, 1995b).

In transparent flow fields, the two constituent motions are already decomposed into a parallel and a radial flow, physically and perceptually. At some level, different populations of neurons seem to be involved in the processing of each flow component, since they are defined independently. The present question is how the subtraction can be achieved. Since we are confining ourselves here to visual mechanisms only, this implies that the subtraction process must consist of a modulation of the relative strength of the response of the two populations. The end result of this process is that the parallel motion signal is effectively removed from the overall representation of the flow field, leaving the behaviorally more important radial component to dominate (Iordanova & von Grünau, 2001).

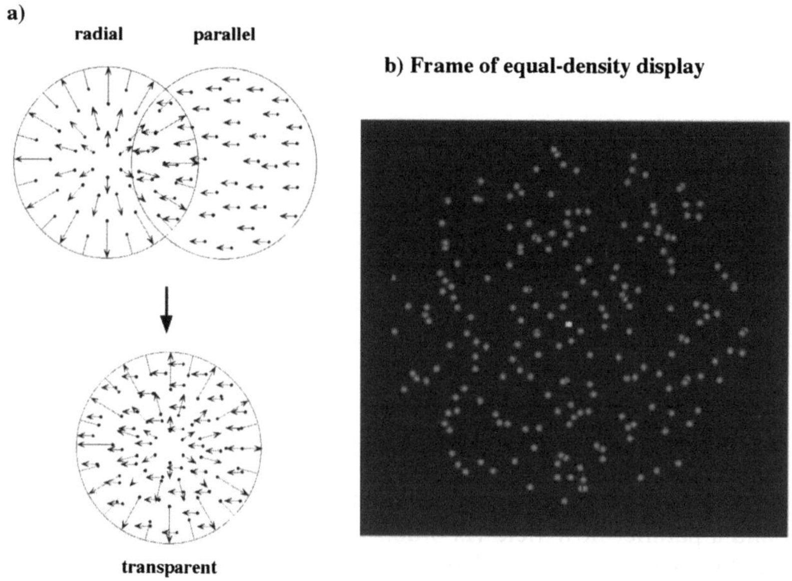

Figure 1. a) Radial and parallel flow fields are combined to give a transparent flow field. b) Static view of a stimulus with equal density of the Gaussian dots.

To investigate the subtraction process, we studied the way in which the detection of each flow component is affected by the presence of the other component. To this end, we developed a masking paradigm where sensitivity to one component (test) is assessed when the other component (mask) is present either as a fully coherent or a fully incoherent motion stimulus. Sensitivity was measured by determining coherence thresholds for the test motion. Coherence refers to the percentage of dots moving in the appropriate direction (radial expanding, radial contracting, parallel left, parallel right). A completely incoherent stimulus would contain the same motion vectors and speeds, but the motions would be jittering back and forth. In all experiments, we found that a radial motion mask suppressed sensitivity to the parallel test motion, while sensitivity to the radial test motion was little affected by the parallel motion mask.

2 EXPERIMENT 1

The goal of this experiment was to establish the nature of the interaction between the two transparent flow fields and to trace its time course. The exposure time of the stimuli was varied. In order to produce a strong effect, the lifetime of the test dots was limited while for the masking dots it was infinite.

2.1 Methods

2.1.1 Observers

Four observers, the two authors and two naïve members of the Vision Lab participated. All had normal or corrected-to-normal vision.

2.1.2 Stimuli

The random dot display consisted of a transparent superposition of a radial and a parallel flow field (200 dots each), (see Figure 1a). This field of 400 bright Gaussian dots (0.3° diameter) was presented on a dark background (Michelson contrast of 0.2). During animation, a specialized algorithm kept dot density constant across all regions of the display (Figure 1b). Parallel motion speed was a constant 9°/sec for all dots across the display. In the radial flow, dot speed increased as a function of eccentricity (from 0° to 70°/sec; illustrated by arrow length, Figure 1a), corresponding to the real-world situation where flow fields are produced by locomotion. The overall extent of the visible stimulus field was 90 x 90 degrees. The masking stimuli (Figure 2) were either parallel (top, with a radial test) or radial (bottom, with a parallel test). In each case, they could have 100% coherence (left) or 0% coherence (right, jitter). The jitter masks (control) were identical to the motion masks, but lacked the coherent motion. Test fields were varied in coherence from 0% to 100% (step size of 20%) with the remaining dots displaying jitter motion. Each test dot had a limited lifetime (moved for 91 ms, then was redrawn at a new random location, consistent with the density control algorithm). Masking dots had infinite lifetime. Exposure duration was varied from 187 to 748 ms with a gradual onset and offset of the stimuli.

2.1.3 Procedure

Observers were seated 70 cm behind a large projection screen and viewed the display through a circular aperture (90 degrees diameter), their head being stabilized by a chin rest. They were fixating in the center of a 5° dot-free central region. No fixation point was present during stimulus presentation. Depending on the experimental condition, observers judged the global

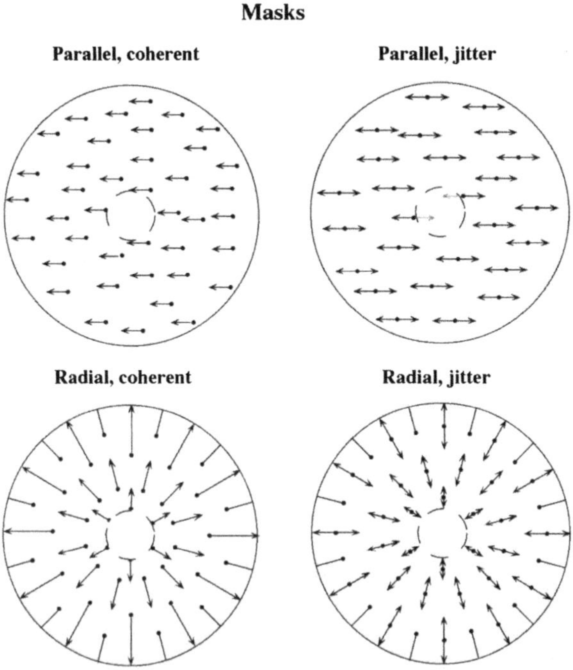

Figure 2. Four kinds of masks were used: They could be coherent (directional) or incoherent (jitter) and either portray parallel or radial flow.

direction of motion of either component of the transparent displays while the other component acted as a mask. Coherence thresholds were determined using the method of constant stimuli (80-100 repetitions) with a 2AFC response (left/right for parallel, expand/contract for radial) for each of the 4 test/mask combinations in separate sessions.

2.1.4 Data Analysis

Coherence thresholds were defined as the percentage of coherently moving dots supporting 82% correct directional judgments (Weibull fit). Each threshold is based on 80 – 100 repetitions of the 6 coherence levels.

2.2 Results

The results are shown for two observers in Figure 3a. Coherence thresholds are graphed as a function of exposure duration for parallel test with

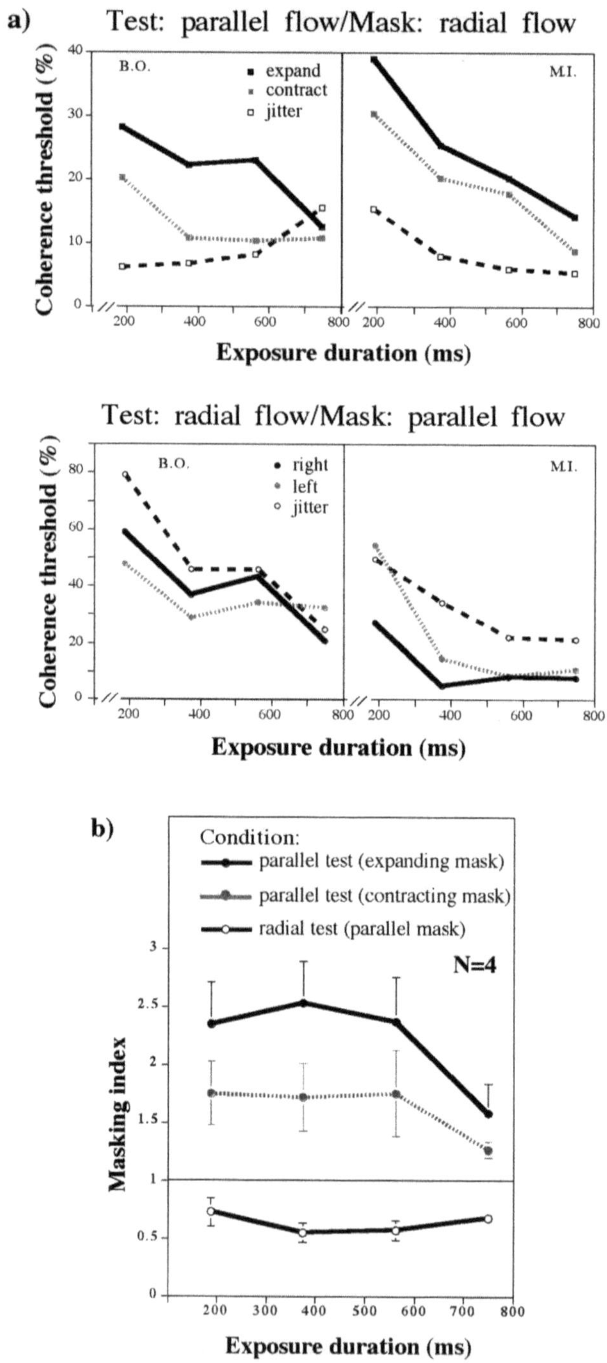

Figure 3. a) Results for two observers for either parallel test with radial mask or radial test with parallel mask. b) Results combined for four observers using a masking index, defined as threshold at motion mask relative to threshold at jitter mask. Index > 1.0 shows threshold increase.

radial mask (expanding, contracting or jitter) and for radial test with parallel mask (left, right or jitter). Coherence thresholds generally declined with increasing exposure duration. As compared to the jitter mask, both radial motion masks elevated thresholds for the parallel test, the expanding mask about twice as strongly as the contracting mask. In contrast, the parallel masks had little consistent effect on the radial thresholds at all exposure durations. The suppressive effect of the radial motion masks was most pronounced at the shortest exposure duration. In order to combine the results of all observers in a meaningful way, we calculated a masking index as the ratio of the threshold at motion mask to the threshold at jitter mask. This is graphed in Figure 3b for four observers. Masking index values above 1.0 indicate threshold increases (parallel test), while values below 1.0 show threshold decreases (radial test).

2.3 Discussion

These results confirm the finding by De Bruyn & Orban (1993), that the radial and parallel components of a transparent flow field interact in an asymmetrical manner, such that the radial flow suppresses sensitivity to the parallel flow, but is itself little affected by parallel flow. This is the kind of modulation one would want in order to subtract the parallel flow from the complex flow to leave the radial flow. It also occurs very early, best at the shortest exposure durations, indicating that it happens before the computation of heading direction, which usually takes over 400 msec (Hooge et al., 1999). The suppression effect seems to decline or disappear for longer durations, so that the parallel flow information becomes available then.

3 EXPERIMENT 2

The flow patterns in the first experiment were designed such as to reflect the characteristics of real-life flow fields. This introduced an important difference between the radial and the parallel flow patterns; namely the radial flow contained a gradient of local speeds (from 0° to 70°/sec), while all parallel flow dots had a constant speed of 9°/sec. The effectiveness of the radial flow as a mask could have been the result of the high speeds and/or the presence of acceleration. The latter also produces a vivid impression of motion-in-depth, which could have contributed to the different findings. In the next experiment, the speed gradient of the radial flow was removed and all dots had the same speed as those of the parallel flow, i.e., 9°/sec. As a further control, thresholds were also measured in the presence of stationary masks.

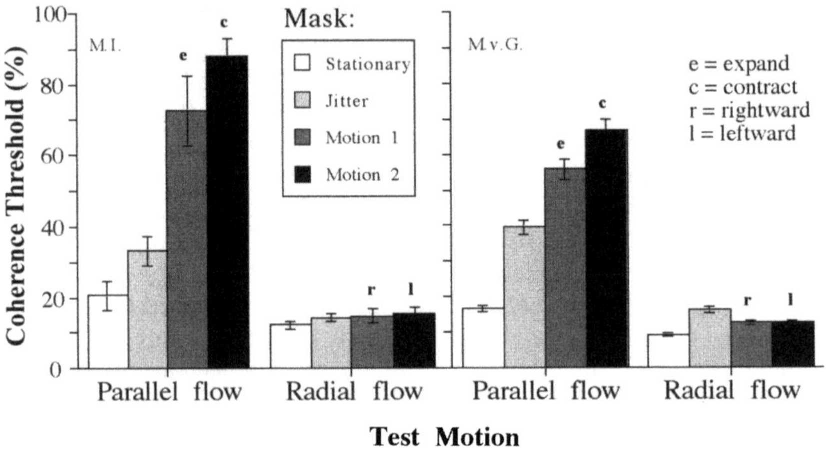

Figure 4. Results for Experiment 2 where all dots had the same velocity and no acceleration.

3.1 Methods

The methods were the same as in Experiment 1, except that the speed of all dots in the parallel and the radial flow was set to be the same, i.e., 9°/sec. Exposure duration was limited to 200 msec to minimize eye tracking. As a further control, thresholds were also determined for stationary masks (in addition to the motion and jitter masks).

3.2 Results

The asymmetric masking effect was again obtained (Figure 4). When the test motion was radial, coherence thresholds were low and roughly the same for all conditions. There was no suppression effect by the parallel flow. When the test motion was parallel, however, radial motion masks led to large increases in threshold, especially when the masking motion was directional.

3.3 Discussion

These results indicate that both accelerating and 'velocity-flat' radial flow fields are equally effective in masking the parallel motion flow. Removing the velocity gradient had one important consequence: Expanding flow, which was a more effective suppressor in Experiment 1, lost this advantage. This often-found asymmetry between radial motion directions (physiologically, there are more cells tuned to expansion than to contraction; Duffy & Wurtz, 1995b for primate MSTd; Rauschecker et al., 1987 for cat PMLS) seems to depend on the correct velocity gradient.

4 EXPERIMENT 3

The next experiments test whether the underlying subtraction mechanism is based on local or global interactions between the radial and parallel motions. When a particular dot is allowed to move over an extended time (or distance), more local information about the trajectories becomes available. With a limited lifetime, this information is curtailed. In the above experiments, lifetime for the test stimuli was set at 91 ms, while that of the masks was infinite. In the present experiment, lifetime was varied and kept the same for tests and masks. Any differential masking effects must be due to interactions at a global level.

4.1 Methods

The methods were the same as before, except that dot lifetime was varied from 52 ms to 104 ms in steps of 13 ms. 52 ms was the shortest time possible, for which motion thresholds could be obtained and motion was still perceived as transparent. Exposure time was 182 ms.

4.2 Results

The results are graphed in Figure 5 for one of the observers. The asymmetrical suppression was present and direction-specific even for the shortest lifetime and did not change much across lifetimes (except for an overall threshold reduction with increasing lifetime). Parallel motion thresholds were elevated by radial motion, especially directional motion, while radial thresholds were little affected. Only one directional motion is plotted, since there was no difference between the masking motion directions. Thus the absence of acceleration again equalized expansion and contraction effects.

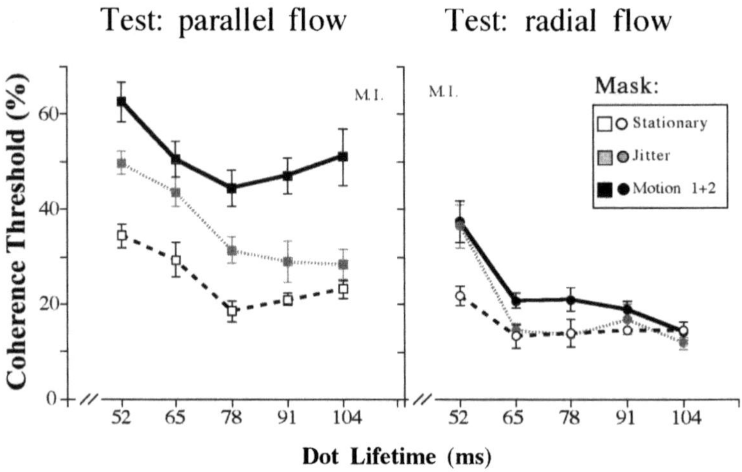

Figure 5. Results for Experiment 3, where dot lifetime was varied. The two opposite motions (expand/contract and left/right) gave similar results and are combined here.

4.3 Discussion

Since integration over time was limited in these displays, the extraction of global motion direction requires the integration of local motions over space. The present results therefore suggest a global mechanism as underlying the subtraction effect.

5 EXPERIMENT 4

We further examined the scale of the global subtraction mechanism by confining the two flow fields to different spatial areas, so that no overlap occurred. Locally, only one motion was present at any location, but globally both motions were contained in the larger stimulus field. This would allow us to examine the spatial limits of the subtraction mechanism.

5.1 Methods

The methods were the same as before, but the circular display area was split into equal segments, the radial and parallel flows filling alternate segments (Figure 6). There were 2 (180 deg each), 4 (90 deg), 8 (45 deg) and 16 (22.5 deg) segments. The absolute position and the type of motion were

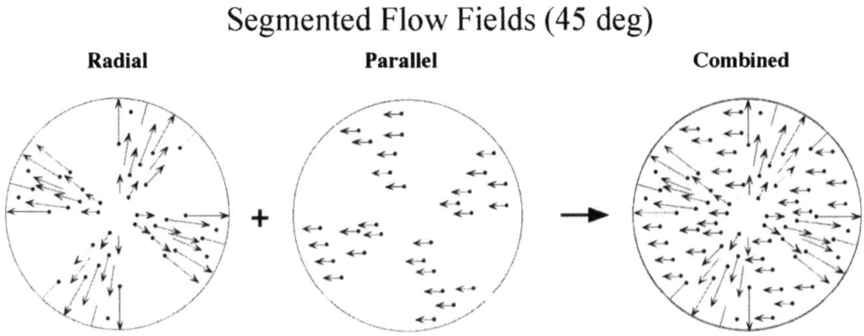

Figure 6. Segmented flow fields (example 45 deg extent).

changed from trial to trial, so that one particular segment could not be used for the task. Only parallel flow was tested and had a dot lifetime of 91 ms. Radial flow was with acceleration and with infinite dot lifetime. Exposure time was 320 ms.

5.2 Results

Results are shown in Figure 7 for two observers. The threshold values were converted to a masking index (threshold at motion mask/threshold at jitter mask). The index is plotted as a function of the segment extent. Included are also the results for the complete overlap condition. Threshold elevation (index > 1.0) was present for all segment extents for both expanding and contracting radial flow. With contraction, the suppression effect was independent of segment size, while with expansion it declined for larger segments. As before, expansion was a much more potent mask than contraction, except for the largest segments (180 deg).

5.3 Discussion

This experiment very clearly shows the global nature of the subtraction mechanism. The presence of the effect did not depend on spatial overlap of the two types of flow, and the magnitude remained constant for overlap and segments up to 45 deg in extent. It was present even when each flow was confined to separate halves of the stimulus field. The spatial extent of the mechanism encoding expansion seems to be larger than that encoding contraction.

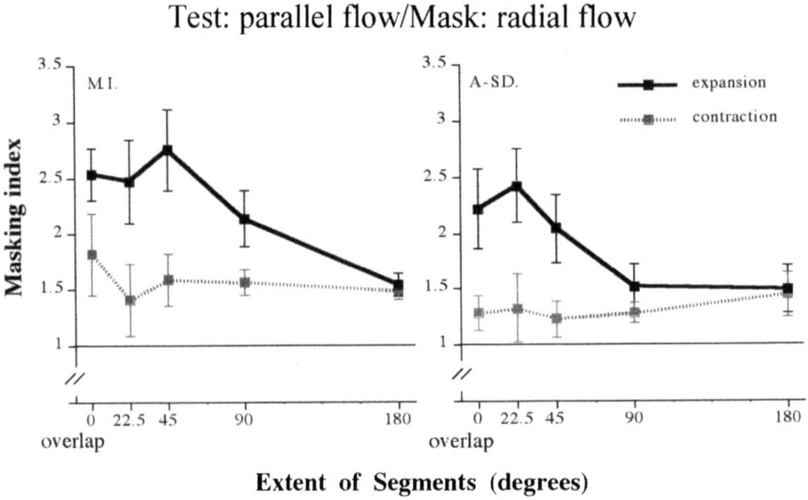

Figure 7. The effect of the extent of the segments for expansion and contraction for two observers. Masking index (as in Figure 3b) is plotted.

6 GENERAL DISCUSSION

We have provided psychophysical evidence for a mechanism that temporarily suppresses the sensitivity to parallel flow when radial motion flow is present. This suppression is asymmetrical in that radial suppresses parallel, but parallel does not suppress radial. This mechanism could perform the task of flow field subtraction, which would enable the visual system for a short time at the beginning of stimulus presentation to extract radial flow information relatively uncontaminated by parallel flow. Thus heading direction could be computed and used for locomotor guidance. This may occur in a separate processing channel, which receives priority for a short time, before the combined flow field becomes dominant.

The dominance of the radial motion that we found depends primarily on the directional structure of the radial flow, not on the presence of acceleration or faster speeds in the radial flow. This dominance could be due to a stronger cooperative integration of radial flow components that overrides local interference, while for parallel flow cooperative links may be weaker and their spatial integration easily disrupted by irrelevant local motions. Alternatively, it is possible that the observed dominance of radial flow appears because of an active suppression of the parallel flow by the radial flow. With different directions of parallel flow, no direction-specific suppression has previously been found (Edwards & Nishida, 1999). Some earlier reports, however, found reciprocal direction-specific suppression in transparent displays (Quian et al., 1994; Snowden, 1989; Mather & Moulden, 1983). In contrast, the asymmetrical suppression that we observed seems to involve interactions at a more global scale (results for non-overlapping components, short lifetime). At that higher level, the perceived motion-in-depth in the radial flow patterns may contribute to the asymmetrical suppression. We found a greatly increased effect for expanding as compared to contracting flow under conditions favoring motion-in-depth. This is consistent with the observations of depth layering in these transparent displays (Grigo & Lappe, 1998) and with the response properties of neurons in primate area MSTd (Geesaman & Andersen, 1996; Duffy & Wurtz, 1995b) and cat area PMLS (Rauschecker et al., 1987).

The conclusions to be drawn from the present experiments are that there is convincing evidence for the existence of a global mechanism of asymmetrical suppression in human vision, which could be used by the visual system to emphasize the behaviorally more relevant information in the radial motion flow (locomotion) over the parallel flow (eye and head movements). The suppression of irrelevant retinal image motions develops very quickly after stimulus presentation, allowing for heading judgments and other behaviorally important tasks to be done early. It remains to be shown that such a mechanism is actually used by the visual system, and under which circumstances it would come into play.

ACKNOWLEDGEMENTS

We thank Peter April for his programming and technical support (www.vpixx.com). This work was supported by NSERC and FCAR research grants to MvG.

REFERENCES

Banks, M. S., Ehrlich, S. M., Backus, B. T. & Crowell, J. A. (1996). Estimating heading during real and simulated eye movements. *Vision. Res.*, *36*, 431-443.

De Bruyn, B. & Orban, G. A. (1993). Segregation of spatially superimposed optic flow components. *J. Exp. Psychol. Hum. Percept. Perform.*, *19*, 1014-1027.

Duffy, C. J. & Wurtz, R. H. (1991). Sensitivity of MST neurons to optic flow stimuli. I. A continuum of response selectivities to large-field stimuli. *J. Neurophysiol.*, *65*, 1329-1345.

Duffy, C. J. & Wurtz, R. H. (1993). An illusory transformation of optic flow fields. *Vision Res.*, *33*, 1481-1490.

Duffy, C. J. & Wurtz, R. H. (1995a). Response of monkey MST neurons to optic flow stimuli with shifted centers of motion. *J. Neurosci., 15*, 5192-5208.

Duffy, C. J. & Wurtz, R. H. (1995b). Medial superior temporal area neurons respond to speed patterns of optic flow. *J. Neurosci., 17*, 2839-2851.

Edwards, M. & Nishida, S. (1999). Global-motion detection with transparent-motion signals. *Vision. Res., 39*, 2239-2249.

Geesaman, B. J. & Andersen, R. A. (1996). The analysis of complex motion patterns by form/cue invariant MSTd neurons. *J. Neurosci., 16*, 4716-4732.

Gibson, J. J. (1950). *The Perception of the Visual World.* Boston: Hougton Mifflin.

Gibson, J. J. (1954). The visual perception of objective motion and subjective movement. *Psychol. Review, 61*, 304-314.

Grigo, A. & Lappe, M (1998). Interaction of stereo vision and optic flow processing revealed by an illusory stimulus. *Vision.Res., 38*, 281-290.

Hooge, I. Th. C., Beintema J. A. & van den Berg A. V. (1999). Visual search of heading direction. *Exp. Brain Res., 129*, 615-628.

Ilg, U. J. (1997). Slow eye movements. *Prog. Neurobiol., 53*, 293-329.

Iordanova M. & von Grünau M.W. (2001). Asymmetrical masking between radial and parallel motion flow in transparent displays. In: *Prog. Brain Res.*, special edition: From Neurons to Cognition, *134*, 333-352.

Kawano, K., Inoue, Y., Takemura, A., Kodaka, Y. & Miles, F. A. (2000). The role of MST neurons during ocular tracking in 3-D space. In: M. Lappe (Ed.), *Neuronal Processing of Optic Flow.* Int. Rev. Neurobiol., 44 (pp. 49-63), San Diego: Academic Press.

Mather, G. & Moulden, B. (1983). Thresholds for movement detection: two directions are less detectable than one. *Q. J. Exp. Psychol. A, 35A*, 513-518.

Meese, T. S., Smith, V. & Harris, M. G (1995). Induced motion may account for the illusory transformation of optic flow fields found by Duffy and Wurtz. *Vision Res., 35*, 981-984.

Pack, C. & Mingolla, E. (1998). Global induced motion and visual stability in an optic flow illusion. *Vision.Res., 38*, 3083-3093.

Quian, N. Andersen, R. A., & Adelson, E. H. (1994). Transparent motion perception as detection of unbalanced motion signals: I. Psychophysics. *J. Neurosci., 14*, 7357-7366.

Rauschecker J.P., von Grünau M.W. & Poulin C. (1987). Centrifugal organization of direction preferences in the cat's lateral suprasylvian visual cortex and its relation to flow field processing. *J. Neurosci., 7*, 943-958.

Royden, C. S., Banks, M. S. & Crowell, J. A. (1992). The perception of heading during eye movements. *Nature, 360*, 583-585.

Royden, C. S., Crowell, J. A. & Banks, M. S. (1994). Estimating heading during eye movements. *Vision. Res., 34,* 3197-3214.

Snowden, R. J. (1989). Motions in orthogonal directions are mutually suppressive. *J. Opt. Soc. Am. A Opt. Image Sci. Vis., 6,* 1096-1101.

van den Berg, A.V. (2000). Human ego-motion perception. In: M. Lappe (Ed.), *Neuronal Processing of Optic Flow.* Int. Rev. Neurobiol., 44 (pp. 3-25), San Diego: Academic Press.

Warren, H. W. Jr. & Hannon, D. J. (1988). Direction of self-motion is perceived from optical flow. *Nature, 336*, 162-163.

Warren, H. W. Jr. & Hannon, D. J. (1990). Eye movements and optical flow. *J. Opt. Soc. Am. A Opt. Image Sci. Vis., 7*, 160-169.

Warren, H. W .Jr, Morris, M. W. & Kalish, M. (1988). Perception of translational heading from optic flow. *J. Exp. Psychol. Hum. Percept. Perform., 14*, 646-660.

Warren, H. W. Jr., Blackwell, A. W., Kurtz, K. J., Hatsopolous, N. G. & Kalish, M. L. (1991). On the sufficiency of the velocity field for perception of heading. *Biol. Cybern., 65,* 311-320.

Warren, H. W. Jr. (1998). The state of flow. In: T. Watanabe (Ed.), *High-level Motion Processing: Computational, Neurobiological, and Psychophysical Perspectives* (pp. 315-358), Cambridge: MIT Press.

4. Eye Movements and an Object-Based Model of Heading Perception

Ranxiao Frances Wang[1] and James E. Cutting[2]

[1] Department of Psychology
University of Illinois, Champaign, IL, USA

[2] Department of Psychology
Cornell University, Ithaca, NY, USA

1 INTRODUCTION

The movement of an observer through an environment of stationary objects creates a complex set of motions called optical flow. Accurately registering these motions with our mobile eyes and acting upon them underlies the everyday commerce and safety of all animals, particularly human beings whose technological means of conveyance have accelerated both speeds and risks. Optical flow is rich and can be parsed in many ways (e.g., Cutting, 1986; Gibson, 1966; Koenderink & van Doorn, 1975; Regan & Beverley, 1982; see Lappe et al., 1999; and Warren, 1998, for reviews). There are many tasks that depend, at least in part, on optical flow — maintaining balance, detecting potential collisions, and monitoring the course of navigation. In this chapter, we discuss a model of heading perception that underlies navigation. It is based on relative motions of object pairs and on eye movements with which observers seek them out prior to heading judgments.

2 MODELS OF HEADING JUDGMENTS

Most approaches to understanding heading judgments generally have two characteristics. First, they focus directly on obtaining *absolute* heading information, the precise location of one's instantaneous aim point. Second, rather than focusing on (a) objects, (b) relative motion, and (c) relative depth, these approaches focus on either (b) motion alone, (b&c) motion and depth, or

L.M. Vaina, S.A. Beardsley and S.K. Rushton (eds.), Optic Flow and Beyond, 61–78.

(a&c) objects and depth. And finally, one approach considers all of this largely irrelevant. Consider each in turn.

2.1 Motion Alone

There are at least two computational schemes that have been posited. Both deal with sectors of the visual field — that is, regions with respect to the fovea. The first is differential motion, proposed by Rieger and Lawton (1985) and elaborated by Hildreth (1992). It squares the length of each motion vector and then adds all vectors within a sector. Second, spatial pooling (Warren & Saunders, 1995) simply adds each vector within a sector. Across sectors heading is determined by comparing the directions and extents of summed vectors. Our implementations of these schemes (Cutting et al., 1999) have shown that the spatial pooling fares better than differential motion. However, we also found that the presence and absence of the invariant cues (as discussed below) predicted heading results of human observers much better than either pooling scheme.

2.2 Motion and Depth

Perrone & Stone (1994, 1998; see Grossberg et al., 1999) presented a model based on their synthesis of anatomy, physiology, and data concerning visual area MST. This scheme is a bit like the models just considered except that motions are also pooled at different depths. This approach can account for virtually all data, but it is also not easily falsifiable. However, since all regions in space and depth are considered essentially equal, there is no role for attention, and we know that attention selects out particular objects of interest. Motter (1993), among others, has shown that responses of single cells in many visual areas are modulated by attention paid to particular objects in motion. These suppress activity that is otherwise present in the classical receptive field. Our notion is that human observers pay attention to particular objects and that it is their relative motion, not the global array of motion, that matters for finding one's way.

2.3 Objects and Depth

Another approach does not directly consider motion or its patterns. Frey and Owen (1999) proposed a measure of information in a heading display, called the *separation ratio*. This ratio, σ, can be expressed as:

$$\sigma = 1 - n/f \tag{1}$$

where n is the distance from the observer to the nearest object at the end of the sequence and f that to the farthest. Its value always falls between 0 and 1. Frey and Owen suggested that this ratio should predict heading performance, the greater the ratio the more accurate observers' responses. Indeed, typically in our displays the value of this ratio is reliably correlated with mean heading judgments. However, in two data sets (Cutting et al., 2000; Cutting & Wang, 2000), a stepwise multiple regression was performed with two predictors of absolute heading performance — the separation ratio, and the presence (coded as 1) or absence (coded as 0) of invariants on any given trial. We found that the invariant code accounted for 35% of the variance in the data, and the separation ratio only an additional 4%. Thus, the separation ratio is correlated with the invariant information in these displays, but by itself does not substantially contribute to the responses. Differential motion and spatial pooling also accounted for no appreciable variance.

2.4 Visual Direction

Rushton and co-workers (1998) and Llewellyn (1971) proposed that optical flow is not paramount for determining one's heading. All one needs to do is rely on visual direction — for example, pointing one's head and body towards a landmark one wishes to attain and simply walking there, allowing the muscles involved in the orientation of one's eye and head to guide one's movements. Indeed, manuals of motorcycle riding (Motorcycle Safety Foundation, 1992) and equitation (Morris, 1990) support this idea and it undoubtedly plays a role in wayfinding. However, Warren and co-workers (2001) showed that both sources are used, with optical flow information dominating when observer-relative velocities increase. Thus, visual direction seems to be a framework within which optical flow makes its major contribution.

3 OUR APPROACH

In analyzing the available information for the perception of one's heading, we have found it useful to start with the relative motions of pairs of objects. These can yield *nominal* information (left or right of a landmark) about heading. When coupled, multiple pairs can yield near-*absolute* information — a heading specification that fall between two landmarks in the visual field. In this section, we discuss three object-based sources of information for heading judgment.

3.1 Object-Based Information

Several classes of pairwise motions occur in the forward visual field: Objects converge or diverge. If the latter, they decelerate apart or accelerate apart. All convergence in the forward field is acceleratory. These three classes of relative motions, and their consequences for heading location, are suggested in Figure 1a. They apply to both straight and curved paths (Wang & Cutting, 1999b). If a moving observer knows the relative motion and the ordinal depth of two objects — which is nearer and which farther away — then several strong statements can be made concerning his or her nominal heading.

First, pairs that converge specify an invariant relation among themselves, a moving observer, and his or her heading. Heading is always to the outside of the nearer member of the pair, regardless of observer velocity or where in the field of view the pair might appear, as shown in Figure 1b. There are no exceptions. As discussed later, convergence is sufficiently potent that Wang and Cutting (1999a) implemented a machine algorithm for heading detection that works without assumptions of depth order; that knows only that heading can never lay between converging stationary objects.

Second, for pairs that decelerate apart, heading is also to the outside of the near member of the pair, but only when both objects are within 45° of the heading. This constraint is not too limiting since the data of Wagner, Baird, and Barbaresi (1981) show that pedestrians look within this limit about 90% of the time (see Cutting et al, 1999).

Third, without other information, pairs that accelerate apart offer no firm statement about heading. Probabilistically, one can state that heading is most often to the outside of the farther member of the pair — 69% of the time as calculated by Wang and Cutting (1999b) and shown in Figure 1c. Thus, accelerating divergence is a heuristic, not an invariant (Cutting & Wang, 2000; see also Gilden & Proffitt, 1989).

Fourth, there are pairs that have crossed over one another in the field of view. If one can remember the crossover of a pair of objects, heading will always be outside its farther member. This too is an invariant. It follows from the fact that converging items will, if the observer continues long enough on a straight path, always meet in the field of view, the nearer occlude the farther, and the two accelerate apart. Thus, what separates these third and fourth sources is, in part, observer memory. However, whereas all crossover pairs accelerate apart, all pairs accelerating apart need not have crossed over. Any

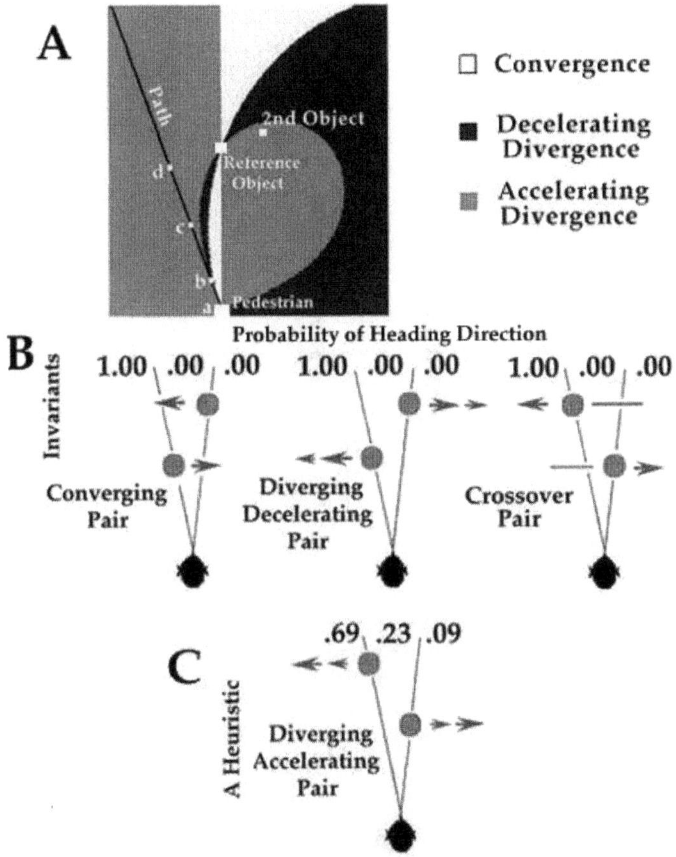

Figure 1. Panel A is a plan view of the layout around a pedestrian, and relative motions around an object 20° off path to the right. All other objects converge, diverge and decelerate, or diverge and accelerate from it. Panels B show invariant relations for pairs specifying heading direction. For pairs converging or decelerating apart, heading is always outside the nearer object; for crossover pairs it is always outside the farther object. Panel C shows that heading direction is probabilistic for pairs that decelerate apart, but most often to the outside of the farther object. A fixed sequence can occur among these four relative motions, suggested in Panel A: At Point (a) the reference and 2nd objects accelerate apart, at Point (b) they decelerate apart, at Point (c) they converge, and at Point (d) they cross over.

change in direction of the moving observer will change the orientation of the patterns as seen in Figure 1a, and create a plethora of new pairs accelerating apart that never crossed over from the point of view of the observer.

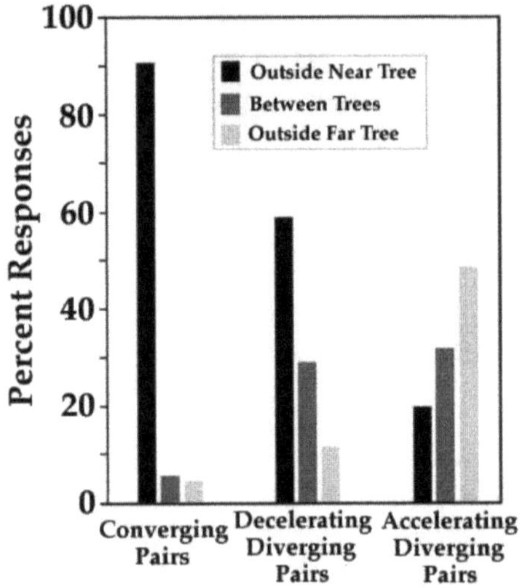

Figure 2. Heading responses for three types of stimuli in environments with only two trees. Responses are divided into three categories — those outside the near and far trees and those between. The information in Figure 1b would predict heading judgments should be most often placed outside the near tree for pairs that converge and decelerate apart, and outside the far tree when they accelerate apart. Results from Cutting & Wang (2000) showed three reliably distinct patterns, and that these predictions were upheld. Convergence is the more salient (and powerful) invariant, and the heuristic of accelerating divergence yields the least diverse pattern of responses. There were no crossover trials in this study. See also Best et al (2002) for a replication.

3.2 Invariants and Heuristic Compared

How effective are our posited sources of information for human observers? To eliminate other factors Cutting and Wang (2000) used minimal stimuli, those with two trees and also no crossovers. They used a pursuit fixation technique (Cutting, 1986; Kim et al., 1996; Royden et al., 1992; Warren & Hannon, 1988), with the dolly and pan of a camera simulating observer translation and eye/head rotation. Maximum rotations were less than 1°/s, and thus eye muscle feedback is not necessary (Royden, et al, 1992). From Figure 1b, observers ought to place their heading to the outside of the near tree of invariant pairs that converge or decelerate apart. Indeed, as shown in Figure 2, observers followed this placement 91% of the time for convergent pairs and 58% for pairs decelerating apart. These results show, as also found by Wang and Cutting (1999b) and Cutting et al (2000), that convergence is the more salient and powerful invariant.

Figure 1c suggests that for heuristic pairs, on the other hand, observers should place their heading to the outside of the far tree. And indeed they did

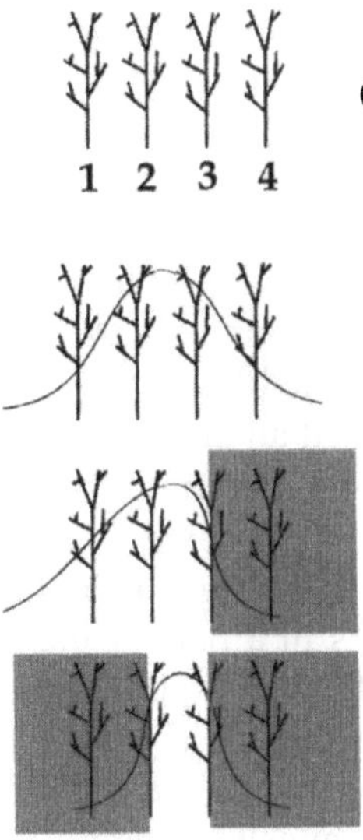

Figure 3. Schematic versions of four-tree stimulus arrays, diagrams of response distributions, and the response constraints in the three cases (in gray). For stimuli with no invariants, responses were distributed normally with a standard deviation (sd) of 3.6°. With one invariant the sd was 2.6° and the distribution highly skewed. An example is shown where an invariant pair, Trees 3 and 4, constrain responses to the left of Tree 3. With two invariants of opposing sign, responses can be constrained to a region between two trees (here Trees 2 and 3). Here again they were normally distributed, but with an sd of only 1.4°. Mean errors also decreased across stimulus types — from 4.1 to 2.2 to 0.8°. Required heading accuracy in this situation is 3.5° (Cutting et al, 1992).

— 47% of the time as shown in Figure 2 (see also Best et al., 2002). After considering these information sources, regression analyses showed no significant contributions of other variables, such as the angular separation of the two trees, or their relative distances from each other or the observer. The role of the heuristic in heading judgments with multiple objects is still not clear. What is clear is that the heuristic information cannot be combined with invariant information in the same manner the invariant information is combined. Thus, if accelerating divergence plays a role in heading judgment with multiple objects, some form of probability measure needs to be used. One such mechanism is illustrated in the next section on the simulations.

3.3 Invariants and Absolute Heading

Accurate judgments of absolute heading can be achieved with invariants of opposing sign. Wang and Cutting (1999b) investigated such situations and

Figure 3 shows a schematic rendering of how multiple invariants operate. In arrays without invariants responses were normally distributed; in arrays with one invariant responses were less varied and skewed away from the boundary specified by the invariant pair; and in arrays with two opposing invariants, with a permissible response region between two trees, responses were normally distributed but with a standard deviation less than half that of arrays without invariant pairs. Perhaps most importantly, absolute heading accuracy was within 0.8°, a value unsurpassed in the literature.

Wang and Cutting (1999a) developed an algorithm based on convergence alone. To compute the horizontal heading direction, the algorithm first divides the visual field into *n* regions of equal size. At the beginning, it assumes that heading can be at any of these regions, with equal probability ($1/n$). Then it computes the relative motion of objects in a pair of the regions to determine whether there is a converging pair. If there is a converging pair, then heading is more likely to be toward the regions outside the pair. Thus, the probability rating of these regions increases, while that of the other regions decreases. If there is no converging pair, then the opposite happens (see Figure 4a). This process continues until the algorithm goes through every pair. The final heading judgment is determined by the region with the highest probability rating (see Figure 4c).

Using this algorithm, Wang and Cutting (1999a) examined the effects of the number of objects in the environment, the depth variations, and the rotation rate on the accuracy of heading judgments. In the simulation, an observer moved toward a dot cloud with different densities of the dots, different depth separation between the nearest dot and the farthest one, and different rotation rates. Then the retinal motion of each dot was calculated. Based on the retinal motion, the algorithm then computed the heading based on the relative velocity of pairs of the dots, by adjusting the probability rating for each region according to the rule of convergence. Finally, the angular error between the computed heading direction and the actual heading direction was calculated.

The simulation results suggest that the algorithm provides both accurate heading judgments and a pattern of errors consistent with that of human observers. The algorithm achieves heading judgment accuracy of 0.1 degree even with the rotation of the eyes, outperforming many field-based models in metric accuracy, even though the model itself is not metric. Moreover, accuracy improves as the number of dots increases, consistent with the observation that human heading judgments are more accurate in richer environments (Li & Warren, 2000; Warren & Hannon, 1988). Accuracy also increases as a function of the depth variation. When the dots were all on a single frontal-parallel plane, heading error was rather large. As the depth separation among the dots increased, heading judgments became more accurate.

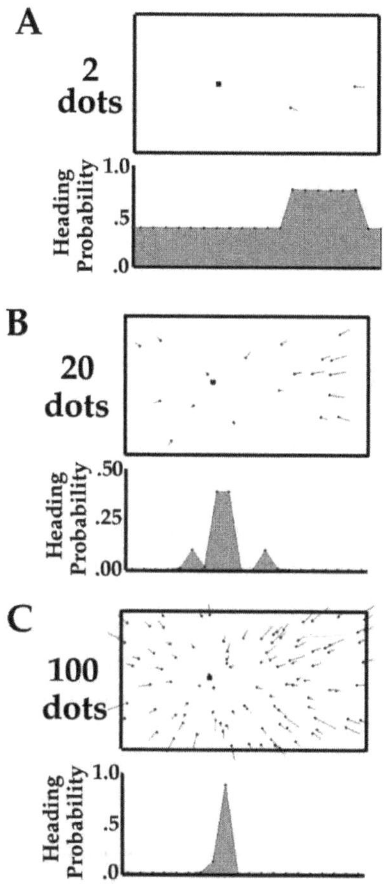

Figure 4. Simulation results of the algorithm based on convergence without depth information. Panel A shows a sample screen and results for two dots; Panel B for 20 dots, and Panel C for 100 dots. The small square shows the actual heading. The algorithm divides the horizontal visual field into 20 regions, and computes the probability of heading for each region. The judged heading is determined by the region with the highest probability. Notice that accuracy increases greatly with the number of dots (objects) in the display using one aspect of a single invariant, i.e., that one's heading can never lay between converging objects.

3.4 Invariants and Detecting a Moving Object

Objects in the environment often move. Such objects could pose a threat to travel, or alternatively they might be food or an object to catch. In either case it would behoove an observer to be able to segregate motion (the independent translation of objects) from movement (object displacements due to observer translation), a distinction first made by Gibson (1954). Consider three ways. First, familiarity with objects will go some way towards specifying what is in motion and what is not. Cars, people, and animals can translate to new positions; trees, buildings, and rocks normally cannot. But, of course, at any given moment cars and other objects can be either translating or be stationary. Thus, familiarity cannot be the sole means by which motion among movement is detected.

Second, binocular disparities might aid the segregation of motion from movement (Kellman & Kaiser, 1995). However, these may not be useful in situations where object motion is very slight, or when it is further away than about 30 m (Cutting & Vishton, 1995).

Third, retinal velocity might predict their difference (Brenner, 1991; Wertheim, 1995). Wagner et al. (1981) found that pedestrians look at stationary objects about 60% of the time, and moving objects 40%. If we fixate a stationary object near our path, any object in motion is likely to be the fastest moving object in the field of view, or at least near the fovea, unless its velocity is quite minimal. If one fixates an object in motion the patterns of movement of stationary objects are complex, but such movements are critical for detections of collisions and bypasses (Cutting, Vishton, & Braren, 1995).

Cutting and Readinger (2002) removed familiarity cues, binocular disparities, and insofar as possible, differences in retinal motion as variables that could aid in the detection of motion during movement. They presented viewers an array of identical poles, one of which (albeit perhaps somewhat mysteriously) was mobile, but only to a very modest degree. A sample stimulus arrangement is shown and explained in Figure 5. Observers indicated which pole was in motion during their simulated translation, and they could see each trial as often as they liked.

Detection of the mobile pole was strongly influenced by the field of invariants on a given trial. That is, if all invariants were *coherent* — they all specified that heading was in the same direction or they all converged uniformly on a central region — then judgments were less accurate (47% correct where chance was 10%) and more difficult (observers chose to see each such trial more often). When the invariants of the stimuli array were *incoherent*, as shown in Figure 5, detection accuracy was greater (60%) and less difficult. Moreover, if the trial was incoherent and the mobile pole misidentified, two thirds of all confusion errors involved a neighboring pole that was involved in the invariant incoherent with the rest of the field. These results cannot be explained by image velocities of moving poles as opposed to the stationary ones or the relative depth of the moving pole within the display.

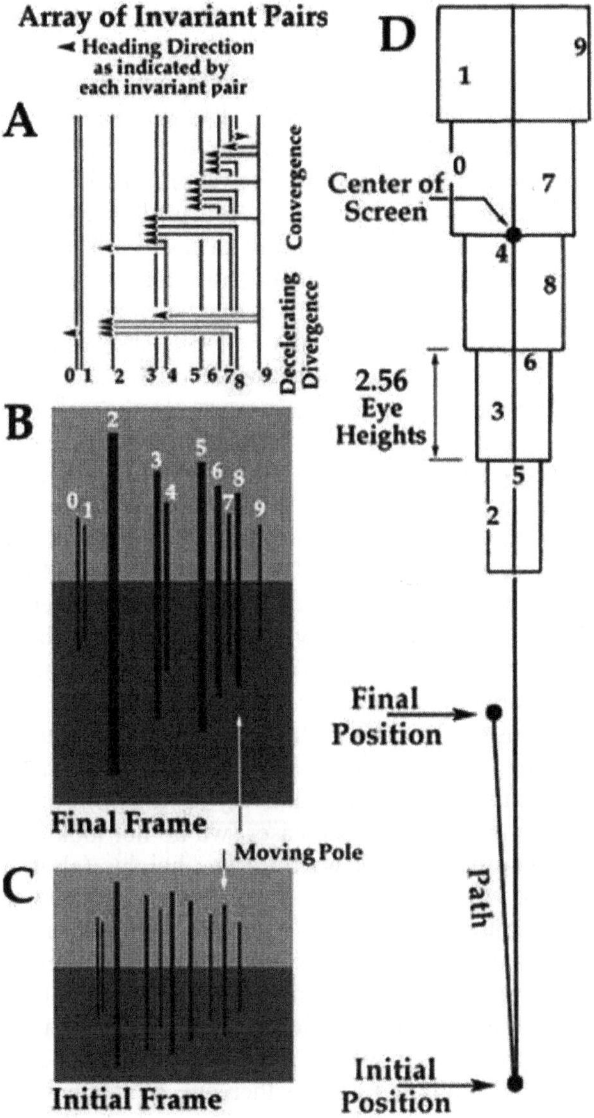

Figure 5. An example trial and its structure. Panel D shows a plan, scaled view of its layout, the locations of the observer, and the regions within which poles could be randomly placed. Panels C and B show the beginning and ending frames, respectively, of the four second sequence simulating observer movement along a path 3° to the left of the poles and with a pan to the right to keep the center of the region of poles in view. The moving pole on this particular trial is Pole 8. Pole numbers and positions correspond in all panels. Panel A shows the 19 pairs of

4 EYE MOVEMENTS AND HEADING JUDGMENTS

Animals and humans move their heads and eyes to reduce blur, particularly when they move (Land, 1999). Eye and head movements assure optimal resolution of objects of interest around them. But what do we look at, and why? Unfortunately, the data from free-ranging pedestrians is scant. Eye movement data from motorists have been discussed at length (e.g., Chapman & Underwood, 1998; Land, 1992; Land & Lee, 1994; see also Readinger et al., 2002). Although valuable, these data show that motorists often look at things that are not pertinent to a pedestrian (e.g. such as the inner tangent of a roadway during a turn). In addition Calvert (1954) observed that, as motorists increase their speed, they scan a smaller sector of the visual field, narrowing in on a region near their aim point. This is due to the fact that, with speed, accurate pursuit fixation off to the side becomes increasingly difficult. Motorists thus face a situation foreign to our evolutionary ecology. With an eye height less than a pedestrian and a translational velocity often an order of magnitude greater, they commonly experience a rapidity of optical flow never seen on foot, and never experienced by human beings before the mid-19th century (Schivelbusch, 1986). Motorists' fixations, eye movements, and skills are a marvel of human adaptability, but they cannot be mapped in a straightforward manner onto situations and constraints in which we evolved.

Important eye movement data also exist for pedestrians whose footfall is unsure (e.g., Hollands et al., 1995; Patla & Vickers, 1997), suggesting ways in which we guarantee our safe progress by looking at a surface of support. However, in most situations we are relatively assured of our locomotion and not endangered of imbalance. In such cases, we do not look down very often; only 25% of the time do we look within 4 eye heights (about 6.4 m) of our feet (Wagner et al., 1981). Thus, we have been interested in the looking behavior of pedestrians who would be secure in their gait and how they pick their paths through environments containing interesting things.

Recently, investigators have begun to consider the relation between eye movements and heading judgments. Some research has focused on whether, and how fast, observers can move their eye towards their heading (Hooge et al., 1999). Others have focused on whether eye movements are needed for

---

Figure 5 cont'd: poles involved in heading invariants, 14 converging and 5 decelerating apart. Each arrow points in the specified heading direction, its stem connects the two poles involved in a particular invariant relation, and the base of the arrowhead delimits the edge of the response region allowed by the particular invariant pair. Eighteen of these invariants specify that heading be to the left; one pair involving Pole 8 yields a heading result incoherent with the rest, specifying that heading be to the right. This discrepancy would contribute to the detection of Pole 8 as a mobile pole, and also the possible confusion of Pole 7 with it. Were all invariant pairs consistent in their specified heading direction the trial would be called coherent.

heading judgments in unstructured environments with brief-display times (Grigo & Lappe, 1999). Neither of these research foci is of direct interest here. Much research has demonstrated that people can find their heading rapidly so it is little surprise that they can do this with their eyes. Also, we have shown that unstructured environments (fields of moving dots) often yield results quite different than those of more structured ones (e.g., arrays of schematic trees; Cutting et al., 1997; Vishton & Cutting, 1995; Li & Warren, 2001). In particular, dot fields may introduce biases not present in more naturalistic stimuli. Given that responses are often different in the two situations, we would expect eye movements to differ as well.

What eye fixation data are relevant? The data of Wagner et al. (1981) derive from an ecological survey of 800 fixations by 16 people walking through a familiar setting (a town and campus) for 90 min. These data suggest that 90% of the time we are looking away from our aim point by more than 5°. This is a conundrum, since the physical constraints of pedestrian locomotion suggest that we must know heading within 3.5° (Cutting et al., 1992). If knowledge of heading is so important to us why do we spend so much of our time looking where it is not? The argument made here, and in our previous research, is that the best information for one's heading is indeed often well off one's path. Since the environment is typically a plenum of attention-grabbing objects, most of which are off one's path, one can well afford looking at them for their interest value while simultaneously gathering information about heading. Thus, we have focused on the observer-relative information off one's path.

4.1 Invariants and Eye Movements

Cutting et al (2000) investigated observers' eye movements during the course of simulated travel towards a grove of four trees and prior to making a heading judgment. Stimulus sequences mimicked pursuit eye fixations on one tree, which varied in screen position as shown in Figure 6a, during forward movement. This manipulation varied initial eye position within the stimulus array allowing for saccades elsewhere, and partly overcame the general tendency for observers to continue to look at midscreen. Important comparisons were between trees across trials in the same array position that were and were not part of invariant pairs.

Wearing an eye-tracker, each of twelve participants looked anywhere on the screen they liked during the course of the trial; they determined their heading during the course of the trial, and at the end of the sequence moved the screen cursor to their apparent heading and pressed the mouse. Thus, two

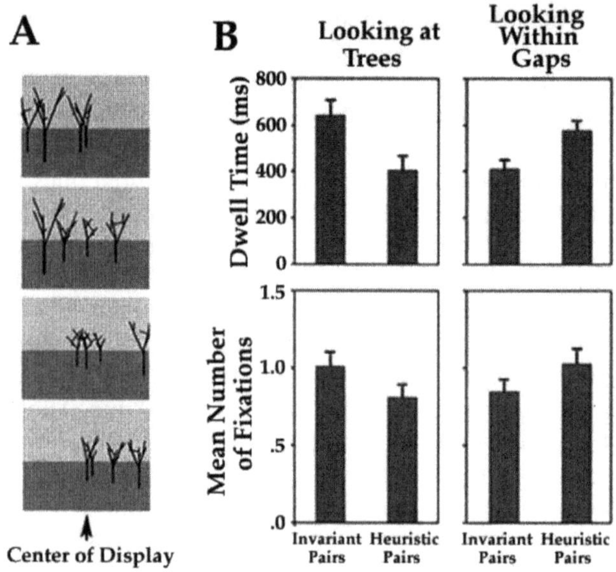

Figure 6. Panel A shows four sample arrangements of the four trees on given stimulus trials. Initial fixation was at screen center and would fall on one tree at the beginning of the trial. This variation allowed placement of invariant pairs on and off the initial fixation point. Panel B shows mean dwell times and mean numbers of fixations on each tree within invariant pairs and heuristic pairs, and within each gap between invariant pairs and heuristic pairs. Error bars indicate one standard error of the mean.

types of response were recorded — heading judgments at the end of each trial and eye fixations throughout each trial. Figure 6b shows the results.

Observers spent most of their time (63%) looking at one of the four trees, and considerably less time (24%) looking in gaps between them. Given the resolution of the eye tracker the residual dwell time (13%) was indeterminate. More importantly, as shown in Figure 6b, observers looked 240 ms longer at, and produced 0.18 more fixations on, each tree that was a member of an invariant pair than each heuristic-pair tree in its same stimulus position. From Figures 1b and 1c such a result makes sense if observers are seeking reliable information about the location of their heading. In addition, mean dwell times were 170 ms shorter in gaps between invariant pairs and there was a mean of 0.17 fewer fixations there as well. This too makes sense. Figure 1b shows that heading can never lie between members of invariant pairs, so there is little need to spend much time looking there.

Finally, consider the relation between fixations and heading responses. For each observer, we tallied a 2X2 table: whether they looked at the near tree of the invariant pair or not, and whether they placed their heading response to the correct side of this tree or not. When observers looked at the pertinent tree they were 75% correct; when they did not look at this tree, they were only 19% correct. All observers showed this pattern. These patterns strongly suggest that observers seek out their heading through the use of invariant information among pairs of objects as they move through the environment.

5 OVERVIEW

Across the studies reported here on the information available in the retinal array during locomotion, we have shown that:

(a) the observer-relative motion of invariant generating pairs of stationary objects specifies nominal heading;
(b) the observer-relative motion of heuristic generating pairs offers probabilistic heading information;
  - the invariants and heuristic can be used by observers in judgments about their heading;
  - couplings of invariant pairs on either side of the heading with opposite sign constrain observer judgments to the narrow region between them, and these constraints are honored;
(c) a collection of many invariant pairs can, when yielding an incoherent heading solution due to the independent motion of one object, guide the detection of that moving object; and
(d) observers seek out invariant pairs through patterns of eye movements and fixations prior to their heading judgments. The rationale for this appears to be that, by fixating objects belonging to invariant pairs, convergent and decelerating divergent motion can be more easily registered, depth order noted, and confirmatory information about heading accrued. We assume making accurate heading judgments is a task that requires scrutiny; no pop out occurs, at least at pedestrian speeds.

In this chapter we have presented an overview of our research program on how observers determine their heading, or the instantaneous location in the visual field of the point towards which they are moving. Most approaches to this problem use the global information available across the entire visual field. Ours, in contrast, is a piecemeal approach which focuses on the information generally available at the fovea and which can accrue over a sequence of fixations. In terms of the information available to the moving observer, there is no difference between these two types of approaches. In terms of information use, however, we believe ours is superior in that it accounts for both correct performance and for errors that observers make in experiments and, by extrapolation, in the real world.

REFERENCES

Best, C. J., Crassini, B., & Ross, R. H.. (2002). The roles of static depth information and object-image relative motion in perception of heading. *J. Exp. Psychol. Hum. Percept. Perform., 28,* 884-901.

Brenner, E. (1991). Judging object motion during smooth pursuit eye-movements: The role of optic flow. *Vision Res., 31*, 1893-1902.

Calvert, E. S. (1954). Visual judgments in motion. *J. Inst. Nav., 7* , 233-251 & 398-402.

Chapman, P. R. & Underwood, G. (1998). Visual search in dynamic scenes: Event type and the role of experience in viewing driving situations. In G. Underwood (ed.) *Eye Guidance in Reading and Scene Perception* . pp. 369-394. Amsterdam: Elsevier Science.

Cutting, J. E. (1986). *Perception with an Eye for Motion.* Cambridge, MA: MIT Press.

Cutting, J. E., Alliprandini, P.M.Z., & Wang, R.F. (2000). Seeking one's heading through eye movements. *Psychon. Bull. Rev., 7*, 490-498.

Cutting, J. E. & Readinger, W. O. (2002). Perceiving motion while moving, or how heading invariants make optical flow cohere. *J. Exp. Psychol. Hum. Percept. Perform.,, 26*, 731-747.

Cutting, J. E., Springer, K., Braren, P. A., & Johnson, S. H. (1992). Wayfinding on foot from information in retinal, not optical, flow. *J. Exp. Psychol. Gen., 121*, 41-72 & 129.

Cutting J. E., & Vishton P. M. (1995). Perceiving layout and knowing distances. in W. Epstein & S. Rogers (Eds.). *Perception of Space and Motion.* pp. 69-117. San Diego, CA: Academic Press.

Cutting, J. E., Vishton, P. M. & Braren, P. A. (1995). How we avoid collisions with stationary and with moving obstacles. *Psychol. Rev., 102*, 627-651.

Cutting, J. E., Vishton, P. M., Flückiger, M., Baumberger, M., & Gerndt, J. (1997). Heading and path information from retinal flow in naturalistic environments. *Percept. Psychophys., 59,* 426-441.

Cutting, J. E. & Wang, R. F. (2000). Heading judgments in minimal environments: The value of a heuristic when invariants are rare. *Percept. Psychophys., 62*, 1146-1159.

Cutting, J. E., Wang, R. F., Flückiger, M., & Baumberger, M. (1999). Human heading judgments and object-based motion information. *Vision Res., 39*, 1079-1105.

Frey, B. F. & Owen, D. H. (1999). The utility of motion parallax information for the perception and control of heading. *J. Exp. Psychol. Hum. Percept. Perform., 25*, 445-460.

Gibson, J. J. (1954). The visual perception of objective motion and subjective movement. *Psychol. Rev., 61*, 304-314.

Gibson, J. J. (1966). *The Senses Considered as Perceptual Systems.* Boston: Houghton Mifflin.

Gilden, D. L. & Proffitt, D. R. (1989). Understanding collision dynamics. *J. Exp. Psychol. Hum. Percept. Perform., 15*, 372-383.

Grigo, A. & Lappe, M. (1999). Dynamical use of different sources of information in heading judgments from retinal flow. *J. Opt. Soc. Am. A, 16*, 2079-2091.

Grossberg, S., Mingolla, E., & Pack, C. (1999). A neural model of motion processing and visual navigation by cortical area MST. *Cereb. Cortex, 9*, 878-895.

Hildreth, E. C. (1992). Recovering heading for visually-guided navigation. *Vision Res., 32*, 1177-1192.

Hollands, M. A. Marplehorvat, D. E., Henkes, S., & Rowan, A. K. (1995). Human eye movements during visually guided stepping. *J. Mot. Behav., 27*, 155-163.

Hooge, I. T. C., Beintema, J. A., & van den Berg, A. V. (1999). Visual search of heading direction. *Exp. Brain Res., 129*, 615-628.

Kellman, P. J. & Kaiser, M. K. (1995). Extracting object motion during observer motion: Combining constraints from optic flow and binocular disparity. *J. Opt. Soc. Am. A, 12*, 623-625.

Kim, N.-G., Turvey, M. T., & Growney, R. (1996) Wayfinding and the sampling of optical flow by eye movements. *J. Exp. Psychol. Hum. Percept. Perform., 22,* 1314-1319.

Koenderink, J. J. & van Doorn, A. J. (1975). Invariant properties of the motion parallax field due to the movement of rigid bodies relative to an observer. *Optica Acta, 22*, 773-791.

Land, M. (1992). Predictable eye-head coordination during driving. *Nature, 359*, 318-320.

Land, M. (1999). Motion and vision: Why animals move their eyes. *J. Comp. Physiol. A, 185*, 341-352.

Land, M. & Lee, D. N. (1994). Where do we look when we steer? *Nature, 369,* 742-744.

Lappe, M., Bremmer, F., & van den Berg, A. V. (1999). Perception of self-motion from visual flow. *Curr. Trends Cognit. Sci., 3*, 329-336.

Li, L. & Warren, W. H. (2001). Perception of heading during rotation: Sufficiency of dense motion parallax and reference objects. *Vision Res., 40,* 3873-3894.

Llewellyn, K. R. (1971). Visual guidance of locomotion. *J. Exp. Psychol., 91*, 245-261.

Morris, G. H. (1990). *Hunter Seat Equitation* (3rd ed.). New York: Doubleday.

Motorcycle Safety Foundation. (1992). *Evaluating, Coaching, and Range Management Instructor's Guide.* Irvine, CA: Author.

Motter, B. (1993). Focal attention produces spatially selective processing in visual area V1, V2, and V4 in the presence of competing stimuli. *J. Neurophysiol., 7*, 2239-2255.

Patla, A. E. & Vickers, J. N. (1997). Where and when do we look as we approach and step over an obstacle in the travel path? *NeuroReport, 8*, 3661-3665.

Perrone, J. & Stone, L. (1994). A model of self-motion estimation within primate visual cortex. *Vision Res., 34,* 1917-1938.

Perrone, J. & Stone, L. (1998). Emulating the visual receptive-field properties of MST neurons with a template model of heading estimation. *J. Neurosci., 18*, 5958-5975.

Readinger, W. O., Chatziastros, A., Cunningham, D. W., Bülthoff, H. H., & Cutting, J. E. (2002). Systematic effects of gaze-eccentricity on steering. *J. Exp. Psychol. Appl.,* in press.

Regan, D. M. & Beverley, K. I. (1982). How do we avoid confounding the direction we are looking with the direction we are going? *Science, 215*, 194-196.

Rieger, J. H. & Lawton, D. T. (1985). Processing differential image motion. *J. Opt. Soc. Am. A, 2*, 354-360.

Royden, C. S., Banks, M. S., & Crowell, J. A. (1992). The perception of heading during eye movements. *Nature, 360*, 583-585.

Rushton, S. K., Harris, J. M., Lloyd, M. R., & Wann, J. P. (1998). Guidance of locomotion on foot uses perceived target location rather than optic flow. *Curr. Biol., 8*, 1191-1194.

Schivelbusch, W. (1988). *The Railway Journey.* Berkeley, CA: University of California Press.

Underwood, G & Radach, R. (1988). Eye guidance and visual information processing: Reading: visual search, picture perception, and driving, in G. Underwood (ed.) *Eye Guidance in Reading and Scene Perception.* pp. 1-27. Amsterdam: Elsevier Science.

Vishton, P. M. & Cutting, J. E. (1995). Wayfinding, displacements, and mental maps: Velocity fields are not typically used to determine one's aimpoint. *J. Exp. Psychol. Hum. Percept. Perform., 21*, 978-995.

Wagner, M., Baird, J. C., & Barbaresi, W. (1981). The locus of environmental attention. *J. Environ. Psychol., 1*, 195-201.

Wang, R. F. & Cutting, J. E. (1999a). A probabilistic model for recovering camera translation. *Comput. Vision. Im. Understanding. 76*, 205-212.

Wang, R. F. & Cutting, J. E. (1999b). Where we go with a little good information. *Psychol. Sci., 10*, 72-76.

Warren, W. H. (1998). The state of flow, in T. Watanabe (ed.) *High-level Motion Processing.* pp. 315-358. Cambridge, MA: MIT Press.

Warren, W. H. & Hannon, D. J. (1988). Direction of self-motion is perceived from optical flow. *Nature, 336*, 162-163.

Warren, W. H., Kay, B. A., Zosh, W. D., Duchon, A. P., Sahuc, S. (2001). Optic flow is used to control human walking. *Nat. Neurosci., 4*, 201-202.

Warren, W. H. & Saunders, J. A. (1995). Perceived heading in the presence of moving objects. *Percept., 24,* 315-331.

Wertheim, A. (1995). Motion perception during self-motion: The direct versus inferential controversy. *Behav. Brain Sci., 17*, 293-355.

5. Short-Latency Eye Movements: Evidence for Rapid, Parallel Processing of Optic Flow

F.A. Miles[1], C. Busettini[1,2], G.S. Masson[1,3] and D.S. Yang[1,4]

[1] Laboratory of Sensorimotor Research
National Eye Institute, Bethesda, MD, USA

[2] Department of Physiological Optics
University of Alabama at Birmingham, Birminham, AL, USA

[3] Institut de Neurosciences Physiologiques et Cognitives
C.N.R.S., Marseille, France

[4] Department of Ophthalmology
Columbus Children's Hospital, Columbus, OH, USA.

1 INTRODUCTION

As we go about our daily activities we view the world from a constantly shifting platform and some visual functions are compromised if the images on the retina are not reasonably stable. For example, visual acuity begins to deteriorate when retinal image speeds exceed a few degrees per second (Westheimer & McKee, 1975). There are a number of visual reflexes that help to stabilize our gaze on particular objects of interest by generating eye movements to offset our head movements. However, it is important to remember that these visual mechanisms normally operate in close synergy with vestibuloöcular reflexes that rely on two types of end-organ embedded in the base of the skull: the *semicircular canals*, which are selectively sensitive to *angular* accelerations of the head, and the *otolith organs*, which are selectively sensitive to *linear* accelerations (Goldberg & Fernandez, 1975). Thus, the vestibular end-organs decompose head movements into their angular and linear components and support two quite independent reflexes, the RVOR and TVOR, that compensate selectively for rotational and translational disturbances of the head respectively with latencies <10 msec. These vestibular reflexes operate open-loop—because their output, eye movement, does not influence their input, head movement—and neither is perfect, hence motion of the observer must often be associated with some

L.M. Vaina, S.A. Beardsley and S.K. Rushton (eds.), Optic Flow and Beyond, 79–107.

residual retinal image motion and this is where the visual stabilization mechanisms become involved. However, the visual end-organs — the two retinas — see *all* visual disturbances, regardless of whether they result from rotation and/or translation of gaze so that if any visual decomposition is to be done it must be by signal processing in the central nervous system (CNS). It is our contention that the visual system does attempt to perform such decomposition, using visual filters to sense the pattern of optic flow and thereby infer the observer's motion and the eye movements that best compensate for that motion.

The traditional approach to visual stabilization of the eyes has ignored translational problems completely and placed the observer inside a rotating drum to simulate the visual events associated with a failure of the RVOR during head turns. This elicits a pattern of tracking eye movements, often termed optokinetic nystagmus (OKN), which appears to be present in all animals with mobile eyes, and the finding that the rabbit's optokinetic system was organized in canal or eye-muscle coordinates strongly endorsed the idea that the adequate stimulus for OKN was rotation (Simpson, 1984; Simpson & Graf, 1985). However, such data are not available for primates, whose OKN has two distinct components: an early component (OKNe) with brisk dynamics and a delayed component (OKNd) with sluggish dynamics (Cohen et al., 1977).

The view that OKNd compensates for rotational disturbances was reinforced by the finding that long-term adaptive changes in the gain of the RVOR—induced by exposure to magnifying or minifying spectacles (Miles & Eighmy, 1980; Miles & Fuller, 1974)—were associated with proportional changes in the gain of OKNd (Lisberger & Miles, 1980). This led to the suggestion that RVOR and OKNd were synergistic, shared the same coordinate system with a common adaptive gain element, and specifically compensated for head rotations. Significantly, there were no changes in the gain of OKNe in those experiments.

More recent studies of OKNe have employed large moving patterns back-projected onto a translucent tangent screen facing the observer (because it offers much better control of the stimulus parameters) and the responses evoked in this situation have been termed "ocular following responses", or OFR (Miles et al., 1986). Of particular interest here is the finding that changes in the gain of the TVOR—resulting from changes in viewing distance (Paige, 1989; Schwarz et al., 1989; Schwarz & Miles, 1991)—were associated with proportional changes in the gain of OFR (Busettini et al., 1991; Busettini et al., 1994; Schwarz et al., 1989; Schwarz & Miles, 1991). Once again it was assumed that this arose from a gain element that was shared by the visual and vestibular mechanisms—this time, OKNe/OFR and the TVOR—which this time functioned to compensate for translational disturbances of the head. (Unfortunately, the gain of OKNd was not tested in these experiments.) From these findings the idea emerged that OKNd and OKNe/OFR operated largely independently as backups to the RVOR and TVOR, respectively.

One of the striking features of OFR is their ultra-short latency: <60 ms in monkeys (Miles et al., 1986) and <85 ms in humans (Gellman et al., 1990). Subsequent studies on the visually driven oculomotor responses have revealed that there are at least three visual reflexes with ultra-short latencies that have special features to deal with the visual disturbances associated with translation of the observer. These seem to constitute a family of machine-like, ultra-rapid reflexes and have a number of features in common. Notably, all utilize relatively low-level, preattentive cortical processing of complex visual stimuli and yet probably function largely independently of perception. Two of these visual reflexes sense the observer's motion by decoding the global pattern of optic flow, one (OFR) dealing with the problems of the observer who looks off to one side and the other (radial-flow vergence) with the problems of the observer who looks straight ahead. The third visual reflex (disparity vergence) complements the second, helping to maintain binocular alignment on the objects that lie ahead, but utilizes a different cue to motion in depth: the change in the relative alignment of the images on the two retinas (binocular disparity). Nonetheless, this last reflex shares much in common with the other two and we think that all three normally work in close synergy to deal with the complex optic flow associated with motion of the observer in a 3-D visual environment.

In reviewing the operation of these three reflexes our focus will be on their very earliest responses that are generated open-loop, i.e., on the eye movements that occur within two reaction times: after one reaction time the eye-movement responses begin to modify the visual stimulus and, after a second reaction time, this reafference will in turn begin to modify the recorded eye movement responses, and so on and so forth. This concentration on the initial open-loop responses hugely simplifies the interpretation of the data and allows one to infer the sensory-motor processing directly from the stimulus-response relationships.

There is one characteristic that is shared by all three visual reflexes and that we feel provides an important insight into the visual context in which they operate: all three show transient *post-saccadic enhancement* whereby a stimulus in the immediate wake of a saccadic eye movement elicits a much larger oculomotor response than the same stimulus applied a few hundreds of milliseconds later (Busettini et al., 1996b, 1997, 2001; Gellman et al., 1990; Kawano & Miles, 1986). Much of this effect is visual in origin and results from the motion of the background across the retina during the saccade (reafference)—in fact, a saccadic-like shift of the visual scene is sufficient to produce enhancement. Thus, all three reflexes are particularly sensitive to changes occurring in the wake of saccadic eye movements, a time when binocular stability and alignment are particularly at risk and when there is an urgent need to stabilize binocular gaze as soon as possible in order to scrutinize the new images that have just been brought into the fovea. Accordingly, we think that these reflexes play a special role in promoting binocular stability immediately after saccadic gaze shifts and all of the experiments that will be reviewed here dealt with the initial ocular responses elicited in the immediate wake of a centering saccade.

There is another visual tracking mechanism—the so-called *smooth pursuit system*—that should be mentioned. This operates as a feature-tracking mechanism—its hallmark is the ability to track small targets that move across a stationary background—but it will not be considered here because, at least so far, it has not provided any insights into the neural processing of global optic flow patterns, which are our major concern in this book. Fortunately, these pursuit responses have quite a long latency and usually do not get under way until after the initial open-loop reflex responses that are our major concern in this review are completed, hence they are not a complicating factor in most of the experiments under consideration here. The deployment of this pursuit system by subjects experiencing continuous global optic flow has been reviewed recently by Lappe & Hoffmann (2000).

2 THE PATTERNS OF OPTIC FLOW ASSOCIATED WITH ROTATIONAL AND TRANSLATIONAL DISTURBANCES OF GAZE

There are some fundamental differences between the optic flow associated with rotational and translational disturbances of gaze: see Koenderink (1986) for review. Let us first consider the pure rotation case. If an eye rotates with respect to the stationary surroundings, there is a coherent motion of the entire retinal image that can be considered to be distributed over the surface of a sphere and created by projection through a vantage point at the center. In the simplest case, the eye rotates about this vantage point and the pattern of optic flow resembles the lines of latitude on a globe (Figure 1A), so that the speed of flow is greatest at the "equator" and decrements as the cosine of angle of latitude. However, because the eye has only a restricted field of view, only a portion of this flow field will be visible at any given time (e.g., Figure 1B). If the image motion resulted from rotation of the head then it could be almost completely eliminated by equal and opposite (i.e., compensatory) rotations of the eyes in the head so that the entire scene would be stabilized on the retina.[1]

---

[1] This ignores the fact that we have two eyes and that the eyes are some distance away from the usual axis of head rotation, which means that the eyes will always undergo some translation during real-world rotations of the head.

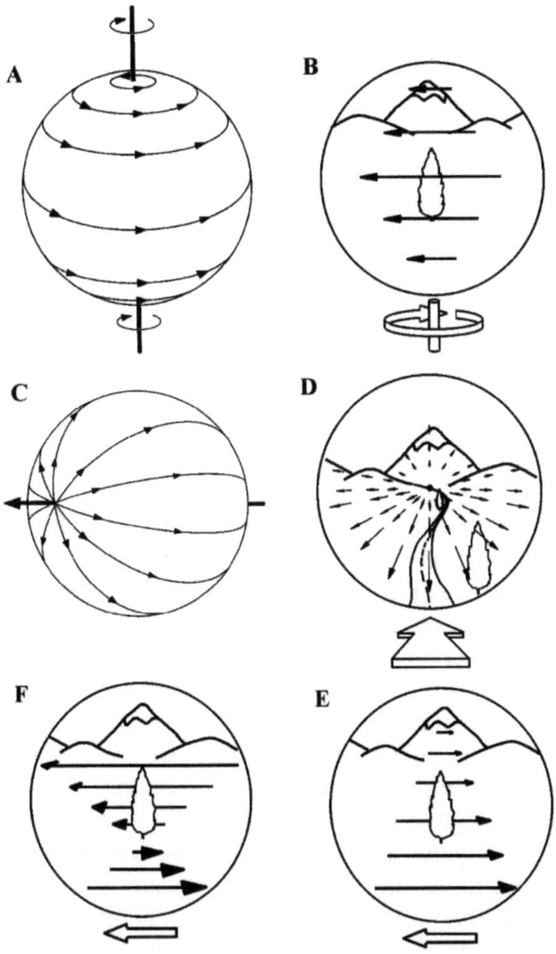

Figure 1. Patterns of optic flow experienced by a moving observer. A. The retinal flow can be considered to be distributed over the surface of a sphere and created by projection through a vantage point at the center. Here, the observer rotates about this vantage point (Miles et al., 1991). B & D-F. Cartoons showing the observer's limited field of view and the pattern of motion experienced. B. Observer rotates about a vertical axis as in A and looks straight out to one side; the speed of optic flow is greatest at "equator" and decrements as cosine of angle of latitude. The pattern and speed of flow are solely determined by the observer's motion (Miles, 1997). C. Optic flow associated with translation of the vantage point (Miles et al., 1991). D. Observer who translates as in C and looks in direction of heading sees expanding flow (Busettini et al., 1997). E. Observer who translates as in C and looks off to the right side (but makes no compensatory eye movements) sees image motion in inverse proportion to the viewing distance so the world appears to pivot about infinity (Miles et al., 1992). F. As in E but here the observer stabilizes the image of the tree in the middle ground by generating compensatory eye movements so that the world now appears to pivot about the tree (Miles et al., 1992).

When an eye undergoes pure translation, the optic flow consists of streams of images emerging from a focus of expansion straight ahead and disappearing into a focus of contraction behind, the overall pattern resembling the lines of longitude on a globe (Figure 1C). As with rotational disturbances, the *direction* of flow at any given point depends solely on the motion of the observer but in contrast, the *speed* of the flow also depends on the 3-D structure of the visual surroundings, being inversely proportional to the viewing distance at that location. Thus, nearby objects move across the field of view much more rapidly than more distant ones due to motion parallax (Gibson, 1950; 1966). Again, given the eye's restricted field of view the pattern of motion actually seen depends on where the eye is pointing relative to the direction of heading. If the eye points exactly in the direction of heading then the retina sees a radially expanding pattern of flow (e.g., Figure 1D), whereas if the eye points off to one side then there will be a laminar pattern of optic flow (Figure 1E).

In the real world, of course, such translations result from movements of the head but it is not always immediately obvious what useful role eye movements might play here. In general, we assume that foveal vision takes precedence, especially in the wake of a saccade when the visual tracking systems that are of most concern to us here are most actively engaged. In the case of the moving observer who looks in the direction of heading, objects of interest that lie ahead are getting closer and, because the observer is not the cyclops implied by Figure 1D but has two eyes with slightly differing viewpoints, he/she must converge his/her two eyes in order to keep the two fovea aligned on those objects. This appears to be the role of the *radial-flow vergence* mechanism that will be our first major concern in this article.

The optic flow depicted in Figure 1E is that experienced by the moving observer who, for example, looks off to one side and is passive, i.e., makes no eye movements, so that retinal image motion is inversely related to the viewing distance. In this case, the only stable retinal images are those whose objects are at infinity. This would be appropriate if the observer wished to scrutinize the mountain in Figure 1E, but not if he/she wished to scrutinize the tree in the middle ground; for this, the observer must become active and his/her eyes must compensate for the motion of the head, rotating in their sockets by an amount that is inversely proportional to the distance between the eye and the tree. In this event, the compensatory eye movements would be conjugate, i.e., the same for both eyes, and would completely transform the pattern of optic flow because the translational motion (due to the motion of the head) is now combined with rotational motion (due to the compensatory eye movements): see Figure 1F. Thus, during translation, at any given time compensatory eye movements can stabilize only the images in one particular depth plane—in our cartoons in Figure 1, this is the plane of the mountain or of the tree but not of both. It is in situations like these that we think OFR have their major role and they will be our second major concern in this article.

If the observer who looks in the direction of heading (as in Figure 1D) fails to converge his/her eyes adequately then the plane of fixation will overtake the object(s) of regard, shifting those objects out of the plane of fixation so that they acquire so-called crossed disparity, a very potent vergence stimulus. Recent experiments indicate that disparities applied to large patterns elicit vergence eye movements at latencies very similar to those of OFR and radial-flow vergence. These *disparity vergence* responses, which will reinforce the radial-flow vergence responses, will be our last major concern in this review.

It is now apparent that we will deal with two kinds of eye movement: *vergence* that alters the angle between the two lines of sight and thereby changes the distance to the plane of fixation, and *version* that alters the eccentricity of the two eyes together and thereby shifts gaze within the plane of fixation (often termed conjugate movements). Vergence (Vg), which is given by the difference in the positions of the two eyes [L-R], and version (Vs), which is given by the average position of the two eyes [(L+R)/2], are orthogonal representations and provide a complete description of binocular eye movements such that the positions of each eye can be reconstructed from them. Thus, adopting the convention that rightward movement is positive, then increases in convergence are also positive, and L=Vs+Vg/2 while R=Vs-Vg/2.

3 RADIAL-FLOW VERGENCE

We now concentrate upon the gaze stability problems of the moving observer who looks in the direction of heading and so experiences the radial pattern of optic flow featured in Figure 1D. Insofar as the radial pattern of flow is associated with a change in viewing distance, the observer must converge his/her eyes if the object of interest in the scene ahead is to stay imaged on both foveas. Recent experiments on primates have indicated that radial optic flow elicits vergence eye movements at ultra-short latencies, ~85 ms in humans (Busettini et al., 1997) and ~60 ms in monkeys (Inoue et al., 1998). Expanding flow, which signals a forward approach and hence a decrease in the viewing distance resulted in increased convergence, and contracting flow, which signals the converse, resulted in decreased convergence. These experiments used a two-frame movie (two slide projectors with fast mechanical shutters) to simulate the visual experience of an observer who undergoes sudden (i.e., step-like) displacement towards or away from a tangent screen covered in random dots. The focus of

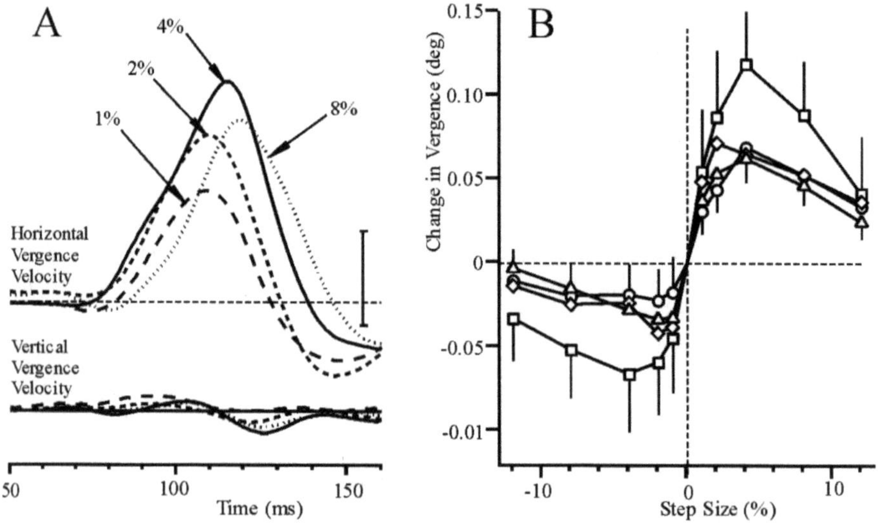

Figure 2. Initial vergence responses to expanding and contracting radial flow (two-frame movie; radial optic flow plus size change). A. Mean horizontal and vertical vergence velocity associated with expanding steps equivalent to a reduction in viewing distance of 1-8% at time zero (one human subject). Upward deflections indicate increased convergence and left sursumvergence. Calibration 2°/s. B. Dependence of mean changes in horizontal vergence (over the period 95-128 ms from stimulus onset) on simulated percentage change in viewing distance for four human subjects. Error bars, ±1 SD. From Busettini et al. (1997).

expansion/contraction was at the center of the screen and the stimulus was specified in terms of the equivalent change in the viewing distance (which was actually fixed throughout the experiment at 33.3 cm). For example, increasing the size and eccentricity (with respect to the screen center) of each dot by 1% simulated a decrease in viewing distance of ~1%. The subjects first brought their eyes to the center of the screen so that the focus of expansion/contraction would be roughly centered on their two foveas. Sample mean vergence velocity profiles—both horizontal and vertical—in response to steps of expanding flow are shown in Figure 2A, and indicate clear horizontal vergence responses, albeit of small amplitude and brief duration. The quantitative dependence of these responses on the magnitude of the steps is shown in Figure 2B, which reveals that the optimal steps were quite small, 4% or less.

The vergence responses resulted solely from the changes in the eccentricity of the dots (i.e., radial flow) and their changes in size made little or no contribution. This is evident from Figure 3, which shows the responses when steps included both radial flow and size change (Figure 3A and

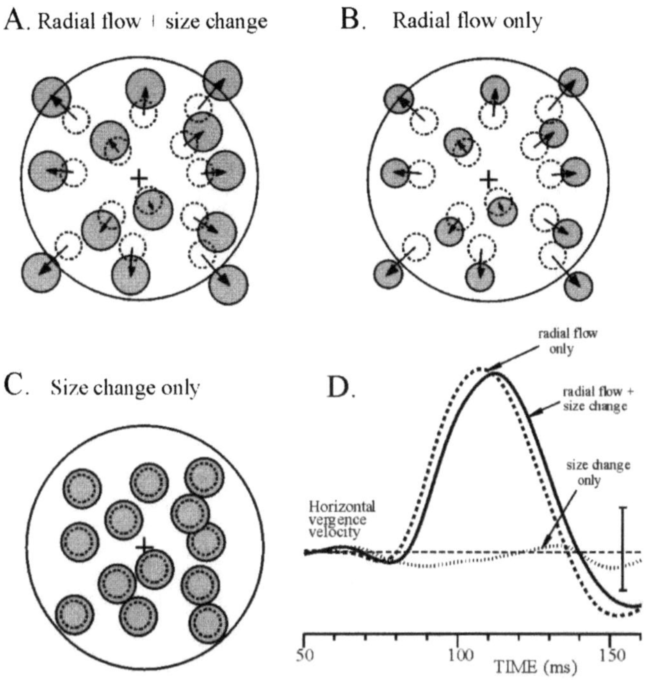

Figure 3. The adequate stimulus for radial-flow vergence: cartoons of the two-frame movies used and sample initial vergence data. A. Radial flow plus size change, each dot increased in size and eccentricity (by a fixed percentage) with respect to the focus of expansion (cross). B. Radial flow alone, each dot remained the same size but increased its eccentricity. C. Size change alone, each dot enlarged in place. D. Mean horizontal vergence velocity associated with the three types of stimuli applied at time zero (one human subject). Upward deflections indicate increased convergence. Calibration 2°/s. Note that A-C are only cartoons to illustrate the principle: the actual patterns started out with 2° dots covering 50% of the screen that occupied almost 90°x90°. From Busettini et al. (1997).

continuous line in Figure 3D), radial flow alone (Figure 3B and dashed line in Figure 3D), and size change alone (Figure 3C and dotted line in Figure 3D). Size changes have been shown to elicit transient vergence eye movements but only at latencies that have been estimated to be more than twice that of radial-flow vergence (Erkelens & Regan, 1986).

A notable feature of these vergence responses is that they are very transient, lasting <100 ms. It was possible that this was because the stimulus was transient—a single step displacement of the random dot patterns—but this was not the case. The responses remained transient even with multiple steps applied in rapid succession ("multi-frame movie", using multiple slide projectors with fast mechanical shutters, at a frame rate of 50Hz). Figure 4

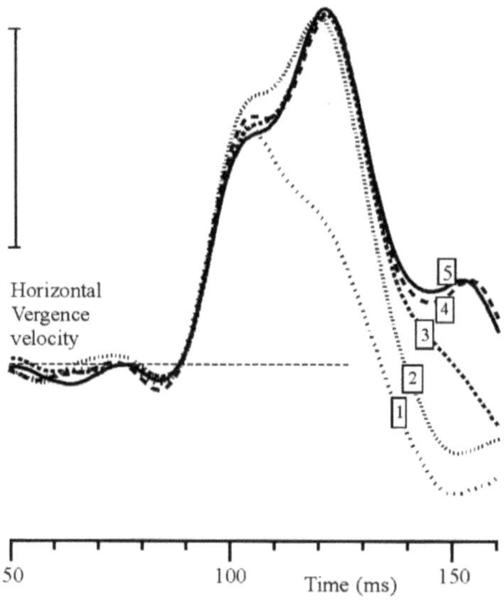

Figure 4. Initial vergence responses to multiple radial-flow steps applied at 20-ms intervals, each producing a 4% expansion (one human subject). Numbers of steps indicated by numbers on traces. Calibration, 2°/s.

shows the initial vergence velocity profiles when 1, 2, 3, 4, or 5 radial-flow steps were applied at 20-ms intervals, each step producing expansion increments of 4%. The profiles in Figure 4 are all synchronized on the first step and it is clear that the initial transient was slightly prolonged by a second step but that a third (or later) step had only minor impact and there was only a very weak sustained component. Thus, the response is intrinsically transient. Note that we consider the responses only up to 160 ms after the radial-flow step. This is because the convergence that commences 80-90 ms after expanding steps, for example, results in the random-dot stimuli acquiring uncrossed disparity that in turn would be expected to initiate antagonistic divergent eye movements about 90 ms later (Busettini et al., 2001). Consequently, the vergence responses to the radial-flow stimuli will start to be "corrupted" by disparity-vergence responses 170-180 ms after the step (end of the "open-loop" period).

The clear suggestion here is that the brain is able to sense the radial pattern of flow and to infer from this that there has been a change in viewing distance that calls for a change in the vergence angle. However, a characteristic of these ocular responses to radial flow patterns is that each eye always moves in the direction of the net motion vector in the nasal hemifield. This raises the possibility that the vergence did not result from the radial pattern of optic flow *per se* but from monocular tracking, in which each eye

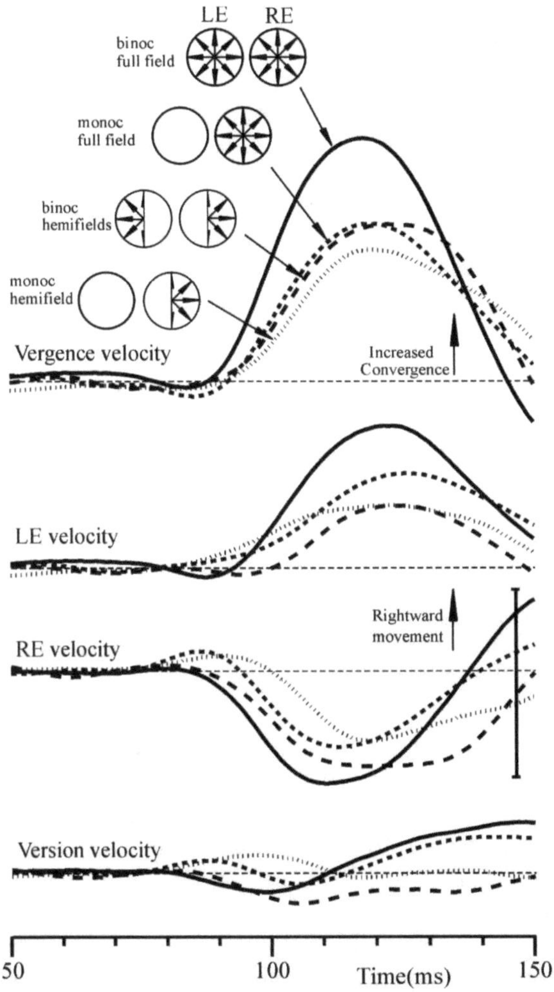

Figure 5. Effects of masking off parts of the field on initial horizontal oculomotor responses to expanding flow (4% change; two-frame movie; one human subject). No mask, 'binoc full field'; left eye masked, 'monoc full field'; both nasal hemifields masked, 'binoc hemifields'; all except right temporal hemifield masked, 'monoc hemifield'. Mean velocity traces in order from above: vergence, left eye, right eye, version. Calibration, 2°/s. From Busettini et al. (1997).

independently tracked the motion that it saw in the nasal hemifield. For example, with expanding flow the net motion vector in the nasal hemifields was "towards the nose", which was also the direction in which each eye moved to produce the convergence. This explanation is predicated upon each eye having its own independent tracking mechanism that receives its visual drive preferentially from the nasal hemifield. That this was *not* the case was

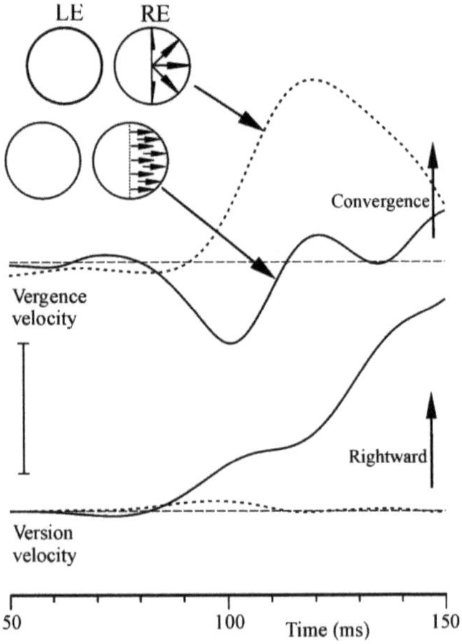

Figure 6. Initial mean horizontal vergence and version responses to radial (dashed line, from Figure 5) and planar flow (continuous line) for one human subject. Two-frame movie. Calibration: 1°/s (vergence), 3°/s (version).

apparent from the observation that vergence responses persisted when both nasal hemifields were masked off: see the traces labeled, "binoc hemifields" in Figure 5. Note that here each eye moved "towards the nose" but the net motion vector was "away from the nose". Thus, Busettini et al. (1997) concluded that the vergence responses resulted from a true parsing of the radial pattern of flow.

Perhaps not surprisingly, radial-flow vergence did not require binocular visual stimulation even though the motor output of the reflex was binocular: see "monoc full field" in Figure 5. Indeed, the vergence responses persisted when the only parts of the radial pattern visible were in the temporal hemifield of one eye: see "monoc hemifield" in Figure 5. This last stimulus was especially interesting because, unlike the other three stimuli in Figure 5, it had a strong net motion vector (to the right), and yet it still elicited almost pure convergence with almost no (rightward) version: see the "Version velocity" trace in Figure 5. Clearly, the mere existence of a net motion vector was not sufficient to generate version responses when orthogonal (i.e., vertical) motion vectors were also present. When these vertical motion vectors were removed, there was now vigorous rightward version/OFR. This effect can be seen in Figure 6, where the condition in which *all* motion was rightward and restricted to one temporal hemifield (see the traces in continuous line) is compared to the "monoc hemifield" condition with radial flow (see traces in dotted line in Figure 6) that was already seen in Figure 5.

Note that there was a transient loss of convergence when planar motion was substituted for radial flow in Figure 6 and we think that this was because the stimulus was unnatural (uniocular): when motion stimuli are presented to one eye only, both eyes move in the direction of the stimulus but the eye that sees the motion usually moves a little more vigorously. In our present case, both eyes moved rightward but the right eye moved a little more vigorously than the left eye so that there was a transient reduction in the vergence angle; when the motion was applied to both retinas, little vergence was seen (not shown).

Figure 6 demonstrates that the brain is able to determine whether the pattern of optic flow calls for version or vergence in <100 ms. The extremely short latency surely suggests that the system depends on parallel processing to arrive at an appropriate response based on the pattern of optic flow. We further suggest that this points to the existence of neurons or networks that act like templates or tuned spatial filters to detect specific patterns of optic flow and generate appropriate oculomotor responses to serve the needs of visual stabilization, cf., the template models of Perrone & Stone (1994; 1998).

Optical geometry indicates that as an observer moves forwards (or backwards) the vergence angle between the two lines of sight must increase (or decrease) at a rate that is inversely proportional to the square of the viewing distance in order for both eyes to remain aligned on the object of regard in the scene ahead (Yang et al., 1999). This means that a moving observer who looks far ahead will not need to converge his/her eyes, at least at first. However, if the environment is cluttered, as in a forest for example, then he/she will experience considerable radial optic flow, which we might expect will result in convergence that would be inappropriate. An engineer modeling a radial-flow vergence mechanism—as in the case of Sandini et al. (2001), for example—might deal with this by modulating the gain of the response so that it is inversely related to the square of the viewing distance. It seems that the brain has adopted a similar solution albeit less than perfect. The vergence responses induced by a given step of radial flow are a linear function of the vergence angle at the time the step is applied: see Figure 7. Thus, under normal circumstances, the gain of the radial-flow vergence mechanism will be inversely related to the viewing distance. This is somewhat short of the ideal performance but will clearly render the moving observer who looks ahead less sensitive to the optic flow resulting from nearby visual clutter.

One curious aspect of the radial-flow vergence mechanism is that it uses the global pattern of flow over large areas of the visual field to generate eye movements that presumably serve to maintain the binocular alignment of the two foveas, which represent only a small fraction of that visual field. In all of

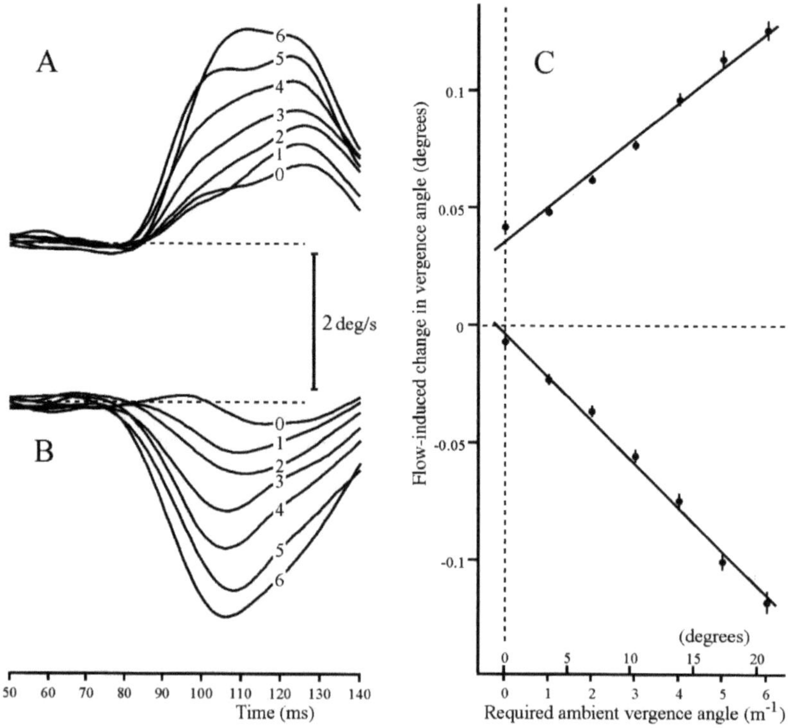

Figure 7. Dependence of radial-flow vergence on the preexisting vergence angle. Mean vergence velocity responses to expanding (A) and contracting (B) flow-steps for one human subject (4% steps; two-frame movie). Numbers on traces: vergence angle in reciprocal meters at the time radial-flow steps were applied. Upward deflections indicate increasing convergence. C. Mean changes in vergence angle (over the time period 90-140 ms from step) for expanding (above) and contracting (below) flow vs. vergence angle when the flow-steps were applied. Lines are best-fit linear regressions. Error bars, ±SE. From Yang et al. (1999).

the experiments mentioned thus far the focus of expansion/contraction was always centered on the fovea(s), simulating the situation in which gaze is in the direction of heading. This raises the question of what would happen if the observer looks off slightly to one side so that this focus is eccentric on the retina.

To address this we investigated the effect of positioning the focus of expansion/contraction so that it was to the right or left of the center of the screen (the approximate location of the fovea). Figure 8 shows the dependence of the horizontal vergence (Figure 8A) and version/OFR (Figure 8B) responses to 4% steps of expansion/contraction on the horizontal retinal eccentricity of the focus of expansion/contraction.

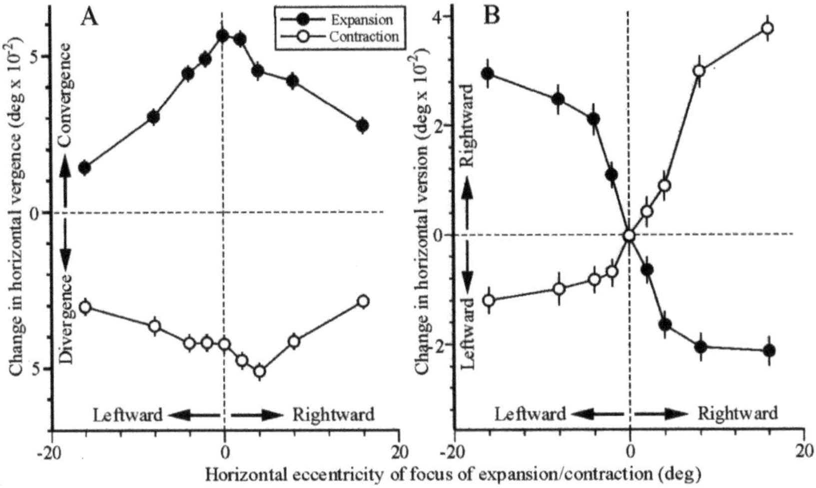

Figure 8. Mean vergence (A) and version (B) responses associated with expanding (filled symbols) and contracting (open symbols) radial-flow steps (two-frame movies): dependence on horizontal retinal eccentricity of focus of expansion/contraction (one human subject). Measures based on changes in vergence and version position over the period 90-157 ms after the time of the step. Error bars, ±SE.

As the focus became more eccentric on the retina, the vergence responses tended to get smaller and version/OFR now became apparent. The direction of the version/OFR responses was commensurate with the direction of motion in the region of the fovea(s). For example, shifting the focus of expansion (contraction) to the right brought out leftward (rightward) version/OFR and vice versa.[2] Further, the sensitivity of version/OFR to changes in eccentricity tended to be greatest in the immediate vicinity of the fovea.

These data are all consistent with the idea that version/OFR gives greater weight to motion inputs in the vicinity of the fovea. The implication is that when the observer who is moving forwards looks slightly off to one side of the direction of heading his/her oculomotor responses will acquire a version/OFR component that works to increase the eccentricity of his/her eyes with respect to the focus of expansion. If the observer looks off to the right then his/her eye movements will have a rightward version component and if he/she looks downward then his/her eye movements will have a downward version component etc. This dependence of the direction of eye tracking on

[2] Of course, this situation prevails only for a short time—these initial eye movement responses will tend to shift the focus of expansion/contraction on the retina towards the fovea, significantly complicating the etiology of subsequent (closed-loop) eye movements.

the direction of the motion in the foveal region is also a feature of the tracking phases of closed-loop OKN when humans and monkeys view a large radial-flow patterns simulating continuous forward or backward motion across a ground plane (Lappe et al., 1998; Niemann et al., 1999). This superiority of the central retina in driving OKN is in accord with the findings of several studies that used uniform motion (Dubois & Collewijn, 1979; Howard & Ohmi, 1984; Murasugi & Howard, 1989). However, all of these OKN studies used prolonged stimuli so that pursuit responses may have contributed to them.

We will now examine the extreme situation in which the observer's gaze is orthogonal to the direction of heading, as in (E) and (F) of Figure 1, calling for pure version.

4 OCULAR FOLLOWING

We saw in the previous section that the version/OFR mechanism gives precedence to images in the foveal region so that we might assume that, when the observer looks orthogonal to the direction of heading, as in Figures 1E and 1F, the resulting eye movements and optic flow patterns are largely determined by the local motion where gaze happens to be directed. Thus, we might expect that in Figure 1E gaze would have to be directed at the mountain in order for its image to be the one that was selectively stabilized by the OFR whereas in Figure 1F gaze would have to be directed at the tree. However, ocular following responds to image motion extending over wide regions of the central retina and is not exclusively responsive to image motion in the foveal region. Given this, the question arises as to how the ocular following mechanism can single out the motion of particular objects in the scene and ignore the competing motion of the objects elsewhere. One way to achieve this would be to use attentional focusing mechanisms to spotlight the target of interest within the complex flow field. Such mechanisms exist and are used by closed-loop OKN (Mestre & Masson, 1997) as well as by the pursuit system (Niemann et al., 1999) but they require high-level executive decisions to select the image whose motion is to be tracked and this of necessity is very time consuming (Keller & Khan, 1986; Kimmig et al., 1992; Mestre & Masson, 1997).

Thus, the pursuit system will not begin to influence eye movements until long after ocular following is well under way. One clue to a potential short-latency mechanism comes from the realization that the compensatory eye movements required to stabilize the images of a given object(s) that lies orthogonal to the observer's direction of heading are inversely proportional to

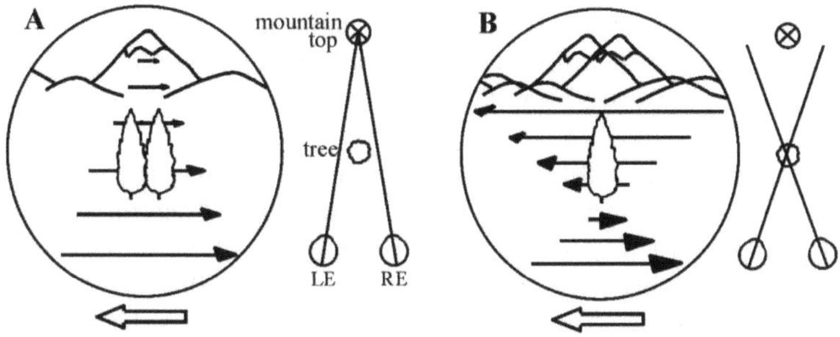

Figure 9. Optic flow experienced by a translating observer (binocular viewing). A. As in Figure 1E, except that with binocular viewing the mountain in the plane of fixation is seen as single and the nearer tree is seen as double (disparate). A plan view of the observer's eyes and the two objects is shown to the right. B. As in Figure 1F, except that with binocular viewing the tree is in the plane of fixation and so is seen as single whereas the distant mountain is now seen as double. Again, the plan view is shown to the right. Note that the dimensions of the eyes and their separations have been exaggerated to show the disparity more clearly. From Busettini et al. (1996a).

the viewing distance.[3] This means that, at any given time, the compensatory eye movements can only stabilize images in one particular depth plane. The question, therefore, now becomes: What determines which depth plane is stabilized? Whatever the mechanism it must be ultra-rapid.

We think that the ocular following system solves this problem using low-level stereomechanisms that perform rapid parallel processing of binocular images, effectively sorting them on the basis of the depth plane that they occupy. This stereo algorithm, which utilizes the fact that we have two eyes with slightly differing viewpoints, is illustrated in Figures 9A and 9B, which show "binocular" renditions of the cartoons in Figures 1E and 1F. The two eyes are assumed to be aligned on the mountain in Figure 9A and on the tree in Figure 9B as a result of prior vergence and saccadic eye movements guided by some high-level process that selects the target to be placed in the two foveas. The net result of this is that the object likely to be of most interest to the observer has been brought into the plane of fixation.

Objects in the plane of fixation are imaged at corresponding positions on the two retinas—they are said to lack binocular disparity—and are perceived as single (the mountain in Figure 9A and the tree in Figure 9B). In contrast, objects that are nearer or farther than the plane of fixation have images that occupy non-corresponding positions on the two retinas—they are said to have

[3] Indeed, the TVOR that generates eye movements to compensate for lateral motions of the head has a gain that is inversely proportional to the viewing distance (Busettini et al., 1994; Paige, 1989; Paige & Tomko, 1991; Schwarz et al., 1989; Schwarz & Miles, 1991).

binocular disparity—and are seen as double (the tree in Figure 9A and the mountain in Figure 9B). It seems reasonable to assume that the objects in the plane of fixation are the ones whose retinal images should receive the top priority from the visual stabilization mechanism in the event that the observer moves.

Clearly, one possible algorithm for stabilizing gaze on objects in the plane of fixation would be to track only those objects whose images lack disparity. Early support for this idea was the finding that OKN is best for images with zero binocular disparity (Howard & Gonzalez, 1987; Howard & Simpson, 1989), but high-level processing, perhaps involving selective attention, may have contributed to these studies, which examined the closed-loop steady-state responses. However, experiments examining the earliest OFR to sudden motion of a large random-dot pattern showed only a weak preference for images in the plane of fixation, i.e., images with zero binocular disparity (Masson et al., 2001). It was argued that such coherent motion stimuli lack the subsidiary cues—like motion parallax, for example—that are normally associated with translation of the observer and might be critical for uncovering a role for disparity.

To simulate the presumed everyday situation more closely, Masson et al. (2001) partitioned the random-dot pattern into two interleaved sets of horizontal bands that suddenly underwent opposite horizontal motion: see the cartoon in Figure 10A. In the example shown, one set of (test) dots moved to the left and the other set of (conditioning) dots moved to the right. Motion of either set of dots alone—when in the plane of fixation—produced vigorous OFR: see the traces labeled, "Test only" and "Conditioning only" in Figure 10B, which shows sample mean version velocity profiles. When both sets of dots were present and in the plane of fixation their competing motions produced very weak OFR: see the dashed trace labeled "0°" in Figure 10B. However, when the conditioning dots were given crossed disparity then OFR now favored the leftward motion of the test dots that were still in the plane of fixation: see traces in Figure 10B labeled, 0.4°, 0.8° etc. Figure 10C shows the dependence of the changes in version on the disparity of the conditioning dots and indicates that responses progressively favored the leftward motion of the test images as more and more disparity—crossed, as in Figure 10B, or uncrossed—was applied to the conditioning images. This indicates that OFR is disparity-sensitive and has a strong preference for zero disparity. These data were obtained from humans but the OFR of monkeys also shows a preference for motion in the plane of fixation (Takemura et al., 2000).

The question arises as to why Masson et al. (2001) found that the initial OFR elicited by coherent motion of a large random-dot pattern are only modestly attenuated when the whole pattern is given disparity, whereas the same disparity applied to the partitioned display seen in Figure 10A had a much more powerful effect. Masson et al. suggested that OFR might rely on

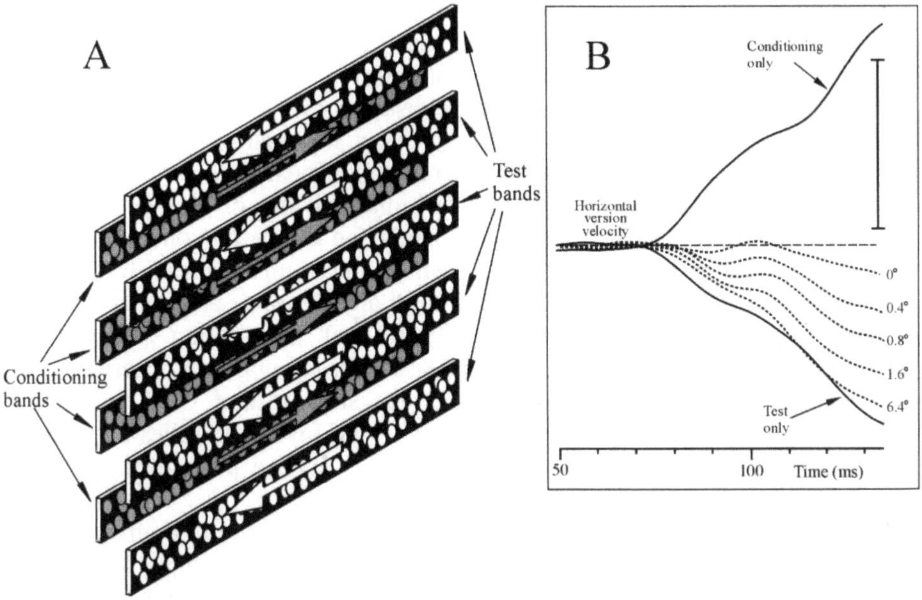

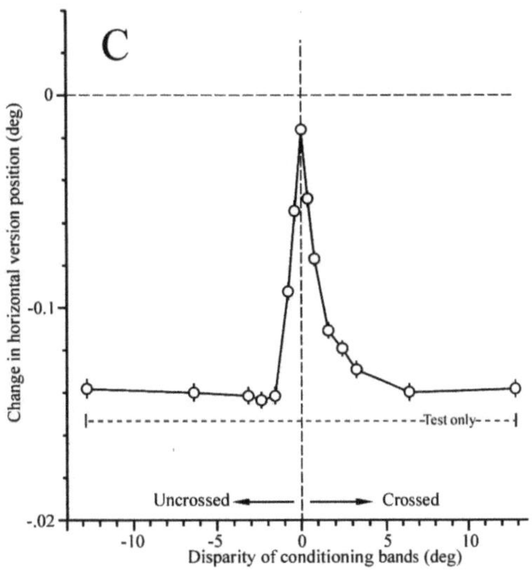

Figure 10. Ocular following responses (OFR): dependence on binocular disparity. A. OFR stimuli were two interleaved sets of dots that suddenly underwent horizontal motion (40°/s) in opposite directions. The test dots always remained in the plane of the screen (plane of fixation) and the disparity of the conditioning dots was varied from trial to trial. B. Time course of initial OFR: dependence on (crossed) disparity of the conditioning dots, whose values are given at the ends of traces. 'Test only' and 'Conditioning only' stimuli were in the plane of fixation. C. Mean change in version (measured over the 50-ms period starting 85 ms after stimulus onset): dependence on the disparity of the conditioning dots. Responses in the (rightward) direction of the conditioning bands are plotted above zero and responses in the (leftward) direction of the test stimulus are plotted below zero. Responses to the 'Conditioning only' stimulus are off scale. Error bars, ±SE. Data for one human subject. From Masson et al. (2001).

motion detectors that are selectively sensitive to relative motion and/or relative disparity. Absolute disparity refers to the slight differences in the positions of the two retinal images of a given object resulting from the differing viewpoint of the two eyes, and relative disparity refers to the differences in the absolute disparities of different objects within the visual scene resulting from differences in their distance to the observer. To obtain definitive evidence for relative disparity it is necessary to show that the disparity tuning in one region is dependent on the disparity in another region.

Two of us have recently used the partitioned display of Masson et al. and found that applying disparity to the test bands had little effect on the disparity at which the conditioning bands had their greatest impact on OFR, i.e., the responses to the test and conditioning stimuli are separable and hence it is the absolute—rather than the relative—disparity of the motion stimuli that determines their impact on OFR (Yang & Miles, unpublished observations). In sum, the experiments reviewed thus far suggest that OFR has some preference for image motion in the vicinity of the fovea and relies on motion detectors that are sensitive to absolute disparity with a preference for zero disparity.

Earlier studies showed that OFR have properties generally attributed to low-level motion detectors (Borst & Egelhaaf, 1993). For example, when sine wave grating patterns are used, provided the patterns are within the spatial frequency bandwidth of the system (<0.5 cycles/deg), the latency is solely a function of contrast and temporal frequency (Miles et al., 1986). These and other data led to the development of a model consisting of a drive mechanism that integrates the motion errors over time and a separate trigger mechanism that responds solely to changes in luminance and acts as a gate with the power of veto over the output of the drive mechanism. The trigger mechanism has a high threshold and functions to improve the signal-to-noise ratio (without impeding the integration of the motion error signals) so that the system is less likely to chase spurious internal noise—a potential problem with an automatic servomechanism with such a short latency.

More recently, it has been shown that human OFR can be elicited at short latency by step-wise displacements of random-dot patterns and that reversing the luminance polarity during the steps reverses the OFR, consistent with the idea that these earliest tracking responses result, at least in part, from the operation of a first-order (luminance) motion-energy detection mechanism (Masson et al., 2002). There is also a built-in safety mechanism that prevents the system from tracking the visual disturbances created by the subject's own saccadic eye movements as they sweep the image of the world across the retina (Kawano & Miles, 1986). This mechanism senses the rapid motion in the peripheral visual field and transiently suppresses any ocular following. In fact, rapid motion in the peripheral field of one eye can prevent the tracking of visual motion presented simultaneously to the central visual field of the other eye (inter-ocular transfer), indicating that this saccadic suppression must take place within the cerebral cortex, beyond the point at which visual inputs from the two eyes are combined.

A motion-detection mechanism that relies solely on local measurements cannot correctly report the direction of motion of elongated luminance edges—this is the so-called "aperture problem" (Movshon et al., 1985). Recent work on the human OFR suggests that the brain can solve this by detecting localized moving 2-D features, such as corners, whose motion is unambiguous, or by extracting higher-order motion signals such as texture cues. When confronted with two sinusoidal gratings differing in orientation by 45°, one moving and the other stationary ("unikinetic" plaid), OFR starts in the direction of the grating motion and only ~20 ms later starts to veer in the direction of the pattern motion (Masson & Castet, 2002). A similar temporal partitioning of OFR is seen when a moving grating pattern is viewed through an elongated aperture oriented 45° to the grating: the very earliest OFR are in the direction of the grating motion but later responses gradually veer in the direction of the elongated aperture, cf. the "barber pole" illusion (Masson et al., 2000). These findings are consistent with the idea that the earliest open-loop OFR rely on local motion detectors—albeit distributed across the visual field—and the later OFR result from processing by higher-order mechanisms.

The ultra-short latencies mean that ocular following gets under way before the subject is even aware of the stimulus that drives it. Such rapid operation is presumably one reason why ocular following is subject to extensive long-term adaptive gain control, which helps to ensure that the amplitude and direction of these ultra-rapid responses are appropriate (Miles & Kawano, 1986). All of these properties are characteristic of a mechanism that operates as an automatic reflex, independently of perception.

5 DISPARITY VERGENCE

When the moving observer looks in the direction of heading, radial optic flow is only one of several cues that indicate the forward rate of progress. Another is binocular disparity and recent experiments have demonstrated that when random-dot patterns are viewed dichoptically and suddenly subjected to small binocular misalignments (disparity steps), corrective vergence eye movements are elicited at latencies closely comparable with those for ocular following and radial flow vergence (Busettini et al., 2001; 1996b). Crossed disparity steps elicited increased convergence and uncrossed steps decreased convergence—provided that they did not exceed more than a few degrees—exactly as expected of a depth-tracking servo mechanism driven by disparity. That these responses could not be the result of monocular tracking was evident from experiments in which the disparity step was confined to one eye: when the crossed disparity step was restricted to the right eye (which saw a leftward step), the result was (binocular) convergence in which the left eye moved rightward even though that eye had seen only a stationary pattern. The (rightward) movement of the left eye here is in the direction expected of a stereoscopic mechanism that responds to a binocular misalignment but is in the opposite direction to the only available motion cues—the leftward motion at the right eye.

The range of disparities over which the system behaves like a servo mechanism, that is, the range over which increases in the disparity vergence error result in roughly linear increases in the vergence response is <2°. Thus, this vergence mechanism can correct only small misalignments of the two eyes, commensurate with a mechanism that performs only local stereo matches and merely attempts to bring the nearest salient images into the plane of fixation. During forward locomotion this mechanism will help to prevent images from leaving the plane of fixation. However, the *primary* function of this mechanism is to eliminate small vergence errors, regardless of their source, as evidenced by the fact that it also operates in the vertical plane using vertical disparity, which is unrelated to depth and translation *per se* in primary gaze positions (Busettini et al., 2001). While the specific involvement with vergence errors resulting from locomotion is clear, this is a *secondary* function, placing it in a slightly different category from OFR and radial-flow vergence whose primary function is to compensate for the observer's translational motion. However, once more we have a mechanism that functions as a low-level automatic servo and is not involved in high-level operations like the transfer of fixation to new images in new depth planes, which requires time consuming target selections and (often) the decoding of large disparity errors (>10°) that necessitate solution of the correspondence problem.

Vergence responses can also be elicited at ultra-short latencies by disparity stimuli applied to dense (50%) *anticorrelated* binocular patterns, in which the two images have opposite contrast so that each black dot in one eye is matched to a white dot in the other eye (Masson et al., 1997). Figure 11A shows sample mean vergence velocity profiles in response to crossed disparity stimuli applied to correlated (continuous line) and anticorrelated (discontinuous line) patterns. Note that the vergence responses to the anticorrelated stimuli are in the reverse direction of those to the correlated stimuli. The disparity tuning curves for these data are shown in Figure 11B, the curve obtained with the normal correlated patterns (filled symbols) having a characteristic s-shape that is well fitted by a Gabor function with offset. The curve for the anticorrelated data (open symbols) is almost a mirror image, and the cosine term for the best-fit Gabor function is phase shifted almost exactly 180 degrees. In two-alternative-forced-choice tests, subjects could readily

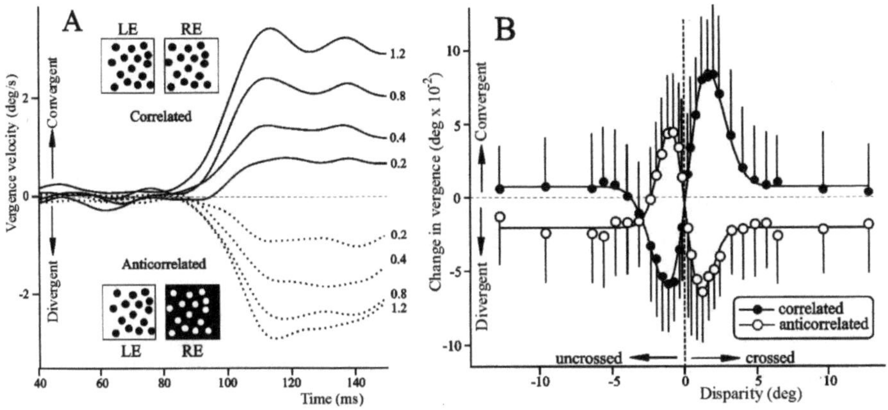

Figure 11. Vergence eye movements elicited by disparity steps applied to random-dot patterns. A. Mean horizontal vergence velocity profiles to crossed-disparity stimuli applied at time zero to correlated (continuous line) and anticorrelated (dotted line) patterns with step amplitudes (in deg) indicated at ends of traces. Patterns shown are cartoons only—actual patterns extended 80°x80° and had 50% density, each dot being 2° diameter. B. Dependence of mean (±SD) change in vergence position (over the 33 ms time period starting 90 ms after step) on disparity step with correlated (filled symbols) and anticorrelated (open symbols) patterns. Curves are best-fitting Gabor functions, the linear (servo) region being restricted to disparities <±2°. Data for one human subject. From Masson et al. (1997).

discriminate between crossed and uncrossed disparities when applied to the correlated patterns but not when applied to the anticorrelated patterns (Masson et al., 1997). This is consistent with the idea that these short-latency vergence responses derive their visual input from an early stage of cortical processing prior to the level at which depth percepts are elaborated. Actually, the large-field stimuli used in all of these disparity vergence studies contain only absolute disparity cues, which are known to be poorly perceived in depth (Erkelens & Collewijn, 1985a,b; Regan et al., 1986).

6 NEURAL MEDIATION

There is extensive evidence from monkeys that the three visual reflexes are all cortically mediated despite their ultra-short latency. Bilateral lesions of the medial superior temporal area of the cortex (MST) result in major impairments of all three visual reflexes (Takemura et al., 2002) and there is extensive data from single unit recordings indicating that neurons in this region discharge in relation to the visual stimuli used to drive all three reflexes. Thus, MST has long been known to contain neurons that are selectively sensitive to radial optic flow patterns such as those now known to evoke vergence eye movements at ultra-short latencies (Duffy, 2000; Duffy & Wurtz, 1991a,b; 1995; 1997a,b,c; Lagae et al., 1994; Saito et al., 1986; Tanaka et al., 1986; Tanaka & Saito, 1989). However, most of these recordings describe neuronal events that have much longer latencies than those that could generate the radial-flow vergence eye movements described here. Kawano and colleagues have shown that there are neurons in MST that discharge in relation to the earliest OFR responses, their temporal profiles even reproducing the irregularities in the temporal profiles for OFR, and sharing OFR's sensitivity to disparity (Takemura et al., 2000). Although there are many neurons in MST that are sensitive to disparity very few of these have disparity tuning curves whose shapes resemble those for short-latency disparity vergence eye movements (Eifuku & Wurtz, 1999; Roy, Komatsu & Wurtz, 1992; Takemura et al., 2001). However, when all of the disparity tuning curves obtained from a given monkey are summed together, they match the disparity tuning curves for the vergence responses very closely, even reproducing the idiosyncrasies of the individual animals, i.e., the vergence eye movements are encoded in the population activity (Takemura et al., 2001; Takemura et al., 2002). In sum, there is accumulating evidence from monkeys that MST plays a significant role in the generation of all three visual reflexes.

7 SUMMARY

Eye movements serve to improve vision. Primates have several reflexes that generate eye movements to compensate for bodily movements that would otherwise disturb their gaze and undermine their ability to process visual information. Two vestibuloöcular reflexes, the RVOR and the TVOR, compensate selectively for rotational and translational disturbances of the head, and each has visual backups that operate as negative feedback tracking mechanisms to reduce any residual disturbances of gaze. Of particular interest here are three visual tracking mechanisms that operate in machine-like fashion with ultra-short latencies (<60 ms in monkeys, <85 ms in humans). These visual reflexes each have special features that are selectively engaged during translational disturbances and use pre-attentive parallel processing of optic flow to extract control signals that initiate eye movements before the observer is even aware that there has been a disturbance, i.e., they operate independently of perception. We suggest that this processing is accomplished by visual filters tuned to different features of the binocular image flow. One of the reflexes uses the radial patterns of flow experienced by the moving observer who looks in the direction of heading and generates vergence eye movements that help maintain binocular alignment on the objects that lie ahead: *radial-flow vergence.* A second reflex deals with the motion parallax experienced by the moving observer who looks off to one side and uses binocular stereo cues to help selectively stabilize the retinal images of objects in the plane of fixation: *ocular following.* A third reflex reinforces the first, using changes in the binocular alignment of the retinal images to generate vergence eye movements: *disparity vergence.* Despite their rapid, reflex nature, there is evidence from monkeys strongly suggesting that all three mechanisms are mediated, at least in part, by the medial superior temporal (MST) area of cortex.

REFERENCES

Borst, A., & Egelhaaf, M. (1993). Detecting visual motion: Theory and models. In F. A. Miles, & J. Wallman (Eds.), *Visual Motion and its Role in the Stabilization of Gaze.* pp. 3-27: Elsevier Science Publishers BV.

Busettini, C., FitzGibbon, E. J., & Miles, F. A. (2001). Short-latency disparity vergence in humans. *J. Neurophysiol., 85*, 1129-1152.

Busettini, C., Masson, G. S., & Miles, F. A. (1996a). A role for stereoscopic depth cues in the rapid visual stabilization of the eyes. *Nature, 380*, 342-345.

Busettini, C., Masson, G. S., & Miles, F. A. (1997). Radial optic flow induces vergence eye movements with ultra-short latencies. *Nature, 390*, 512-515.

Busettini, C., Miles, F. A., & Krauzlis, R. J. (1996b). Short-latency disparity vergence responses and their dependence on a prior saccadic eye movement. *J. Neurophysiol., 75*, 1392-1410.

Busettini, C., Miles, F. A., & Schwarz, U. (1991). Ocular responses to translation and their dependence on viewing distance. II. Motion of the scene. *J. Neurophysiol., 66*, 865-878.

Busettini, C., Miles, F. A., Schwarz, U., & Carl, J. R. (1994). Human ocular responses to translation of the observer and of the scene: dependence on viewing distance. *Exp. Brain Res., 100*, 484-494.

Cohen, B., Matsuo, V., & Raphan, T. (1977). Quantitative analysis of the velocity characteristics of optokinetic nystagmus and optokinetic after-nystagmus. *J. Physiol. Lond., 270*, 321-344.

Dubois, M. F. W., & Collewijn, H. (1979). Optokinetic reactions in man elicited by localized retinal stimuli. *Vision Res., 19*, 1105-1115.

Duffy, C. J. (2000). Optic flow analysis for self-movement perception. *Int. Rev. Neurobiol., 44*, 199-218.

Duffy, C. J., & Wurtz, R. H. (1991a). Sensitivity of MST neurons to optic flow stimuli. I. A continuum of response selectivity to large-field stimuli. *J. Neurophysiol., 65*, 1329-1345.

Duffy, C. J., & Wurtz, R. H. (1991b). Sensitivity of MST neurons to optic flow stimuli. II. Mechanisms of response selectivity revealed by small-field stimuli. *J. Neurophysiol., 65*, 1346-1359.

Duffy, C. J., & Wurtz, R. H. (1995). Response of monkey MST neurons to optic flow stimuli with shifted centers of motion. *J. Neurosci., 15*, 5192-5208.

Duffy, C. J., & Wurtz, R. H. (1997a). Medial superior temporal area neurons respond to speed patterns in optic flow. *J. Neurosci., 17*, 2839-2851.

Duffy, C. J., & Wurtz, R. H. (1997b). Multiple temporal components of optic flow responses in MST neurons. *Exp. Brain Res., 114*, 472-482.

Duffy, C. J., & Wurtz, R. H. (1997c). Planar directional contributions to optic flow responses in MST neurons. *J. Neurophysiol., 77*, 782-796.

Eifuku, S., & Wurtz, R. H. (1999). Response to motion in extrastriate area MSTl: disparity sensitivity. *J. Neurophysiol., 82*, 2462-2475.

Erkelens, C. J., & Collewijn, H. (1985a). Eye movements and stereopsis during dichoptic viewing of moving random-dot stereograms. *Vision Res., 25*, 1689-1700.

Erkelens, C. J., & Collewijn, H. (1985b). Motion perception during dichoptic viewing of moving random-dot stereograms. *Vision Res., 25*, 583-588.

Erkelens, C. J., & Regan, D. (1986). Human ocular vergence movements induced by changing size and disparity. *J. Physiol. Lond., 379*, 145-169.

Gellman, R. S., Carl, J. R., & Miles, F. A. (1990). Short latency ocular-following responses in man. *Vis. Neurosci., 5*, 107-122.

Gibson, J. J. (1950). *The Perception of the Visual World.* Boston: Houghton Mifflin.

Gibson, J. J. (1966). *The Senses Considered as Perceptual Systems.* Boston: Houghton Mifflin.

Goldberg, J. M., & Fernandez, C. (1975). Responses of peripheral vestibular neurons to angular and linear accelerations in the squirrel monkey. *Acta Otolaryngol., 80*, 101-110.

Howard, I. P., & Gonzalez, E. G. (1987). Human optokinetic nystagmus in response to moving binocularly disparate stimuli. *Vision Res., 27*, 1807-1816.

Howard, I. P., & Ohmi, M. (1984). The efficiency of the central and peripheral retina in driving human optokinetic nystagmus. *Vision Research, 24*, 969-976.

Howard, I. P., & Simpson, W. A. (1989). Human optokinetic nystagmus is linked to the stereoscopic system. *Exp. Brain Res., 78*, 309-314.

Inoue, Y., Takemura, A., Suehiro, K., Kodaka, Y., & Kawano, K. (1998). Short-latency vergence eye movements elicited by looming step in monkeys. *Neurosci. Res., 32*, 185-188.

Kawano, K., & Miles, F. A. (1986). Short-latency ocular following responses of monkey. II. Dependence on a prior saccadic eye movement. *J. Neurophysiol., 56*, 1355-1380.

Keller, E. L., & Khan, N. S. (1986). Smooth-pursuit initiation in the presence of a textured background in monkey. *Vision Res., 26*, 943-955.

Kimmig, H. G., Miles, F. A., & Schwarz, U. (1992). Effects of stationary textured backgrounds on the initiation of pursuit eye movements in monkeys. *J. Neurophysiol., 68*, 2147-2164.

Koenderink, J. J. (1986). Optic flow. *Vision Res., 26*, 161-179.

Lagae, L., Maes, H., Raiguel, S., Xiao, D.-K., & Orban, G. A. (1994). Responses of macaque STS neurons to optic flow components: A comparison of areas MT and MST. *J. Neurophysiol., 71*, 1597-1626.

Lappe, M., & Hoffmann, K. P. (2000). Optic flow and eye movements. *Int. Rev. Neurobiol., 44*, 29-47.

Lappe, M., Pekel, M., & Hoffmann, K. P. (1998). Optokinetic eye movements elicited by radial optic flow in the macaque monkey. *J. Neurophysiol., 79*, 1461-1480.

Lisberger, S. G., & Miles, F. A. (1980). Role of primate medial vestibular nucleus in long-term adaptive plasticity of vestibuloocular reflex. *J. Neurophysiol., 43*, 1725-1745.

Masson, G. M., Yang, D.-S., & Miles, F. A. (2002). Reversed short-latency ocular following. *Vision Res., 42*, 2081-2087.

Masson, G. S., Busettini, C., & Miles, F. A. (1997). Vergence eye movements in response to binocular disparity without depth perception. *Nature, 389*, 283-286.

Masson, G. S., Busettini, C., Yang, D.-S., & Miles, F. A. (2001). Short-latency ocular following in humans: sensitivity to binocular disparity. *Vision Res., 41*, 3371-3387.

Masson, G. S., & Castet, E. (2002). Parallel motion processing for the initiation of short-latency ocular following in humans. *J. Neurosci., 22*, 5149-5163.

Masson, G. S., Rybarzyck, Y., Castet, E., & Mestre, D. R. (2000). Temporal dynamics of motion integration for the initiation of tracking eye movements at ultra-short latencies. *Vis. Neurosci., 17*, 753-767.

Mestre, D. R., & Masson, G. S. (1997). Ocular responses to motion parallax stimuli: the role of perceptual and attentional factors. *Vision Res., 37*, 1627-1641.

Miles, F. A. (1997). Visual stabilization of the eyes in primates. *Curr. Opin. Neurobiol., 7*, 867-871.

Miles, F. A., & Eighmy, B. B. (1980). Long-term adaptive changes in primate vestibuloocular reflex. I. Behavioral observations. *J. Neurophysiol., 43*, 1406-1425.

Miles, F. A., & Fuller, J. H. (1974). Adaptive plasticity in the vestibulo-ocular responses of the rhesus monkey. *Brain Res., 80*, 512-516.

Miles, F. A., & Kawano, K. (1986). Short-latency ocular following responses of monkey. III. Plasticity. *J. Neurophysiol., 56*, 1381-1396.

Miles, F. A., Kawano, K., & Optican, L. M. (1986). Short-latency ocular following responses of monkey. I. Dependence on temporospatial properties of visual input. *J. Neurophysiol., 56*, 1321-1354.

Miles, F. A., Schwarz, U., & Busettini, C. (1991). The parsing of optic flow by the primate oculomotor system. In A. Gorea (Ed.) *Representations of Vision: Trends and Tacit Assumptions in Vision Research, vol.* pp. 185-199. Cambridge: Cambridge University Press.

Miles, F. A., Schwarz, U., & Busettini, C. (1992). The decoding of optic flow by the primate optokinetic system. In A. Berthoz, W. Graf, & P.P. Vidal (Eds.), *The Head-Neck Sensory-Motor System, vol.* pp. 471-478. New York: Oxford University Press.

Movshon, J. A., Adelson, E. H., Gizzi, M. S., & Newsome, W. T. (1985). The analysis of moving visual patterns. In C. Chagas, R. Gattass, & C. Gross (Eds.), *Pattern Recognition Mechanisms, vol.* New York: Springer Verlag.

Murasugi, C. M., & Howard, I. P. (1989). Up-down asymmetry in human vertical optokinetic nystagmus and afternystagmus: contributions of the central and peripheral retinae. *Exp. Brain Res., 77*, 183-192.

Niemann, T., Lappe, M., Büscher, A., & Hoffmann, K.-P. (1999). Ocular responses to radial optic flow and single accelerated targets in humans. *Vision Res., 39*, 1359-1371.

Paige, G. D. (1989). The influence of target distance on eye movement responses during vertical linear motion. *Exp. Brain Res., 77*, 585-593.

Paige, G. D., & Tomko, D. L. (1991). Eye movement responses to linear head motion in the squirrel monkey. I. Basic characteristics. *J. Neurophysiol., 65*, 1170-1182.

Perrone, J. A., & Stone, L. S. (1994). A model of self-motion estimation within primate extrastriate visual cortex. *Vision Res., 34*, 2917-2938.

Perrone, J. A., & Stone, L. S. (1998). Emulating the visual receptive-field properties of MST neurons with a template model of heading estimation. *J. Neurosci., 18*, 5958-5975.

Regan, D., Erkelens, C. J., & Collewijn, H. (1986). Necessary conditions for the perception of motion in depth. *Invest. Ophthal. Vis. Sci., 27*, 584-597.

Roy, J. P., Komatsu, H., & Wurtz, R. H. (1992). Disparity sensitivity of neurons in monkey extrastriate area MST. *J. Neurosci., 12*, 2478-2492.

Saito, H., Yukie, M., Tanaka, K., Hikosaka, K., Fukada, Y., & Iwai, E. (1986). Integration of direction signals of image motion in the superior temporal sulcus of the macaque monkey. *J. Neurosci., 6*, 145-157.

Sandini, G., Panerai, F., & Miles, F. A. (2001). The role of inertial and visual mechanisms in the stabilization of gaze in natural and artificial machines. In J.M. Zanker, & J. Zeil (Eds.), *Motion Vision: Computational, Neural, and Ecological Constraints, vol.* pp. 189-218). Berlin: Springer Verlag.

Schwarz, U., Busettini, C., & Miles, F. A. (1989). Ocular responses to linear motion are inversely proportional to viewing distance. *Science, 245*, 1394-1396.

Schwarz, U., & Miles, F. A. (1991). Ocular responses to translation and their dependence on viewing distance. I. Motion of the observer. *J. Neurophysiol., 66*, 851-864.

Simpson, J. I. (1984). The accessory optic system. *Ann. Rev. Neurosci., 7*, 13-41.

Simpson, J. I., & Graf, W. (1985). The selection of reference frames by nature and its investigators. In A. Berthoz, & G. Melvill Jones (Eds.), *Adaptive Mechanisms in Gaze Control: Facts and Theories, vol.* pp. 3-16). Amsterdam: Elsevier/North-Holland.

Takemura, A., Inoue, Y., & Kawano, K. (2000). The effect of disparity on the very earliest ocular following responses and the initial neuronal activity in monkey cortical area MST. *Neurosci. Res., 38*, 93-101.

Takemura, A., Inoue, Y., & Kawano, K. (2002). Visually driven eye movements elicited at ultra-short latency are severely impaired by MST lesions. *Ann. N. Y. Acad. Sci., 956*, 456-459.

Takemura, A., Inoue, Y., Kawano, K., Quaia, C., & Miles, F. A. (2001). Single-unit activity in cortical area MST associated with disparity-vergence eye movements: evidence for population coding. *J. Neurophysiol., 85*, 2245-2266.

Takemura, A., Kawano, K., Quaia, C., & Miles, F. A. (2002). Population coding in cortical area MST. *Ann. N. Y. Acad. Sci., 956*, 284-296.

Tanaka, K., Hikosaka, K., Saito, H., Yukie, M., Fukada, Y., & Iwai, E. (1986). Analysis of local and wide-field movements in the superior temporal visual areas of the macaque monkey. *J. Neurosci., 6*, 134-144.

Tanaka, K., & Saito, H. (1989). Analysis of motion of the visual field by direction, expansion/contraction, and rotation cells clustered in the dorsal part of the medial superior temporal area of the macaque monkey. *J. Neurophysiol., 62*, 626-641.

Westheimer, G., & McKee, S. P. (1975). Visual acuity in the presence of retinal-image motion. *J. Opt. Soc. Am., 65*, 847-850.

Yang, D., Fitzgibbon, E. J., & Miles, F. A. (1999). Short-latency vergence eye movements induced by radial optic flow in humans: dependence on ambient vergence level. *J. Neurophysiol., 81*, 945-949.

6. Functional Neuroanatomy of Heading Perception in Humans

Lucia M. Vaina[1,2] and Sergei Soloviev[1]

[1] Brain and Vision Research Laboratory, Department of Biomedical Engineering
Boston University, Boston, MA, USA

[2] Department of Neurology
Harvard Medical School, Boston, MA, USA

1 INTRODUCTION

As we move through the environment, the pattern of visual motion on the retina provides rich information about our passage through the scene. There is an abundant physiological and psychophysical evidence that this information, termed "optic flow" (Gibson, 1950), is an essential component of navigation in the three-dimensional space, since it is critical for encoding self-motion, for the perception of object movement and for controlling posture and locomotion. Psychophysical experiments have demonstrated that human observers can recover with high accuracy the direction of heading from optic flow patterns, even in the absence of actual self-motion (Warren, 1998, 1999). Electrophysiological studies on the 'motion system' in the dorsal extrastriate cortex in monkeys have identified cortical areas that might provide the neural substrate for the analysis of optic flow patterns and for the direction of self-motion (heading). In particular, the dorsal medial division of the macaque middle superior temporal area (dMST) has been proposed to be specialized for the analysis of complex optic flow information (Albright, 1993; Albright & Stoner, 1995; Andersen, 1997; Andersen et al., 2000; Wurtz & Duffy, 1992) and recent studies (Britten & Van Wezel, 1998, 2002) have shown that neurons in this area are also involved in recovering self-motion direction from optic flow cues.

Particularly relevant for this chapter are the results from recent research on electrical stimulation of MST neurons in monkeys trained to indicate the side of the display containing the focus of optic flow expansion (Britten & Van Wezel, 1998, 2002). These studies reported that the monkey's indication

L.M. Vaina, S.A. Beardsley and S.K. Rushton (eds.), Optic Flow and Beyond, 109–137.

of heading location was influenced by the activity of the MST neurons altered by electrical stimulation and the shift of responses was in the direction expected from the visual stimulus and the preference of the activated neurons — thus when electrically stimulated neurons had preference for left heading, the monkey's response was also biased toward left heading. This is the first strong evidence for the fact that MST neurons directly contribute to the computations underlying the direction of self-motion (heading) from optic flow. Is MST the only cortical area significantly involved in computing heading? A number of physiological studies showed that several other motion responsive cortical areas, especially parietal area 7a may provide additional neural substrate for mechanisms involved in optic flow computation ((Siegel & Read, 1997) and see Chapter 1, for a review). Does each of these areas compute optic flow and perhaps heading, or do they operate in concert in a network of neural mechanisms to mediate different aspects of these computations? In this chapter we partially address this question by examining some of the recent functional neuroimaging literature on optic flow and heading direction discrimination in healthy human subjects. However, as pointed out by Wurtz (1998), just the perceptual investigation will not give us the full answer since in the psychophysical tasks observers are asked to indicate to which side of a reference mark they are heading, and it is the experimenter's inference that the responses reflect a judgment of heading. However, this is the best approximation of the problem we now have on our hands, before quantitative studies of heading estimation in subjects navigating in the real everyday environment will emerge with some consistency.

2 CORTICAL AREAS RESPONSIVE TO OPTIC FLOW REVEALED BY FMRI IN HUMANS

In humans, functional neuroimaging studies (PET and fMRI) have repeatedly reported that many regions in the human brain respond selectively to moving stimuli (Braddick et al., 2001; Dupont et al., 1994; Grezes et al., 2001; Grossman et al., 2000; Grossman & Blake, 2002, Servos et al., 2002; Sunaert, 2001; Sunaert et al., 1999; Tootell et al., 1997; Vaina et al., 2001). Most of the studies published thus far have investigated the anatomical areas of activation and the extent to which cortical activation, as measured by stimulus-evoked changes in blood flow and tissue oxygenation, are characteristic of specific motion tasks.

A large number of studies have concentrated on a region in the ascending limb of the inferior temporal sulcus referred to as hMT+ because it functionally represents the human homologue of the macaque areas MT and MST (Beauchamp & DeYoe, 1996; Braddick et al., 2001; Cheng et al., 1995;

Dumoulin et al., 2003; Dupont et al., 1994, McKeefry et al., 1997; Reppas et al., 1997; Tootell et al., 1993, 1995a, 1995b, 1998; Watson et al., 1993; Zeki et al., 1991). In particular, when Tootell et al. (1995b) mapped BOLD (blood oxygenation level-dependent) responses in the striate and extrastriate visual areas to motion stimuli portraying radial gratings, they found that the hMT+ strongly responded to low contrast stimuli (it saturated already at 4% contrast). Orban and collaborators conducted several combined functional neuroimaging (both PET and fMRI) and psychophysical studies of direction and speed discrimination of frontoplanar random dot motions (Dupont et al., 1994, 1997; Orban, 2001; Sunaert et al., 1999, 2000; Van Oostende et al., 1997), which strengthens the evidence that several areas other than hMT+ are involved in motion processing.

The posterior part of area hMT+ is bound by area V3a, which is also quite sensitive to motion (Tootell et al., 1997). An area anterior to V3a has been designated as the kinetic occipital (KO) area, to reflect sensitivity to the form of structured kinetic stimuli (Dupont et al., 1997; Van Oostende et al., 1997). Area KO responds strongly to motion defined borders in complex motion displays. This area, referred to as V3b by Smith and collaborators (Smith et al., 1998), has been shown to also respond to certain types of second order motion.

Recently, several fMRI studies aimed to further subdivide the area hMT+ into two distinct regions suggested as corresponding to the macaque areas MT and MST (Dukelow et al., 2001; Huk et al. 2002; Morrone et al., 2000). In particular, the Huk et al. (2002) study cleverly exploits the physiological evidence from the macaque for a distinguishable retinotopic map in MT and a much coarser retinotopic organization in MST, together with the larger extension into the ipsilateral hemifield of the receptive fields of MST neurons compared to those in MT. Thus, using stimuli specifically devised for assessing retinotopic organization and receptive field size within hMT+, they found two distinct but adjacent regions of activity: one, suggested to be the human homologue of the macaque MT, exhibited retinotopic organization and smaller receptive fields, and the other, corresponding to area MST in the macaque, did not show clear retinotopy but responded to peripheral ipsilateral stimulation consistent with large receptive sizes of its neurons.

3 FUNCTIONAL NEUROIMAGING OF OPTIC FLOW AND OF DIRECTION OF HEADING JUDGMENT

A PET study (de Jong et al., 1994) of the cortical areas responsive to optic flow stimuli simulating forward motion in depth over a flat horizontal surface found bilateral foci of activity in the fusiform and temporal gyri, the right dorsal cuneus (area V3), and the latero-posterior cuneus (or superior parietal lobe). No significant activity was found in hMT+. The optic flow field in this study consisted of small bright dots on a dark background (viewed binocularly) and comparisons were made between displays with 100% coherent radial expansion motion from a virtual horizon and 0% coherent motion in which all the dots moved in random directions. In a follow up study (Howard et al., 1996), the stimulus was modified to activate both the inferior and superior visual field during fMRI scanning. Significant activation was found in both the dorsal and ventral V3, within the area hMT+ and a region within the superior temporal gyrus (STG), hypothesized by these authors as the probable human equivalent of the macaque superior polysensory area (STP) known to respond to optic flow stimuli (Siegel, 1998; Siegel & Read, 1997).

Using a block design paradigm alternating static dots with random dot kinematograms portraying optic flow, Greenlee (2000) found clusters of activation in the striate (V1) and extrastriate (V2, V3/V3a) cortices, in ventral area V3, in areas KO/V3b and in the hMT+. This interesting study also demonstrated that the hMT+ activation did not vary significantly with the type of optic flow (e.g., rotation or radial). However, area KO/V3b appeared to respond selectively to the disparity gradient present in the optic flow field stimulus In particular, within KO/V3b the disparity effect was measurable in the rotation condition and somewhat less in the expansion condition. Rutchsmann and colleagues (Rutschmann et al., 2000) investigated BOLD responses to optic flow in a paradigm comparing either monoptic or dichoptic presentations of either expansion or expanding-spiral motions with a random-walk stimulus. In the monoptic presentation subjects viewed the optic flow binocularly in plane, with both eyes viewing the same stimulus. In the dichoptic presentation the stimuli were presented with stereo depth, using speed gradients combined with varying amounts of disparity. In a block design experimental paradigm, alternating fixation with one type of optic flow, significant responses to all types of stimuli were found in Brodmann areas (BA) 17, 18, 19 and 37. However, selective response to the direction components of the flow (expansion > spiral > rotation) were found only in the middle portion of area 19, labeled 19m and postulated as equivalent to the functionally defined area KO/V3b (Dupont et al., 1997; Smith et al., 1998; Van Oostende et al., 1997). This provides further support for the involvement of areas KO/V3b in the processing of optic flow. However, there was no stimulus selective response in area hMT+, except under the dichoptic viewing conditions (in the presence of binocular disparity). Furthermore, the dichoptic stimuli also consistently elicited a small increase in response (percent signal increase) in BA 19m (KO/V3b) and BA 19d (putative V3a) which was coupled with the directional component of the flow, possibly representing a correlate of the additional processing associated with the neural analysis of optic flow in the 3-D conditions (Rutschmann et al., 2000).

These and other studies not discussed here compared moving dots with static stimuli. However, as pointed out by Braddick and collaborators (Braddick et al., 2001) by doing this, cortical areas specifically activated by direction of motion cannot be distinguished from areas activated by temporal frequency (flicker). To address specifically the cortical substrate of motion coherence (direction discrimination) these authors used a block design stimulus paradigm consisting of a uniform moving field of random dots alternating with dynamic noise dots, but keeping throughout the same temporal and spatial frequencies. The results of this study reveal that in the occipital lobe, V1 was not specifically activated by coherent motion, but strong activation was seen in several extrastriate areas, among which most prominently in the areas hMT+, V3a and V3. The unexpected outcome of this study is the strong response to coherent motion in area V3, suggesting that the ventral occipital region is also sensitive to this type of motion. In addition, this study is somewhat at odds with the findings of McKeefery and colleagues (McKeefry et al., 1997) who reported that hMT+ was more strongly activated by incoherent than by coherent motion. One possible explanation for this discrepancy might be that in this latter study (McKeefry et al., 1997) the stimuli had very low dot density, which prevented summation within the receptive field, and thus the directional activation could not predominate in hMT+. Sensitivity to motion coherence, assessed as the increased stimulus-related blood flow, was also reported in the intraparietal sulcus (IPS) and in the superior temporal sulcus (STS).

In a recent PET imaging study Beer and colleagues (Beer et al., 2002) stimulated different types of observer's continuous movements in depth using a very large visual display in order to identify brain areas selective to these movements. They concluded that in order to detect direction of heading and analyze its parameters the brain engages a network of widely distributed regions, particularly KO and several additional sites in the temporal, temporoparietal and occipital areas. Similar to studies discussed above, these authors also found that hMT+ was selectively activated by incoherent motion, but was not selective for the continuous coherent, wide-field motion simulating self-motion.

Peuskens and colleagues (Peuskens et al., 2001) used PET and fMRI to determine the cerebral activation pattern elicited when subjects performed an active task of heading direction discrimination when viewing a ground plane of optic flow. In the PET study the main effect of heading in the occipital and occipital-temporal cortices was a strong activation in the cuneus, with the local maxima in the presumed areas V2 and V3a, and in an area located posterior to the reported location of the hMT+ (Talairach coordinates –44, -80, 4; z-value; z = 7.2 and 40, -82, 4; z = 5.3), (Talairach & Tournoux, 1988).

More anterior, bilateral activation was noted in the superior parietal lobule, in a region located dorsally in the intraparietal sulcus corresponding to the areas DIPSM/DIPSL (medial and dorsal intraparietal sulcus regions), (Sunaert et al., 1999). In a subsequent fMRI study these authors used a passive viewing task for hMT+ localization, in addition to the heading discrimination task. Comparison in the Talairach space of the areas activated by these two tasks revealed that hMT+ was significantly involved in heading discrimination. Further significant activation was also documented in several posterior frontal regions, both in the dorsal and ventral premotor areas, suggesting the linking of heading information to motor plans.

D'Avossa and colleagues (D'Avossa et al., 1998) interested in the spatial coding of heading representation, investigated whether different values of a relevant parameter were represented in different cortical regions. Their approach is quite interesting as it directly addresses the concept of neural maps that have been shown to contain specific representations within different regions of cortex (Knudsen et al., 1987). They used fMRI to determine whether homologous areas of each hemisphere encode heading direction towards the opposite side of the space and to map the azimuthal and elevational components of heading (D'Avossa & Kersten, 1996). The stimulus consisted of random dot kinematograms portraying optic flow which simulated heading toward a 3-D cloud of stationary dots either in the gaze direction or in an eccentric direction (i.e. 3 deg above, below, right or left of the fixation point). In each block of trials optic flow simulating heading straight ahead was paired with optic flow simulating one of the eccentric headings. In accord with previously discussed studies, the cerebral activations during the perception of optic flow simulating self-motion in an eccentric direction activated regions in the occipital and parietal lobes presumably corresponding to the areas V3 and 7a. The lateralization of activity in these areas co-varied with the direction of heading, consistent with a map of heading in either or both of these areas. The anterior regions showed a right hemisphere lateralization irrespective of direction, which may be explained by a possible involvement of attentional mechanisms. Further data analysis of activations points to the involvement in the left hemisphere of the cuneus and the superior parietal lobule, and the cuneus and the medial occipital gyrus bilaterally.

Similar to the results reported by de Jong and colleagues (de Jong et al., 1994), there were no significant activations in the hMT+ to any of the heading stimuli. However, unlike in de Jong's study, here the heading stimuli did not elicit any activation in the ventral occipital-temporal areas. Wunderlich and collaborators (Wunderlich et al., 2002) used fMRI to investigate the neural substrate of perception of objects that appear to move in depth (toward or away from the observer). Such stimuli were alternated in a block design paradigm with a static random dots field. The major question in this study was whether brain activation elicited by the perception of motion towards the observer is different from activity resulted from motion away from the observer. They found that regions in the lateral inferior occipital cortex bilaterally and the right lateral superior occipital cortex were quite specifically implicated in processing motion toward the observer. The activated region in the superior lateral occipital cortex corresponded to the cortical area KO. The area hMT+ was not differentially activated by dots moving towards the observer.

Different studies used different stimuli parameters and consequently found somewhat different areas of activity. Surprisingly, there was no agreement on the specific sensitivity of hMT+ to the various heading stimuli. The detailed study of Peuskens and collaborators conjectured that if the activation pattern in DIPSM/L can be considered suggestive for this area's central role in heading perception, these parietal regions may correspond, functionally, to the macaque area 7a, which is believed to be the apex of the motion pathway. Neurophysiological studies of Siegel and colleagues (Chapter 1, for a review) suggest that area 7a might be involved in the extrapersonal representation of space and the representation of self-motion, or direction of heading and of object motion. D'Avossa and colleagues (D'Avossa et al., 1998) reported activity in the probable homologue of 7a for rightward and leftward heading, but it is puzzling that no activation was found in the parietal lobes for upward heading.

To further determine the neuro-anatomical substrate of heading perception in humans, here we compare results from fMRI and behavioral studies in three related heading tasks. In the first two, *Heading RDK* and *Heading RDK Objects*, the stimulus consisted of a dynamic random dot patterns simulating heading towards a stationary 3-D cloud of dots towards which the observer moved on a straight path. The *Heading RDK Objects* contained four square objects defined by random dots translating in plane and all crossing the focus of expansion. In the third task, *Heading Landmarks*, the stimulus portrayed a large room populated with different static objects that served as landmarks. In particular, we were interested to determine whether cortical activation in three heading tasks was strongly sensitive to different image cues or to the task itself. Our focus was on comparing the brain regions activated by each of the tasks and on the correlation of behavioral results with activation (BOLD signal percent change) in particular regions of interests (ROIs) identified in the basic fMRI data analysis.

3.1 Subjects and Procedures

Eight naïve, healthy subjects with normal vision (ranging in age between 22-35 years; 4 women) were paid to participate in the study of visual heading. All subjects participated in the psychophysical and the fMRI studies of the tasks described above. The entire study was approved by the ethics committees at the Boston University and the NMR Center at Massachusetts General Hospital and all subjects gave informed consent. Displays were generated by a Power Macintosh G4 and for psychophysics were presented on the Macintosh screen. During fMRI scanning, visual stimuli were rear-projected onto an acrylic screen (DaTex, Da-Lite Corp.) providing an activated visual field up to 40×25 degrees. Stimuli were projected onto the screen by a Sharp 2000 color LCD projector, through a collimating lens (Buhl Optical). Details on the fMRI data acquisition are given in the next section.

Two-three days prior to the fMRI study each subject first underwent a practice session until they felt comfortable with all the tasks (usually 10-15 repetition trials) and then additional thresholds were obtained on each experimental condition from each subject. Feedback was given only during the practice trials. Subjects were instructed to maintain fixation throughout each trial, but the examiner controlled fixation only informally. During the psychophysical testing the room was completely dark except for the display. To obtain estimates of the threshold at which subjects can judge their heading accurately, the angle between the heading and the target line was varied according to an adaptive staircase procedure. A button press on a keypad indicated the response. For the psychophysical tasks, threshold was calculated as the average of the last six reversal values, during which stimulus presentation followed a classical staircase (one error - one step up, three correct response - one step down). The variable in all three tasks was the heading angle defined as the visual angle between the actual heading and the probe. Subjects completed 5-8 repeats of the staircase procedure, until their performance appeared to asymptote three times in a row. A final threshold was calculated for each subject as the mean of the three consecutive similar thresholds. In Figure 1 (d), for each heading task we report the mean of the final thresholds from all subjects.

3.2 The Heading Stimuli

3.2.1 Heading Perception in Random Dot Kinematograms

The first experiment tests the ability of naive observers to judge heading from motion fields that simulate pure translation of the observer in two conditions: (a) the stimulus portrays only the observer's motion toward the stationary scene. The focus of expansion is visible throughout a trail; (b) the scene contains four additional translating objects that cross the observer's path in each trial (and hence the FOE was not visible at all times).

(a) Subjects viewed binocularly optic flow patterns, centered at the fovea, simulating self-motion relative to a 3D cloud of stationary dots. The stimulus consisted of a dynamic random dot field displayed in a square aperture subtending 35.5 deg^2 at 30° viewing distance. Dots were randomly generated from the super volume including the visible part at the beginning to obtain constant dot density projection of the environment on the screen. Dot size was 6 $arcmin^2$ and dot density (2 dots/deg^2) was algorithmically kept constant. The simulated observer speed was 200 cm/sec, which corresponds to a fast walk of 4.2 miles/hour, or 7.2 km/hour. The direction varied with uniform probability between extreme values of 2.5° and 12° to the left and right of the center of the display.

(b) In this condition we tested how observers' heading judgments were affected by the presence of moving objects. Previous precise psychophysical studies of Royden and Hildreth (1996) and Warren and Saunders (1995) reported that in the presence of a single moving object that crossed the observer's path, there was a small directional bias in the heading judgments. Our focus was on characterizing and comparing the neuroanatomical substrates of these two test conditions, simple straight trajectory heading and heading in the presence of several moving objects crossing the focus of expansion (FOE).

Superimposed on the display described above there were four small objects defined as dynamic occluding planes defined by the difference in speed and direction from the heading pattern defined above. The objects had diameters ranging between 1.75°- 4°, and a constant dot density of 2 dots/deg^2. The density of the entire display was held constant at 2 dots/deg^2. The dots within the illusory borders of the objects translated relative to the observer within the XY plane at a constant speed of 100 cm/sec at the viewing distance of 30 cm. The motion of the objects was independent relative to the observers' simulated motion. The objects kept a constant linear planar trajectory (two diagonal in opposite directions, one upward and the other downward) and each object intersected once and occluded temporarily the focus of expansion. Throughout the test the starting positions and direction motions of the objects were kept constant while the observer heading direction was varied from trial to trial by the staircase procedure.

In both experimental conditions, (a) and (b), observers were instructed to fixate on a central fixation mark. They watched a motion sequence for 440 msec at the end of which a vertical white probe was displayed on a newly generated static frame of random dots with the same image characteristics as in the RDK (except motion) at a given horizontal angle from the true heading (FOE). The observers' task was to judge whether it appeared that they were heading to the left or right of the vertical probe line. A button press on a keypad indicated the observer's responses. Prior to task (b), subjects were told that moving objects would be present and they were asked to ignore them as much as possible and to base their response to their perceived movement toward the 3D cloud of dots.

3.2.2 Results

Figure 1d (conditions (a) and (b)) shows the results on these tasks from eight observers. Consistent with previously published literature, for all subjects the straight-line heading (Heading RDK) without objects was slightly easier than the Heading-RDK with objects, as reflected by the thresholds on the two tasks which referred to the judgment error between the true heading (FOE) and the probe.

3.2.3 Heading in the Presence of Landmarks

(c) The stimulus (shown schematically in Figure 1c) was a colored naturalistic display with texture mapped objects and complex 3-D structures portraying a large room with furniture, walls, and objects. The movement of the display simulated the direction of translation of an observer. The starting position was randomly generated from the central 2/3 of the room width and then the target of heading was chosen randomly at the right or left of the central fixation mark. The stimulus subtended 44×44 deg from a viewing distance of 30 cm. For each trial the first frame of the motion sequence portraying the room was displayed on the screen before the trial began. Observers were asked to fixate on a cross in the center of the display and watch the motion sequence for 1.9 seconds at the end of which the fixation mark disappeared. A vertical probe in the form of an hourglass was shown on a static view of the room and observers where asked to judge whether they were heading to the left or to the right of the probe via a button press. As in

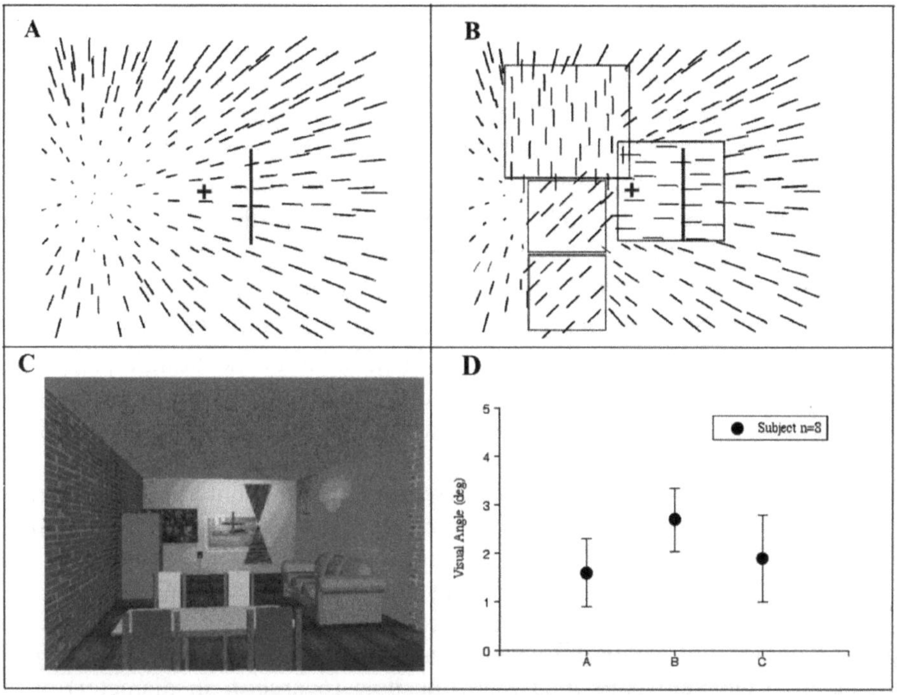

Figure 1. Heading stimuli and psychophysical performance of all subjects for each heading task. a) *Heading RDK.* b) *Heading RDK Objects.* c) *Heading Landmarks.* d) The mean final thresholds from all subjects reported for each heading task.

the previous heading task, an adaptive staircase was used to vary the angle between the true heading and the probe.

3.2.4 Results

Figure 1d (condition (c)) shows mean results from six staircases of eight observers. As before, the y-axis portrays the difference in degrees of visual angle between the true FOE and the probe. The difference among the results from the three tests was not statistically significant (t-test, $p>0.05$).

3.3 FMRI Data Acquisition and Experimental Design

Data were acquired in a 1.5-T whole-body MRI system (Magnetom Vision, Siemens, Germany) equipped with a head volume coil (with 40 mT/m maximum gradient strength, a slew rate of 200 (T/m)/sec and a FOV of 40 cm). For fMRI, echo planer imaging (EPI), sensitive to blood oxygen level dependent (BOLD) effects, was used (repetition time TR = 2.53 sec, echo time TE = 70 ms, flip angle 90°, field of view 200 mm). To cover the whole brain, we used twenty two 5 mm thick slices with 1 mm gap with an axial orientation and an image size of 64×64 pixels. The slices covered the whole brain with a voxel size of 3.13×3.13×6 mm for the EPI images. For anatomical localization, we used 3-D gradient echo T1-weighted images (TR = 11.1 ms, TE = 4.3 ms, flip angle 8°, FOV 256 mm). Anatomical image size was 256×256×128 pixels, with a slice thickness of 1.33 mm, and voxel size of 1×1×1.33 mm. The stimuli were presented in a block design paradigm, alternating "fixation" as a baseline condition with one of the three stimuli, depending on the particular time series acquired. The subjects indicated their response by a button press on a magnet compatible keypad. The order of the tests was pseudo randomized and presentation was counterbalanced across subjects. The switching between the heading tasks was transmitted verbally. For every subject, each time series was repeated 3 times and for each of the eight subjects all data was obtained in a single scanning session.

Two additional time series were acquired from all subjects in which a passive viewing of a radially moving random dot kinematogram (10 deg in diameter, 6 deg/sec) embodied by white dots (6 arcmin in diameter) on a black background was interleaved with stationary dots of the same characteristics (except motion). A small white fixation mark was shown in the middle of the display and as previously, subjects were asked to maintain fixation. The motion alternated from trial to trial (1 sec each) between expansion and contraction. This condition was used for localizing the hMT+ complex in every subject.

3.4 FMRI data analysis

FMRI data were post processed using the MEDx 3.41 software package (Sensor Systems Inc., Sterling, VA) and complementary scripts (MEDx TCL, MATLAB (The Mathworks Inc., Natick, MA), and PERL) developed in our laboratory. Details of the initial steps of fMRI data analysis have been published (Vaina et al., 2001). Briefly, all individual functional images were motion-corrected (Woods, Grafton, Holmes et al., 1998; Woods, Grafton, Watson et al., 1998; Woods et al., 1993), spatially smoothed using a Gaussian filter (FWHM of two times the voxel size). Global intensity normalization was performed to normalize the average of each volume to the same mean value and linear signal intensity drift not related to the task under study was estimated for each voxel and removed from the time series data. Active brain regions were determined by means of a t-test comparison of the rest and active conditions of each experimental paradigm and the timing of each paradigm condition were automatically recorded during each fMRI scan. The first four frames were removed from each acquisition to compensate for the effects of adaptation of a subject to the start of a new experimental run, and a five second delay was introduced into each paradigm file to account for the effects of hemodynamic delay in the fMRI responses. A statistical significance threshold of $P<0.05$ (Resel corrected) was applied to the data with an extent threshold of a minimum cluster size of four voxels (Worsley et al., 1992, 1996). For group analysis, statistical maps (z-values) for each subject in each condition were used to form a group statistical model by calculating the sum of individual z-values (over each voxel) divided by the square root of the number of subjects. For each subject, EPI images were registered to the high-resolution deskulled structural volume, the same transformation was applied to statistical volumes, registration in Talairach space (Talairach & Tournoux, 1988) was performed for the structural and statistical volumes, and thresholded statistical maps ($z \geq 3.0$ for individual subjects) were superimposed onto the high resolution structural volume.

A "goodness of data" test was performed to evaluate the similarity of statistical maps coming from the same paradigm acquisitions. The measure of similarity between statistical maps was based on the normalized cross-correlation commonly used in template matching. For a pair of statistical maps f and w the coefficient of normalized cross-correlation is given by

$$r = \frac{\sum_x \sum_y (f(x,y) - \bar{f})(w(x,y) - \bar{w})}{\left(\sum_x \sum_y (f(x,y) - \bar{f})^2 \sum_x \sum_y (w(x,y) - \bar{w})^2\right)^{1/2}} \tag{1}$$

summations are carried out over all pixels in the image. The measure of similarity defined above is based on the Schwartz inequality

$$\int fw \leq \left(\int f^2 \int w^2\right)^{1/2} \tag{2}$$

where f and w are two integrable, real-valued functions. It follows from the inequality that $-1 \leq r \leq 1$. Therefore, if statistical maps f and w coincide then $r = 1$, if they are not correlated $r = 0$, and if f and w are anti-correlated $r = -1$. After averaging the similarity measure over several slices, the result was compared to a threshold value set to 0.4. Data were considered to be good if similar acquisitions had the value of parameter r exceeding the threshold value set for the subject. Two weakly correlated acquisitions were found and they were removed from further analysis. In one subject the data was inconsistent from one time series to another across every task and furthermore this subject had significant head movement. This subject was discarded from the further analysis of fMRI data.

To achieve a statistically sensitive analysis in a larger number of brain regions, we defined several regions of interests (ROIs) based on activations obtained in the initial statistical analysis of the data and previously reported results on brain areas involved in visual processing in humans, including tasks directly relevant to our study such as topographical navigation and judgment of heading (Aguirre & D'Esposito, 1997; Aguirre et al., 1998; D'Esposito et al., 1998; Dupont et al., 1994, 1997; Howard et al., 1996; Mendola et al., 1999; Peuskens et al., 2001; Sunaert et al., 1999; Tootell et al., 1997; Vaina et al., 2001; Van Oostende et al., 1997; Zeki et al., 2003). It is important to note that the estimated percent BOLD signal change values for each functionally defined cortical region were obtained on the basis of ROIs defined for each individual subject in a group and not on the basis of the average statistical activation map obtained in the group analysis for an fMRI task. This approach is somewhat more complex than the group analysis frequently seen in the fMRI literature, but it has a net advantage in that it is significantly less prone to inter-subject variability in the Talairach space of brain areas with similar functional properties.

For each ROI defined from our own fMRI data we calculated the Talairach coordinates of its center, mean and standard deviation of fMRI signal percent change, activation volume (total number of voxels having $z \geq 3.0$), and asymmetry index (AI) (Binder et al., 1996; Desmond et al., 1995; Thulborn et al., 1999). To reduce the contribution of voxels exhibiting sustained negative BOLD responses associated with reductions in blood flow and neuronal activity (Shmuel et al., 2002), only voxels in which BOLD signal percent change was positive were included in calculation of the mean percent change and standard deviation. For each ROI, we estimated the variation of the mean percent change $P = \mu \pm \Delta\mu$ by the following formula

$$\Delta\mu \approx P_{\max} + \frac{S_p}{S}(\sigma - P_{\max}) \tag{3}$$

where $P_{\max}$ is a maximal percent change with the ROI, S_p is the area of the ROI adjusted to contain only voxels with positive signal percent change, and S is the area of the original ROI. For each experimental condition, bar plots of mean fMRI signal percent change together with error bars (of estimated variation) were plotted by ROIs. Furthermore, for each ROI we performed a three-way ANOVA to compare mean values of fMRI signal percent change with respect to the three factors: different tasks, subjects, and hemispheric sides (Hogg & Ledolter, 1987), and a multiple comparison test (using Bonferroni adjustment) of means of fMRI signal percent change was used to determine in which task the fMRI response change was significantly different from that in other tasks (Hochberg & Tamhane, 1987).

To study the relationship of fMRI activation to the subject's psychophysical performance, we performed an analysis of the correlations between fMRI responses and behavioral data. For all subjects (n=7) scatter plots showing normalized fMRI activation vs. normalized psychophysical performance for the individual subjects and specific tasks were generated for each ROI (Gilaie-Dotan et al., 2001) – the high correlation value ($r > 0.4$) indicates involvement of an ROI in processing of the visual task.

3.4.1 Results

The data from seven subjects was analyzed following all stages described above. First, statistical analysis of the fMRI data was conducted to determine the brain regions that may be involved in some or all three heading tasks used in this study. Group activation maps for the seven subjects on the three heading tasks are shown in axial brain slices registered in the Talairach space (Figure 2). By and large, the activation maps were similar for all heading tasks. The activated voxels in the occipital lobe were located in the fusiform gyrus (BA 17, 18), the cuneus (BA 18) and middle occipital gyrus (BA 18, 19). Activated voxels in the temporal lobe having z-values below threshold value of 5.7 (not visible in Figure 2), were located in the superior temporal gyrus (BA 22), middle and superior temporal gyrus, (BA 21, 22). Activated voxels in the parietal lobe (not shown) were located in inferior parietal lobule (BA 40), precuneus and superior parietal lobule (BA 7). The main focus of activation in the frontal lobe was located in precentral gyrus (BA 6). For reference, the hMT+ complex was localized in each subject and the mean area of activation for the seven subjects is outlined with concentric circles.

A subsequent analysis at a finer scale of resolution was carried out in order to reveal any significant differences across tasks in the level of fMRI responses. We defined separately for each task and subject several ROIs as described in the Methods section. The average values (over the seven subjects) of Talairach coordinates of the ROIs defined, standard deviations, corresponding maximal Z values, and tentative functional names are shown in Table 1.

After labeling each ROI with a tentative functional name, we estimated BOLD percent signal change within the functional regions defined for each task and for every subject. The estimated signal change values were averaged over all the subjects (n=7, shown in Figure 3).

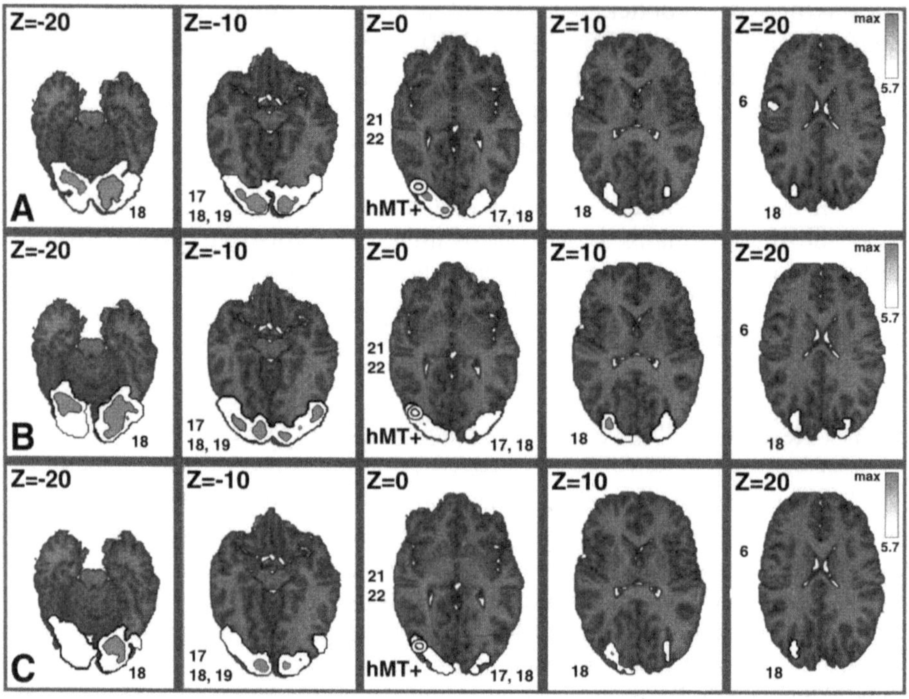

Figure 2. Activation regions for control subjects thresholded at z = 5.7 (n = 7). a) Averaged statistical map for the *Heading RDK* experiment. b) Averaged statistical map for the *Heading RDK Objects* experiment. c) Averaged statistical map for the *Heading Landmarks* experiment. Data was collected in a 1.5T scanner. Area hMT+ is outlined in a white circle.

To determine whether there were significant differences in activation levels within ROI's (Table 1) among the three tasks, subjects, and side of activity, we performed a three-way ANOVA and multiple comparisons analysis (see methods section). By and large the activations levels were similar across tasks and most ROI's. Significant differences were found only in V2 (p = 1.7e-09; F = 62.68) due to higher activation level in *Heading Landmarks* than in the other two tasks, V3a (p = 5.93e-05; F = 21.11) and KO/V3b (p = 0.02; F = 5.49) due to higher activation in the *Heading RDK-Objects* and *Heading Landmarks* than in *Heading RDK*, and PreCs (p = 0.02; F = 9.64,) due to higher activation level in *Heading RDK Objects* than in the other two tasks. As one would expect, these results illustrate that the more complex experimental situations portrayed by *Heading RDK-Objects* and *Heading Landmarks* stimuli were selectively associated with a higher signal change in several ROI's than the simpler *Heading RDK* task that simulates the flow field of an observer traveling on straight path in a rigid environment.

Table 1. For each ROI defined based on the initial statistical analysis of fMRI data for all heading tasks for several subjects (n=7) participated in the study, the table contains the ROI's tentative functional name, its stereotaxic coordinates (mean, std) and activation level (max Z value, std).

Stereotaxic coordinates and activation level for each ROI defined

ROI	Stereotaxic coordinates	Z values
V1	12.5 (0.9); -90.2 (2.5); -12.0 (0.0)	7.4 (2.9)
	-12.7 (1.8); -94.9 (1.3); -12.3 (0.7)	5.6 (3.4)
V2	29.4 (0.6); -84.5 (1.5); -11.0 (0.0)	6.1 (2.8)
	-26.6 (1.3); -86.8 (1.0); -11.0 (0.0)	5.6 (3.8)
VP	40.7 (3.3); -71.0 (1.7); -11.0 (0.0)	6.2 (3.3)
	-42.7 (1.8); -73.0 (2.3); -10.9 (0.4)	6.2 (2.6)
V3	15.7 (0.5); -93.9 (0.8); 7.3 (0.7)	4.7 (3.0)
	-18.0 (0.0); -92.7 (0.5); 7.0 (0.0)	3.9 (2.8)
V4	28.8 (5.2); -64.0 (4.2); -17.0 (0.0)	6.4 (3.1)
	-25.6 (3.7); -67.5 (1.4); -19.0 (0.0)	5.9 (3.8)
V3a	26.0 (0.3); -81.5 (1.9); 15.0 (0.0)	5.3 (3.8)
	-24.6 (1.4); -80.6 (3.0); 15.0 (0.0)	3.7 (2.0)
KO	28.2 (1.0); -84.7 (1.8); 2.0 (0.0)	5.4 (3.2)
	-27.6 (0.7); -84.2 (1.0); 2.0 (0.0)	4.4 (2.3)
MT	41.4 (3.1); -69.1 (1.3); -2.1 (0.4)	5.1 (2.5)
	-41.4 (2.4); -72.0 (2.9); -1.9 (0.8)	4.9 (2.1)
STS	54.6 (0.5); -37.3 (0.7); 5.0 (0.0)	1.3 (2.1)
	-53.3 (2.2); -39.5 (0.2); 5.5 (0.9)	1.4 (2.2)
STG	57.2 (0.2); -0.6 (1.4); 5.0 (0.0)	1.8 (2.3)
	-52.2 (1.8); -4 .0 (4.0); 5.5 (1.4)	1.4 (1.1)
PreCs	44.4 (2.1); -1.9 (3.6); 31.5 (3.0)	5.1 (2.5)
	-46.2 (1.3); -2.2 (2.4); 30.0 (0.0)	4.5 (3.2)
PostCs	28.8 (1.6); -51.4 (1.2); 37.3 (1.6)	4.9 (3.3)
	-32.2 (3.4); -49.3 (2.4); 36.1 (2.5)	4.7 (2.4)
DIPSM/L	13.4 (2.3); -72.9 (2.7); 41.8 (0.7)	6.7 (3.4)
	-16.9 (0.9); -71.9 (1.2); 42.0 (0.0)	7.3 (3.0)
SFS	23.0 (2.6); -13.3 (2.5); 52.3 (1.3)	5.0 (4.2)
	-26.0 (1.5); -15.5 (1.3); 50.5 (2.3)	4.1 (2.8)
PhG	26.6 (3.3); -48.0 (3.8); -20.0 (2.1)	3.2 (2.2)
	-29.2 (1.5); -51.3 (5.2); -19.9 (1.8)	2.9 (2.6)

3.5 Direct Correlation Between FMRI Signal and Performance on the Heading Tasks

We determined quantitatively the extent to which there was a direct correlation between the recorded fMRI signals and subjects' psychophysical performance measured by their responses during the fMRI scanning session. Figure 4 shows the relationship between normalized fMRI signal and subjects' normalized performance on the three tasks for several ROI's (Gilaie-Dotan et al., 2001). We considered the correlation to be significant if its correlation value (r) exceeded 0.4. For the *Heading RDK* task, the highest correlation between performance and normalized signal was in the functional area DISPM/L (r = 0.83) and a weaker correlation in the PostCS (r = 0.44) and VP (r = 0.41). This result strengthens Peusken's et al. (2001) suggestion that, compared with other parietal motion areas, the DISPM/L region of the posterior portion of the intraparietal sulcus is specifically tuned to heading estimation.

Behavioral responses in the *Heading RDK Objects* task were significantly correlated in area KO (r = 0.74), STS (r = 0.63), hMT+ (r = 0.53), V3 (r = 0.53) and V3a (r = 0.49), suggesting that the subjects might have computed, both motion and object borders when doing this heading task. Activity in both the PreCs (r = 0.76), and PostCs (r = 0.72) were also highly correlated with behavior. These areas are part of the frontoparietal network related to visual attention and oculomotor processing (Corbetta, 1998). Considering the three areas that were most correlated with behavior suggests that in order to compute heading direction in the presence of multiple moving objects, the visual system uses the relative motion system to segment and eliminate the objects. We suggest that the extraction of the kinetic objects occurs under the control of spatial attention (Corbetta, 1998; Vaina et al., 2001).

In the *Heading Landmarks* task the most significant correlations with behavior was found in STG (r = 0.64) followed by somewhat weaker correlations in V3 (r = 0.51), V3a (r = 0.69), and KO (r = 0.40). The Talairach coordinates of the center of area STG here, corresponds to a region suggested by previous functional neuroimaging studies (Howard et al., 1996, Vaina et al., 2001) as a human homologue of the macaque Superior Temporal Sulcus (Stp). It has been proposed that STP integrate motion information from the dorsal visual stream and object information from the ventral visual stream (Racine et al., 1996; Vaina & Gross, (submitted)). Both types of information coexist in this task and they might be used in concert to discriminate direction of heading in a scene populated with objects. Alternatively, object information may be extracted but not necessarily used in performing this task,

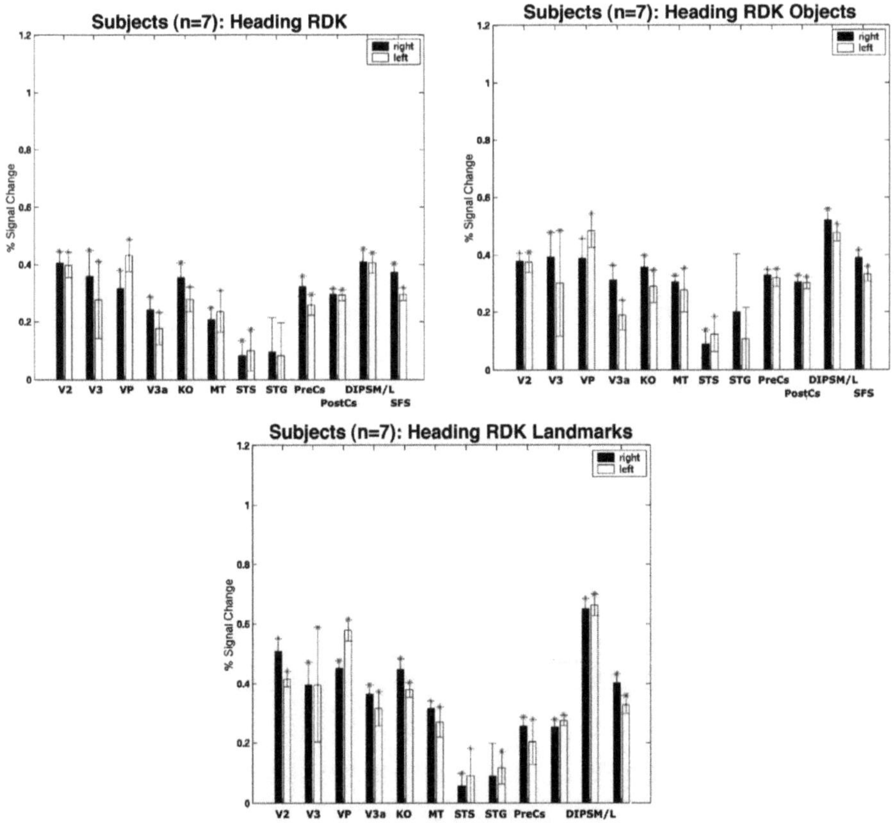

Figure 3. ROI analysis for control subjects (n=7). The bar graphs show the percent change in the fMRI signal within each ROI. Each pair of bars shows signal change in both the right and left hemispheres. Error bars indicate error estimates calculated as described in the methods section. Results from heading RDK (left), heading objects (right), and heading landmarks experiments (bottom) are shown.

since information from optic flow that is still present in the scene might be sufficient for judging the direction of heading.

The results of heading discrimination presented here are consistent with previous data on the same tasks from neurological patients with discrete lesions. For example, the suggestion of the important role in heading estimation of area DIPSM/L compared to other parietal areas ((Peuskens et al., 2001) and the data presented here), is strongly supported by our previous reports on neurological patients with bilateral posterior parietal lesions suffering of the Balint-Holmes syndrome, whose low level motion perception was normal, but perception of optic flow patterns, especially heading perception (Jornales et al., 1997; Vaina, 1998; Vaina & Rushton, 2000) was

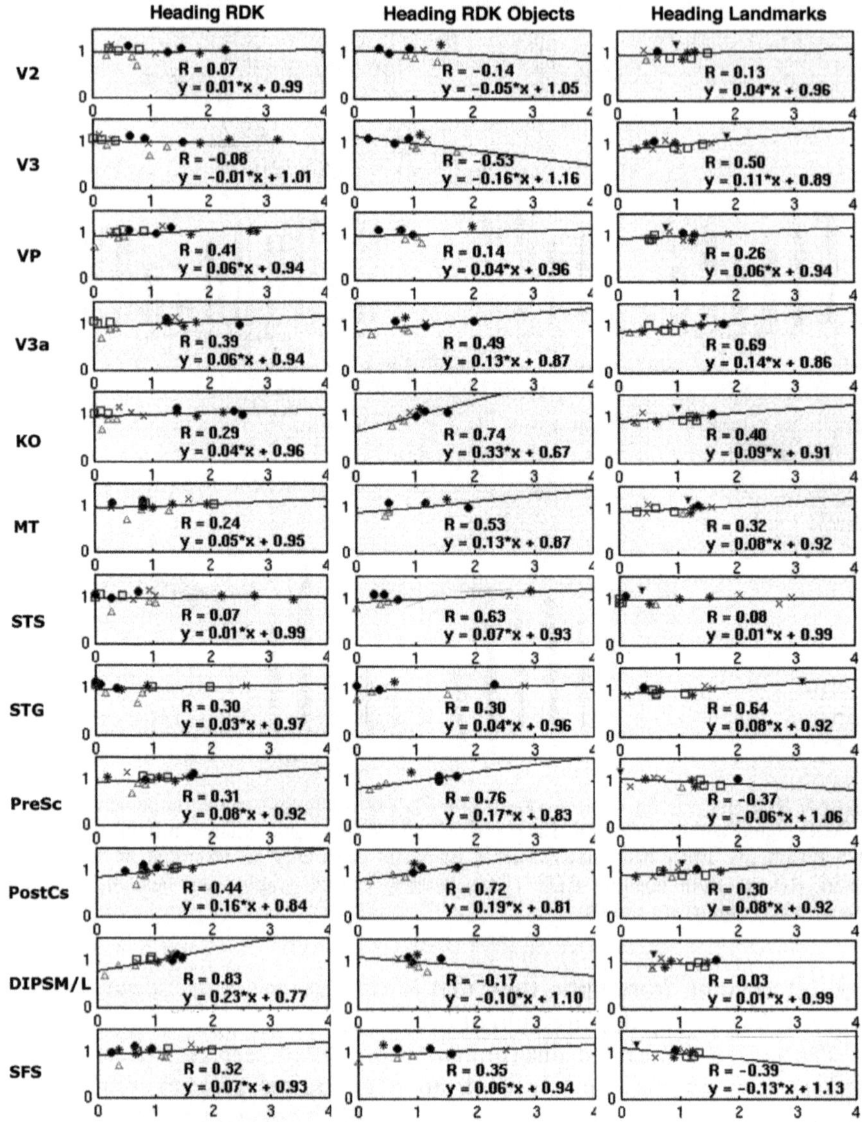

Figure 4. Correlation between subjects' behavioral performance and fMRI activation for the different ROIs. Scatter plots showing normalized fMRI activation (x-axis) vs. normalized recognition (y-axis) rate for the individual subjects (each subject is shown by different marker style) and specific heading tasks, *Heading RDK* (left column), *Heading RDK Objects* (middle column), and *Heading Landmarks* (right column). Results from each ROI are shown separately, in different rows. The regression line (solid), its equation and correlation value (R) are given for each plot.

severely impaired. These patients were first studied with the straight line *Heading RDK* stimulus that we also used in fMRI. Interestingly, while they remained unable to discriminate heading from random dot kinematograms, two of these patients had normal performance on the discrimination of *Heading Landmarks* task. (The third patient was not tested). Revisiting the specific brain areas correlated with behavior in the fMRI study described here, we note that they overlap very little, suggesting that different neuronal mechanisms may be used to judge direction of heading in the two tasks. Moreover, the areas significantly correlated with behavior in the *Heading Landmarks* task were not involved in these patients' lesions.

But still, how can the perception of self-motion occur without using optic flow? Several psychophysical studies showed (Cutting et al., 1992, 1997; Vaina & Rushton, 2000; Vishton & Cutting, 1995) that heading perception was possible even when instead of smooth motion observers view discrete samples, suggest that velocity information is not necessary for heading computation.

We explored this alternative by modifying the stimulus presentation in the *Heading Landmark* task (Figure 5). We removed the motion information by displaying three different static frames extracted from the motion of the room, each shown for 665 msec with interleaved blank intervals of 285 msec each. Five subjects, naïve to task and conditions were presented with this *Choppy Heading Landmarks* display (Figure 5b) and, as previously, were asked to determine whether their direction of heading was to the left or to the right of a probe. Figures 5c and d show psychophysical results for several subjects (n=5) and fMRI activation from a representative subject. As expected, subjects were unable to perceive heading in the sequence of static frames extracted from the heading RDK in which the only cue to calculate heading was provided by motion. In the Heading Landmark tasks, there was the motion cue but also the landmarks information. Eliminating motion information subjects presumably still could use the landmark cues to compute their heading. Figure 5d revealed a strong and stimulus specific activation of the parahippocampal gyrus in the *Choppy Heading Landmarks* condition.

This activation is consistent with results from a recent study by Gron and colleagues (Gron et al., 2000) that found when objects were used as specific landmarks for navigation during encoding of a maze there was a significant activation in the parahippocampal gyrus. Aguirre and D'Esposito (Aguirre & D'Esposito, 1997; Aguirre et al., 1998) also showed activity in the parahipocampal gyrus during subjects' acquisition of spatial layout for navigation. These results, taken together with our data from the *Choppy Heading Landmarks* suggest that when optic flow is not available for navigation, the computation of heading direction may be based on the knowledge of the spatial layout of the environment.

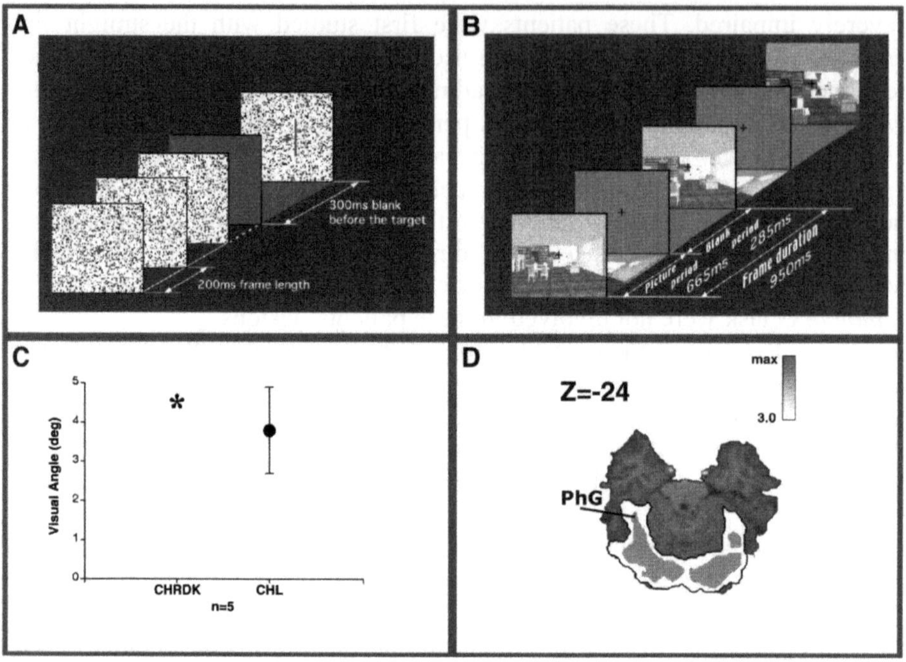

Figure 5. Choppy heading tests stimuli and results. a) Schematic view of the *Choppy Heading RDK* test. b) Schematic view of the *Choppy Heading Landmarks* test. In both *Choppy Heading RDK* and *Choppy Heading Landmarks* the probe was displayed in the last frame and subjects were asked to report whether they were heading to the left or right of the probe. c) Psychophysical data from five subjects on the *Choppy Heading RDK* (CHRDK) and *Choppy Heading Landmarks* (CHL) tasks. The asterisk denotes that subjects could not perform the task. The stimulus was presented with central fixation. d) Representative subject's brain slice indicating activity during presentation of the CHL test in the parahippocampal gyrus (PhG), an area that did not elicit activation by any other heading task described in this paper. (*Choppy Heading RDK* was not used in the fMRI).

4 WHERE ARE WE GOING?

Figure 6 is an illustrative summary of the major cortical regions we found activated by the three major heading tasks carried out in our laboratory, in relation with the functional neuroimaging studies of heading judgment briefly summarized at the beginning of the chapter. The different experimental paradigms, stimuli and imaging modalities used by the various studies might have contributed to the difference in results. However, all concur to support the view that judgment of heading direction engages a network of widely distributed neural regions which may be, at least in part, determined by the nature of the stimulus and the task.

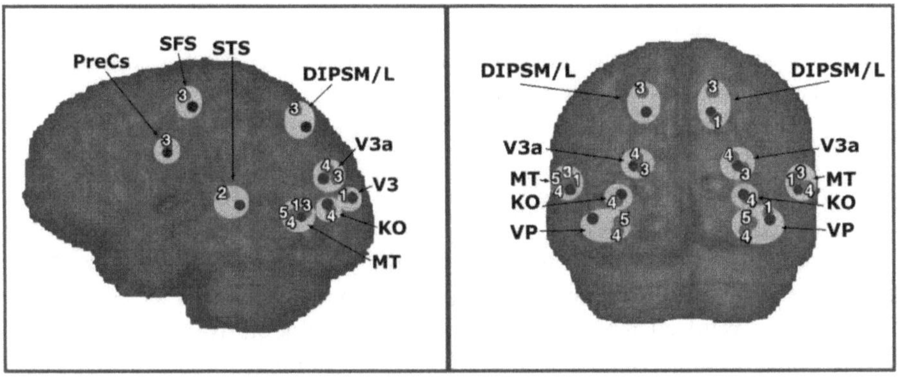

Figure 6. Location of brain areas involved in the visual heading estimation superimposed onto the brain of a typical normal subject. Lateral view of the left hemisphere (left) and posterior view of the back of the brain (right) is shown. Activations are shown in the Talairach coordinates (Talairach & Tournoux, 1988). Black disks show location of the activated areas according to the present study, dark gray disks show activated areas location according to 1 = (de Jong et al., 1994), 2 = (Howard et al., 1996), 3 = (Peuskens et al., 2001), 4 = (Rutschmann et al., 2000), 5 = (Wunderlich et al., 2002). Light gray disks summarize results of the present and previous studies into several clusters of areas corresponding to the following ROIs: VP, V3, V3a, KO, MT, STS, PreCs, DIPSM/L, STS.

In view of the ongoing controversies regarding the mechanisms involved in computing self- motion, and because of the importance of heading computation for locomotion and ultimately for action, further studies are needed to pin down the specific functional neuroanatomy of motion for navigation. The current psychophysical theories of self motion perception complemented by predictions from emerging computational models constrained by the newer results from monkey physiology, offer a great opportunity for well designed, hypothesis driven, functional neuroimaging studies of the possible underlying neural mechanisms of heading.

However, limiting the inquiry to just the perceptual investigation will not give us the full answer, because in the psychophysical tasks the human subjects are asked to indicate to which side of a reference mark they are heading, and it is an inference that their responses reflect a judgment of heading (Wurtz, 1998). But this is the best approximation of the problem we now have on our hands, before quantitative studies of heading estimation in subjects navigating in the real everyday environment emerge with some consistency. There are encouraging signs that this is already beginning to happen. Several laboratories are using head mounted displays and virtual reality to study heading and navigation in normal (Tarr & Warren, 2002) and neurological human subjects (Wann, 1996). Before long these displays will be MR compatible and we will be able to quantitatively measure what the brain is doing when it navigates through a busy world, compare walking in a desert to walking in the busy New York City at rush hour, or to running on the most crooked street in San Francisco.

4.1 Acknowledgments

We thank Martin Kopcik for programming the psychophysical and fMRI stimuli used here. We thank Rosalyn Ano for help with the fMRI data analysis, and Winfred Kao for technical support in acquiring the fMRI data. This work was supported by a grant from the National Institutes of health, EY-2R01-07861-13 to LMV.

5 REFERENCES

Aguirre, G. K., & D'Esposito, M. (1997). Environmental knowledge is subserved by separable dorsal/ventral neural areas. *J. Neurosci., 17* (7), 2512-2518.

Aguirre, G. K., Zarahn, E., & D'Esposito, M. (1998). Neural components of topographical representation. *Proc. Natl. Acad. Sci. USA., 95* (3), 839-846.

Albright, T.D. (1993). Cortical processing of visual motion. In: F.A. Miles, & J. Wallman (Eds.), *Visual Motion and its Role in the Stabilization of Gaze* (pp. 177-201): Elsevier Science Publishers B.V.

Albright, T.D., & Stoner, G.R. (1995). Visual Motion Perception. *Proc. Natl. Acad. Sci. USA., 92*, 2433-2440.

Andersen, R.A. (1997). Neural mechanisms of visual motion perception in primates. *Neuron, 18*, 865-872.

Andersen, R. A., Shenoy, K. V., Crowell, J. A., & Bradley, D. C. (2000). Neural Mechanisms for Self-Motion Perception in Area MST. In: M. Lappe (Ed.) *Neuronal Processing of Optic Flow,* 44 (pp. 219-233). New York: Academic Press.

Beauchamp, M., & DeYoe, E. (1996). Brain areas for processing motion and their modulation by selective attention. *NeuroImage, 3*, 245.

Beer, J., Blakemore, C., Previc, F. H., & Liotti, M. (2002). Areas of the human brain activated by ambient visual motion, indicating three kinds of self-movement. *Exp. Brain Res., 143* (1), 78-88.

Binder, J. R., Swanson, S .J., Hammeke, T. A., Morris, G. L., Mueller, W. M., Fischer, M., Benbadis, S., Frost, J. A., Rao, S. M., & Haughton, V. M. (1996). Determination of language dominance using functional MRI: a comparison with the Wada test. *Neurology, 46* (4), 978-984.

Braddick, O. J., O'Brien, J. M., Wattam Bell, J., Atkinson, J., Hartley, T., & Turner, R. (2001). Brain areas sensitive to coherent visual motion. *Perception, 30* (1), 61-72.

Britten, K. H., & Van Wezel, R. J. (1998). Electrical microstimulation of cortical area MST biases heading perception in monkeys. *Nat. Neurosci., 1* (1), 59-63.

Britten, K. H., & Van Wezel, R. J. (2002). Area MST and heading perception in macaque monkeys. *Cereb. Cortex, 12* (7), 692-701.

Cheng, K., Fujita, H., Kanno, I., Mura, S., & Tanaka, K. (1995). Human Cortical Regions Activations by Wide Field Visual Motion: An H2(15)O PET Study. *J. Neurophysiol., 74* (1), 413-426.

Corbetta, M. (1998). Frontoparietal cortical networks for directing attention and the eye to visual locations: Identical, independent, or overlapping neural systems? *Proc. Natl. Acad. Sci. USA, 95* (3), 831-838.

Cutting, J. E., Springer, K., Braren, P. A., & Johnson, S. H. (1992). Wayfinding on foot from information in retinal, not optical, flow. *J. Exp. Psychol. Gen., 121* (1), 41-72.

Cutting, J. E., Vishton, P. M., Fluckiger, M., Baumberger, B., & Gerndt, J. D. (1997). Heading and path information from retinal flow in naturalistic environments. *Percept. Psychophys., 59* (3), 426-441.

D'Avossa, G., & Kersten, D. (1996). Evidence in human subjects for independent coding of azimuth and elevation for direction of heading from optic flow. *Vision Res., 36* (18), 2915-2924.

D'Avossa, G., Yacoub, E., Kersten, D., & Hu, X. (1998). Optic flow simulating eccentric ego-motion cause "activation" in occipito parietal areas of the contralateral hemisphere. *Invest. Opthalmol. Vis. Sci., 39* (4), 467.

D'Esposito, M., Aguirre, G. K., Zarahn, E., Ballard, D., Shin, R.K., & Lease, J. (1998). Functional MRI studies of spatial and nonspatial working memory. *Cogn. Brain Res., 7* (1), 1-13.

de Jong, B. M., Shipp, S., Skidmore, B., Frackowiak, R. S., & Zeki, S. (1994). The cerebral activity related to the visual perception of forward motion in depth. *Brain, 117*, 1039-1054.

Desmond, J. E., Sum, J. M., Wagner, A. D., Demb, J. B., Shear, P. K., Glover, G. H., Gabrieli, J. D., & Morrell, M. J. (1995). Functional MRI measurment of language lateralization in Wada-tested patients. *Brain, 118* (6), 1411-1419.

Dukelow, S. P., DeSouza, J. F., Culham, J. C., Van Den Berg, A. V., Menon, R. S., & Vilis, T. (2001). Distinguishing subregions of the human MT+ complex using visual fields and pursuit eye movements. *J. Neurophysiol., 86*, 1991-2000.

Dumoulin, S. O., Baker, C. L., Hess, R. F., & Evans, A. C. (2003). Cortical specialization for processing first- and second-order motion. *Cereb. Cortex, 13* (12), 1375-1385.

Dupont, P., De Bruyn, B., Vandenberghe, R., Rosier, A., Michiels, J., Marchal, G., Mortelmans, L., & Orban, G.A. (1997). The kinetic occipital region in human visual cortex. *Cereb. Cortex, 7*, 283-292.

Dupont, P., Orban, G. A., De Bruyn, B., Verbruggen, A., & Mortelmans, L. (1994). Many areas of the human brain respond to visual motion. *J. Neurophysiol., 72* (3), 1420-1424.

Gibson, J. J. (1950). *The perception of the visual world*. Boston: Houghton Mifflin.

Gilaie-Dotan, S., Ullman, S., Kushnir, T., & Malach, R. (2001). Shape-selective stereo processing in human object-related visual areas. *Hum. Brain Mapp., 15*, 67-79.

Greenlee, M. W. (2000). Human Cortical areas underlying the perception of optic flow: brain imaging studies. *Int. Rev. Neurobiol., 44*, 269-292.

Grezes, J., Fonlupt, P., Bertenthal, B., Delon-Martin, C., Segebarth, C., & Decety, J. (2001). Does perception of biological motion rely on specific brain regions? *NeuroImage, 13*, 775-785.

Gron, G., Wunderlich, A. P., Spitzer, M., Tomczak, R., & Riepe, M. W. (2000). Brain activation during human navigation: gender-different neural networks as substrate of performance. *Nat. Neurosci., 3* (4), 404-408.

Grossman, E., Donnelly, M., Price, R., Pickens, D., Morgan, V., Neighbor, G., & Blake, R. (2000). Brain Areas Involved in Perception of Biological Motion. *J. Cogn. Neurosci., 12* (5), 711-720.

Grossman, E. D., & Blake, R. (2002). Brain Areas Active during Visual Perception of Biological Motion. *Neuron, 35* (6), 1167-1175.

Hochberg, Y., & Tamhane, A. (1987). Multiple Comparison Procedures. New York: Wiley.

Hogg, R. V., & Ledolter, J. (1987). Engineering Statistics. MacMilan Publishing Company.

Howard, R. J., Brammer, M., Wright, I., Woodruff, P. W., Bullmore, E. T., & Zeki, S. (1996). A direct demonstration of functional specialization within motion-related visual and auditory cortex of the human brain. *Curr. Biol., 6* (8), 1015-1019.

Huk, A. C., Dougherty, R. F., & Heeger, D. J. (2002). Retinotopy and functional subdivision of human areas MT and MST. *J. Neurosci., 22* (16), 7195-7205.

Jornales, V. E., Jakob, M., Zamani, A., & Vaina, L. M. (1997). Deficits on complex motion perception, spatial discrimination, and eye-movements in a patient with bilateral occipital-parietal lesions [ARVO Abstract]. *Invest. Ophthalomol. Vis. Sci., 38* (4), S72.

Knudsen, E. I., DuLac, S., & Esterly, S. (1987). Computational maps in the brain. *Annu. Rev. Neurosci., 10*, 41-65.

McKeefry, D. J., Watson, J. D. G., Frackowiak, R. S. J., Fong, K., & Zeki, S. (1997). The activity in human areas V1/V2, V3, and V5 during the perception of coherent and incoherent motion. *NeuroImage, 5* (1), 1-12.

Mendola, J., Dale, A., Fischl, B., Liu, A., & Tootell, R. (1999). The representation of illusory and real contours in human cortical visual areas revealed by functional magnetic resonance imaging. *J. Neurosci., 19* (19), 8560-8572.

Morrone, M. C., Tosetti, M., Montanaro, D., Fiorentini, A., Cioni, G., & Burr, D. C. (2000). A cortical area that responds specifically to optic flow, revealed by fMRI. *Nat. Neurosci., 3* (12), 1322-1328.

Orban, G. A. (2001). Neural coding in area MT/V5 and satellites: from antagonistic surround to the extraction of 3D structure from motion. In: W. Backhaus (Ed.) *Neural Coding of Perceptual Systems,* 9 (pp. 188-202). London: World Scientific Publishing Company.

Peuskens, H., Sunaert, S., Dupont, P., Van Hecks, P., & Orban, G. A. (2001). Human brain regions involved in heading estimation. *J. Neurosci., 21* (7), 2451-2461.

Racine, C., Vaina, L. M., Diaz, J., Zamani, A., & Gross, C. G. (1996). Are there specific anatomical correlates of biological perception in the human visual system? *Society for Neuroscience 26th Annual Meeting. Abstracts, 22* (1), 400.

Reppas, J. B., Niyogi, S., Dale, A. M., Sereno, M. I., & Tootell, R. B. (1997). Representation of motion boundaries in retinotopic human visual cortical areas. *Nature, 388* (6638), 175-179.

Royden, C. S., & Hildreth, E. C. (1996). Human heading judgments in the presence of moving objects. *Percept. Psychophys., 58* (6), 836-856.

Rutschmann, R., Schrauf, M., & Greenlee, M. W. (2000). Brain activation during dichoptic presentation of optic flow stimuli. *Exp. Brain Res., 134* (4), 533-537.

Servos, P., Osu, R., Santi, A., & Kawato, M. (2002). The neural substrates of biological motion perception: an fMRI study. *Cereb. Cortex, 12*, 772-782.

Shmuel, A., Yacoub, E., Pfeuffer, J., Van de Moortele, P., Adriany, G., Hu, X., & Ugurbil, K. (2002). Sustained negative BOLD, blood flow and oxygen consumption response and its coupling to the positive response in the human brain. *Neuron, 36*, 1195-1210.

Siegel, R. M. (1998). Representation of visual space in area 7a neurons using the cetner of mass equation. *J. Comput. Neurosci., 5* (4), 365-381.

Siegel, R.M., & Read, H.L. (1997). Analysis of optic flow in the monkey parietal area 7a. *Cerebral Cortex, 7* (4), 327-346.

Smith, A. T., Greenlee, M. W., Singh, K. D., Kraemer, F. M., & Hennig, J. (1998). The processing of first- and second-order motion in human visual cortex assessed by functional magnetic resonance imaging (fMRI). *J. Neurosci., 18*, 3816-3830.

Sunaert, S. (2001). Functional magnetic resonance imaging: Studies of visual motion processing in the human brain. *ACTA Biomedica Lovaniensia, 226*, 1-123.

Sunaert, S., Van Hecke, P., Marchal, G., & Orban, G. A. (1999). Motion-responsive regions of the human brain. *Exp. Brain Res., 127* (4), 355-370.

Sunaert, S., Van Hecke, P., Marchal, G., & Orban, G. A. (2000). Attention to speed of motion, speed discrimination, and task difficulty: An fMRI study. *NeuroImage, 11* (6), 612-623.

Talairach, J., & Tournoux, P. (1988). Co-Planar Stereotaxic Atlas of the Human Brain. New York: Thieme Medical Publishers.

Tarr, M. J., & Warren, W. H. (2002). Virtual reality in behavioral neuroscience and beyond. *Nat. Neurosci., 5 (Supplement)*, 1089-1092.

Thulborn, K. R., Carpenter, P. A., & Just, M. A. (1999). Plasticity of language-related brain function during recovery from stroke. *Stroke, 30*, 749-754.

Tootell, R. B. H., Mendola, J. D., Hadjikhani, N. K., Ledden, P. J., Liu, A. K., Reppas, J. B., Sereno, M. I., & Dale, A. M. (1997). Functional analysis of V3a and related areas in human visual cortex. *J. Neurosci., 17* (18), 7060-7078.

Tootell, R. B. H., Mendola, J. D., Hadjikhani, N. K., Liu, A. K., & Dale, A. M. (1998). The representation of the ipsilateral visual field in human cerebral cortex. *Proc. Natl. Acad. Sci. USA, 95* (3), 818-824.

Tootell, R. B. H., Kwong, K. K., Belliveau, J. W., Baker, J. R., Stern, C. E., Hockfield, S. J., Breiter, H. C., Born, R., Benson, R., Brady, T. J., & Rosen, B. R. (1993). Mapping human visual cortex: evidence from functional MRI and histology. *Invest. Opthalmol. Vis. Sci., 34* (4), 813.

Tootell, R. B. H., Reppas, J. B., Dale, A. M., Look, R. B., Sereno, M. I., Malach, R., Brady, T. J., & Rosen, B. R. (1995a). Visual motion aftereffect in human cortical area MT revealed by functional magnetic resonance imaging. *Nature, 375*, 139-141.

Tootell, R. B. H., Reppas, J. B., Kwong, K. K., Malach, R., Born, R. T., Brady, T. J., Rosen, B. R., & Belliveau, J. W. (1995b). Functional analysis of human MT and related visual cortical areas using magnetic resonance imaging. *J. Neurosci., 15* (4), 3215-3230.

Vaina, L. M. (1998). Complex motion perception and its deficits. *Curr. Opin. Neurobiol., 8* (4), 494-502.

Vaina, L. M., & Gross, C. G. ((submitted)). Perceptual deficits in patients with impaired recognition of biological motion after temporal lesions.

Vaina, L. M., & Rushton, S. K. (2000). What neurological patients tell us about the use of optic flow. *Int. Rev. Neurobiol., 44*, 293-313.

Vaina, L. M., Solomon, J., Chowdhury, S., Sinha, P., & Belliveau, J. W. (2001). Functional neuroanatomy of biological motion perception in humans. *Proc. Natl. Acad. Sci. USA, 98* (20), 11656-11661.

Van Oostende, S., Sunaert, S., Van Hecke, P., Marchal, G., & Orban, G. A. (1997). The kinetic occipital (KO) region in man: an fMRI study. *Cereb. Cortex, 7*, 690-701.

Vishton, P. M., & Cutting, J. E. (1995). Wayfinding, displacements, and mental maps: Velocity fields are not typically used to determine one's heading. *J. Exp. Psychol. Hum. Percept. Perform., 21*, 978-995.

Wann, J. P. (1996). Virtual reality environments for rehabilitation of perceptual-motor disorders following stroke. *The First European Conference on Disability, Virtual Reality and Associated Technologies* (pp. 233-238). Maidenhead, U.K.

Warren, W. H. (1998). The state of flow. In: T. Watanabe (Ed.) *High-level motion processing* (pp. 315-358). Cambridge, MA: MIT Press.

Warren, W. H. (1999). Visually controlled locomotion, 40 years later. *Ecol. Psychol., 10* (3-4), 177-219.

Warren, W. H., & Saunders, J. A. (1995). Perceiving heading in the presence of moving objects. *Perception, 24* (3), 315-331.

Watson, J. D., Myers, R., Frackowiak, R. S., Hajnal, J. V., Woods, R. P., Mazziotta, J. C., Shipp, S., & Zeki, S. (1993). Area V5 of the human brain: evidence from a combined study using positron emission tomography and magnetic resonance imaging. *Cereb. Cortex, 3* (2), 79-94.

Woods, R. P., Grafton, S. T., Holmes, C. J., Cherry, S. R., & Mazziotta, J. C. (1998). Automated image registration: I. General methods and intrasubject, intramodality validation. *J. Comput. Assist. Tomogr., 22* (1), 139-152.

Woods, R. P., Grafton, S. T., Watson, J. D., Sicotte, N. L., & Mazziotta, J. C. (1998). Automated image registration: II. Intersubject validation of linear and nonlinear models. *J. Comput. Assist. Tomogr., 22* (1), 153-165.

Woods, R. P., Mazziotta, J. C., & Cherry, S. R. (1993). MRI-PET registration with automated algorithm. *J. Comput. Assist. Tomogr., 17* (4), 536-546.

Worsley, K. J., Evans, A. C., Marrett, S., & Neelin, P. (1992). A three-dimensional statistical analysis for CBF activation studies in human brain. *J. Cereb. Blood Flow Metab., 12* (6), 900-918.

Worsley, K. J., Marrett, S., Neelin, P., Vandal, A. C., Friston, K. J., & Evans, A. C. (1996). A unified ststistical approach for determining significant signals in images of cerebral activation. *Hum. Brain Mapp., 4*, 58-73.

Wunderlich, G., Marshall, J. C., Amunts, K., Weiss, P. H., Mohlberg, H., Zafiris, O., Zilles, K., & Fink, G. R. (2002). The importance of seeing it coming: a functional magnetic resonance imaging study of motion-in-depth towards the human observer. *Neuroscience, 112* (3), 535-540.

Wurtz, R. H. (1998). Optic flow: A brain region devoted to optic flow analysis? *Curr. Biol., 8* (16), 554-556.

Wurtz, R. H., & Duffy, C. J. (1992). Neuronal correlates of optic flow stimulation. *Ann. N. Y. Acad. Sci., 656*, 205-219.

Zeki, S., Perry, R. J., & Barteis, A. (2003). The processing of kinetic contours in the brain. *Cereb. Cortex, 2003* (13), 189-202.

Zeki, S., Watson, J. D. G., Lueck, C. J., Friston, K. J., Kennard, C., & Frackowiak, R. S. J. (1991). A direct demonstration of functional specialization in human visual cortex. *J. Neurosci., 11(3)*, 641-649.

7. The Event Structure of Motion Perception

Martin H. Fischer[1] and Heiko Hecht[2]

[1]Psychology Department
University of Dundee, Dundee, Scotland, UK

[2]Psychologisches Institut
Johannes-Gutenberg Universität, Mainz, Germany

1 INTRODUCTION

Motion perception on the basis of optic flow is often studied using purely perceptual response paradigms such as forced choice preferences, and using straightforward motor responses such as simple stereotypical reaction times. Here we argue for a more complex perspective that takes into account the event structure of ecological motion perception. In particular, we hope to convince the reader that the would-be perceptual response to a motion stimulus is noticeably modified be the type of response that is required from the actor. We will argue that our actions modify our perception and more precisely, that the planning component of intended actions influence processing of time critical motion. To make our argument, we first discuss action-perception interactions as well as time-to-contact (TTC) judgments. Then we present some experimental evidence.

2 EVENTS AND ACTION

The conceptual segregation of perception from action has a long tradition. Experimental psychologists have typically minimized the response requirements of their tasks. They have recorded button press latencies rather than the temporal and spatial extent of more natural movements. This widely popular approach has led to an understanding of perception that may not generalize to realistic situations in everyday environments.

First steps away from static stimuli for perception toward a more integrative event-based view of perception have been taken, among others by

L.M. Vaina, S.A. Beardsley and S.K. Rushton (eds.), Optic Flow and Beyond, 139–156.

Michotte (1946) and Johansson (1950). Later, perception of object attributes was shown to depend on the capacities of our motor system. For example, the perception of climbability of stairs depends on the body size of the observers (Warren, 1984): Stairs that we cannot physically climb look higher to us than they are, whereas stairs that are climbable do not. Likewise do hills look steeper to us when we are too tired to walk up their slope (Proffitt et al., 1995), again illustrating how potential actions can influence our perceptual assessments?

The impact of action requirements on visual perception can be traced to the earliest levels of encoding. For example, in a seminal paper Tipper, Lortie, and Baylis (1992) used manual pointing responses to reveal that object selection occurs relative to the responding hand: Task-irrelevant distracters interfered with pointing more when they were near the hand, or between the start and target location of the movement rather than behind the target, regardless of movement direction. Importantly, such configurable effects on overt attention were only obtained with pointing actions and not with verbal responses (Meegan & Tipper, 1999), thus highlighting the task-specificity of visual selection.

Bekkering and Neggers (2002) recently compared attentional adjustments to different action intentions. Their participants pointed at or grasped a target object with a defined orientation and color from among a large set of similar distracter objects. They made more eye fixations on distracters with the relevant orientation before initiating grasping compared to pointing responses. Thus, orientation selection (but not color selection) improved when participants intended to grasp the target. This intriguing finding is in agreement with the notion of separate visuomotor channels that process visual information as required by either the transport or manipulation components of goal-directed hand movements (e.g., Paulignan & Jeannerod, 1996) and underlines the impact of action intention on overt visual selection. This observation has recently been extended to covert visual selection by Fischer and Hoellen (2003) who investigated the impact of different motor demands on space- and object-based attention. Participants detected visual targets placed on objects. They responded by either lifting a finger, pointing, or grasping. More space-based attention was found for pointing than for finger lifting and more object-based attention was found for grasping than for pointing. These results support the view that visual selectivity is tuned to specific motor intentions.

There is also evidence from neurological patients that perception is influenced by action possibilities. When presenting static object pairs to neurological patients with double simultaneous extinction, Riddoch and colleagues (2003) recently found that the presence of an action relationship between the two objects (e.g., wine bottle and cork screw) can reduce the patients' tendency to neglect the object on the side contralateral to their brain damage. Finally, and related to the current study on perceived motion, when presenting dynamic visual objects to healthy observers Viviani and Stucchi (1992) showed that perceived object velocity is biased toward biologically plausible values: Constant velocities were judged as discontinuous while discontinuous velocities such as those generated by a human actor were perceived as smooth.

These and several other examples of action-constrained visual perception (e.g., Fischer, 1997, 1999; Shiffrar & Freyd, 1993; Tucker & Ellis, 1998) fit within theoretical perspectives that allow for novel views on the relationship between visual perception and action when compared to the traditional viewpoint of dissociable processing stages. One such viewpoint is that visual information is processed along two largely independent cortical processing streams; a dorsal stream that connects the primary visual areas in the occipital lobes to frontal motor areas of the brain along the parietal lobes; and a ventral stream that relays the same visual input along the temporal lobes (e.g., Milner & Goodale, 1995). The two visual pathways are believed to serve distinct functions, with the ventral stream leading to object identification and conscious perception and the dorsal stream largely involved in visuomotor coordination for action control. This idea has been supported by a large body of work, ranging from lesion studies in animals and behavioral studies of brain-damaged patients to the observation of dissociations between perception and action in normal observers under certain laboratory conditions.

Alternatively, findings of direct links between perception and action are often summarized under the more traditional concept of "affordances" that is meant to illustrate the higher order-functionality of the perception-action system (Gibson, 1979). An affordance can be considered as an object attribute that signals to an observer a particular use of that object, such as its climbability, graspability etc. The present paper suggests that some such affordances can more parsimoniously be explained as perceptual biases resulting from the anticipated response requirements of the observer. Without having to use the controversial notion of affordance, we hold that motion perception encompasses the entire event, its structure being a function of the visual stimulus and the motor capabilities, and less obviously of the intentions of the observer.

3 TIME-TO-CONTACT AND EVENT SAMPLING

The question arises whether the above mentioned importance of action and action planning for perception is of such general nature that it would affect basic mechanisms of motion processing, as reflected in time-to-contact (TTC) judgments. We hold that this is indeed the case.

Lee and his colleagues examined intercept timing behavior (Lee, 1976; Lee et al., 1983) and concluded that observers exploit an optical variable that directly specifies arrival time (tau). Tau capitalizes on the relative rate of optical expansion undergone by the retinal image of the approaching object or the equivalent rate of gap constriction in the case of motion lateral to the observer. TTC judgments have been found to be fairly accurate and reliable so long as approach velocity is constant and the object is continuously visible before the beginning of an extrapolation period (e.g., Kaiser & Mowafy, 1993). However, a growing number of studies suggests that on the one hand observers are often considerably less accurate that predicted by tau theory, and on the other hand they often continue to make good estimates in the absence of valid tau information.

TTC estimates are often compromised even under good viewing conditions, suggesting that observers use simpler cue-based strategies, rather than deriving tau (Tresilian, 1993, 1994; Wann, 1996). For example, object size biases TTC judgments (DeLucia, 1999; Michaels, 2000), and size, distance, and velocity/acceleration estimates might be necessary for TTC judgments after all (Stewart et al., 1993). These findings among others have reopened the discussion whether tau-theory adequately explains TTC estimation. If TTC estimates can be shown to depend as well on the action context, the notion of an invariant computation of tau would receive another critical challenge.

In order to manipulate the response encoding with a TTC paradigm we made use of the recent finding that spatiotemporal sampling of a moving stimulus leads to a systematic overestimation of TTC (Hecht et al., 2002). A prediction motion (PM) paradigm was used, that is observers were presented with an object moving in the frontoparallel plane until it disappeared from view. A button had to be pressed at the time the object would have reached a designated location farther along its trajectory had it continued to move. When the object appeared and disappeared intermittently during the visible portion of its trajectory TTC was overestimated, corresponding to an underestimation of the object's speed when its motion was sampled. Spatiotemporal sampling is suitable to investigate potential effects of the response on TTC judgments because equivalent sampling can be applied to the stimulus and to the response (continuous movement vs. tapping), or to both.

Spatiotemporal sampling refers to the intermittent display of a moving visual stimulus, repeatedly appearing and disappearing in non-adjacent locations, as if moving behind a picket fence (e.g., Helson, 1930; Dannemiller et al., 1997). Perceived motion speed is typically assessed with paired speed comparisons or button presses in a PM paradigm. This presentation mode often leads to an underestimation of perceived motion speed, that is, the button is pressed too late (e.g., Nijhawan, 1994; but see Castet, 1995; see also

Hecht & Hoffman, 2003 for a differentiation between aperture and sudden onset sampling). However, considering merely the temporal relation between the stimuli and responses, it would be more appropriate to say that sampled motion produces delayed responses when compared to continuous motion. Importantly, both perceptual and response-related processes are involved in the measurement of perceived speed, and the quality of perceptual sampling is thought not to interact with the particular response used to obtain the subsequent judgment of perceived movement speed. In contrast to this widely held belief, we recently found that the response has a strong impact on perceived motion speed and must thus be factored into an analysis of spatiotemporal sampling. Response planning appears to be a fundamental part of motion encoding, thus further strengthening the case for a tight relationship between visual perception and action.

The goal of the current experiment was to assess whether a change of the response requirements (continuous vs. discrete responses) would affect the perceived duration of a linear motion event that was itself either discrete or continuous. It has been well established that observers respond better (faster and more accurately) to stimuli when their responses are congruent with the stimuli in space or time. For example, when observers have to select a left or right button in response to a green or red color patch that randomly appears on the left or right side, they will perform better when the color is presented near the location of the correct response, indicating a physical correspondence benefit. Similar performance benefits are also obtained when the experimental situation allows for a conceptual congruency between stimuli and responses, e.g. responding to the larger of two numbers when that number is also physically bigger (see Hommel & Prinz, 1997, for a review).

4 AN EXPERIMENTAL CASE FOR THE ROLE OF RESPONSE PLANNING

We investigated how discrete vs. continuous manual responding affects visual motion encoding and hypothesized that the manual response requirements would induce systematic perceptual biases. In particular, a match between stimulus and response attributes (both discrete or both continuous) should support accurate speed perception, whereas a mismatch should lead to more variability or possibly to directional biases. We employed a motion extrapolation (PM) paradigm: An object moved laterally from the left side of a computer screen toward its center and then disappeared. Starting from the center of the screen and continuing to the right, the object motion had to be extrapolated by tracing the index finger to the right along the path that the object would have taken, had it continued its path. The screen also acted as a touch screen such that the finger movements could be recorded in space and time.

We fully crossed several stimulus motion conditions with two response types. The stimulus was either continuously visible or it was sampled. In the continuous case, the stimulus flashed on and off at immediately adjacent locations in space and time. In the discrete condition, the stimulus flashed on and off in non-adjacent locations, thus producing an apparent motion percept. Movement speed of the object was also varied (see Methods – Section 4.1). The response was also either continuous or discrete. In the continuous case, subjects traced with their index finger the entire extrapolated path of the object. In the discrete condition the response consisted of tapping at four reference locations along the extrapolated trajectory, such that each tap would be synchronous with the extrapolated stimulus passing through that reference location.

If motion encoding is in fact responsible for the typical sampling effects described above, then two outcomes are possible. First, the effect of delayed responses could persist in all cases of the sampled stimulus, suggesting that the perceptual stimulus was encoded independent of the response requirements. Alternatively, if a commonality or an interaction between perceptual encoding and motor responses exists, then a congruency between stimulus and response should lead to more accurate performance, whereas a discrepancy (in particular sampled stimulus and continuous responses) should lead to more variability of judgments or possibly an underestimation of speed in the motor response.

4.1 Method

Observers: Twelve students (6 men, 6 women) were paid for their participation. They were unfamiliar with the purpose of the study and reported no neurological deficits. Their vision was either normal or corrected to normal. Handedness was not assessed.

Apparatus: Observers were run individually in a quiet laboratory room with a dim background light. They were seated on a height-adjustable office chair in front of a Philips 4 CM 2299 Autoscan Professional Color monitor with 20-inch diagonal screen size. The screen was pitched backward at an angle of 45°. It was equipped with a permanent touch interface (ELO-Touch, controlled with ELO-Graphics MonitorMouse 2.0). The spatial resolution of this touch interface corresponded to the pixel resolution of the screen, which was set to 1024 horizontal x 768 vertical pixels. 26 pixels corresponded to 10 mm. An Apple 4400/200 PowerMac controlled stimulus presentation and

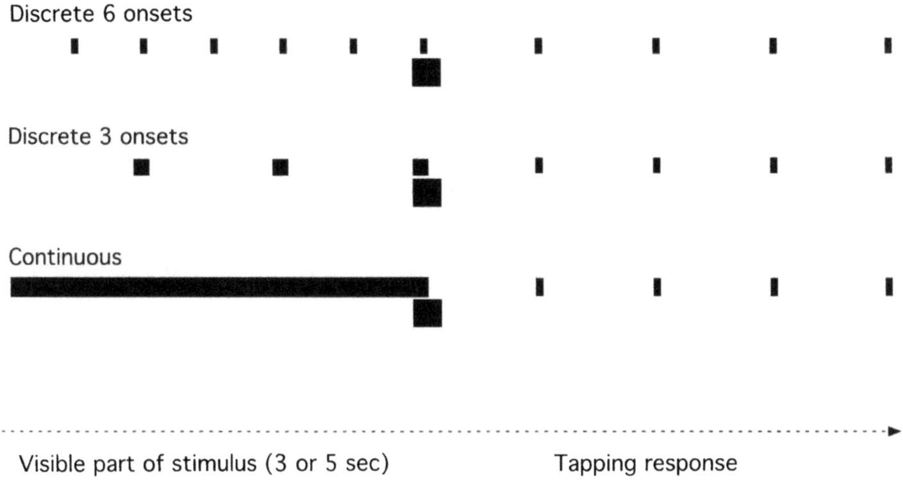

Figure 1. Schematic drawing of the stimulus presentation. The visible stimulus was either discrete or continuous with the total duration being 3 or 5 sec. The response was either a series of taps at the marked positions or a continuous tracing. In the latter case the four positions were also marked such that the display on the right side of the screen remained identical for all trials. The visible stimulus covered 18.5 cm on the screen while the extrapolation distance was about 18 cm, depending on exact start location.

response collection. Viewing distance was approximately 50 cm but no head restraint was used to ensure natural movements (cf. Biguer et al., 1985).

Display: Response instructions for each trial were given in the top portion of the screen and remained visible throughout each trial. Stimuli were presented horizontally from the left side of the screen moving to the right on a light gray background. The moving object that acted as a pacing stimulus for the response was a small black box (a rectangle measuring 3 x 6 pixels). It was presented dynamically on the left side of the screen to induce an apparent motion percept. To create the stimuli we created a horizontal array that consisted of 60 virtual boxes separated by 5 pixels from one another. The array extended over 18.5 cm. Two conditions of discrete stimulus encoding were used. First, each tenth of 60 horizontally arranged boxes was successively visible for 167 ms. This yielded 6 onsets, as is schematically represented in the top panel of Figure 1. Second, each twentieth box was successively visible for 334 ms, yielding 3 onsets (see middle panel of Figure 1). Thus, the pacing stimulus had a cumulative visibility of 1 sec in both discrete encoding conditions. Similarly, in continuous encoding trials (bottom panel) each one of 60 boxes was shown for 17 ms, or each other box was shown for 34 ms, also yielding 1000 ms of cumulative visibility. Intervals between successive onsets were adjusted to yield overall stimulus encoding times of either 3 or 5 seconds, corresponding to two different simulated object speeds.

Note that, in all conditions, perceptual encoding terminated at the display center (box number 60), where the stimulus appeared one last time and in red, thus acting as a go signal. On the right side of the screen four similar reference boxes, separated by 96 pixels (corresponding to 37 mm), were permanently visible in each trial. The number of these reference boxes differed from the number of discrete pacing signals (3 or 6) to prevent observers from using time estimation or rhythmization strategies to optimize their responses. The start location for the motor response was a 40 x 40 pixel green square in the center of the screen just below the onset location of the last pacing signal. Responses were made with the right index finger directly on the screen. The total extrapolation length from the start location to the last reference box subtended 18 cm.

Design: Crossing the two stimulus encoding conditions (discrete, continuous) with two response conditions (discrete, continuous) yielded four main experimental conditions. These conditions were applied to each observer in four separate blocks in counterbalanced sequences. Within each block, stimuli were presented at two speeds (from left side to center in 3 sec vs. 5 sec) and with either few or many onsets (discrete perception: 3 vs. 6 onsets; continuous perception: 30 vs. 60 onsets). Each of these 4 trial types was repeated 5 times in randomized order, yielding 20 trials per block.

Procedure: Observers signed an informed consent form prior to their participation. They were instructed to match their responses to the perceived speed of the pacing signal, such that the finger would be superimposed on the moving target had it continued its motion at its initial constant speed. In other words, the collision times of the (then invisible) pacing signal with the (visible) successive reference points on the right side of the screen should match the observer's contact times. Observers were informed that neither time estimation nor rhythm encoding would provide information for accurate responses. They practiced each condition briefly prior to data collection without receiving feedback about their timing errors.

Each trial began with the presentation of a two-word instruction indicating the stimulus and response conditions in the top area of the display. At the same time, four reference boxes on the right side of the screen, and the start button in the center of the screen, also appeared. Upon contact with the right index finger the start button disappeared and the black pacing signal was presented on the left side of the screen. When the pacing stimulus reached the screen center just above the observer's index finger, it appeared once in red (instead of black) for the assigned onset time and then disappeared. This last offset was the signal for the observer to respond.

In discrete response conditions, the index finger was lifted from the screen and consecutively moved to each of the four reference boxes on the right side of the screen. In these discrete response conditions, reaction time (RT) was the time from offset of the red stimulus to liftoff of the right index finger from the start location, and movement times (MTs) were defined as the times between successive liftoffs and touchdowns on the screen. Dwell times on the screen were also recorded but were not separately analyzed.

In continuous response conditions, the index finger remained in contact with the screen while observers smoothly moved it along the line of reference points. Thus, RT was the time from offset of the red stimulus to the moment when the index finger moved out of the start area (defined as 40 pixels or 15 mm around the initial screen contact), and successive MTs were the times of passing the leftmost horizontal screen coordinates of each of the reference points. Dwell times in these continuous response conditions were zero.

Several precautions were taken to ensure that observers did not obtain information about stimulus speed other than through retinal and extraretinal information. First, the program monitored that the observer's index finger remained in contact with the screen throughout stimulus presentation, to prevent the hand from moving in synchrony with the pacing signal. Whenever the screen contact was interrupted during stimulus presentation, a trial was immediately aborted and repeated later in the block. The same was true when screen contact was lost during continuous responding, or when more than 2,500 ms elapsed between successive contacts during discrete responding. No other performance feedback was given. Finally, the experimenter monitored that the observer's other hand was not used to obtain pacing information.

4.2 Results

Data were accepted for statistical analysis when the following criteria were met: RTs were within 100 and 1000 ms, MTs between successive contacts (excluding dwell times) were within 100 and 2000 ms, times to complete the responses were within 500 ms and 10 seconds, and the discrete contacts were located within +/- 50 pixel horizontally and +/-75 pixel vertically around the correct response marker. These criteria led to an exclusion of 17% of the trials. The remaining observations were used to determine four dependent measures for each trial:

(1) Reaction time, measuring response initiation delay after the red pacing stimulus had disappeared (RT). This measure was thought to reflect both the ease of stimulus encoding and the resulting difficulty of mapping this stimulus information onto the timing parameters of a motor plan.
(2) Movement execution time (MET), which reflects the time from response onset to making the fourth contact, including all MTs and dwell times.

This measure was thought to indicate possible compensatory strategies with respect to movement planning processes, as reflected in RT.

(3) Temporal error (TE), which is the difference between actual response completion time and the time an ideal observer/actor would have produced. A positive score would indicate that the overall response (up to the moment when the final marker was reached) was too slow and a negative value that it was too fast.

(4) Summed temporal deviation (STD), which reflects the unsigned differences between actual movement times and appropriate movement times accumulated across the four intervals of each response. This measure indicated indirectly the degree to which participants corrected their response velocities on-line, given that we could not measure tangential hand velocities directly.

Each of these dependent measures was analyzed with separate repeated measures ANOVA that evaluated the effects of stimulus encoding (discrete, continuous) and response requirements (discrete, continuous), as well as instructed movement speed (fast, slow) and number of onsets during encoding (many, few) on performance.

Overall, the data reveal a strong effect of response mode. Discrete responses led to an overestimation of movement time (= underestimation of speed), whereas continuous responses were associated with an underestimation of movement time.

Reaction Times (RT): Consider first the times required for response initiation. Average times to initiate responses in the discrete-discrete, discrete-continuous, continuous-discrete, and continuous-continuous conditions were 328, 568, 334, and 420 ms, respectively. As depicted in Figure 2, responses were initiated significantly faster toward continuous compared to discrete stimuli, ($F(1, 11) = 5.35$, $p < 0.05$, $MSE = 44{,}945$). Continuous response requirements, on the other hand, led to significantly slower response initiation compared to discrete response requirements, ($F(1, 11) = 30.46$, $p < 0.001$, $MSE = 42{,}049$). Average RTs in the slow and fast movement speed conditions were 450 ms and 375 ms respectively, showing reliably faster response initiation with the faster pacing signal, ($F(1, 11) = 13.60$, $p < 0.001$, $MSE = 19{,}992$). The number of onsets did not reliably affect reaction times, ($F(1, 11) = 3.50$, $p > 0.09$, $MSE = 2{,}353$), with somewhat slower response initiation after few compared to many onsets (419 ms and 406 ms, respectively).

There were two significant interactions, with all other p values > 0.22. Stimulus encoding and response requirements interacted reliably, ($F(1, 11) = 16.34$, $p < 0.001$, $MSE = 17{,}439$), showing that the condition with discrete encoding and continuous responding was associated with the slowest response initiation times (see Figure 2). Finally, the triple interaction of response

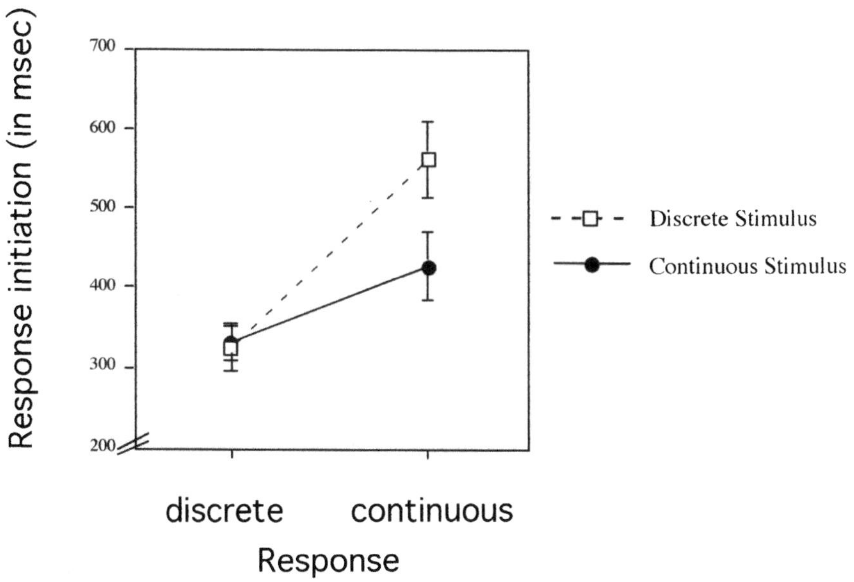

Figure 2. Average response initiation times (RT) in the four main stimulus-response pairings. RT was measured in msec after the moving stimulus had disappeared. Error bars indicate standard errors of the means.

requirements, number of onsets, and movement speed was reliable, (F (1, 11) = 4.82, p = 0.05, MSE = 4,896). In the case of continuous responses, the overall slower reaction times associated with slow instructed movement speed were facilitated by increasing the number of onsets. In the case of discrete responses, on the other hand, the overall slower reaction times associated with slow instructed movement speeds were further prolonged by increasing the number of onsets. Faster required movement speed led to faster response initiation, indicating that observers were aware of the two levels of stimulus speed. They were able to extract this information about equally well from few and many onsets, thus signaling that sampled vs. continuous motion percepts delivered similar speed information.

With respect to our question of how effectively perception is mapped onto action in the compatible and incompatible conditions, the reaction time results show that both encoding and response requirements determine the speed of response initiation. As expected, RT reflected both ease of stimulus encoding and ease of motor planning, and a reliable cost was incurred when discrete stimuli had to be mapped onto a continuous response. Note, however, that with our method RTs for continuous responses were only registered after the index finger had left a critical region of 15 mm around the start location, whereas RTs for discrete responses reflected interruption of the screen contact due to the index finger's lift-off. This procedural difference might account for the main effect of response requirements on reaction times. But irrespective of the nature of the required response, continuous stimuli led to the fastest responses, suggesting that continuous stimuli require the least amount of visuomotor planning.

Movement Execution Times (MET): Consider next the times required to reach the fourth and final response location. Average times to complete responses in the discrete-discrete, discrete-continuous, continuous-discrete, and continuous-continuous conditions were 3,475 ms, 2,575 ms, 3,753 ms, and 2,358 ms, respectively. Manipulating stimulus encoding did not affect METs, ($F(1, 11) = 0.01$, $p > 0.91$, MSE = 3,196,450). However, continuous responses were completed reliably faster than discrete responses, ($F(1, 11) = 53.17$, $p < 0.001$, MSE = 1,188,458). Instructed movement speed also affected completion times reliably, ($F(1, 11) = 54.43$, $p < 0.001$, MSE = 781,223), with 3,511 ms and 2,569 ms in the slow and fast conditions respectively. Completion times were shorter with many onsets compared to few onsets, ($F(1, 11) = 9.47$, $p < 0.01$, MSE = 335,732), with 2,911 ms and 3,169 ms respectively. There were no reliable interactions among these experimental factors, all p values > 0.09. The main effect of instructed movement speed on completion times indicates that observers were well aware of the different required speed levels.

METs were unaffected by the manipulation of stimulus encoding conditions, but continuous response requirements and multiple onsets both led to faster task execution. The absence of interactions indicates that these two manipulations had independent contributions to perceived speed of the pacing signal. Participants were clearly sensitive to the two different stimulus speeds.

Temporal Error (TE): Consider now the signed deviations of actual arrival times from ideal arrival times at the final response location on the right side of the screen. The average differences between actual and required completion times in the discrete-discrete, discrete-continuous, continuous-discrete, and continuous-continuous conditions were 410, -489, 819, and -577 ms, respectively. There was no significant main effect of stimulus encoding, ($F(1, 11) = 0.39$, $p > 0.55$, MSE = 3,196,451). Response requirements reliably affected TE, ($F(1, 11) = 53.17$, $p < 0.001$, MSE = 1,188,459). With discrete responses, completion times were about half a second too long on average, whereas continuous responses were about half a second too fast (see Figure 3). Temporal error was larger with slow compared to fast instructed movement speed, ($F(1, 11) = 19.05$, $p < 0.001$, MSE = 781,224), with +308 and -238 ms respectively. Finally, temporal error was smaller with many compared to few onsets, ($F(1, 11) = 9.47$, $p < 0.01$, MSE = 335,732), with -88 ms and +170 ms on average respectively. The possibility to obtain visual information from multiple onsets helped observers to somewhat calibrate their

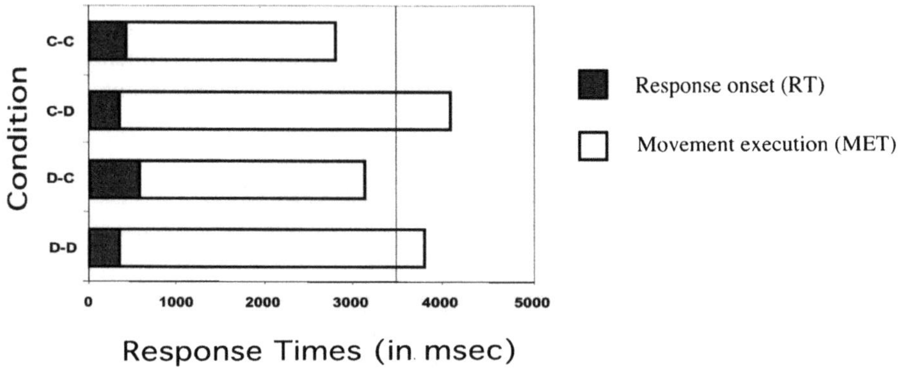

Figure 3. Comparison of response initiation times (RT) and movement execution times (MET) in an aggregated drawing. C-C, C-D, D-C, and D-D: Continuous and Discrete Stimulus-Response pairings. The abscissa depicts time measured from the offset of the last stimulus. The vertical gray line represents the average response completion time for an ideal observer.

temporal response accuracy. Again, there were no reliable interactions among these experimental factors, all p values > 0.09.

Summed Temporal Deviation (STD): Consider finally the absolute deviations between ideal arrival times and actual arrival times, summed across all four reference points on the right side of the screen. Average STDs in the discrete-discrete, discrete-continuous, continuous-discrete, and continuous-continuous conditions were 1,195 ms, 1,154 ms, 1,193 ms, and 1091 ms, respectively. Neither stimulus encoding nor response requirements had a significant impact on temporal accuracy, with ($F(1, 11) = 0.08$, $p > 0.79$, MSE = 685,205), and ($F(1, 11) = 1.10$, $p > 0.32$, MSE = 222,447), respectively. STDs were significantly larger with slow compared to fast response speed (1,450 vs. 866 ms), ($F(1, 11) = 16.90$, $p < 0.001$, MSE = 967,227). The slower response speed conditions presumably left more time for on-line velocity adjustments, as reflected in the larger variability of performance under these conditions. There was no effect of number of onsets, ($F(1, 11) = 0.44$, $p > 0.52$, MSE = 171,975). Average STDs were 1,178 ms and 1,138 ms with many and few onsets respectively. The only significant interaction was between response requirements and response speed, ($F(1, 11) = 17.78$, $p < 0.001$, MSE = 61,833). The improvement in temporal accuracy with increased response speed was larger with discrete responses (735 ms) than with continuous responses (433 ms). The remaining interactions were not significant, all p values > 0.15.

Fast movement speed conditions reduced absolute temporal error, but contrary to the global results, there was no benefit from multiple onsets at the level of individual deviations, possibly due to a competition between on-line correction mechanisms with top-down pacing as specified in the initial motor plan. Figure 3 depicts the response time data averaged across slow and fast stimulus speeds. It is clearly visible that long RTs were overcompensated by shorter METs in the conditions with continuous response requirements.

5 RESPONSE PLANNING IS PART OF THE PERCEPTUAL EVENT

To investigate the influence of response planning on motion perception, observers extrapolated a discrete or continuous pacing signal with either discrete or continuous responses. Temporal performance was the dependent measure, and a surprisingly strong effect of response mode on temporal accuracy was the main result. Contrary to our initial hypothesis derived from the stimulus-response compatibility literature (Hommel & Prinz, 1997), a congruent relationship between motion encoding and response mode did not generally improve temporal accuracy compared to the incongruent stimulus-response mappings. Instead, there was an interaction between response demands and the resulting accuracy of judgments: Although discrete responses were initiated faster they led to slower responding in all cases.

A more detailed inspection of the results reveals clear evidence for two effects. First, we replicated the effect of response delay that accompanies spatiotemporal stimulus sampling within TTC scenarios (Hecht et al., 2002). The fewer sampling steps (onsets) we presented, the larger were all response latencies. Second, the type of required response strongly influenced our results. Continuous motion extrapolation led to faster perceived motion judgments when compared to the discrete tapping responses. This novel finding suggests that the anticipated response requirements modulate perceptual parameters: For a given stimulus, a required continuous response leads to earlier response completion, which is then interpreted as evidence for overestimation of perceived motion speed.

Also, discrete responses were initiated considerably earlier than continuous responses. This finding may be interpreted as indicating an overcompensation strategy: Subjects may have initiated their responses too late in the discrete response conditions and may have tried to compensate their sluggishness during response initiation by increasing their movement speed during response execution. As an alternative explanation, however, we must consider the fact that RT was defined differently in continuous compared to discrete response conditions, which may also account for this finding. Future studies should record effector trajectories to address this issue.

The present findings help to clarify conflicting results obtained with sampled motion stimuli. When merely perceptual judgments are collected (Nijhawan, 1994) or when a single discrete motor response is required (such as in the TTC judgment tasks used by Hecht et al., 2002), sampled motion appears slower than continuous motion (but see Castet, 1995). Our results show that the perceptual effect of slowing of perceived motion due to sampling is modified by the response mode. Specifically, a continuous response can overcome such perceptual bias over the course of an extended motor action. While response initiation delays were larger for continuous than for discrete responses, the continuous responses alone were able to catch up with the speed of the object as originally displayed.

The findings contribute to the demise of tau theory. Moving objects are not processed invariantly but TTC estimates are modulated by the intended response type. Thus, time-critical motion cannot be treated as an isolated stimulus but rather has to be taken into consideration as a structured event. The motion stimulus is just as much part of the event as is the intended action in which it is embedded.

The present results could be interpreted from within the novel theoretical frameworks that were mentioned in the introduction. First, a case can be made that our response requirements engaged the observers' dorsal streams to a larger extent than typical responses in studies of perceived motion that only require single key presses. Moreover, the key presses are typically considered to be reflections of a conscious perceptual judgment, thus tapping into the ventral stream representations of the visual stimulus. From this point of view it is interesting to note that performance in our most "action"-oriented continuous tasks was not necessarily less biased than responses in the more typical discrete response tasks (see Figure 3). Thus, we did not find evidence for a dissociation between perception and action that is currently considered as key support for the two visual pathways notion. On the other hand, we did not include a response condition that required a single key press to indicate time to contact with the last marker on the right hand side of the display.

Second, we can consider the present results from the viewpoint of the affordance concept. We did not find better temporal accuracy when the task had a congruent relationship between motion encoding and response mode, compared to the incongruent stimulus-response mappings. Thus, we cannot say that our results revealed any sensitivity in our observers for the presence of spatiotemporal affordances in the visually presented objects. This may, however, at least in part be due to the relatively abstract and meaningless object displays that were chosen for the present study. It is conceivable that dynamically animated displays of more realistic objects could well reveal observers' sensitivity to spatiotemporal affordances in our paradigm. One particular idea would be to show real-life objects with an inherent direction, such as cars or planes that are known to induce directional biases in similar tasks (e.g., Nagai & Yagi, 2001).

Finally, the present results might also be interpreted in terms of switching costs between task sets that arise when encoding and response differ (e.g.,

Rogers & Monsell, 1995). Switching costs seem to be limited to response initiation. In the case of discrete stimuli our observers produced particularly long response initiation times when continuous responses were required (see the interaction depicted in Figure 2). However, no such interaction occurred when discrete responses were required. Continuous responses might be superior to discrete responses because initiation delays, which are correlated to perceptual lag, are better noticed and better compensated with continuous responses.

With respect to the interplay of perception and action, our observations replicate and extend previous findings (e.g., Viviani & Stucchi, 1992). There is a strong influence of the forthcoming response requirements on the visual perception of dynamic displays. Clearly, the way we perceive our environment is tuned to our action dispositions (Gibson, 1979). Thus, perceptual spatiotemporal pooling effects (Geldard, 1975; Helson, 1930) may be overcome when observes engage in a continuous motor response. The results also suggest that the seemingly basic skill of motion perception cannot be fully understood within frameworks, such as tau theory, that neglect the complete structure of the perceptual-motor event and in particular the intended response.

ACKNOWLEDGEMENTS

This study was supported by an equipment grant from the Max-Planck Institute for Psychological Research (Munich) to MHF. He was also supported by a travel grant from the Carnegie Trust. The study was conducted while MHF was at the Ludwig-Maximilians-Universität, Munich. We thank Birgitt Aßfalg for data collection.

REFERENCES

Bekkering, H., & Neggers, F.W. (2002). Visual search is modulated by action intentions. *Psychol. Sci.*, *13*, 370-374.

Biguer, B., Jeannerod, M., & Prablanc, C. (1985). The role of position of gaze in movement accuracy. In M.I. Posner & O.S.M. Marin (Eds.), *Attention and Performance XI.* pp. 407-424. Hillsdale, N. J. Lawrence Erlbaum Associates.

Castet, E. (1995). Apparent speed of sampled motion. *Vision Res.*, *35*, 1375-1384.

DeLucia, P. R. (1999). Size-arrival effects: The potential roles of conflicts between monocular and binocular time-to-contact information, and of computer aliasing. *Percept. Psychophys.*, *61*, 1168-1177.

Dannemiller, J. L., Heidenreich, S. M., & Babler, T. (1997). Spatial sampling of motion: Seeing an object moving behind a picket fence. *J. Exp. Psychol. Hum. Percept. Perform., 23,* 1323-1342.

Fischer, M. H. (1997). Attention allocation during manual movement preparation and execution. *Eur. J. Cognit. Psychol., 9*, 17-51.

Fischer, M. H. (1999). An investigation of attention allocation during sequential eye movement tasks. *Q. J. Exp. Psychol., 52A*, 649-677.

Fischer, M. H, & Hoellen, N. (2003). Space-based and object-based attention depend on action intention. *(submitted).*

Geldard, F. A. (1975). *Sensory Saltation: Metastability in the Perceptual World.* Hillsdale, NJ: Erlbaum.

Gibson, J. J. (1979). *The Ecological Approach to Visual Perception.* Hillsdale, NJ: Lawrence Erlbaum.

Hecht, H., & Hoffman, D. D. (2003). Extrapolating visual motion: Effects of spatial and temporal sampling. *(submitted).*

Hecht, H., Kaiser, M. K., Savelsbergh, G. J. P., & van der Kamp, J. (2002). The impact of spatio-temporal sampling on time-to-contact judgments. *Percept. Psychophys., 64*, 650-666.

Helson, H. (1930). The *tau*-effect: An example of psychological relativity. *Science, 71*, 536-537.

Hommel, B., & Prinz, W. (1997). *Theoretical Issues in Stimulus-Response Compatibility.* Amsterdam: North-Holland.

Johansson, G. (1950). *Configuration in Event Perception.* Uppsala: Almqvist & Wiksell.

Kaiser, M. K., & Mowafy, L. (1993). Optical specification of time-to-passage: Observers' sensitivity to global tau. *J. Exp. Psychol. Hum. Percept. Perform., 19*, 1028-1040.

Lee, D. N. (1976). A theory of visual control of braking based on information about time-to-collision. *Perception, 5*, 437-459.

Lee, D. N., Young, D. S., Reddish, P. E., Lough, S., & Clayton, T. M. H. (1983). Visual timing in hitting an accelerating ball. *Q. J. Exp. Psychol., 35A*, 333-346.

Meegan, D. V., & Tipper, S. P. (1999). Visual search and target-directed action. *J. Exp. Psychol. Hum. Percept. Perform., 25,* 1347-1362.

Michaels, C. F. (2000). Information, perception, and action: What should ecological psychologists learn from Milner and Goodale (1995)? *Ecol. Psychol., 12*, 241-258.

Michotte, A. (1946/1963). *The Perception of Causality.* (T. R.. Miles & E. Miles, Trans.). London: Methuen.

Milner, D., & Goodale, M. (1995). *The Visual Brain in Action.* Oxford: Oxford University Press.

Nagai, M., & Yagi, A. (2001). The pointedness effect on representational momentum. *Mem. Cognit., 29*, 91-99.

Nijhawan, R. (1994). Motion extrapolation in catching. *Nature, 370*, 256-257.

Paulignan, Y., & Jeannerod, M. (1996). Prehension movements: The visuomotor channels hypothesis revisited. In A.M. Wing, P. Haggard, & J. R. Flanagan (Eds.), *Hand and Brain.* Chapter 13, pp. 265-286. Academic Press.

Proffitt, D. R., Bhalla, M., Gossweiler, R., & Midgett, J. (1995). Perceiving geographical slant. *Psychon. Bull. Rev., 2*, 409-428.

Riddoch, M. J., Humphreys, G. W., Edwards, S., Baker, T., & Wilson, K. (2003). Seeing the action: Neuropsychological evidence for action-based effects on object selection. *Nat. Neurosci., 6*, 82-89.

Rogers, R. D., & Monsell, S. (1995). Costs of a predictable switch between simple cognitive tasks. *J. Exp. Psychol. Gen., 124*, 207-231.

Shiffrar, M., & Freyd, J. J. (1993). Timing and apparent motion path choice with human body photographs. *Psychol. Sci., 4*, 379-384.

Stewart, D., Cudworth, C. J., & Lishman, J. R. (1993). Misperception of time-to-collision by drivers in pedestrian accidents. *Perception, 22*, 1227-1244.

Tipper, S. R., Lortie, C., & Baylis, G. C. (1992). Selective reaching: Evidence for action-centered attention. *J. Exp. Psychol. Hum. Percept. Perform., 18*, 891-905.

Tresilian, J. R. (1993) Four questions of time to contact: A critical examination of research on interceptive timing. *Perception, 22*, 653-680.

Tresilian, J. R. (1994) Approximate information sources and perceptual variables in interceptive timing. *J. Exp. Psychol. Hum. Percept. Perform., 20*, 154-173.

Tucker, M., & Ellis, R. (1998). On the relations between seen objects and components of potential actions. *J. Exp. Psychol. Hum. Percept. Perform., 24*, 830-846.

Viviani, P., & Stucchi, N, (1992). Biological movements look uniform: Evidence of motor-perceptual interactions. *J. Exp. Psychol. Hum. Percept. Perform., 18*, 603-623.

Wann, J.P. (1996). Anticipating arrival: Is the tau margin a specious theory? *J. Exp. Psychol. Hum. Percept. Perform., 22*, 1031-1048.

Warren, W. H. (1984). Perceiving affordances: Visual guiding of stair climbing. *J. Exp. Psychol. Hum. Percept. Perform., 10*, 683-703.

Section 2

Optic Flow Processing and Computation

8. Modeling Observer and Object Motion Perception

Constance S. Royden

Department of Mathematics and Computer Science
College of the Holy Cross, Worcester, MA, USA

1 INTRODUCTION

When a person moves about the world, images of the surrounding scene move across his or her retinas, providing a rich source of information about the environment. The motion of the images arises from numerous sources. The observer's own locomotion causes the images of all stationary items to move in a pattern. This pattern of image velocities on the retina is known as optic flow or retinal flow. This pattern can be a fairly simple, radial pattern (Figure 1a) if the observer moves in a straight line, or more complex (Figure 1b) if the observer moves on a curved path. In addition, the observer may make eye or head movements that also affect the flow field. Finally, objects in the world may themselves be moving, creating additional complexity in the scene (Figure 1c). Somehow the brain is able to process this motion information adeptly to ascertain the direction of motion of the observer as

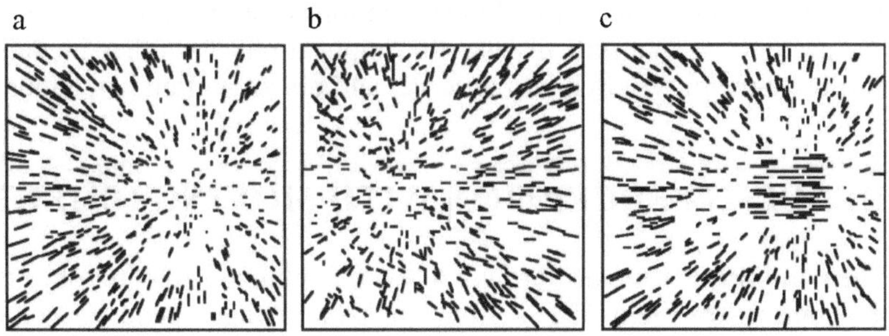

Figure 1. Optic flow fields. Each line represents the image velocity for a point in a 3-D cloud. a) Observer moving in a straight line. b) Observer moving on a curved path. c) Scene containing a moving object.

L.M. Vaina, S.A. Beardsley and S.K. Rushton (eds.), Optic Flow and Beyond, 159–181.

well as the position and direction of motion of moving objects in the scene. This ability allows soccer players to follow a moving ball while maneuvering past other running players, and drivers on busy roads to avoid moving cars and pedestrians while driving straight or negotiating turns.

How the brain accomplishes these tasks has been the subject of much research over the last several decades. One perplexing problem is how the brain sorts out the different sources of motion information, e.g. locomotion, eye movements and moving objects, from a complex flow field. Numerous computational theories (e.g. Bruss & Horn, 1983; Longuet-Higgins & Prazdny, 1980; Heeger & Jepson, 1992; Hildreth, 1992; for review see Hildreth & Royden, 1998) have been developed to compute an observer's translational direction of motion in the presence of rotations, such as eye movements or curved path motion. In addition several physiological models (Perrone, 1992; Perrone & Stone, 1994; Lappe & Rauschecker, 1993; Royden, 1997; Beintema & van den Berg, 1998; Cameron et al., 1998) have been proposed that explain how neurons in the motion areas of visual cortex (the middle temporal area (MT) and the medial superior temporal area (MST)) might process optic flow fields to compute observer translation in the presence of eye movements. While several theoretical approaches have been proposed to address the problem of moving objects in the scene (e.g. Adiv, 1985; Heeger & Hager, 1988; Ragnone et al., 1992; Hildreth, 1992; Thompson & Pong, 1990; for review see Hildreth & Royden, 1998), few of the physiological models have addressed this problem.

In this chapter we will review the psychophysical results regarding an observer's judgment of the translational direction of self-motion (or heading) in the presence of moving objects. We will then discuss one approach for processing flow fields to determine an observer's direction of motion. We will show how this approach is able to account for the psychophysical results for moving objects, both in theory and in simulations of a physiological model of heading perception. We will discuss the implications for other models and further directions for research into observer and object motion perception.

2 HEADING PERCEPTION WITH MOVING OBJECTS

When an observer moves through a scene containing moving objects, the resulting optic flow field differs from the pattern that would be generated from a stationary scene. For example, when an observer moves in a straight line, the optic flow field forms a radial pattern (Figure 1a). The center of this pattern, known as the focus of expansion (FOE), coincides with the observer's direction of motion, or heading (Gibson, 1950). If there is a moving object in the scene (Figure 1c), this object generates additional image velocities in the scene that do not conform to the radial pattern. If one tries to compute the observer's direction of translation using all the image velocities in the scene, the velocities from the moving object will interfere with the computation, leading to errors in the estimate of heading. Several theoretical approaches (e.g. Adiv, 1985; Heeger & Hager, 1988; Ragnone et al., 1992; Thompson & Pong, 1990; Hildreth, 1992; see Hildreth & Royden, 1998 for more discussion of these models) suggested ways of identifying the location of the moving object before or during the heading computation. Once the object's location is known, one can remove from the heading computation the image velocities due to the object. However, it is possible that the brain does not compensate for the errors caused by the moving objects, but instead shows errors in heading estimation when moving objects are present. This is suggested by the following results of psychophysical experiments, and has broad implications for computational models of heading.

People judge their heading well when viewing simulated straight-line motion through stationary scenes (Rieger & Toet, 1985; Warren & Hannon, 1988, 1990; Crowell et al., 1990; van den Berg, 1992; Crowell & Banks, 1993; Cutting et al., 1992). However, in the presence of moving objects, human ability to judge heading depends on the position and the direction of motion of the object. Several researchers have shown that a moving object has little effect on heading judgments when the object is not crossing the observer's path (Royden & Hildreth, 1996; Warren & Saunders, 1995; Cutting et al., 1995). When the object is crossing the observer's path, so that the FOE is obscured, people show a small bias in their heading judgments that depends on the object's 3-dimensional motion direction. When the object is moving toward the observer, the object itself has an FOE that is separate from the FOE for the stationary part of the scene. In this case there is a small bias in the direction of the FOE of the object (Figure 2a), as if the visual system is

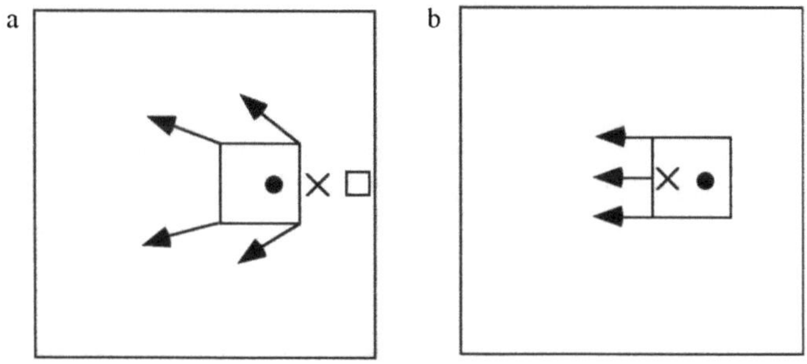

Figure 2. Different object motions produce opposite biases in heading judgments. a) An object moving toward the observer produces a bias in the direction of its FOE. b) An object moving laterally produces a bias in the lateral direction of motion. Filled circles indicate true observer heading, crosses indicate perceived heading, and the small square indicates the object's FOE.

computing an average between the positions of the FOEs from the moving object and the stationary part of the scene. Note that this bias is in the direction opposite that of the lateral component of motion for the object (Warren & Saunders, 1995; Royden & Hildreth, 1996). On the other hand, if the object is moving laterally with respect to the observer, so that it stays at a constant distance from the observer (Figure 2b), people show a small bias in the direction of object motion (Royden & Hildreth, 1996).

The seemingly contradictory results for the different object conditions have several possible explanations. It is possible that the differing biases are the result of separate mechanisms for computing observer heading. For example, Warren and Saunders (1995) suggested that the visual system might average the FOE positions when the FOEs are close to one another, as when the object is approaching the observer, but not when they are far apart, as when the object is moving laterally. They and others (Pack & Mingolla, 1998) suggest that the bias caused by the laterally moving object is the result of a mechanism to compensate for eye movements. In this case, they argue that the lateral motion of the object stimulates a visual mechanism for estimating eye rotations. This mechanism causes the visual system to compensate for this rotation by a shift in the perceived location of the observer's heading. An eye rotation causes the actual image of the FOE to shift in the direction opposite the lateral image motion generated by the eye movement. Therefore, the visual system compensates by shifting the perceived location in the same direction as this lateral motion.

This is similar to an explanation given for a visual illusion described by Duffy and Wurtz (1993). In this illusion, a field of dots moving in a radial pattern overlaps a field of dots moving with lateral motion. Duffy and Wurtz showed that this causes a shift in the perceived FOE of the radial pattern in the direction of the lateral motion. They suggested that this is the result of an eye movement compensation system such as described above. It has been suggested that the laterally moving object acts like a small version of the field of laterally moving dots in the Duffy and Wurtz illusion (Pack & Mingolla, 1998; Warren & Saunders, 1995).

Another explanation of the bias seen with moving objects is that it is the result of a motion subtraction that is performed by the visual system when processing optic flow fields (Royden, 2002). This type of model can account for the biases seen both when the object is moving toward the observer and when the object moves laterally with respect to the observer. It can also account for the perceived shift in bias seen in the Duffy and Wurtz (1993) illusion (Royden & Conti, 2003). The next section describes the theoretical basis of this model, which shows how motion subtraction predicts the direction of the bias in the heading computations when moving objects are present. We also describe a physiologically based model that uses motion subtraction to compute heading, and show how it accounts for both the magnitude and direction of the biases seen with moving objects. Finally, we discuss how this analysis relates to other models and directions for future research in regard to moving objects and moving observers.

3 COMPUTING HEADING BY MOTION SUBTRACTION

3.1 Mathematical Underpinnings

The original idea for using motion subtraction in heading computations comes from a mathematical analysis by Longuet-Higgins and Prazdny (1980), who showed how one could use subtraction of image velocity vectors to compute translational heading in the presence of rotations. For any observer motion, one can instantaneously describe the motion in terms of three translational components of motion (Tx, Ty, Tz), along the X, Y, and Z coordinate axes, and three rotational components of motion (Rx, Ry, Rz), around the coordinate axes. One can then compute the image velocity for a point P = (X, Y, Z) in the world when projected onto an image plane at a distance of 1 unit from the observer. The resulting image velocity is given as follows (Longuet-Higgins & Prazdny, 1980; Royden, 1997):

$$
\begin{aligned}
v_x &= \frac{-T_x + xT_z}{Z} + xyR_x - \left(1 + x^2\right)R_y - yR_z \\
v_y &= \frac{-T_y + yT_z}{Z} + \left(1 + y^2\right)R_x - xyR_y - xR_z
\end{aligned}
\qquad (1)
$$

where (x, y) is the position of the projection of point P onto the image plane, and x = X/Z, y = Y/Z.

Longuet-Higgins and Prazdny (1980) noticed that these equations are separable into two components. One component is dependent on observer translation and not rotation. The other is dependent on observer rotation but not translation. Furthermore, the translation component depends on the distance, Z, of point P from the observer, while the rotation component is independent of this distance. Therefore, if one could measure the image velocities of two points at different distances along the same line of sight, such as immediately on either side of a depth edge, they would have the same rotation component, but their translation components would differ because of the factor 1/Z. Thus, if one subtracts one image velocity vector from the other the rotation components will be eliminated, leaving a difference vector that depends only on observer translation. These difference vectors are given as:

$$v_{xd} = (-T_x + xT_z)(\frac{1}{Z_1} - \frac{1}{Z_2})$$
$$v_{yd} = (-T_y + yT_z)(\frac{1}{Z_1} - \frac{1}{Z_2}) \tag{2}$$

where v_{xd} and v_{yd} represent the x and y components of the difference vectors. These difference vectors point directly toward or away from the observer's direction of translation (Longuet-Higgins & Prazdny, 1980). Thus, if one computes difference vectors across the entire visual scene, they will form a radial pattern, whose center coincides with the direction of observer translation. One can easily locate this center in order to compute the observer's heading.

To show that such a mechanism could be used by the visual system, one must first show that it is sufficient to perform a motion subtraction in a physiologically plausible way, using cells that have receptive field types that are similar to those found in the visual system. Rieger and Lawton (1985) took one step toward this goal by showing that if one performs the subtraction between velocity vectors that are spatially separated by a small amount one can still compute heading fairly accurately. Royden (1997) showed how this motion subtraction could be performed by motion-opponent operators based on the receptive field properties of the cells in the Middle Temporal visual area (MT) of primates, as described below.

3.2 A Physiological Model Using Motion Subtraction

The cells in MT respond to moving stimuli in their receptive fields. They have an excitatory region of their receptive fields that is tuned to direction and speed of the motion within this region (Maunsell & van Essen, 1983). In addition, many of these cells have an inhibitory region, the "surround", adjacent to this excitatory region (Allman et al., 1985; Xiao, Raiguel et al., 1995; Raiguel et al., 1995). The spatial relationship between the excitatory and inhibitory regions varies from cell to cell, with some exhibiting an asymmetric, side-by-side, arrangement (Figure 3a), and others showing more symmetrical patterns, such as a center-surround structure (Figure 3b), (Raiguel et al., 1995). The inhibitory surrounds also show a direction tuning similar to that of the adjacent excitatory region (the "center"). The direction of motion causing the maximum inhibition in the surround is generally the same as the direction of motion that causes the maximum excitation in the center (Allman et al, 1985). Thus, these cells respond poorly to large regions of uniform motion. They instead respond best to a change in motion between the

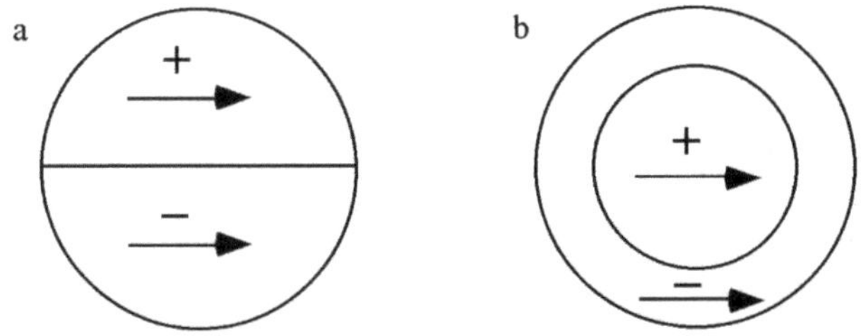

Figure 3. Motion-opponent operators used to model MT cells. The arrow indicates the preferred direction of motion. The plus indicates the excitatory region; the minus indicates the inhibitory region. a) Asymmetric receptive field organization. b) Center-surround receptive field organization.

center and surround regions. In a sense, they are performing a subtraction between the excitatory and inhibitory regions. To test whether these cells might be capable of computing heading in the presence of rotations, Royden (1997) developed a model based on their receptive field properties.

The Royden (1997) model uses "motion-opponent" operators that have receptive fields that are highly simplified versions of the cells in MT. The operators have either asymmetric receptive fields (Figure 3a) or center-surround receptive fields (Figure 3b) with an excitatory region and an adjacent inhibitory region. Each region is tuned to a preferred direction of motion, such that it responds best to motion in that direction and the response falls off with the cosine of the angle between the preferred direction and the actual direction of stimulus motion. The overall response of each operator is computed as the difference between the responses in the excitatory and inhibitory regions. The operators are balanced, so that uniform motion across the receptive fields generates zero response.

The operators are arranged into a network as follows. Each region of the visual field is processed by a group of operators (Figure 4), which vary in their preferred direction of motion. The asymmetric operators also vary in the angle of the line that separates the excitatory from the inhibitory regions. The operator within this group that responds most strongly to a given retinal flow field will have a preferred direction of motion that points approximately toward or away from the observer's direction of motion (Royden, 1997). These operators project to a second layer of cells (Figure 4). Cells in this second layer act as templates for radial patterns of responses from cells in the previous, motion-opponent, layer. Each cell is tuned for a preferred position of the center of the radial pattern. The cell in this layer that responds most strongly will have a preferred center that coincides with the observer's direction of translation. These cells are similar to cells found in MST of the primate. MST cells have large receptive fields and many of them respond well

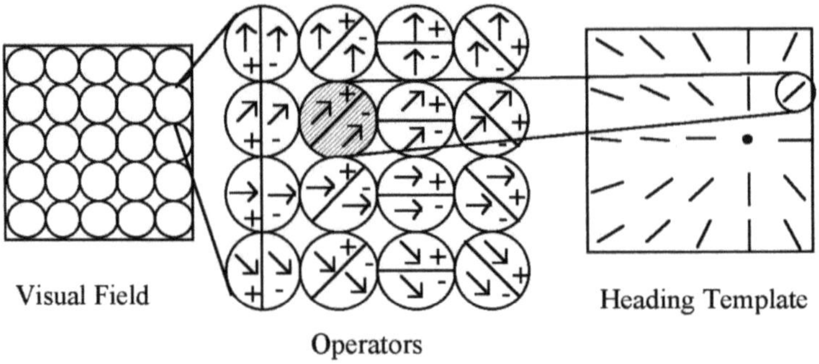

Figure 4. Model using motion-opponent operators. Each region of the visual field is processed by a group of operators. The operators in the first layer project to a second layer of templates for radial patterns.

to expanding or contracting radial patterns (Duffy & Wurtz, 1991; Graziano et al., 1994; Tanaka & Saito, 1989; Saito et al., 1986). In addition, the cells are tuned to the position of the center of the radial pattern (Duffy & Wurtz, 1995).

Royden (1997) showed that this model computes heading well in the presence of observer rotations. Models using either the asymmetric or center-surround operators perform well, although the center-surround operators are less sensitive to gradual depth changes. The model is also robust to added noise in the input velocity field, small changes in receptive field size or the width of the tuning curve (Royden, 1997). This leads to the question of how the model would perform in the presence of moving objects.

3.3 Motion Subtraction with Moving Objects

The presence of moving objects in the scene causes observers to make small but consistent errors in heading judgments under some conditions. By calculating the direction of the difference vectors at the borders between the moving object and the stationary parts of the scene, we can predict how moving objects affect models that use motion subtraction. We can test these predictions by running simulations with the physiologically based model and we can also compare the performance of the model with that of humans.

Consider an observer moving toward a scene at distance Z_1 containing a moving object at a distance Z_2. Suppose the observer motion relative to the stationary part of the scene is (T_{x1}, T_{y1}, T_{z1}) and the observer motion relative to the object is (T_{x2}, T_{y2}, T_{z2}). In that case, the difference vectors at the borders of the object are given by:

$$v_{xd} = \frac{(-T_{x1} + xT_{z1})}{Z_1} - \frac{(-T_{x2} + xT_{z2})}{Z_2}$$
$$v_{yd} = \frac{(-T_{y1} + yT_{z1})}{Z_1} - \frac{(-T_{y2} + yT_{z2})}{Z_2} \tag{3}$$

These vectors form a radial pattern. Lines through the difference vectors intersect at a location given by

$$x = \frac{(Z_2 T_{x1} - Z_1 T_{x2})}{(Z_2 T_{z1} - Z_1 T_{z2})}$$
$$y = \frac{(Z_2 T_{y1} - Z_1 T_{y2})}{(Z_2 T_{z1} - Z_1 T_{z2})} \tag{4}$$

Figure 5 shows the location of this intersection for the two different types of moving objects that cause the different directions of bias in heading judgments. For an object moving laterally to the left (Figure 5a), the intersection of these vectors is to the left of the observer's heading. Thus one would expect that these difference vectors would cause a bias to the left, in the same direction as the object motion. For an object moving in depth toward the observer (Figure 5b) with a lateral component of motion to the left, the difference vectors intersect at a point to the right of the observer's heading. So one would expect these vectors to cause a rightward bias in this condition, in the direction opposite from the object's lateral motion. These are the same directions of bias as are seen with human observers.

Although the direction of the difference vectors generated by the moving

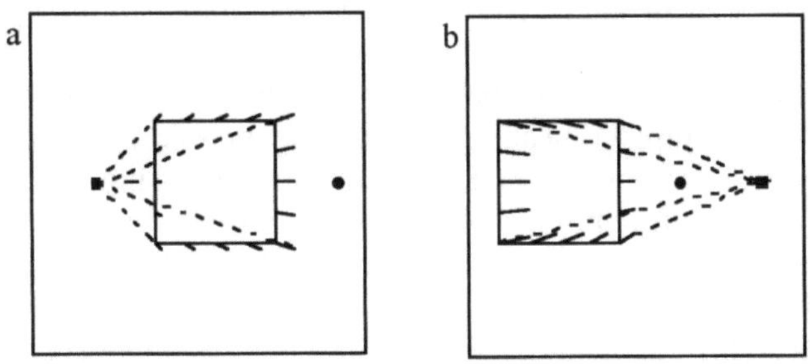

Figure 5. Intersection of difference vectors at the border of a moving object. a) Object is moving laterally to the left. b) Object is moving in depth with a lateral component of motion to the left.

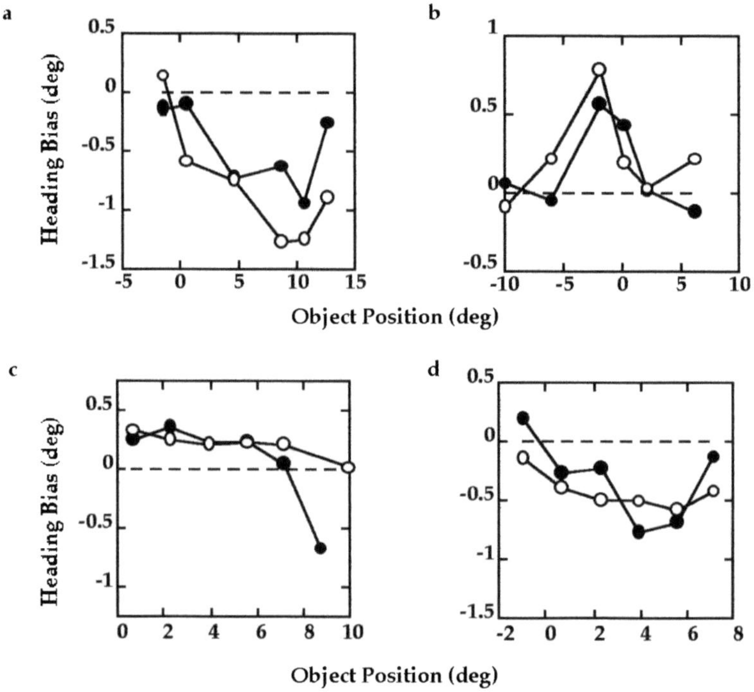

Figure 6. Model and human biases generated by a moving object at different starting positions in the scene. Simulated headings were at 4, 5, 6 and 7 deg to the right of center. Bias is computed as the response when the object is present minus the response when the object is absent. a) Object moving laterally to the left. b) Object moving laterally to the right. c) Object moving in depth, with a lateral component of motion to the left. d) Object moving in depth, with lateral component of motion to the right. Open symbols indicate model biases and filled symbols indicate human biases.

object is consistent with the direction of bias seen for human observers, the position of the point of intersection of these vectors is shifted much more to the left or right than the shift in heading perception by human observers. However, if the visual system makes use of these motion differences, one would expect the computed heading to be somewhere in between the FOE for the stationary part of the scene and the point of intersection for the difference vectors generated by the moving object. We can estimate the position of this computed heading by running model simulations for the motion-opponent heading model described above.

Royden (2002) ran simulations of the motion-opponent model using conditions identical to those used in psychophysical experiments (Royden & Hildreth, 1996). Figure 6 replots data from the psychophysical experiments (Royden & Hildreth, 1996) and from the model simulations (Royden, 2002).

The results of the simulations show that the model exhibits biases in the same direction and of the same magnitude as human observers for both the laterally moving object and the object moving in depth. For an object moving laterally to the left, the model shows a small bias to the left (Figure 6a). When the object moves laterally to the right, the bias is reversed (Figure 6b). For an object moving in depth, the bias is opposite the direction of the lateral component of motion (and toward the object's FOE), (Figures 6c and 6d). This demonstrates that a physiological model that computes motion differences can account for the biases in human heading judgments when moving objects are present.

The model responses are also dependent on the starting position of the object, as are the human responses. This position dependence for the model arises from weighting the inputs from the first layer to the second layer with a Gaussian function, such that inputs from regions of the visual field that are close to the template cell's preferred radial center are weighed more heavily than inputs from more distant regions (Royden, 2002). This weighting system was initially proposed by Warren and Saunders (1995) to account for the position dependence of the biases caused by moving objects. The Gaussian weighting may also serve to increase the influence of the most informative velocities in the scene for computing heading, i.e., those near the observer's heading direction (Crowell & Banks, 1996). The fact that it also helps to minimize the effects of moving objects that are not crossing the observer's path may be a fortuitous consequence of a mechanism to minimize noise in the computation.

4 ALTERNATIVE NEURAL ARCHITECTURES

The above discussion shows that a physiologically based model using motion subtraction can compute heading accurately even in the presence of eye movements or other observer rotations. In addition, the model accounts for biases in observer heading judgments caused by the presence of moving objects in the scene. It should be pointed out, however, that this neural architecture is not the only one that could lead to these results. The critical feature of the model presented above is the local subtraction of neighboring image velocities in the computation of heading. While the motion-opponent operators seem like the logical functional units for performing this subtraction, there are other plausible neural architectures that could accomplish the same result. Figure 7 diagrams three such possible architectures, including the motion-opponent strategy (Figure 7a).

One such alternative neural architecture (Figure 7b), performs the subtraction at the level of the connections between the first (MT) and second

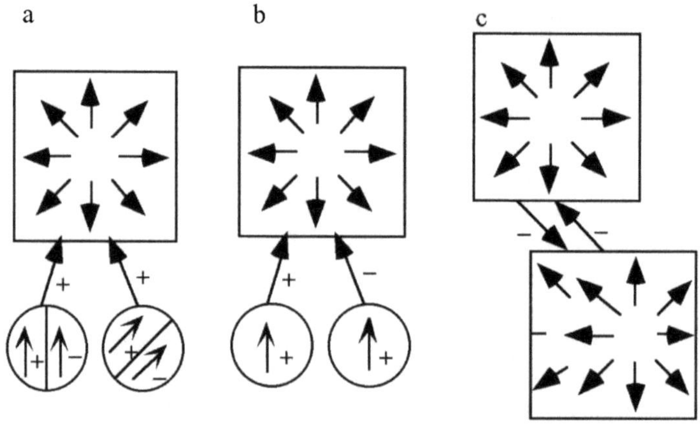

Figure 7. Alternative neural architectures for performing motion subtraction. a) Motion subtraction performed by motion-opponent operators in the first layer (e.g., MT). b) Motion subtraction performed by inhibitory connections between the first and second layers. c) Motion subtraction performed by lateral inhibition between cells in the second layer (e.g., MST).

(MST) layers of the neural network. In this scenario, cells in MT that do not have inhibitory surrounds act as the input layer and the weights of the connections to the second layer are excitatory or inhibitory. The inhibition could be a direct inhibition of the second layer cell or, to make the subtraction more locally specific, it could be accomplished through presynaptic inhibition of one cell by a neighbor with similar direction preference.

A third possibility for the neural architecture might be one that subtracts by lateral inhibition between cells in MST (Figure 7c). In this case, template cells in the MST layer would again receive input from direction selective cells in MT that do not have the inhibitory surround. The templates would be tuned to radial patterns of motion and would inhibit other template cells with shifted centers of motion. Such lateral inhibition of cells with similar preferred centers of radial motion can serve to sharpen the tuning curves for observer heading. This method of subtraction is less specific and less local than the other architectures described here. It would therefore be important to thoroughly test such a model to determine whether this global form of subtraction would show the same biases as do people and the motion-opponent model in the presence of moving objects.

Finally, it is possible that the motion subtraction is carried out at a processing level before MT, such as the primary visual cortex (V1). It has recently been shown that cells in V1 respond in different ways depending on the image context surrounding the classical receptive field. For example, a cell's response may change depending on the figure-ground relationship between the image in the receptive field and the surrounding scene (Lamme,

1995). It would not be surprising if these cells showed contextual effects in their motion responses, some of which could amount to subtraction of adjacent image velocities similar to that in MT cells. The advantage of performing motion subtraction at the level of V1 is that the receptive field sizes are generally smaller than those of MT cells and so the subtraction could be localized to smaller image areas. This would eliminate some of the noise in the heading computation that arises due to spatial separation of the image velocity vectors being subtracted (Rieger & Lawton, 1985).

While all of the above architectures are possibilities for computing differences in motion across an optic flow field, the Royden (1997) model that uses motion-opponent operators based on MT cells is appealing for several reasons. First, it uses cell types that are known to exist in the visual system, while the other architectures are as yet hypotheses that need experimental verification. The fact that cells in MT exhibit motion opponent properties that are capable of computing heading and accounting for observer biases from moving objects makes these cells obvious candidates for this function. Second, it provides a reasonable hypothesis as to the function of these motion-opponent cells in MT, i.e., the purpose of the inhibitory surrounds of these cells is to process motion differences to facilitate the computation of heading in the presence of rotations.

5 CONSEQUENCES FOR OTHER MODELS

As mentioned above, there are several models based on physiological properties of cells in MT and MST that compute heading. Most of these have been developed to compute heading for an observer moving through stationary scenes and have not been tested for their response when a moving object is present. Therefore, one cannot compare these models based on their known responses. However, one can make predictions as to how these models may be influenced by moving objects based on the architecture of the model and the underlying theory for the computation of heading.

Several models have been proposed that use templates to compute heading. These template cells are tuned to particular optic flow patterns that arise from specific combinations of observer translation and rotation. Hatsopoulos and Warren (1991) developed a neural network that used back propagation to set connection weights between an input layer consisting of direction tuned motion cells and an output layer that was trained to compute heading for observers moving in a straight line. The resulting pattern of connections in the trained network resembled the connections for a set of templates for radial patterns of motion.

Perrone (1992) and Perrone & Stone (1994) developed a more general model for computing heading that allowed for observer rotations. As in the Hatsopoulos and Warren model the inputs were direction and speed selective cells based on MT cells without the inhibitory surrounds. The second layer consisted of templates tuned to the patterns of motion that would be generated by a combination of rotation and translation. To limit the large number of templates that would be required in such a model, the Perrone & Stone (1994) version limited rotations to those generated by an observer tracking a stationary object in the scene.

Both the Hatsopoulos and Warren (1991) and the Perrone and Stone (1994) models use the image velocities from the optic flow field directly to compute heading using optic flow template cells. In other words, no motion subtraction occurs in these models. One would therefore predict that both of these models would show a bias similar to humans in the case of an object approaching the observer. In this case, the human bias is in the direction that would be predicted by averaging the position of the FOE due to the moving object and the FOE due to the stationary part of the scene. Warren and Saunders (1995) report that a template model similar to Hatsopoulos and Warren (1991) does indeed show such a bias. However, one would not expect these models to show the same bias as humans for the condition in which the object moves laterally with respect to the observer. In this condition, the FOE of the object is in the direction opposite the object's lateral direction of motion. Thus one would predict that the template models would show a bias in the direction opposite the direction of motion of the object, which is inconsistent with the bias shown by human observers. If this prediction is correct, then these template models would not be able to account for the biases shown by humans in the presence of moving objects. Some second mechanism, such as an eye movement compensation mechanism described above, would have to be invoked.

Another physiologically based model (Lappe & Rauschecker, 1993) is based on a computational approach developed by Heeger and Jepson (1992). Heeger and Jepson showed that one can compute observer translation and rotation parameters from the image velocities at five or more different locations in the scene. The known image velocities are put into a set of simultaneous equations that can then be solved for the unknown observer motion parameters as well as the distance to the points in the scene. Heeger and Jepson showed that these equations could be solved by finding the observer translation and rotation parameters that minimize a residual function. Lappe and Rauschecker developed a physiological model based on this approach. Their model uses a layer of direction selective cells, as do the template models described above, and the residual function is implemented through the variation of the connection weights between the first and the second layers. They showed that this model computes observer heading well in the presence of eye rotations, and that the cells in the second layer have receptive field properties similar to those of MST cells in the primate.

Because the connection weights of the Lappe and Rauschecker model are more complex than the weights in the template models or the motion-opponent model, it is not entirely clear how the Lappe and Rauschecker model would respond in the presence of moving objects in the scene. It is possible to make an educated guess, however. First, the connection weights in the Lappe and Rauschecker model can be positive or negative, and each positive connection is balanced with a negative (or inhibitory connection) between the two layers. This architecture could be likened to the architecture described above (Figure 7b) in which the motion subtraction is carried out through inhibition between the first and second layers of the model. If this is true, then one would expect the presence of moving objects to affect this model in much the same way as they affect the motion-opponent operators, with the motion subtraction leading to the biases seen.

One piece of evidence suggesting this is indeed the case is that the Lappe and Rauschecker model exhibits a shift in the computed FOE for the stimulus that generates the visual illusion described by Duffy and Wurtz (Duffy & Wurtz, 1993; Lappe & Rauschecker, 1995). In this illusion, as described above, when a lateral plane of moving dots overlaps a plane of dots moving in a radial pattern, observers experience an illusory shift of the perceived location of the FOE of the radial pattern. This shift may be caused by the same mechanism as the shift seen for laterally moving objects, and Royden and Conti (2003) have shown that motion subtraction can explain the shifts seen with this stimulus. The fact that the Lappe and Rauschecker model shows the same shift in the Duffy and Wurtz illusion suggests it would show a similar shift in the presence of moving objects, which would be consistent with that seen with human observers. If this shift is due to a motion subtraction through the excitatory and inhibitory connections of the model, then one might also expect the Lappe and Rauschecker model to show a shift in the direction of the object's FOE for the objects moving toward the observer, as does the Royden (1997) model and as do people. This prediction remains to be tested with model simulations.

One final physiological model that deserves discussion here uses velocity gain fields to compute observer heading in the presence of eye rotations (Beintema & van den Berg, 1998). In this model, the responses of template cells tuned to retinal flow are multiplied by a 'rate-coded' measure of eye velocity, producing a layer of cells that have a preferred flow field that changes dynamically to compensate for eye movements. While this model generally assumes the rate-coded eye movement signal arises from extra-retinal sources, it is possible to gain an estimate of eye velocity from the visual input. In this case, the model compensates for the estimated eye movement by a shift in the tuning of the receptive fields. It seems likely that this model would show a bias similar to humans in the case of the laterally moving objects, since it is an implementation of the eye movement compensation theory described above. However, it is unclear that it would show the same biases as humans in the case of the object approaching the observer. Model simulations would be necessary to test this.

6 FUTURE RESEARCH DIRECTIONS

The above discussion considers how the brain might compute heading in the presence of moving objects. However, to navigate successfully while avoiding collisions one must also be able to detect moving objects and judge their 3-D direction of motion. There currently is little known about these abilities for human observers and there has been little in the way of physiologically plausible computational modeling of these functions. Therefore, these are both areas that are ripe for future research.

6.1 Detection of Moving Objects in the Scene

To determine the neural mechanisms for the detection of moving objects in the scene, it will be necessary both to perform psychophysical experiments and to build computational models that are based on both electrophysiological and psychophysical results. Initial psychophysical results indicate that the perception of egomotion from visual or vestibular cues impairs the ability to detect moving objects in the scene (Probst et al., 1986; Brandt et al., 1991; Niemann & Hoffmann, 1997). However, there has not been a thorough examination of the stimulus parameters that allow for moving object detection.

One line of research may shed some light on this problem. Royden and colleagues used a visual search paradigm to investigate observers' abilities to detect a moving object among "stationary" distractors (Royden et al., 1996, 2001). In this investigation the question was whether observers search efficiently (or perform a "parallel" search) for a moving item among stationary items when the observer is in motion. Because the observer's motion causes the images of all items in the scene to move, the target item is recognizable only because it is moving differently from the others. For example, if the observer moves in a straight line, all the items that are stationary in the scene will move in a radial pattern. The moving item will move in a direction inconsistent with this pattern (Figure 8a). The observer's task is to detect this item. The efficiency of a search is determined by measuring the reaction time for determining whether or not the target item is

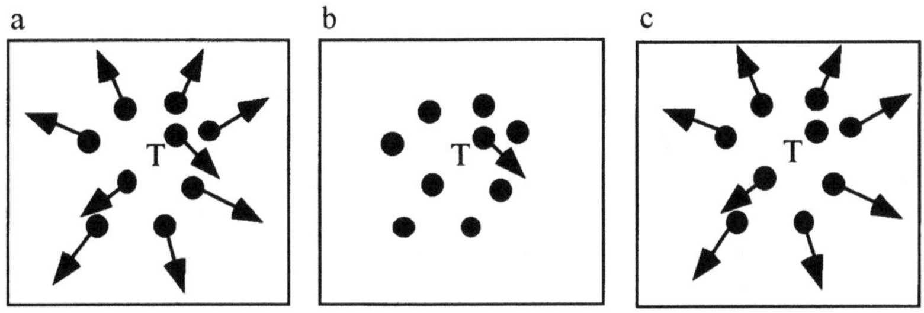

Figure 8. Stimuli used for visual search experiments. a) Target moves perpendicular to radial flow lines and distractors move in radial pattern. b) Moving target and stationary distractors. c) Stationary target and distractors moving in radial pattern. T indicates the target.

present in the scene as the number of distracting items increases. If the reaction time remains constant as the number of distracting items increases, this denotes an efficient (or "parallel") search. In these cases, the target appears to "pop out" of the surrounding scene. Inefficient (or "serial") searches are characterized by a reaction time that increases as the number of distractors in the scene increases, as if the observer must consider each item individually to find the target. The results showed that the target moving object "popped out" for some conditions, such as when the item moved at 90 deg to the radial flow lines (Figure 8a), but not for others, such as when the angle between the target's motion and that of the radial flow lines was reduced to 15 deg (Royden et al., 1996). It seems likely that the difference is related to the detectability of the moving object within the optic flow field. The closer the object's motion to the radial flow lines, the more difficult it is to detect. As the detectability gets closer to a threshold level, the search becomes less efficient. There is still much to be done to fully analyze the conditions that allow observers to detect moving objects within the optic flow field.

One interesting result from the search literature relates to the difference between the search for a moving target among stationary distractors versus the search for a stationary target among moving distractors. It is well known that a moving item among stationary distractors (Figure 8b) "pops out", i.e., the search is efficient. It is also known that search for a stationary item among distractors moving in random directions is inefficient (Verghese & Pelli, 1992). Royden et al. (2001) showed that this asymmetry holds when searching for a stationary item among distractors moving in a radial optic flow pattern (Figure 8c), although the reaction time does not increase as quickly for the optic flow condition as for the random motion condition. This is interesting because, in the optic flow case, the target item with a stationary image would correspond to an item that is moving relative to the other items in the scene. One might have expected that the target would pop out in this case, if the visual system were able to discount the radial motion in the optic flow field as being the result of observer motion. The fact that it does not pop-out suggests that the detection of moving objects by a moving observer requires considerable allocation of computational resources.

There has been some modeling, mainly in the area of computer vision research, on the detection of moving objects in the scene. If one knows that the observer is moving in a straight line, then one can detect moving objects by identifying items whose image motion is inconsistent with the radial pattern generated by observer translation (Adiv, 1985; Heeger & Hager, 1988; Ragnone et al., 1992; Thompson & Pong, 1990; see Hildreth & Royden, 1998 for more discussion). Hildreth (1992) extended this idea so that moving objects could be detected in the presence of observer rotations as well. This model computes difference vectors by local velocity subtractions. Observer heading is then computed as the position consistent with the majority of the difference vectors. The borders of moving objects can then be detected by identifying difference vectors that do not correspond to the radial pattern of difference vectors associated with the computed heading. This model could be put into physiological form using the Royden (1997) model with motion-opponent operators. The Royden (1997) model already computes heading in the presence of moving objects (Royden, 2002) and so it would only be necessary to add a mechanism to identify operator responses that are inconsistent with the computed heading in order to locate the borders of the moving object. This is the subject of ongoing research.

A different approach was taken by Zemel and Sejnowski (1998). They created an artificial neural network whose input layer consisted of cells modeled on MT neurons and whose output layer was based on properties of MST cells. They presented this network with stimuli generated from scenes containing multiple independent object motions and applied an unsupervised optimization technique to set the connection strengths of the network. They showed that the resulting network could segment the scene into regions of coherent 3-D object motion, suggesting that one role of MST cells is to signal different 3-D motions within the visual scene.

The above discussion highlights some of the research that has been done to determine how the brain identifies moving objects when the observer is in motion. While the results are intriguing, much more work must be done both in the psychophysical area and in the modeling area before we gain a thorough understanding of moving object detection.

6.2 Determining 3-D Direction of Object Motion

Once a moving object is detected, the observer must judge its direction of motion in 3-D well enough to avoid a collision with the object. This judgment would not necessarily have to be a well-refined estimate of the 3-D motion. It would be sufficient for the observer to identify collision trajectories versus non-collision trajectories. Recent work has shown that the presence of optic flow in the background impairs an observer's ability to judge the velocity of moving objects in the scene (Brenner, 1993; Brenner & van den Berg, 1996). However, there has been little work done examining human ability to judge 3-D object trajectories in the presence of optic flow fields, so it is difficult to speculate how well people are able to perform this task.

One study has examined how shifting attention affected judgments of heading and of 3-D object motion (Royden & Hildreth, 1999). Observers viewed a simulated scene in which they approached a location to the right of center and in which a moving object was present to the left of center. Observers were asked to judge their heading as well as to determine whether the object was approaching them (and therefore its image increased in size) or moving with pure lateral motion (and therefore its image stayed the same size). While observers were able to make judgments about the moving object's 3-D motion very well when they did not have another task to do, when they attended the heading task their ability to judge the 3-D object motion deteriorated dramatically (Royden & Hildreth, 1999). Interestingly the same was not true about heading judgments. Observers' heading judgments remained fairly stable whether or not they were attending to the heading task. Further research is needed to study the parameters that allow accurate judgment of 3-D object motion in the context of optic flow fields.

Intuitively one might expect that modeling the computation of 3-D object motion should be very similar to the modeling of observer heading. Both involve the computation of the motion of the observer relative to a surface or a set of surfaces. However, there are some difficulties in the case of moving objects that may make the modeling less straightforward than modeling for observer heading. First, the moving objects are generally smaller than the surrounding scene, so there is less information to use in computing their 3-D motion direction. Second, there may be little texture on the surface of the object, so the velocity measurements may be noisier on average than those for the scene as a whole. Third, if the visual system uses motion differences to compute observer motion relative to a surface, then the primary information regarding 3-D object motion will be at the borders between the moving object and the stationary parts of the scene. The difference vectors generated at these borders point to neither the scene FOE nor the moving object FOE, but rather to some other location that depends on the relative motions of the object and observer (Royden, 2002). It is unclear whether this will bias judgments of 3-D object motion, or whether difference vectors are used at all in judgments of 3-D object motion. Much more psychophysical work must be done to address these questions before realistic models can be implemented. This area of research clearly has many questions left to be answered and is ready for new experiments and modeling to address them.

7 CONCLUSIONS

Moving observers often must navigate through a dynamic world containing multiple moving objects that must be avoided. Observers must therefore judge their own motion direction as well as judging the 3-D direction of motion of the moving objects. The presence of moving objects in the scene affects how well people judge their heading. Models that use motion subtraction to compute heading theoretically should show similar biases in the computed heading direction. In particular, a model that uses motion-opponent operators similar to cells found in area MT of the primate visual system shows the same direction and magnitude of heading biases as do humans when faced with moving objects in the scene.

To avoid collisions, a moving observer must be able to detect moving objects and judge their 3-D direction of motion. While there has been some research in the detection of moving objects, there are still many questions left to be answered regarding human abilities in this area. Furthermore, modeling this ability with physiologically plausible models still needs much more work. The problem of judging an object's 3-D direction of motion when the observer is moving still has much to be investigated and is an area that is open for future research. The combination of psychophysical, electro-physiological and computational research has already proved fruitful in our understanding of human heading perception. In the future this combination should provide new insights into the neural mechanisms for processing observer and object motion.

ACKNOWLEDGMENTS

This work was funded by NSF grant #IBN-0196068.

REFERENCES

Adiv, G. (1985). Determining three-dimensional motion and structure from optical flow generated by several moving objects, *IEEE Trans. Pat. Anal. Mach. Intel., PAMI-7,* 384-401.

Allman, J., Miezin, F. & McGuiness, E. (1985). Direction- and velocity-specific responses from beyond the classical receptive field in the middle temporal visual area (MT), *Perception, 14,* 105 - 126.

Beintema, J. A. & van den Berg, A. V. (1998). Heading detection using motion templates and eye velocity gain fields, *Vision Res., 38,* 2155-2179.

Brandt, T., Dieterich, M. & Probst, T. (1991). Self-motion and oculomotor disorders affect motion perception, in Paillard, J. (ed.), *Brain and Synapse.* Oxford, U.K.: Oxford Science Publications.

Brenner, F. (1993). Judging an object's velocity when its distance changes due to ego-motion, *Vision Res., 33,* 487 - 504.

Brenner, F., and van den Berg, A. V. (1996). The special role of distant structures in perceived object velocity, *Vision Res., 36,* 3805 - 3814.

Bruss, A. R. & Horn, B. K. P. (1983). Passive navigation, *Comput. Vis. Graph. Im. Process., 21,* 3-20.

Cameron, S., Grossberg, S. & Guenther, F. H. (1998). A self-organizing neural network architecture for navigation using optic flow, *Neural Computat., 10,* 313 - 352.

Crowell, J. A. & Banks, M. S. (1993). Perceiving heading with different retinal regions and types of optic flow, *Percept. Psychophys., 53,* 325 - 337.

Crowell, J. A. & Banks, M. S. (1996). Ideal observer for heading judgments, *Vision Res., 36,* 471 - 490.

Crowell, J. A., Royden, C. S., Banks, M. S., Swenson, K. H. & Sekuler, A. B. (1990). Optic flow and heading judgments, *Invest. Ophthalmol. Vis. Sci. 31 (Suppl),* 522 (Abstract).

Cutting, J. E., Springer, K., Braren, P. A., & Johnson, S. H. (1992). Wayfinding on foot from information in retinal, not optical, flow, *J. Exp. Psychol. Gen., 121,* 41-72.

Cutting, J. E., Vishton, P. M. & Braren, P. A. (1995). How we avoid collisions with stationary and moving objects, *Psychol. Rev., 102,* 627 - 651.

Duffy, C. J. & Wurtz, R. H. (1991). Sensitivity of MST neurons to optic flow stimuli. I. A continuum of response selectivity to large field stimuli, *J. Neurophysiol., 65,* 1329 - 1345.

Duffy, C. J. & Wurtz, R. H. (1993). An illusory transformation of optic flow fields, *Vision Res., 33,* 1481 - 1490.

Duffy, C. J. & Wurtz, R. H. (1995). Response of Monkey MST neurons to optic flow stimuli with shifted centers of motion, *J. Neurosci., 15,* 5192 - 5208.

Gibson, J. J. (1950). *The Perception of the Visual World.* Boston, Mass.: Houghton Mifflin.

Graziano, M. S. A., Andersen, R. A. & Snowden, R. (1994). Tuning of MST neurons to spiral motions, *J. Neurosci., 14,* 54 - 67.

Hatsopoulos, N. G. & Warren, W. H. (1991). Visual navigation with a neural network, *Neural Netw., 4,* 303 - 317.

Heeger, D. J. & Hager, G. (1988). Egomotion and the stabilized world, *Proceedings of the 2nd International Conference on Computer Vision*, Tampa, FL, 435-440.

Heeger, D. J. & Jepson, A. D. (1992). Subspace methods for recovering rigid motion I: Algorithm & implementation, *Int. J. Comput. Vis.*, *7*, 95-117.

Hildreth, E. C. (1992). Recovering heading for visually-guided navigation, *Vision Res., 32*, 1177 - 1192.

Hildreth, E. C. & Royden, C. S. (1998). Computing observer motion from optic flow, in Watanabe, T. (ed.), *High-level Visual Motion Processing. Computational, Neurobiological and Psychophysical Perspectives.* Cambridge, MA: MIT Press.

Lamme, V. A. F. (1995). The neurophysiology of figure-ground segregation in primary visual cortex, *J. Neurosci.*, *15*, 1605-1615.

Lappe, M. & Rauschecker, J. P. (1993). A neural network for the processing of optic flow from ego-motion in man and higher mammals, *Neural Computat.*, *5*, 374-391.

Lappe, M. & Rauschecker, J. P. (1995). An illusory transformation in a model of optic flow processing, *Vision Res., 35*, 1619 - 1631.

Longuet-Higgins, H. C. & Prazdny, K. (1980). The interpretation of a moving retinal image, *Proc. R. Soc. Lond. B, 208,* 385 - 397.

Maunsell, J. H. R. & van Essen, D. C. (1983). Functional properties of neurons in middle temporal visual area of the macaque monkey. I. Selectivity for stimulus direction, speed and orientation, *J. Neurophysiol., 49,* 1127 - 1147.

Niemann, T. & Hoffmann, K.-P. (1997). Motion processing for saccadic eye movements during visually induced sensation of ego-motion in humans, *Vision Res., 37*, 3163 - 3170.

Pack, C. & Mingolla, E. (1998). Global induced motion and visual stability in an optic flow illusion, *Vision Res., 38*, 3083 - 3093.

Perrone, J. A. (1992). Model for the computation of self-motion in biological systems, *J. Opt. Soc. Am., A*, *9*, 177 - 194.

Perrone, J. A. & Stone, L. S. (1994). A model of self-motion estimation within primate extrastriate visual cortex, *Vision Res., 34,* 2917 - 2938.

Probst, T., Brandt, T. & Degner, D. (1986). Object-motion detection affected by concurrent self-motion perception: psychophysics of a new phenomenon, *Behav. Brain Res., 22*, 1 - 11.

Ragnone, A., Campani, M. & Verri, A. (1992). Identifying multiple motions from optical flow, *Second European Conference on Computer Vision*, Santa Margerita Ligure, Italy, 258-266.

Raiguel, S., Van Hulle, M. M., Xiao, D. K., Marcar, V. L. & Orban, G. A. (1995). Shape and spatial distribution of receptive fields and antagonistic motion surrounds in the middle temporal area (V5) of the macaque, *Eur. J. Neurosci., 7,* 2064 - 2082.

Rieger, J. H. & Lawton, D. T. (1985). Processing differential image motion, *J. Opt. Soc. Am. A, 2,* 354 - 360.

Rieger, J. H. & Toet, L. (1985). Human visual navigation in the presence of 3-D rotations, *Biol. Cybern., 52*, 377-381.

Royden, C. S. (1997). Mathematical analysis of motion-opponent mechanisms used in the determination of heading and depth, *J. Opt. Soc. Am. A, 14*, 2128 - 2143.

Royden, C. S. (2002). Computing heading in the presence of moving objects: A model that uses motion-opponent operators, *Vision Res., 42*, 3043 - 3058.

Royden, C. S. & Conti, D. (2003). A model using MT-like motion-opponent operators explains an illusory transformation in the optic flow field. *Vision Res., (submitted).*

Royden, C. S. & Hildreth, E. C. (1996). Human heading judgments in the presence of moving objects, *Percept. Psychophys., 58*, 836 - 856.

Royden, C. S. & Hildreth, E. C. (1999). Differential effects of shared attention on perception of heading and 3-D object motion, *Percept. Psychophys., 61*, 120-133.

Royden, C. S., Wolfe, J. & Klempen, N. (2001). Visual Search Asymmetries in Motion and Optic Flow Fields, *Percept. Psychophys., 63*, 436-444.

Royden, C. S., Wolfe, J. M., Konstantinova, E. & Hildreth, E. C. (1996). Search for a moving object by a moving observer. *Invest. Ophthalmol. Vis. Sci. 37 (suppl),* 299 (Abstract).

Saito, H., Yukie, M., Tanaka, K., Hikosaka, K., Fukada, Y. & Iwai, E. (1986). Integration of direction signals of image motion in the superior temporal sulcus of the macaque monkey, *J. Neurosci., 6,* 145 - 157.

Tanaka, K. & Saito, H. (1989). Analysis of motion in the visual field by direction, expansion/contraction, and rotation cells clustered in the dorsal part of the medial superior temporal area of the macaque monkey, *J. Neurophysiol., 62*, 626 - 641.

Thompson, W. T. & Pong, T. C. (1990). Detecting moving objects, *Int. J. Comput. Vis., 4*, 39-57.

van den Berg, A. V. (1992). Robustness of perception of heading from optic flow, *Vision Res., 32*, 1285-1296.

Verghese, P., & Pelli, D. G. (1992). The information capacity of visual attention, *Vision Res., 32*, 983 - 995.

Warren, W. H. & Hannon, D. J. (1988). Direction of self-motion is perceived from optical flow, *Nature, 336*, 162-163.

Warren, W. H. & Hannon, D. J. (1990). Eye movements and optical flow, *J. Opt. Soc. Am. A, 7*, 160-169.

Warren, W. H. & Saunders, J. A.(1995). Perceiving heading in the presence of moving objects, *Perception, 24* , 315 - 331.

Xiao, D. K., Raiguel, S., Marcar, V., Koenderink, J. & Orban, G. A. (1995). Spatial heterogeneity of inhibitory surrounds in the middle temporal visual area, *Proc. Nat. Acad. Sci. USA, 92,* 11303 - 11306.

Zemel, R. S. & Sejnowski, T. J. (1998). A model for encoding multiple object motions and self-motion in area MST of primate visual cortex, *J. Neurosci., 18*, 531 - 547.

9. Linking Perception and Neurophysiology for Motion Pattern Processing: The Computational Power of Inhibitory Connections in Cortex

Scott A. Beardsley[1] and Lucia M. Vaina[1,2]

[1] Brain and Vision Research Laboratory, Department of Biomedical Engineering
Boston University, Boston, MA, USA

[2] Department of Neurology
Harvard Medical School, Boston, MA, USA

1 INTRODUCTION

The motion of the visual scene across the retina, termed optic flow (Gibson, 1950), contains a wealth of information about our dynamic relationship within the environment. Perceptual information regarding heading, time to contact, object motion and object segmentation can all be recovered to various degrees by analyzing the complex motion components of optic flow; for review see (Andersen, 1997, Lappe, et al., 1999). While the usefulness of such information for visually guided actions and navigation is clear, the complex neural mechanisms underlying its processing and extraction remain, for the most part, poorly understood.

As our understanding of the ecological importance of visual motion processing has increased over the last 40 years, so to has research into the perceptual mechanisms and cortical areas where optic flow processing may occur (Andersen, et al., 2000; Bremmer, et al., 2000; Duffy, 2000; Lappe, 2000; van den Berg, 2000); for review see (Vaina, 1998). Studies of the anatomical and physiological pathways in cortex have identified a coarse hierarchy of visual motion processing in which the primary visual cortex (V1) and adjacent visual areas, such as V2 and V3, send afferent projections to motion sensitive cells in the middle temporal (MT) cortex. From MT, cells in turn project to later visual motion areas including the medial superior temporal (MST) cortex and ventral intraparietal (VIP) cortex among others (Boussaoud, et al., 1990; DeYoe & Van Essen, 1988; Felleman & Van Essen,

L.M. Vaina, S.A. Beardsley and S.K. Rushton (eds.), Optic Flow and Beyond, 183–221.

1991; Maunsell & Van Essen, 1983; Van Essen & Maunsell, 1983). Within this coarse hierarchy, neurons exhibit a progressive increase in receptive field size and in their ability to encode more complex forms of visual information that facilitates the ability of visual motion areas to deal more directly with the complexity of the visual scene. Understanding the computational structures that underlie this progressive increase in information complexity and their relationship to our perception of the visual scene has been a primary goal of visual motion research and motivates much of the work presented here.

A current challenge in computational neuroscience is to elucidate the architecture of the cortical circuits for sensory processing and their effective role in mediating behavior. In the visual motion system, biologically constrained computational models are playing an increasingly important role in this endeavor by providing an explanatory substrate linking psychophysical performance and the visual motion properties of single cells.

Early in the visual motion pathway, research examining the neural structures within cortical areas such as V1 and MT has begun to probe the computational and functional role of neural connectivity between and within cortical regions (Carandini & Ringach, 1997; Chey, et al., 1998; Grossberg & Williamson, 2001; Koechlin, et al., 1999; Lund, et al., 1993; Malach, et al., 1997; Stemmler, et al., 1995; Teich & Qian, 2002). Recently, neural models based on the motion properties of these areas have been developed to examine the link between neural structures and perceptual performance on psychophysical tasks (Chey et al., 1998; Grossberg & Williamson, 2001; Koechlin et al., 1999).

In the work presented here, we apply a similar methodology to examine the structure and function of optic flow-based processing of the wide-field motion patterns encountered during self-motion. By combining human perceptual performance with biologically constrained neural models our aim is to elucidate the neural structures and computational mechanisms associated with optic flow processing of the visual scene. Specifically what neural structures within visual cortex are sufficient to encode and process the perceptually relevant motion patterns typically encountered during self-motion through the environment?

1.1 Visual Motion Processing in Cortex

Given the wide variety of visually perceived motions we experience as we move through the world, one might expect that the visual system should contain specialized detectors sensitive to the motion components (e.g. radial, circular, planar, etc.) associated with optic flow. Human psychophysical studies support this form of specialization. Perceptual performance in tasks examining motion pattern detection and discrimination indicate the existence of specialized detectors sensitive to radial, circular, and planar motion patterns across a wide range of psychophysical techniques (Burr, et al., 1998; Freeman & Harris, 1992; Meese & Harris, 2001ab, 2002; Morrone, et al., 1995, 1999; Regan & Beverley, 1978, 1979; Snowden & Milne, 1996, 1997; Te Pas, et al., 1996). Perceptually, these motion pattern mechanisms have been shown to integrate local motions along complex trajectories to obtain global motion percepts over wide visual fields (Burr et al., 1998; Morrone et al., 1999).

1.1.1 Motion Pattern Processing

Neurophysiological studies in non-human primates support the existence of such mechanisms and, together with anatomical studies, indicate a coarse visual motion hierarchy that begins in V1 and extends into posterior parietal cortex (Boussaoud et al., 1990; DeYoe & Van Essen, 1988; Felleman & Van Essen, 1991; Maunsell & Van Essen, 1983; Van Essen & Maunsell, 1983). Within posterior parietal cortex several cortical areas have been identified, including the dorsal division of MST (MSTd), VIP and area 7a, whose neurons exhibit preferred responses to motion pattern stimuli and contain qualitatively similar visual motion properties (Duffy & Wurtz, 1991a,b, 1995, 1997b; Graziano, et al., 1994; Schaafsma & Duysens, 1996; Siegel & Read, 1997; Tanaka, et al., 1989). In light of these similarities and given the relative abundance of motion pattern studies in MSTd we focus our subsequent discussion of the physiology on this region with the understanding that the underlying computational mechanisms and visual motion properties are not necessarily unique to this area.

Single cell studies in MSTd have identified cells that respond over large regions of the visual field (~60°) and exhibit preferred responses to simple motion pattern components of optic flow characterized by coherent radial, circular, and planar motions (Duffy & Wurtz, 1991a,b, 1995, 1997b; Geesaman & Andersen, 1996; Graziano et al., 1994; Lagae, et al., 1994; Orban, et al., 1992; Saito, et al., 1986; Tanaka et al., 1989, 1989). The distribution of preferred motion patterns represented within MSTd spans a continuum in the stimulus space formed by radial, circular, and spiral motions that is biased in favor of expanding motions (Geesaman & Andersen, 1996; Graziano et al., 1994). Moreover, many cells in this area also respond to planar motions, suggesting a more extensive set of preferred motions that includes the four planar directions of motion (up/down, left/right) (Duffy & Wurtz, 1991a,b, 1997b).

Within this multi-dimensional plano-radial-circular space, cells respond across a wide range of stimulus speeds and exhibit speed tuning profiles best characterized by their filtering properties (i.e., low-pass, linear, high-pass), (Duffy & Wurtz, 1997a; Orban, et al., 1995). The preferred pattern responses of these neurons are scale and position invariant to small and moderate variations in the stimulus size, location of the center-of-motion (COM), and the visual cues conveying the motion (Geesaman & Andersen, 1996; Graziano et al., 1994; Tanaka et al., 1989). For larger variations and with tests using non-optimal stimuli, cell responses degrade continuously (Duffy & Wurtz, 1995; Graziano et al., 1994). This sensitivity to the global speed and motion pattern information contained within optic flow has led to speculation that cells in MSTd could be used to encode flow based heading through the visual scene.

Combined neurophysiological and psychophysical studies in the MT/MST complex provide indirect support for such a representation, suggesting a strong link between the patterns of neural activity and motion based perceptual performance. Microstimulation of local neural clusters has been shown to reliably bias perceptual judgments in favor of the stimulated neural cluster's preferred visual motion properties (Britten & van Wezel, 1998; Celebrini & Newsome, 1995; Salzman, et al., 1990, 1992). Analyses of psychophysical performance versus single cell responses support these findings and indicate a correlation between the intensity of a cell's response and the corresponding perceptual judgment (Britten, et al., 1992, 1996; Celebrini & Newsome, 1994). While such studies do not necessarily imply that perception occurs within the affected visual areas *per se*, they do suggest that the motion processing on which perception is based occurs in these areas.

1.1.2 Neural Structures within Cortical Areas

In early visual motion areas, such as the primary visual cortex (V1) and MT, single cell studies have identified a variety of intrinsic neural structures (Gilbert, et al., 1996; Gilbert & Wiesel, 1985, 1989; Gilbert, 1983, 1985, 1992; Grinvald, et al., 1986; Kisvarday, et al., 1997; Lund et al., 1993; Malach et al., 1997; McGuire, et al., 1991; Ts'o, et al., 1986). Extensive modeling of the local horizontal connections inherent in these structures, both in V1 (Ben-Yishai, et al., 1995; Stemmler et al., 1995; Worgotter, et al., 1991) and MT (Koechlin et al., 1999; Liu & Hulle, 1998), suggest the existence of complex interconnected architectures whose most basic connections can impart considerable computational power to simulated neural populations. Combined psychophysical and computational studies support these findings and further suggest that horizontal connections may play a significant role in encoding the visual motion properties associated with various psychophysical tasks (Adini, et al., 1997; Chey et al., 1998; Koechlin et al., 1999; Nowlan & Sejnowski, 1995; Stemmler et al., 1995). As the importance of such local interactions has become clearer, attention has begun to focus on their likely structure and function in higher visual areas as well (Amir, et al., 1993; Edelman, 1996; Miikkulainen & Sirosh, 1996; Sakai & Miyashita, 1991; Taylor & Alavi, 1996; Wiskott & von der Malsburg, 1996).

1.2 Neural Coding of Visual Information

In the cortex, the interpretation of psychophysical and physiological data within a common neural framework is dependent on how the information is represented; for review see (deCharms & Zador, 2000). Although precise temporal codes are capable of transmitting large amounts of information between individual neurons, there is good evidence that cortical neurons represent information using a coarse population code of redundant units (Shadlen & Newsome, 1998, 1994; Softky, 1995; Softky & Koch, 1993).

Indirect support for information transfer via populations of noisy cortical units has come from computational studies across a wide range of processing modalities including motor planning for reaching (Georgopoulos, et al., 1986, 1988; Lukashin & Georgopoulos, 1993, 1994; Lukashin, et al., 1996; Salinas & Abbott, 1994, 1995), orientation discrimination, and motion direction discrimination (Sundareswaran & Vaina, 1996; Vaina, et al., 1995; Zemel, et al., 1998; Zohary, 1992). In each case, the sensory information encoded across neural populations has been shown to be computationally sufficient to extract perceptually useful information from the underlying neural representation. Theoretical studies support these empirical results, demonstrating near optimal encoding of perceptually relevant stimulus properties for a variety of population coding techniques (Seung & Sompolinsky, 1993; Snippe, 1996).

While there is wide spread support for population codes in biological systems, optimal methods for decoding the neural information remain unclear. A variety of decoding techniques have been proposed including Bayesian inference (Foldiak, 1993; Oram, et al., 1998), population vector analysis (Seung & Sompolinsky, 1993), maximum likelihood (Pouget, et al., 1998), center of mass (Snippe, 1996), and probability density estimation (Sanger, 1996; Zemel et al., 1998). However, the computational efficiency and biological plausibility of these methods are often at odds with one another, preventing general acceptance for any one technique.

1.3 Computational Models of Motion Pattern Processing

The use of theoretical models and computer simulations in conjunction with other experimental modalities provides a powerful approach that can be used to probe the types of neural structures that may exist in the brain and their role in mediating perception. Biologically constrained models can, for example, be used to establish whether the hypothesized neural structures are computationally sufficient to account for the experimental results derived from psychophysical and neurophysiological studies. More importantly, the assumptions built into these models (both implicit and explicit) often provide predictions regarding the underlying computational mechanisms that can, in turn, be used to guide experimental research.

The existence of qualitative similarities between the visual motion properties observed in MT and MST and the computational mechanisms believed to underlie visual motion perception has motivated the development of a wide range of biologically constrained models intended to link perception, neural structure, and function within a common computational framework. While some of these models have been developed to quantify the characteristics of motion pattern tuning and their emergent properties (Beardsley & Vaina, 1998; Beardsley, et al., 2003; Pitts, et al., 1997; Wang, 1995, 1996; Zhang, et al., 1993), others have focused on identifying the computational mechanisms sufficient to extract perceptually meaningful motion attributes, particularly those associated with estimates of self-motion (Grossberg, et al., 1999; Hatsopoulos & Warren, 1991; Lappe & Rauschecker, 1993, 1995; Lappe et al., 1996; Perrone & Stone, 1994, 1998; Royden, 1997; Zemel & Sejnowski, 1998). For example, using a winner-take-all template model of self-motion estimation, Perrone and Stone (1994, 1998) obtained heading estimates under gaze-stabilized conditions that were well matched to equivalent measures of human performance. Moreover, they observed visual motion properties that were consistent with cells in MSTd across a wide range of conditions, suggesting that MSTd is computationally sufficient to extract estimates of heading.

Throughout these models, neural structures that parallel the function of the visual motion pathway have been implemented to examine how the visual motion properties of individual cells develop and to quantify the computational mechanisms required to extract perceptually useful information from optic flow. Within this scope, such models have improved our understanding of visual motion processing in cortex. However, they have generally done little to address the functional role of neural connections *within* visual motion areas, such as MST and VIP, typically associated with flow specific heading and navigation tasks.

In the work outlined in this chapter we investigate the computational and functional role of simple horizontal connections *within* a population of MSTd-like units. Based on the known visual motion properties of cells within MSTd we ask what neural structures are computationally sufficient to encode human perceptual performance on a motion pattern discrimination task? Is a population of independently responsive units, indicative of a simple feed-forward pooling of local motion estimates, computationally sufficient to encode the perceptual task? Or are specific neural structures necessary to extract equivalent measures of human perceptual performance?

2 DISCRIMINATING PERTURBATIONS IN WIDE-FIELD MOTION PATTERNS

Much of the research examining motion pattern perception has focused on an observer's ability to detect coherent motions in the presence of a masking background motion. A variety of masking conditions, including both local random motion and conflicting global motion patterns, have been used to quantify the existence of motion pattern mechanisms (Burr et al., 1998; Freeman & Harris, 1992; Morrone et al., 1995). The qualitative similarities between these mechanisms and motion pattern sensitive cells found in MSTd of non-human primates has lead to speculation that similar neural structures may exist in the human visual motion pathway. Such speculation has received support from recent functional imaging studies of motion direction and motion pattern perception that have reported significant activation in an area referred to as MT+, the human homologue of the MT/MST complex (de Jong, et al., 1994; Greenlee, 2000; Heeger, 1999; Morrone, et al., 2000; Rees, et al., 2000; Rutschmann, et al., 2000; Tootell, et al., 1995; Vaina, et al., 2000).

Using stimuli consistent with previous motion pattern experiments (Burr et al., 1998; Freeman & Harris, 1992; Morrone et al., 1995), we developed a unique psychophysical task designed to facilitate a more direct comparison between human perceptual performance and the visual motion properties in cortical areas, such as MSTd, believed to underlie the perceptual task. We hypothesize that if neural substrates similar to those reported in MSTd do play a role in motion pattern processing, then the bias for expanding motions should result in measurable and predictable artifacts in psychophysical performance that can provide additional insight into the underlying neural mechanisms.

If the human homologue of MSTd does play a significant role in motion pattern processing, then we would expect significant variations psychophysical performance to arise as a function of the tested motion pattern and speed in ways that are correlated with the underlying visual motion

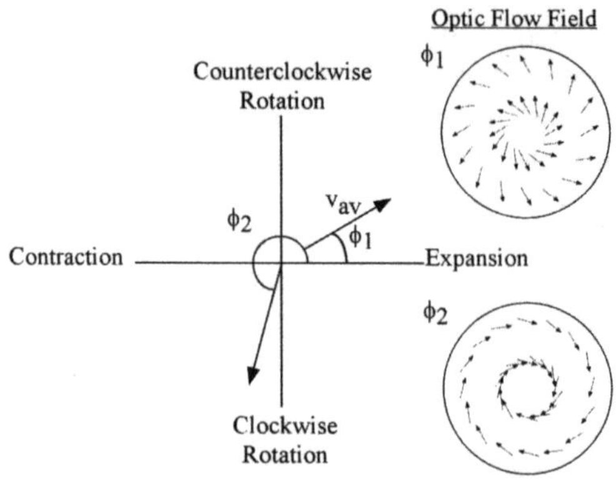

Figure 1. Radial, circular, and spiral motions can be represented as vectors in a 2-D stimulus space. The vector magnitude (v_{av}) corresponds to the average dot speed across the motion field and the flow angle (ϕ) defines the type of motion pattern relative to a 0° baseline expansion. Off-axis regions correspond to intermediate degrees of spiral motion. [From Beardsley & Vaina, 2001; used by permission]

properties reported in MSTd. Specifically is there a preference for radially increasing speed gradients and does there exist a perceptual bias favoring the discrimination of expanding motions?

2.1 Methods

In a temporal two-alternative-forced-choice (2TAFC) graded motion pattern (GMP) task we measured discrimination thresholds to global changes in the patterns of complex motion (Beardsley & Vaina, 2001). Stimuli consisted of dynamic random dot displays presented in a 24° annular region, with the central 4° removed, in which each dot moved coherently from frame-to-frame such that its spatial location (x,y) in the n+1 frame was given by:

$$\begin{pmatrix} x_{n+1} \\ y_{n+1} \end{pmatrix} = e^{\omega\cos\phi}\begin{pmatrix} x_n\cos(\omega\sin\phi) - y_n\sin(\omega\sin\phi) \\ x_n\sin(\omega\sin\phi) - y_n\cos(\omega\sin\phi) \end{pmatrix} \qquad (1)$$

Within this representation, motion patterns could be uniquely described by their 'flow angle', ϕ, and speed, ω, in the 2-D stimulus space formed by cardinal (radial and circular) and spiral motions (Figure 1). Throughout the task the motion pattern remained centered in the visual display.

Prior to the start of the experiment, observers adapted for 10 sec to the background display in a darkened room. During testing observers were required to fixate a small central square and discrimination pairs of stimuli,

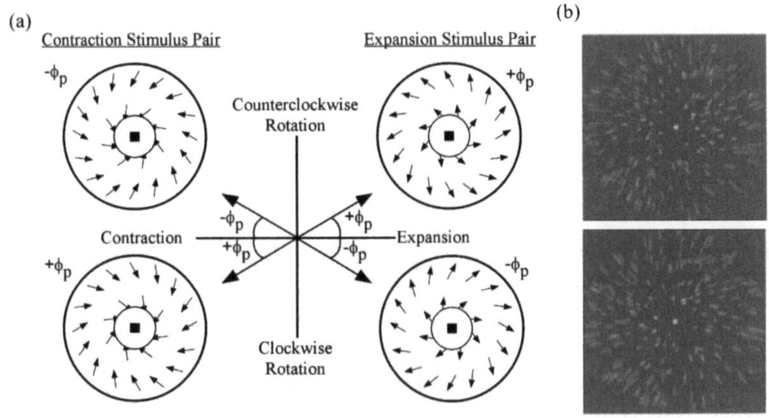

Figure 2. Schematic representation of the graded motion pattern (GMP) task for radial motions. All stimulus apertures were *illusory* as defined by an absence of dots. a) Pairs of stimuli were created by perturbing the flow angle (ϕ) of each 'test' motion by $\pm\phi_p$ in the stimulus space. For the radial 'test' motions shown here, the perturbations are equivalent to the addition of a circular motion component to the 'test' motion (expansion or contraction). b) A scaled time-lapsed version of an expansion stimulus pair. [From Beardsley & Vaina, 2001; used by permission]

formed by perturbing the 'test' motion by $\pm\phi_p$ (Figure 2), were presented following a constant stimulus paradigm (440±40 msec stimulus duration). To minimize the build-up of adaptation to specific motion patterns (expansion, clockwise rotation, etc.), opposing motions (e.g. expansion/contraction) were interleaved across paired presentations.

For each paired stimulus presentation observers were required to select the stimulus, via key press, containing a negative perturbation relative to the 'test' motion (Figure 2). During testing the task of discriminating negative/positive perturbations was facilitated through the presentation of radial, circular and spiral 'test' motions in separately interleaved blocks of trials. Within each block of trials observers were required to discriminate motion pattern perturbations using a conceptual simplification to the negative perturbation judgment; e.g. for radial 'test' motions select the stimulus containing a CCW motion component; for circular 'test' motions select the stimulus containing an expanding motion component, etc.

For each observer, discrimination thresholds were obtained at two average dot speeds (8.4 and 30°/sec) for each of eight 'test' motions (expansion, contraction, clockwise (CW) and counterclockwise (CCW) rotation, and the four intermediate spiral motions). Within each constant stimulus session discrimination thresholds were estimated using a weighted two-parameter Weibull fit to a minimum of four perturbation levels. A χ^2 goodness-of-fit measure with (v) degrees of freedom was used to exclude data sets with poor curve fits from further analysis ($\chi^2(\nu) < \chi_R^2(\nu)$; p<0.1), (Bevington, 1969).

2.2 Results

Discrimination thresholds are reported here from a subset of the observer population consisting of three experienced psychophysical observers; one of which was naïve to the purpose of the psychophysical task (Beardsley & Vaina, 2001). For each condition performance is reported as the mean and standard error averaged across 8-12 threshold estimates.

Across observers and dot speeds perceptual performance followed a distinct trend in the stimulus space with discrimination thresholds for radial motions (expansion/contraction) significantly lower than those for circular motions (CW/CCW rotation), (Figure 3). Within these sub-pairs, thresholds for radial and circular motions were well matched across observers. Figure 3 also shows that while the individual thresholds for the intermediate spiral motions were not significantly different from those for circular motions, the trends across 'test' motions were well fit by sinusoids whose period and phase were approximately $196 \pm 10^{\circ}$ and $-72 \pm 20^{\circ}$ respectively.

Comparable trends in performance were obtained for stimuli containing 50% flicker noise (data not shown). In this condition, dot motions were randomly assigned as 'signal' or 'noise' from frame-to-frame such that the proportion of signal dots in each frame was 50%. This resulted in a decreasing probability of uninterrupted dot motion that was structurally similar to the 2-frame motion used previously to quantify the existence of specialized motion pattern mechanisms (Burr et al., 1998; Morrone et al., 1995).

The consistency of the cyclic threshold profile in stimuli that restrict the temporal integration of individual dot motions, and simultaneously contain all directions of motion, generally argues against a primary role for local motion mechanisms in the perceptual task. While there exist a wide variety of "local" motion direction anisotropies within the psychophysical literature whose properties are reminiscent of the threshold variations observed here, (Ball & Sekuler, 1987; Coletta, et al., 1993; Edwards & Badcock, 1993; Gros, et al., 1998; Matthews & Qian, 1999; Matthews & Welch, 1997; Raymond, 1994; Zhao, et al., 1995), all predict equivalent thresholds for radial and circular motions for a set of uniformly distributed and/or spatially restricted motion direction mechanisms. Together with the need for a significant spatial integration across dispirit directions to offset the reduced temporal integration in the 50% flicker noise stimulus, this suggests the presence of a global mechanism that spatially integrates across local motions to encode more complex representations of visual motion.

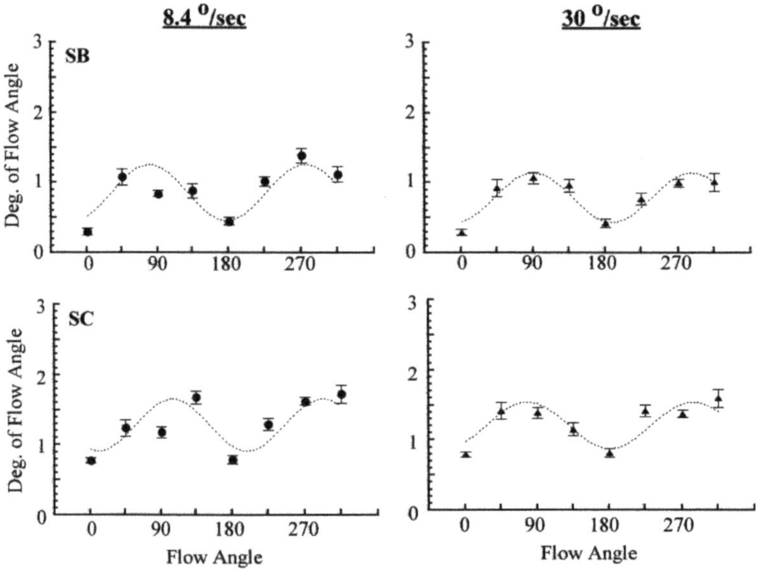

Figure 3. GMP thresholds across eight 'test' motions at two mean dot speeds (8.4 and 30°/s) for two observers (SB and SC). Performance varied continuously as a function of the 'test' motion with thresholds for radial motions (ϕ = 0,180) significantly lower than those for circular motions (ϕ = 90,270), ($p<0.001$; $t(37) = 3.39$). Thresholds for spiral motions were not significantly different from circular motions ($p=0.223$, $t(60) = 0.75$), however, the trends across 'test' motions were well fit (SB: $r>0.82$, SC: $r>0.77$) by sinusoids whose period and phase were $196 \pm 10°$ and $-72 \pm 20°$ respectively.

Additional support for the role of a global motion mechanism in the perceptual task comes from a control experiment designed to assess the computational impact of the structured speed information in the stimulus. Three observers were tested using a modified set of radial and circular motions in which the radial speed gradient was removed. This condition, referred to as random speed, resulted in stimuli whose spatial distribution of dot speeds was randomized while preserving the local and global trajectory information.

In the random speed control, threshold performance increased significantly across observers, particularly for circular motions (Figure 4). Such performance suggests a measurable perceptual contribution associated with the presence of the speed gradient and is particularly interesting given the spatial symmetry implicit in the stimulus design.

Since the speed gradient was a function of the spatial distance from the stimulus center, which was itself fixed in the center of the visual display, its distribution for each 'test' motion was radially symmetric. Within this limited subset of conditions the speed gradient did not contribute computationally

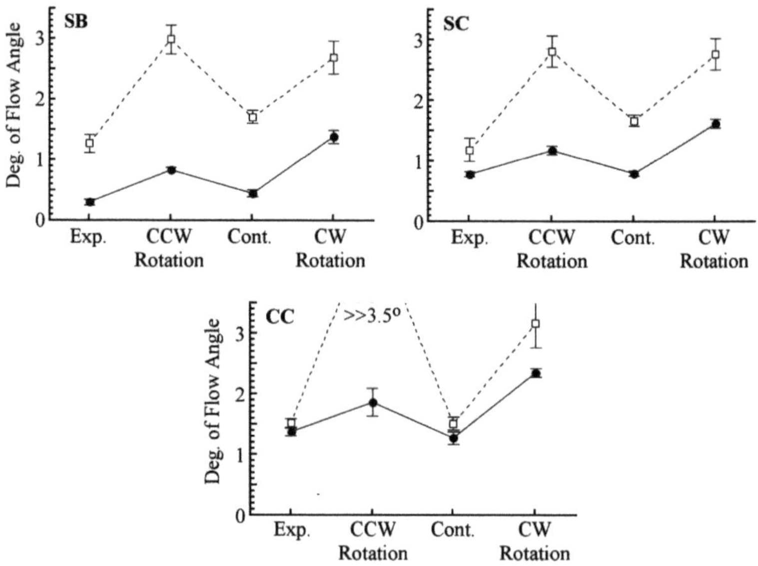

Figure 4. Radial and Circular GMP thresholds for stimuli with (filled circles) and without (open squares) a radial speed gradient (average dot speed = 8.4 deg/s). Across observers (SB, SC, and CC) thresholds for radial motions remained significantly lower than those for circular motions in the random speed condition (i.e., no speed gradient). Relative to stimuli containing speed gradients, random speed thresholds increased significantly across observers ($p<0.05$; $t(17)=1.91$), particularly for circular motions ($p<0.005$; $t(25)=3.31$).

relevant structural information to the task. However, the speed gradient did convey information regarding the integrative structure of the global motion field and as such suggests a preference in the underlying motion mechanisms for spatially structured speed information. Together with similar perceptual studies reporting the existence of wide-field motion pattern mechanisms (e.g. Burr et al., 1998; Meese & Harris 2001, 2002), these results suggest that the threshold differences observed here may be associated with variations in the computational properties across a series of specialized motion pattern mechanisms.

3 A COMPUTATIONAL MODEL OF MOTION PATTERN PROCESSING

3.1 GMP Discrimination via a Population Code in MSTd

The similarities between the visual motion patterns used to quantify human perception and the tuning properties of cells in visual motion areas such as MSTd and VIP suggest that such cortical regions may mediate the perceptual task. If, for example, motion pattern perception were based on a simple population code of complex motion detectors drawn from the anisotropic cell distribution reported in MSTd (Geesaman & Andersen, 1996; Graziano et al., 1994), one might expect that GMP discrimination should follow a distribution symmetric about the radial motion axis with thresholds for expansions lowest and thresholds for contractions among the highest. Within this framework, deviations in human performance from that of a simple population code could suggest the presence of horizontal connections within the population that serve to modify and refine the underlying neural representation.

To examine this hypothesis, we constructed a population of MSTd-like units whose visual motion properties were consistent with those reported in the neurophysiology (Duffy & Wurtz, 1991a,b; Geesaman & Andersen, 1996; Graziano et al., 1994; Orban et al., 1995), (Figure 5; see Beardsley & Vaina, 2001 for details). Within the model, simulations were categorized according to the underlying distribution of preferred motions represented across the population (2 reported in MSTd - Figure 5b, and a uniform control). The first distribution simulated an expansion bias in which the density of preferred motions decreased symmetrically from expansions to contractions across the motion pattern space (Graziano et al., 1994). The second distribution was similar in structure but contained a higher percentage of cells tuned to contracting motions (Geesaman & Andersen, 1996). The third distribution simulated a uniform preference for all motions and was used as a control to examine the effects of a non-homogeneous distribution on perceptual performance. Throughout the remainder of the chapter we refer to simulations containing these distributions as Unimodal, Bimodal, and Uniform respectively.

In MSTd receptive fields are large, spanning up to one-half of the visual field (Duffy & Wurtz, 1991a,b; Tanaka & Saito, 1989). For neurons preferring similar motions, the resultant overlap in visual input can introduce correlation in the neural output that varies with the relative spatial positions of the cells' receptive fields. The structure associated with such correlations implies a redundant representation of neural information that can be used to aid information extraction across populations of noisy units. To simulate this effect in the model we assumed that units had comparable receptive fields such that the responses of units with similar preferred motions were moderately correlated ($r = [0.4,0.8]$). Since the model did not explicitly contain a feed-forward layer of spatially localized inputs, we correlated neural responses by imposing a maximum preferred motion response, R_{max}, that varied as a Poisson process ($\lambda=28$) for each stimulus presentation (Figure 5a).

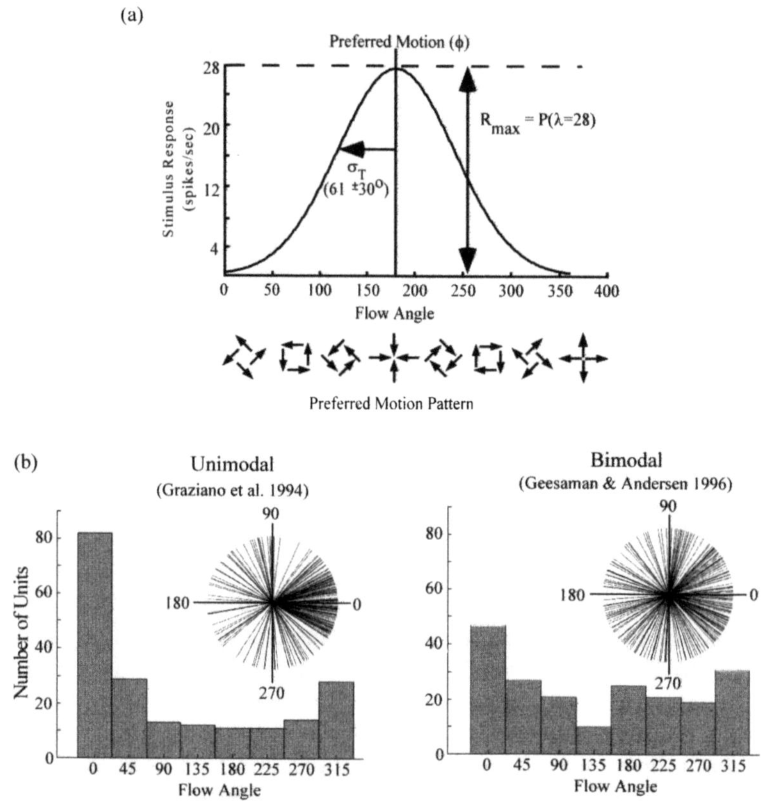

Figure 5. Motion pattern properties of the MSTd-like units used in the model. a) Motion pattern tuning followed a gaussian profile ($\sigma_T = 61^o \pm 30^o$) in the stimulus space centered on each unit's preferred motion. The preferred motion responses, R_{max}, varied as a Poisson process based on the reported range of firing rates in MSTd. b) The preferred motion for each unit was selected randomly from a continuous distribution consistent with one of two distributions reported in MSTd. [From Beardsley & Vaina, 2001; used by permission]

The baseline neural output of each unit (N_i) was simulated as an uncorrelated Poisson process (λ=12).

3.1.1 Extracting Perceptual Estimates from the Neural Code

The psychophysical task was simulated in the model using a "population vector" to represent the net motion pattern estimated in the 2-D stimulus space (Figure 1). A structural diagram of the model is shown in Figure 6. For each stimulus, the response of the ith unit was represented as the average firing rate, R_{av}, estimated from the unit's motion pattern tuning profile (Figure 5a),

$$R_{av_i} = R_{max}\, e^{-(\phi - \phi_i)^2 / 2\sigma_i^2} \qquad (2)$$

where R_{max} is the maximum preferred stimulus response (spikes/s), ϕ is the stimulus flow angle and ϕ_i and σ_i correspond to the preferred motion and standard deviation of the ith unit's tuning profile respectively. The net response, R_i, of each unit was formed by combining R_{av_i} with the baseline neural output N_i.

For each stimulus, units 'voted' for their preferred motions with a weight equal to their net firing rate. Across the population weighted responses were represented as vectors in the units' preferred motion directions and estimates of the stimulus flow angle were decoded as the "population vector" formed from the vector sum of responses across all units.

Using the population vector estimates obtained for each set of paired stimuli, the model's performance was quantified according to the negative perturbation ($-\phi_p$) discrimination of flow angle outlined in the psychophysical task. As with the perceptual task, a least-squares fit to percent correct performance across constant stimulus levels was used to estimate discrimination thresholds.

3.2 Simulation 1: An Independent Neural Code

In the first series of simulations, we quantified GMP performance across populations of independently responding units. For each class of models (Unimodal, Bimodal, and Uniform), threshold performance was examined across five populations (100, 200, 500, 1000, and 2000 units). Both the range in thresholds and their trends across 'test' motions were compared across simulations to identify combinations of population size and preferred motion distributions that yielded quantitatively similar performance relative to the psychophysical task.

It is important to note that within this implementation the population vector does *not* provide an unbiased estimator of the stimulus flow angle (ϕ). The bias in the distribution of preferred motions (Figure 5b) introduces a corresponding bias into the vector representation that is skewed toward expanding motions. As a result, increasing populations do not yield vector representations that converge to the proper 2-D motion pattern. This makes accurate interpretation of individually decoded stimuli problematic without

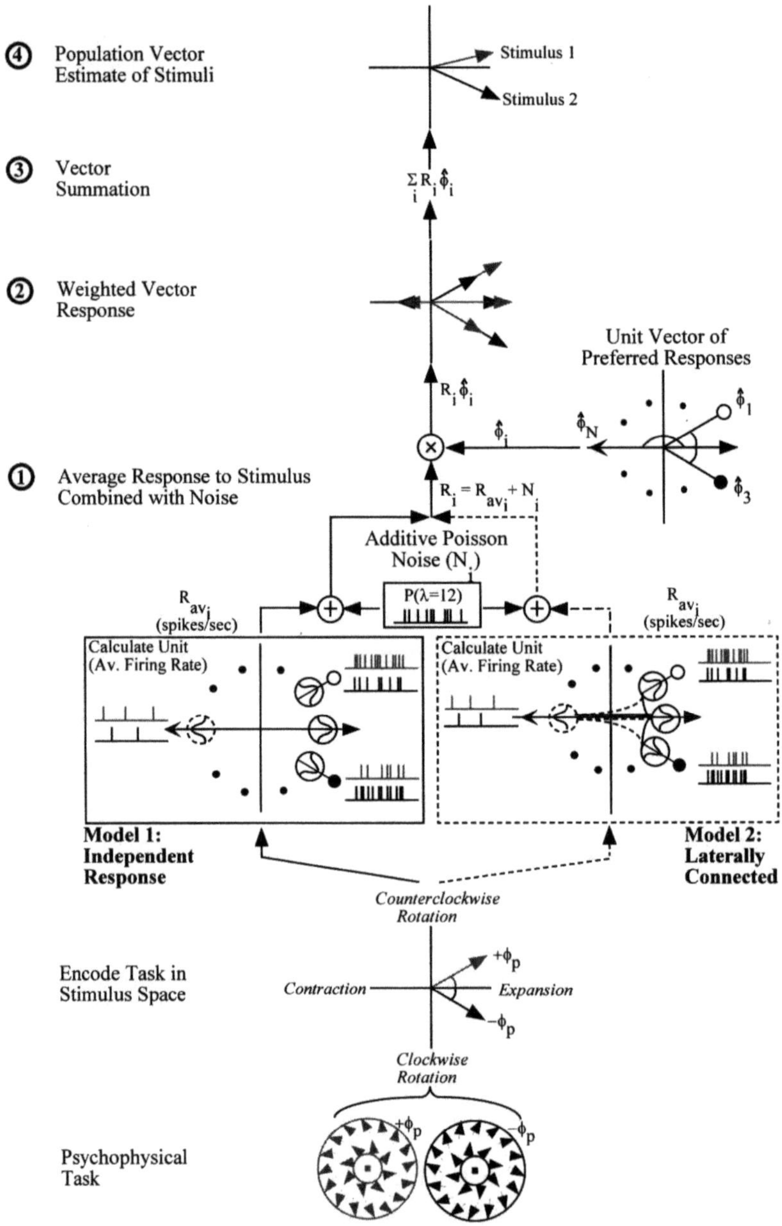

Figure 6. Schematic representation of the model structure. The psychophysical task was simulated using a vector representation of the stimuli in the 2-D motion pattern space. For each stimulus, units 'voted' (2) for their preferred motion with a weight equal to their net firing rate (1). Stimuli were decoded from the neural representation as the population vector formed by the vector sum of weighted responses across units (3). As in the psychophysical task, "perceptual" judgments were based on the negative perturbation ($-\phi_p$) discrimination of flow angle (4). [From Beardsley & Vaina, 2001; used by permission]

a prior knowledge of the nonlinear transformation associated with the bias in the vector representation.

To compensate for this, the model's performance was based on the relative spatial locations *between* the pairs of motion patterns ($\pm\phi_p$) presented in the 2TAFC task and not the absolute location of each decoded stimulus in the stimulus space. In this way, the population vector could assume any smooth nonlinear mapping without imposing *a prior* assumptions regarding the underlying transformation and thus was not constrained to linearly map the stimulus space onto the internal neural representation.

3.2.1 Simulation 1: Results

Performance varied considerably as a function of the population size and distribution of preferred motions (Figure 7). Over the psychophysical range of interest ($\phi \pm 7^{\circ}$), discrimination thresholds for contracting motions were at chance across all Unimodal populations (100-2000 units) while thresholds for circular motions remained significantly higher than those reported in the perceptual task (Figure 7a). Across the eight 'test' motions, discrimination thresholds showed no significant improvement with increasing population size.

Discrimination thresholds for Bimodal populations were more consistent with human performance, particularly as population size increased. Contraction ($\phi = 180^{\circ}$) thresholds for larger populations steadily decreased to those obtained for expansions ($\phi = 0^{\circ}$), (Figure 7b), however, even with the largest populations discrimination for contracting spiral motions remained poor.

Unlike the Unimodal populations, Bimodal performance also improved asymptotically with population size. This was particularly true for contractions ($\phi = 180^{\circ}$), which decreased by a factor of six as the size of the population was increased. While this decrease made the overall range of thresholds more comparable with human performance, Bimodal populations remained unable to accurately reproduce psychophysical thresholds in the asymptotic limit observed through 2000 units.

For simulations containing a Uniform distribution of preferred motions, the range of discrimination thresholds was consistent with human performance, however, the trend in thresholds was generally flat. What variability did occur, for small populations in particular, was due primarily to the discrete sampling of preferred motions from the uniform distribution. In subsequent simulations (not shown here), more extensive averaging reduced the observed variability, resulting in threshold profiles that were well fit by a line whose slope was near zero.

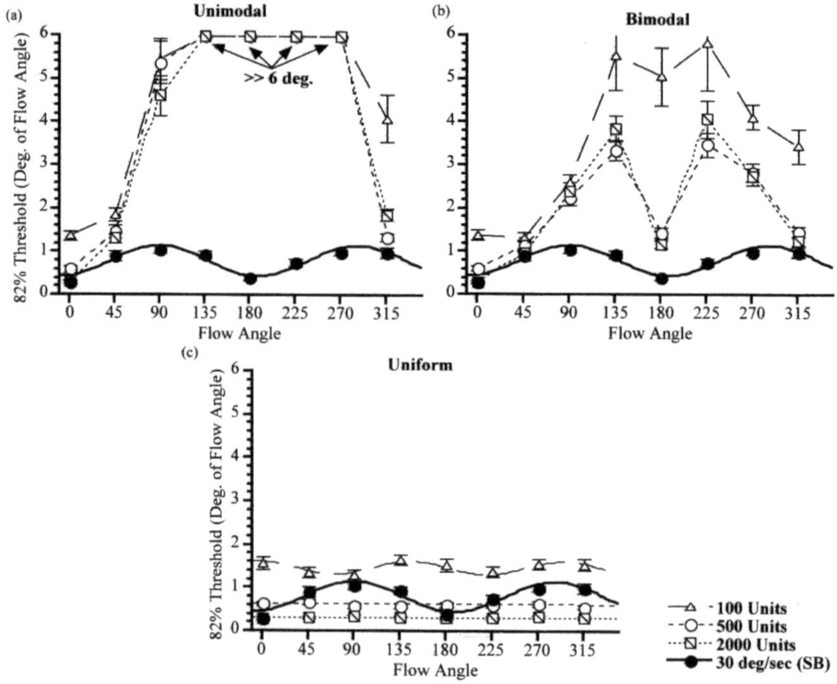

Figure 7. Model vs. psychophysical performance on the GMP task for three simulated populations of independently responding units. GMP thresholds (±1 standard error) were averaged across five independently generated populations for each population size. a) Across all populations, Unimodal thresholds were highest for contracting motions and lowest for expanding motions. b) Bimodal thresholds were more consistent with human performance (SB) particularly for moderately large populations (>500 units). Although discrimination thresholds for contracting motions ($\phi = 180^{o}$) steadily decreased as the number of units increased, Bimodal populations remained unable to accurately reproduce psychophysical thresholds in the asymptotic limit observed through 2000 units. c) Across Uniform populations the range of discrimination thresholds was well-matched with human performance, however, the flat trend across motion patterns was inconsistent with the sinusoidal trend observed in the psychophysical task. [From Beardsley & Vaina, 2001; used by permission]

Examination of the pooled information within the Unimodal and Bimodal models suggests that the threshold differences resulted from the combination of background noise and the population bias for expanding motions. For non-expanding sub-populations, such as those preferring contractions, the proportion of units responsive to a non-expanding stimulus was typically a small percentage of the total population. In these cases, the population vector contained a significant amount of noise associated with the non-preferred background response. As population size increased, the non-uniform effect of the expansion bias on the signal-to-noise ratio across the population caused discrimination thresholds to closely track the degree of preferred motion bias within each model class.

3.3 Simulation 2: An Interconnected Neural Structure

In a second series of simulations, we examined the computational effect of adding lateral (horizontal) connections between neural units. If the distribution of preferred motions in MSTd is in fact biased towards expansions as the neurophysiology suggests, it seems unlikely that independent estimates of the visual motion information would be sufficient to yield the perceptual profiles obtained in the psychophysical task. Instead we hypothesize that a simple fixed architecture of excitatory and/or inhibitory connections between neural units may be necessary to yield the observed psychophysical thresholds.

In this series of simulations, units were fully interconnected with fixed excitatory and/or inhibitory connections. Within this architecture, each neural response consisted of the summed output from three nonlinear sources corresponding to (a) the visual motion stimulus, (b) internal neural noise (not associated with the visual stimulus), and (c) a modulating input from other units within the population.

Across the population, connection strength varied as a function of the similarity in preferred motions between neural units. Using the flow angle representation for preferred motion patterns, excitatory connections from the ith unit (with preferred motion ϕ_i) were made to units with similar preferred motions. The strength of excitation followed a Gaussian profile centered at ϕ_i with a fixed standard deviation ($\sigma_E = 30^o$) across the population. Similarly, units with anti-preferred motions received inhibitory connections whose strength followed a Gaussian profile centered at $180+\phi_i$. To examine the effects of inhibitory spread and connection strength in the model, the standard deviation of the inhibition profile, σ_I, and the level of excitatory/inhibitory activity, S_A, were considered free parameters.

3.3.1 Simulation 2: Results

Within the parameter space used to define connectivity between units (i.e., S_A and σ_I), both the overall level of the GMP thresholds and their trends across 'test' motions varied widely. Monte Carlo simulations across the parameter space yielded regions of high correlation (with respect to the psychophysical thresholds) that were consistent across independently

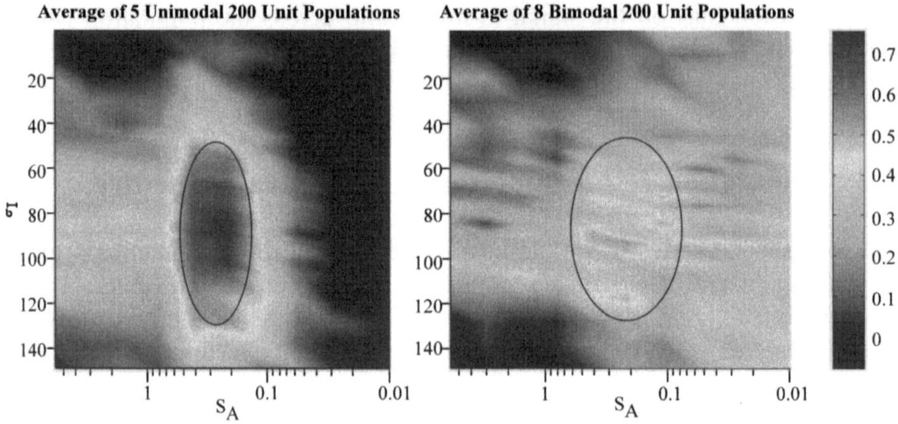

Figure 8. Correlation maps of model vs. psychophysical performance as a function of the strength of the horizontal connections, S_A, and the spread of inhibition, σ_I. For Unimodal/Bimodal populations the regions of interest (ROIs) denoted by the ellipses, (≥80% of maximum correlation), corresponded to areas within the parameter space where model thresholds were well matched to human perceptual performance. These regions were generally robust extending over a broad range of σ_I [60,120°] and S_A [0.15,0.5]. [From Beardsley & Vaina, 2001; used by permission]

simulated populations for both the Unimodal and Bimodal models. Typically these regions were well defined over a wide but limited range of σ_I and S_A. Figure 8 illustrates this result for averages of five Unimodal and eight Bimodal populations containing 200 units. In both cases the regions-of-interest (ROIs), corresponding to the areas of best correlation (80% of maximum), were generally robust, extending over a broad range of σ_I [60,120°] and S_A [0.15,0.5].

While the maximum level of correlation within these regions typically accounted for only a fraction of the psychophysically observed variance across motion patterns (25-50%), simulations containing 100 units indicated an increase in correlation as a function of population size. Together these results suggested that larger populations would yield regions of increasingly higher correlation, however, the computational cost precluded an exhaustive search of the parameter space for populations of more than 200 units. To offset this limitation, we confined simulations of larger populations to a set of 'optimal' parameters estimated from each ROI, (Figure 8).

Using a fixed σ_E (=30°), ROI center-of-mass estimates of [σ_I = 106°, S_A = 1.77] and [σ_I = 76°, S_A = 1.93] for 100-unit populations and of [σ_I = 88°, S_A = 0.37] and [σ_I = 91°, S_A = 0.36] for 200-unit populations were obtained for Unimodal and Bimodal models respectively. Given the lack of significant correlation across Uniform populations and the similarities in center-of-mass

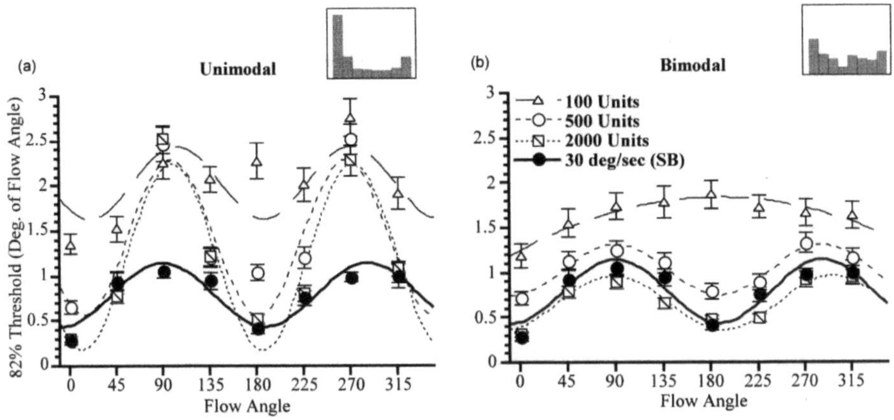

Figure 9. Model vs. psychophysical performance on the GMP task for populations containing excitatory and inhibitory horizontal connections. As the number of units increased, discrimination thresholds (averaged across five simulations) decreased to psychophysical levels and the sinusoidal trend in thresholds began to emerge for both the a) Unimodal and b) Bimodal models. Sinusoidal trends were established for as few as 500 units and were well fit (r>0.9) by sinusoids whose periods and phases were (173.7 ± 6.9°, -67.21 ± 16.78°) and (208.3 ± 10.2°, -60.68 ± 15.74°) for the Unimodal and Bimodal models respectively.

estimates for the Unimodal/Bimodal populations, we set σ_I and S_A to 80° and 1.5 respectively for all 100-unit populations. For larger populations the connection strength, S_A, was scaled inversely with the number of units to maintain an equivalent level of horizontal activity and facilitate a comparative analysis across populations.

Figure 9 shows the averaged threshold performance for three population sizes. The incorporation of horizontal connections between units decreased the range of discrimination thresholds to psychophysically observed levels. More importantly, as population size increased the sinusoidal trend in threshold performance emerged for both the Unimodal and Bimodal models. The threshold trends were well established for as few as 500 units and were well fit (r>0.9) by sinusoids whose periods and phases were consistent with the fitted psychophysical trends (196 ± 10° and -72 ± 20°).

Unlike the Unimodal/Bimodal models, Uniform populations were not significantly affected by the presence of horizontal connections. The range of thresholds were typically well matched to those from Simulation 1 and asymptotically decreased to levels well below those of human observers for large populations (i.e., 2000 units).

Throughout the Unimodal and Bimodal simulations presented here, threshold performance decreased asymptotically with increasing population size for all 'test' motions. The decrease followed an inverse power law whose exponent varied as a function of the 'test' motion across an order of magnitude

[0.05,0.5]. For all 'test' motions the threshold slope decreased quickly with population size such that for 2000 units the change in thresholds was 0.008° for each 100-unit increase in the population. Together with the computational robustness to large variations in the underlying population properties (i.e. size and the distribution of preferred motions) the model's asymptotic performance makes this form of highly interconnected architecture appealing as a method for encoding visual motion information consistently across a neural population.

3.4 Simulation 3: Thresholding Neural Responses

In a third series of simulations we examined the effect of imposing a threshold on the unit responses contributing to the population vector. Here, sub-threshold responses were rectified to zero using a Heaviside function, $H(T_R)$, prior to weighting the unit's preferred motion vector

$$\begin{aligned} H(T_R) &= 0, \qquad R_i < T_R \\ &= 1, \qquad R_i \geq T_R \end{aligned} \tag{3}$$

where T_R is the response threshold. GMP performance was examined over a range of rectifying thresholds (T_R=[1,61] spikes/s) for each class of models. Like the previous simulations, the model's performance was correlated with the psychophysical thresholds to estimate the range of T_R over which the population code was consistent with human performance. Threshold performance was then examined as a function of population size for each of the three model classes (Unimodal, Bimodal, and Uniform) using an 'optimal' T_R selected from the composite regions of highest correlation.

3.4.1 Simulation 3: Results

While the addition of horizontal connections resulted in discrimination thresholds that were in good agreement with human performance, it does not preclude other methods of noise reduction within the neural structures used to encode the visual motion information. When population vector estimates were based on a subset of thresholded (T_R) responses within a population of unconnected units, the resulting thresholds were similar to those obtained with populations of interconnected units.

With a T_R of 35 spikes/s imposed prior to calculating the population vector, the range of discrimination thresholds decreased to psychophysically observed levels for populations containing 500 units. As with the interconnected populations, the sinusoidal trend in threshold performance emerged as the number of units increased. Here the cyclic trends were established for 1000 units and were well fit (r>0.85) by sinusoids whose periods and phases were (218.5 ± 15.7°, -37.98 ± 21.35°) and (216.9 ± 21.9°, -42.17 ± 30.34°) for Unimodal and Bimodal models respectively.

Closer examination of the response characteristics within these "thresholded" populations indicates that the proportion of units responding consistently above threshold (>80% of the time) varied with the 'test' motion, from a low of 1.4% for contractions to a high of 11% for expansions (Unimodal populations). The subsequent exclusion of minimally responsive units reduced the variability of the 'predicted' motion pattern in a manner consistent with the presence of inhibitory horizontal connections between anti-preferred motion pattern units. As a result, for Unimodal and Bimodal models in particular, the proportion of units contributing to the population vector varied as a function of the 'test' motion and typically comprised only a small fraction of the total population. Such performance is generally consistent with a winner-take-all strategy of neural coding and the pre-synaptic inhibitory architecture it implies (Haykin, 1999; Hertz, et al., 1991).

3.5 Simulation 4: The Role of Inhibitory Connections

The development of a sinusoidal trend in GMP performance for large (>1000 units) "thresholded" populations, together with the inhibitory winner-take-all structures implied by their sparse coding, suggest that similar performance could be obtained from populations containing *only* inhibitory connections. Although early simulations using small populations to search for regions of significant correlation across the (σ_I, S_A) parameter space were inconclusive, the inability of small "thresholded" populations in Simulation 3 to reproduce the psychophysical trend suggests that larger populations might be required. Using the 'optimal' structural parameters obtained previously from Simulation 2, we examined the model's GMP performance for mutually inhibiting populations of up to 2000 units, (Figure 10).

With the inclusion of anti-preferred inhibition between units, discrimination thresholds for large populations asymptotically approached those for the interconnected excitatory/inhibitory populations (i.e. Simulation 2). As population size increased, the sinusoidal trend in thresholds emerged for both the Unimodal and Bimodal models. Typically the trend was established for as few as 1000 units and was well fit (r>0.9) by sinusoids whose periods and phases were consistent with the fitted psychophysical trends (196 ± 10° and -72 ± 20°). These results are in good agreement with both the fitted psychophysical trends and the performance of the

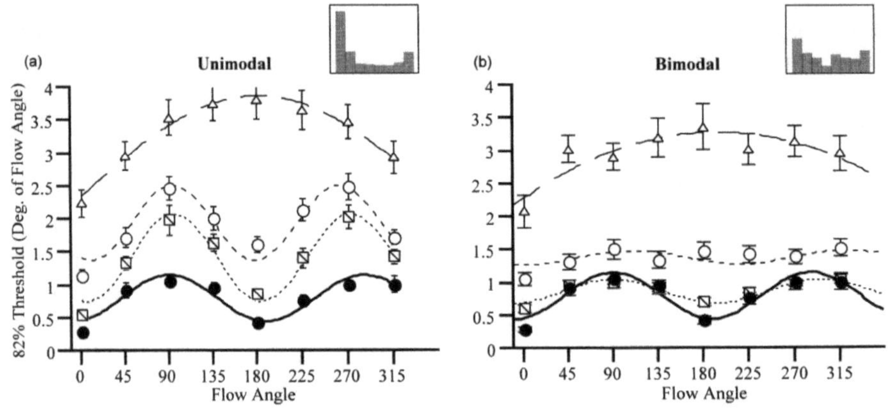

Figure 10. Model vs. psychophysical performance on the GMP task for populations containing only inhibitory connections. As population size increased the sinusoidal trend in thresholds emerged for both the a) Unimodal and b) Bimodal models. The trends were established for as few as 1000 units and were well fit (r>0.9) by sinusoids whose periods and phases were (178.0 ± 6.0°, -79.48 ± 13.93°) and (205.8 ± 16.2°, -65.77 ± 25.88°) for Unimodal and Bimodal models respectively.

interconnected excitatory/inhibitory models (Simulation 2) for moderate population sizes (~500 units).

As with the excitatory/inhibitory architecture simulated previously, discrimination thresholds decreased asymptotically with increases in population size. Subsequent simulations for even larger populations (~5000 units) showed little change in the threshold amplitudes or their trends across 'test' motions. Together with the robust encoding of motion pattern information across σ_E (see Beardsley & Vaina, 2001 for details), these results suggest that perceptual performance in the GMP task may be mediated primarily through signal enhancement associated with the strength and spread of inhibitory connections within a population of motion detectors selective for the simple motion pattern components of optic flow.

4 DISCUSSION

In the graded motion pattern (GMP) task presented here (see also Beardsley & Vaina, 2001) we have shown that human discrimination varies consistently with the pattern of complex motion over a range of thresholds (ϕ = 0.5-2°). For each observer, the trend across psychophysical thresholds consistently demonstrated a preference for radial motions that is qualitatively similar to the bias for expanding motions reported in MSTd. Together with the preferences for global motion patterns reported in MSTd, these results suggest that much of the perceptual information, if not the underlying neural circuitry, involved in the discrimination of motion patterns may be represented in the human homologue to MST.

Using biologically plausible neural units we have shown that the motion information encoded across a population of MSTd neurons is sufficient to obtain discrimination thresholds consistent with human performance. Theoretically, the computations required to perform the perceptual task can be obtained by comparing the nonzero responses between pairs of motion pattern cells that adequately span the 2-D stimulus space. However, in practice neural populations are typically required to extract the desired visual motion information given the presence of neural noise not associated with the stimulus and the lack of *a prior* knowledge concerning the variability in stimulus tuning profiles across cells.

In the neural populations simulated here, the 2-D stimulus representation facilitated the choice of a "population vector" decoding strategy to extract perceptually meaningful motion pattern information. With this implementation, the bias for expanding motions reported in MSTd introduced a computational bias in the vector estimate of the flow angle (ϕ) that was skewed toward expanding motions. Under these conditions, increases in population size did not yield vector representations that converged to the proper 2-D motion pattern, making an accurate interpretation of individually decoded stimuli problematic without *a prior* knowledge of the nonlinear transformation associated with the bias.

To compensate for this, the model's performance was based on the *relative* locations between the pairs of stimuli presented in the psychophysical task and not the proper 2-D location of each decoded motion in the stimulus space. In this way, the population vector could assume any smooth nonlinear mapping without imposing *a prior* assumptions regarding the underlying transformation, and thus was not constrained to linearly map the stimulus space onto the internal neural representation.

While sufficient for the psychophysical task outlined here, this limitation in the decoding strategy restricts the model's general application to perceptual tasks consisting of single stimulus presentations. This deficiency could be overcome by using computationally more robust decoding schemes such as Bayesian inference (Oram et al., 1998) or probability density estimation (Sanger, 1996; Zemel et al., 1998) that do not require an explicit representation of the stimulus space. Within such decoding schemes the addition of inhibitory connections between units, while mathematically more complex, would seem likely to affect GMP discrimination in comparable ways by minimizing the stimulus uncertainty associated with overlapping probability distributions within a multi-dimensional representation.

Similar but less extreme conditions may also occur in models of free arm reaching (Georgopoulos et al., 1986, 1988; Lukashin & Georgopoulos, 1993, 1994; Lukashin et al., 1996) for which the monkey physiology suggests the presence of statistically significant deviations from uniformity in both motor cortex (Schwartz, et al., 1988) and parietal area 5 (Kalaska, et al., 1983). Within each of these models there has been an implicit assumption of uniformity in the neural representation that, for large populations, predicts population vector estimates will converge to the proper reaching vector representation. However if this uniformity assumption is not met, as in the model we have presented, the theoretical convergence breaks down in proportion to the degree of non-uniformity and the locations of stimulus vectors relative to the bias. Depending on the actual degree of anisotropy within cortical motor areas, asymptotic reaching errors, such as those reported by Georgopoulos et al. (1988), might in fact be artifacts associated with an underlying non-uniformity in the 3-D reaching representation.

4.1 Motion Pattern Discrimination in a Population of Independently Responding Units

Using a population of MSTd-like neural units we have demonstrated that the population vector information decoded from independently responding units *is not* computationally sufficient to extract perceptual estimates that are consistent with human performance. For anisotropic neural populations containing a strong bias for expanding motions (Unimodal and Bimodal models), large amounts of noise are introduced into the population vector estimate for non-expanding stimuli. In the extreme case of contracting motions, only 20% of the population typically responded to the stimulus. For those units that did respond significantly, the encoded signal-to-noise ratio was typically offset by a significantly larger noise contribution encoded across the sub-population of non-responsive units. This introduced considerable variability in the motion pattern estimated from the population, causing discrimination thresholds to increase.

In contrast the control simulations, containing a uniform distribution of preferred motions across the population, were able to approximate the psychophysical range of discrimination thresholds. However, they did not develop the sinusoidal trend in thresholds observed across 'test' motions. If, as we have hypothesized, the trend in threshold performance were due to a combination of neural properties (such as the distribution of preferred motion patterns) inherent to the cells underlying the perceptual task, then we would not have expected the Uniform models to deviate significantly from the flat trend obtained across all conditions.

Taken together these results suggest that a population code of independent neural responses is capable of extracting perceptual information relevant to the motion pattern task. However, the model's strong dependence on population size and the distribution of preferred motions prevent this scheme from producing the robust trends in psychophysical performance reported across observers.

4.2 The Computational Role of Horizontal Connections

The addition of horizontal connections that minimized neural noise within the population produced discrimination thresholds that were well matched with human psychophysical performance. Neural structures containing both excitatory and inhibitory connections were able to accurately simulate psychophysical thresholds for populations spanning a 20-fold increase in size (100-2000 units). More interestingly, equivalent levels of performance were also obtained using larger neural populations that contained *only* inhibitory connections.

Together with the model's robust performance as a function of the spread of excitatory connections (σ_E), these results suggest that the sinusoidal trend in perceptual thresholds may be mediated primarily through signal enhancement associated with the strength (S_A) and spread (σ_I) of inhibitory connections within an anisotropic distribution of preferred motion patterns. The ability of preferred stimulus units to inhibit non-preferred units greatly reduced the level of noise across the population, thereby increasing discrimination. This helped offset the strong bias for expanding motions reported in the physiology and produced the cyclic trend in which radial discrimination thresholds were lowest.

Complimentary support for such inhibitory structures can be found in the performance of neural populations whose responses were thresholded prior to decoding the motion pattern estimate (Simulation 3). As with the various interconnected architectures examined here, such populations were able to accurately approximate human performance using as few as 500 units.

Although this strategy for reducing noise is computationally appealing, the simulations suggest that to be effective it would require a relatively high threshold ($T_R \approx$ 35 spikes/s). At this level, much of the visual motion information encoded across the population would be sacrificed to extract the perceptually relevant information. Such an effect cannot be easily reconciled with the correlations reported between perceptual performance and single cell responses in visual motion areas such as MT and MST (Britten et al., 1992; Celebrini & Newsome, 1994). Together with the necessity of *a priori* knowledge regarding what constitutes noise vs. signal implicit in this structure, the use of such a mechanism in the perceptual task presented here seems unlikely.

However, the similarities between the sparse neural representation and a winner-take-all strategy for encoding motion pattern stimuli does provide some computational insights. In biologically plausible neural structures, winner-take-all strategies are often implemented using mutually inhibiting horizontal connections that suppress all but the strongest neural responses. Taken in the less extreme case of a winner-take-all strategy across neural sub-populations (i.e. contraction preferred units etc.), the resultant inhibitory structures would be qualitatively similar to the mutually inhibiting connection profiles explicitly simulated here.

4.3 Inhibitory Influences in MST

In the work presented here we have proposed a specific set of local interactions within a population MSTd-like units whose visual motion properties are consistent with the available physiological evidence. Specifically the model predicts an 80-100° spread of inhibition (σ_I), as defined in the 2-D motion pattern space, whose profile is continuous across preferred motions. While such structures have not been explicitly examined in higher visual motion areas, such as MSTd, there is anecdotal support for the presence of inhibitory connections within the neurophysiological literature. Several studies have reported inhibitory effects in the motion pattern responses of MSTd neurons, although the degree and specificity of inhibition has not been characterized in detail (Duffy & Wurtz, 1991b; Lappe, et al., 1996).

Multi-cellular studies in MST would be well suited to examine this result in greater detail and further quantify the extent of inhibitory influence across motion pattern sensitive cells. By correlating neural responses across cell pairs as a function of the relative distances between their preferred motions and the retinotopic positions of their receptive field centers, multi-cellular studies could be used to quantify a) the spread of inhibition as a function of preferred motion patterns, b) the extent to which horizontal interactions span a continuous profile of relative strength and c) the retinotopic extent of such interactions.

It is worth noting that while the performance of the neural models suggests the presence of specific inhibitory connections between motion pattern detectors, the inhibitory effects implied by the model *do not* require the massive degree of interconnections implemented here. Equivalent interactions could be achieved in more biologically plausible architectures containing small populations of inhibiting interneurons whose pre-synaptic inhibition of preferred motion sub-populations follows an anti-hebbian rule of connection strength. Discrete inhibitory profiles whose strength is constant within sub-populations could also be implemented without seriously degrading the visual motion information stored across the population. For equivalent representations of σ_I and/or half-width in continuous and discrete profiles respectively, the influence of the anti-preferred inhibitory structures would likely be similar such that discrimination thresholds continued to be well matched to human perceptual performance.

4.4 Neural Interactions within and Across Visual Motion Areas

The gross increase in computational complexity observed along the visual motion pathway has typically been associated with a feed-forward pooling of visual motion information *between* visual areas. In this scheme, neural responses to simple visual motion components are combined to encode more complex and perceptually relevant properties of the visual scene.

The model presented here does not explicitly preclude the existence of inhibitory feed-forward structures in the projection of motion information from MT to MSTd. While it is possible to obtain similar computational results using feed-forward mechanisms that inhibit "global" motion pattern responses via convergent "local" motion inputs, the mechanisms underlying the development of such a system remain unclear. In a previous computational model (Beardsley & Vaina, 1998; Beardsley et al., 2003) we illustrated the development of horizontal inhibition between MSTd-like units under a supervised learning rule. We also reported evidence of inhibitory feed-forward projections whose development appeared to be linked to the underlying learning rule. It seems likely that the visual motion pathway contains a combination of both inhibitory feed-forward and horizontal connections that act to maintain the visual information encoded across neural populations. Future extensions of this model to include a feed-forward layer of direction selective MT inputs and neurophysiological studies by others incorporating multi-cellular recording techniques within and across visual motion areas will likely provide the insights necessary to address these issues in greater detail.

4.5 Relation to Existing Models of MSTd and Heading

As we alluded to in the introduction, numerous biologically constrained feed-forward models of visual motion pooling from MT to MSTd have been developed to extract perceptually relevant visual motion properties. Three in particular, Perrone & Stone (1994, 1998), Zemel & Sejnowski (1998), and Lappe et al. (1996), contain many of the underlying visual motion properties assumed here and could, in principle, also be used to model GMP performance. In practice, the methods employed in each model may in some ways restrict their extension to graded motion pattern discriminations, but in doing so they are likely to provide useful insights into the computational and structural requirements of higher-dimensional encoding and 'perceptual' decoding across multiple visual motion tasks.

In the template model of heading estimation proposed by Perrone and Stone (1994, 1998), perceptual estimates are obtained according to the gaze-stabilized translation properties of the most active template unit. In this case much of the visual motion information encoded across the population is discarded, removing a potential source of correlated signal that could be used to compensate the effects of internal and/or external noise. While such methods also have the advantage of removing much of the population noise, the trade-off in perceptual performance is strongly dependent on the underlying computational structure and physiological properties. As a result it is unclear how 'perceptual' performance for small perturbations in the overall motion would degrade as a function of neural and visual motion noise. Specifically, can maximal unit responses accurately decode the percept under noisy conditions or is it necessary to make more explicit use of the visual signal encoded across the population as a whole?

Other models, including Zemel & Sejnowski (1998) and Lappe et al. (Lappe & Rauschecker, 1993, 1995; Lappe et al., 1996; Lappe & Duffy, 1999), have decoded 'perceptual' estimates of heading based on the maximal activity obtained across neural sub-populations. While the signal integration afforded such methods is likely to provide a computationally more robust 'perceptual' estimate under noisy conditions, in the case of the GMP task it is not immediately clear how many neural sub-populations might be required, nor how many units each should contain.

Together with the extension of the model presented here to include a feed-forward layer of MT projections, these models could be readily extended to examine GMP discrimination across a wide range of stimulus conditions. In doing so each would provide additional insight into the relative roles of feed-forward versus horizontal connections and the range of decoding schemes sufficient to obtain equivalent measures of human perceptual performance. Specifically, to what extent might inhibitory feed-forward projections be used to mediate the decoded motion patterns? And what limitations do existing decoding strategies place on the extraction of equivalent measures of perceptual performance in the presence of noise (internal and external) and across multi-dimensional parameter spaces associated with dispirit perceptual tasks (e.g., Heading vs. GMP discrimination)?

5 SUMMARY

Within the visual cortex, the use of experimental and computational techniques to investigate neural connectivity continues to refine the role of intrinsic horizontal connections in the emergent computational and perceptual properties of the visual system. The model we have presented here builds on and extends these concepts to cortical areas later in the visual motion pathway. Through simulated populations of MSTd-like units we have identified a set of anti-preferred inhibitory structures whose computational effects on the encoded visual motion information are well matched to equivalent measures of human psychophysical performance. These structures are qualitatively similar to the inhibition of anti-preferred direction tuned cells reported in earlier visual motion areas such as MT and suggest a more a complex neural architecture throughout the visual motion system.

ACKNOWLEDGMENTS

The authors wish to thank Colin Clifford and Peter Foldiak for their helpful suggestions during the early stages of this work. This work was supported by National Institutes of Health grant EY-2R01-07861-13 to LMV.

REFERENCES

Adini, Y., Sagi, D., & Tsodyks, M. (1997). Excitatory-inhibitory network in the visual cortex: Psychophysical evidence. *Proc. Natl. Acad. Sci., 94*, 10426-10431.

Amir, Y., Harel, M., & Malach, R. (1993). Cortical hierarchy reflected in the organization of intrinsic connections in macaque monkey visual cortex. *J. Comp. Neurology, 334*, 19-46.

Andersen, R. A. (1997). Neural mechanisms of visual motion perception in primates. *Neuron, 18*, 865-872.

Andersen, R. A., Shenoy, K. V., Crowell, J. A., & Bradley, D. C. (2000). Neural Mechanisms for Self-Motion Perception in Area MST. In: M. Lappe (Ed.). *Neuronal Processing of Optic Flow*, 44 (pp. 219-234). New York: Academic Press.

Ball, K., & Sekuler, R. (1987). Direction-specific improvement in motion discrimination. *Vision Res., 27* (6), 953-965.

Beardsley, S. A., & Vaina, L. M. (1998). Computational modeling of optic flow selectivity in MSTd neurons. *Network: Comput. Neural Syst., 9*, 467-493.

Beardsley, S. A., & Vaina, L. M. (2001). A laterally interconnected neural architecture in MST accounts for psychophysical discrimination of complex motion patterns. *J. Comput. Neurosci., 10*, 255-280.

Beardsley, S. A., Ward, R. L., & Vaina, L. M. (2003). A feed-forward network model of spiral-planar tuning in MSTd. *Vision Res., 43*, 577-595.

Ben-Yishai, B., Bar-Or, R. L., & Sompolinsky, H. (1995). Theory of orientation tuning in visual cortex. *Proc. Natl. Acad. Sci., 92*, 3844-3848.

Bevington, P. (1969). *Data Reduction and Error Analysis for the Physical Sciences,* (p. 336). New York: McGraw-Hill.

Boussaoud, D., Ungerleider, L. G., & Desimone, R. (1990). Pathways for motion analysis: cortical connections of the medial superior temporal and fundus of the superior temporal visual areas in the macaque. *J. Comp. Neurol., 296* (3), 462-495.

Bremmer, F., Duhamel, J.-R., Ben Hamed, S., & Werner, G. (2000). Stages of Self-Motion Processing in Primate Posterior Parietal Cortex. In: M. Lappe (Ed.) *Neuronal Processing of Optic Flow,* 44 (pp. 173-198). New York: Academic Press.

Britten, K. H., Newsome, W. T., Shadlen, M. N., Celebrini, S., & Movshon, J. A. (1996). A relationship between behavioral choice and the visual responses of neurons in macaque MT. *Vis. Neurosci., 13 (1),* 87-100.

Britten, K. H., Shadlen, M. N., Newsome, W. T., & Movshon, J. A. (1992). The analysis of visual motion: a comparison of neuronal and psychophysical performance. *J Neurosci., 12 (12),* 4745-4765.

Britten, K. H., & van Wezel, R. J. A. (1998). Electrical microstimulation of cortical area MST biases heading perception in monkeys. *Nat. Neurosci., 1*, 59-63.

Burr, D. C., Morrone, M. C., & Vaina, L. M. (1998). Large receptive fields for optic flow detection in humans. *Vision Res., 38 (12),* 1731-1743.

Carandini, M., & Ringach, D. L. (1997). Prediction of a recurrent model of orientation selectivity. *Vision Res., 37*, 3061-3071.

Celebrini, S., & Newsome, W. T. (1994). Neuronal and psychophysical sensitivity to motion signals in extrastriate area MST of the macaque monkey. *J. Neurosci., 14 (7),* 4109-4124.

Celebrini, S., & Newsome, W. T. (1995). Microstimulation of extrastriate area MST influences performance on a direction discrimination task. *J. Neurphysiol., 73 (2),* 437-448.

Chey, J., Grossberg, S., & Mingolla, E. (1998). Neural dynamics of motion processing and speed discrimination. *Vision Res., 38*, 2769-2786.

Coletta, N. J., Segu, P., & Tiana, C. L. (1993). An oblique effect in parafovial motion perception. *Vision Res., 33 (18)*, 2747-2756.

de Jong, B. M., Shipp, S., Skidmore, B., Frackowiak, R. S. J., & Zeki, S. (1994). The cerebral activity related to the visual perception of forward motion in depth. *Brain, 117*, 1039-1054.

deCharms, R. C., & Zador, A. (2000). Neural representation and the cortical code. *Annu. Rev. Neurosci.,* (23), 613-647.

DeYoe, E. A., & Van Essen, D. C. (1988). Concurrent processing streams in monkey visual cortex. *Trends Neurosci., 11* (5), 219-226.

Duffy, C. J. (2000). Optic Flow Analysis for Self-Movement Perception. In: M. Lappe (Ed.) *Neuronal Processing of Optic Flow,* 44 (pp. 199-218). New York: Academic Press.

Duffy, C. J., & Wurtz, R. H. (1991a). Sensitivity of MST neurons to optic flow stimuli. I. A continuum of response selectivity to large-field stimuli. *J. Neurophysiol., 65 (6),* 1329-1345.

Duffy, C. J., & Wurtz, R. H. (1991b). Sensitivity of MST neurons to optic flow stimuli. II. Mechanisms of response selectivity revealed by small field stimuli. *J. Neurophysiol., 65 (6),* 1346-1359.

Duffy, C. J., & Wurtz, R. H. (1995). Response of monkey MST neurons to optic flow stimuli with shifted centers of motion. *J. Neurosci., 15 (7),* 5192-5208.

Duffy, C. J., & Wurtz, R. H. (1997a). Medial superior temporal area neurons respond to speed patterns in optic flow. *J. Neurosci., 17 (8),* 2839-2851.

Duffy, C. J., & Wurtz, R. H. (1997b). Planar directional contributions to optic flow responses in MST neurons. *J. Neurophysiol., 77,* 782-796.

Edelman, S. (1996). Why Have Lateral Connections in the Visual Cortex? In: J. Sirosh, R. Miikkulainen, & Y. Choe (Eds.), *Lateral Interactions in the Cortex: Structure and Function,* Electronic Book (http://www.cs.utexas.edu/users/nn/web-pubs/htmlbook96/edelman/). Austin: The UTCS Neural Networks Research Group.

Edwards, M., & Badcock, D. R. (1993). Asymmetries in the sensitivity to motion in depth: a centripetal bias. *Perception, 22,* 1013-1023.

Felleman, D. J., & Van Essen, D. C. (1991). Distributed hierarchical processing in the primate cerebral cortex. *Cereb. Cortex, 1,* 1-47.

Foldiak, P. (1993). The 'Ideal Homunculus': Statistical Inference from Neural Population Responses. In: F.H. Eeckman, & J.M. Bower (Eds.), *Computation and Neural Systems* (pp. 55-60). Norwell: Kluwer Academic Publishers.

Freeman, T. C., & Harris, M. G. (1992). Human sensitivity to expanding and rotating motion: Effects of complementary masking and directional structure. *Vision Res., 32 (1),* 81-87.

Geesaman, B. J., & Andersen, R. A. (1996). The analysis of complex motion patterns by form/cue invariant MSTd neurons. *J. Neurosci., 16 (15),* 4716-4732.

Georgopoulos, A. P., Kettner, R. E., & Schwartz, A. B. (1988). Primate motor cortex and free arm movements to visual targets in three-dimensional space. II. Coding of the direction of movement by a neuronal population. *J. Neurosci., 8 (8),* 2928-2937.

Georgopoulos, A. P., Schwartz, A. B., & Kettner, R. E. (1986). Neuronal population coding of movement direction. *Science, 233* (26), 1416-1419.

Gibson, J. J. (1950). *The Perception of the Visual World.* (Boston: Houghton Mifflin).

Gilbert, C., Das, A., Ito, M., Kapadia, M., & Westheimer, G. (1996). Spatial integration and cortical dynamics. *Proc. Natl. Acad. Sci. USA, 93,* 615-622.

Gilbert, C., & Wiesel, T. (1989). Columnar specificity of instrinsic horizontal and corticocortical connections in cat visual cortex. *J. Neurosci., 9 (7),* 2432-2442.

Gilbert, C. D. (1983). Microcircuitry of the visual cortex. *Ann. Rev. Neurosci, 6,* 217-247.

Gilbert, C. D. (1985). Horizontal integration in the neocortex. *Trends Neurosci.,* (April), 160-165.

Gilbert, C. D. (1992). Horizontal integration and cortical dynamics. *Neuron, 9,* 1-13.

Gilbert, C. D., & Wiesel, T. N. (1985). Intrisic connectivity and receptive field properties in visual cortex. *Vision Res., 25 (3),* 365-374.

Graziano, M. S., Anderson, R. A., & Snowden, R. (1994). Tuning of MST neurons to spiral motions. *J. Neurosci., 14 (1),* 54-67.

Greenlee, M. W. (2000). Human cortical areas underlying the perception of optic flow: brain imaging studies. *Int. Rev. Neurobiol., 44,* 269-292.

Grinvald, A., Lieke, E., Frostig, R. D., Gilbert, C. D., & Wiesel, T. N. (1986). Functional architecture of cortex revealed by optical imaging of intrinsic signals. *Nature, 324,* 361-364.

Gros, B. L., Blake, R., & Hiris, E. (1998). Anisotropies in visual motion perception: a fresh look. *J. Opt. Soc. Am. A, 15 (8),* 2003-2011.

Grossberg, S., Mignolla, E., & Pack, C. (1999). A neural model of motion processing and visual navigation by cortical area MST. *Cereb. Cortex, 9 (8),* 878-895.

Grossberg, S., & Williamson, J. R. (2001). A neural model of how horizontal and interlaminar connections of visual cortex develop into adult circuits that carry out perceptual grouping and learning. *Cereb. Cortex, 11,* 37-58.

Hatsopoulos, N., & Warren, W. J. (1991). Visual navigation with a neural network. *Neural Netw., 4,* 303-317.

Haykin, S. (1999). *Neural Networks: A Comprehensive Foundation,* (p. 842). Upper Saddle River, NJ: Prentice Hall.

Heeger, D. J. (1999). Linking visual perception with human brain activity. *Curr. Opin. Neurobiol., 9 (4),* 474-479.

Hertz, J., Krogh, A., & Palmer, R. G. (1991). Introduction to the Theory of Neural Computation. *A Lecture Notes Volume in the Santa Fe Institute Studies in the Sciences of Complexity* (p. 327). New York: Addison-Wesley Publishing Company.

Kalaska, J. F., Caminiti, R., & Georgopoulos, A. P. (1983). Cortical mechanisms related to the direction of two-dimensional arm movements: relations in parietal area 5 and comparison with motor cortex. *Exp. Brain Res., 51,* 247-260.

Kisvarday, Z., Toth, E., Rausch, M., & Eysel, U. (1997). Orientation-specific relationship between populations of excitatory and inhibitory lateral connections in the visual cortex of the cat. *Cereb. Cortex, 7 (7),* 605-618.

Koechlin, E., Anton, J., & Burnod, Y. (1999). Bayesian interference in populations of cortical neurons: A model of motion integration and segmentation in area MT. *Biol. Cybern., 80,* 25-44.

Lagae, L., Maes, H., Raiguel, S., Xiao, D.-K., & Orban, G. A. (1994). Responses of macaque STS neurons to optic flow components: a comparison of areas MT and MST. *J. Neurophysiol., 71 (5),* 1597-1626.

Lappe, M. (2000). Computational mechanisms for optic flow analysis in primate cortex. In: M. Lappe (Ed.) *Neuronal Processing of Optic Flow,* 44 (pp. 235-268). New York: Academic Press.

Lappe, M., Bremmer, F., Pekel, M., Thiele, A., & Hoffmann, K. (1996). Optic flow processing in monkey STS: A theoretical and experimental approach. *J. Neurosci., 16 (19),* 6265-6285.

Lappe, M., Bremmer, F., & van den Berg, A. V. (1999). Perception of self-motion from visual flow. *Trends Cogn. Sci., 3* (9), 329-336.

Lappe, M., & Duffy, C. (1999). Optic flow illusion and single neuron behavior reconciled by a population model. *Eur. J. Neurosci., 11*, 2323-2331.

Lappe, M., & Rauschecker, J. P. (1993). A neural network for the processing of optic flow from ego-motion in man and higher mammals. *Neural Comput., 5*, 374-391.

Lappe, M., & Rauschecker, J. P. (1995). Motion anisotropies and heading detection. *Biol. Cybern., 72 (3),* 261-277.

Liu, L., & Hulle, V. (1998). Modeling the surround of MT cells and their selectivity for surface orientation in depth specified by motion. *Neural Comput., 10*, 295-312.

Lukashin, A. V., & Georgopoulos, A. P. (1993). A dynamical neural network model for motor cortical activity during movement: population coding of movement trajectories. *Biol. Cybern., 69*, 517-524.

Lukashin, A. V., & Georgopoulos, A. P. (1994). A neural network for coding of trajectories by time series of neuronal population vectors. *Neural Comput., 6*, 19-28.

Lukashin, A. V., Wilcox, G. L., & Georgopoulos, A. P. (1996). Modeling of directional operations in the motor cortex: a noisy network of spiking neurons is trained to generate a neural-vector trajectory. *Neural Netw., 9 (3),* 397-410.

Lund, J., Yoshioka, T., & Levitt, J. (1993). Comparison of intrinsic connectivity in different areas of the macaque monkey cerebral cortex. *Cereb. Cortex, 3 (2),* 148-162.

Malach, R., Schirman, T., Harel, M., Tootell, R., & Malonek, D. (1997). Organization of intrinsic connections in owl monkey area MT. *Cereb. Cortex, 7 (4),* 386-393.

Matthews, N., & Qian, N. (1999). Axis-of-motion affects direction discrimination, not speed discrimination. *Vision Res., 39*, 2205-2211.

Matthews, N., & Welch, L. (1997). Velocity-dependent improvements in single-dot direction discrimination. *Percept. Psychophys., 59 (1),* 60-72.

Maunsell, J. H., & Van Essen, D. C. (1983). The connections of the middle temporal visual area (MT) and their relationship to a cortical heirarchy in the macaque monkey. *J. Neurosci., 3 (12),* 2563-2586.

McGuire, B., Gilbert, C., Rivlin, P., & Wiesel, T. (1991). Targets of horizontal connections in macaque primary visual cortex. *J. Comp. Neurol., 305*, 370-392.

Meese, T., & Harris, M. (2001a). Independent detectors for expansion and rotation, and for orthogonal components of deformation. *Perception, 30*, 1189-1202.

Meese, T. S., & Harris, M. G. (2001b). Broad direction bandwidths for complex motion mechanisms. *Vision Res, 41* (15), 1901-1914.

Meese, T. S., & Harris, S. J. (2002). Spiral mechanisms are required to account for summation of complex motion components. *Vision Res., 42*, 1073-1080.

Miikkulainen, R., & Sirosh, J. (1996). Introduction: The Emerging Understanding of Lateral Interactions in the Cortex. In: J. Sirosh, R. Miikkulainen, & Y. Choe (Eds.), *Lateral Interactions in the Cortex: Structure and Function,* Electronic Book (http://www.cs.utexas.edu/users/nn/web-pubs/htmlbook96/miikkulainen/). Austin: The UTCS Neural Networks Research Group.

Morrone, C., Burr, D., & Vaina, L. (1995). Two stages of visual processing for radial and circular motion. *Nature, 376*, 507-509.

Morrone, M. C., Burr, D. C., Di Pietro, S., & Stefanelli, M. A. (1999). Cardinal directions for visual optic flow. *Curr. Biol., 9*, 763-766.

Morrone, M. C., Tosetti, M., Montanaro, D., Fiorentini, A., Cioni, G., & Burr, D. C. (2000). A cortical area that responds specifically to optic flow, revealed by fMRI. *Nat. Neurosci., 3 (12),* 1322-1328.

Nowlan, S. J., & Sejnowski, T. J. (1995). A selection model for motion processing in area MT of primates. *J. Neurosci., 15 (2),* 1195-1214.

Oram, M. W., Foldiak, P., Perrett, D. I., & Sengpiel, F. (1998). The 'ideal homunculus': decoding neural population signals. *Trends Neurosci., 21 (8),* 365-371.

Orban, G. A., Lagae, L., Raiguel, S., Xiao, D., & Maes, H. (1995). The speed tuning of medial superior temporal (MST) cell responses to optic-flow components. *Perception, 24 (3),* 269-285.

Orban, G. A., Lagae, L., Verri, A., Raiguel, S., Xiao, D., Maes, H., & Torre, V. (1992). First-order analysis of optical flow in monkey brain. *Proc. Natl. Acad. Sci. USA, 89*, 2595-2599.

Perrone, J., & Stone, L. (1994). A model of self-motion estimation within primate extrastriate visual cortex. *Vision Res., 34 (21),* 2917-2938.

Perrone, J. A., & Stone, L. S. (1998). Emulating the visual receptive-field properties of MST neurons with a template model of heading estimation. *J. Neurosci., 18 (15),* 5958-5975.

Pitts, R. I., Sundareswaran, V., & Vaina, L. M. (1997). A model of position-invariant, optic flow pattern-selective cells. In: *Computational Neuroscience: Trends in Research 1997* (pp. 171-176). New York: Plenum Publishing Corporation.

Pouget, A., Zhang, K., Deneve, S., & Latham, P. E. (1998). Statistically efficient estimation using population coding. *Neural Comput., 10*, 373-401.

Raymond, J. E. (1994). Directional anisotropy of motion sensitivity across the visual field. *Vision Res., 34 (8),* 1029-1039.

Rees, G., Friston, K., & Koch, C. (2000). A direct quantative relationship between the function properties of human and macaque V5. *Nat. Neurosci., 3 (7),* 716-723.

Regan, D., & Beverley, K. (1978). Looming detectors in the human visual pathway. *Vision Res., 18*, 415-421.

Regan, D., & Beverley, K. I. (1979). Visually guided locomotion: Psychophysical evidence for a neural mechanism sensitive to flow patterns. *Science, 205*, 311-313.

Royden, C. S. (1997). Mathematical analysis of motion-opponent mechanisms used in the determination of heading and depth. *J. Opt. Soc. Am. A, 14*, 2128-2143.

Rutschmann, R. M., Schrauf, M., & Greenlee, M. W. (2000). Brain activation during dichoptic presentation of optic flow stimuli. *Exp. Brain Res., 134*, 533-537.

Saito, H.-a., Yukie, M., Tanaka, K., Hikosaka, K., Fukada, Y., & Iwai, E. (1986). Integration of direction signals of image motion in the superior temporal sulcus of the macaque monkey. *J. Neurosci., 6 (1),* 145-157.

Sakai, K., & Miyashita, Y. (1991). Neural organization for the long-term memory of paired associates. *Nature, 354*, 152-155.

Salinas, E., & Abbott, L. (1994). Vector reconstruction from firing rates. *J. Comput. Neurosci., 1*, 89-107.

Salinas, E., & Abbott, L. (1995). Transfer of coded information from sensory to motor networks. *J. Neurosci., 15*, 6461-6474.

Salzman, C. D., Britten, K. H., & Newsome, W. T. (1990). Cortical microstimulation influences perceptual judgements of motion direction. *Nature, 346*, 174-177.

Salzman, C. D., Murasugi, C. M., Britten, K., & Newsome, W. T. (1992). Microstimulation in visual area MT: effects on direction discrimination performance. *J. Neurosci., 12 (6)*, 2331-2355.

Sanger, T. D. (1996). Probability density estimation for the interpretation of neural population codes. *J. Neurophysiol., 76 (4)*, 2790-2793.

Schaafsma, S. J., & Duysens, J. (1996). Neurons in the ventral intraparietal area of awake macaque monkey closely resemble neurons in the dorsal part of the medial superior temporal area in their responses to optic flow patterns. *J. Neurophysiol., 76 (6)*, 4056-4068.

Schwartz, A. B., Kettner, R. E., & Georgopoulos, A. P. (1988). Primate motor cortex and free arm movements to visual targets in three-dimensional space. I. Relations between single cell discharge and direction of movement. *J. Neurosci., 8 (8)*, 2913-2827.

Seung, H. S., & Sompolinsky, H. (1993). Simple models for reading neuronal population codes. *Proc. Natl. Acad. Sci. USA, 90*, 10749-10753.

Shadlen, M., & Newsome, W. (1998). The variable discharge of cortical neurons: implications for connectivity, computation, and information coding. *J. Neurosci., 18 (10)*, 3870-3896.

Shadlen, M. N., & Newsome, W. T. (1994). Noise, neural codes, and cortical organization. *Curr. Opin. Neurobiol., 4*, 569-579.

Siegel, R. M., & Read, H. L. (1997). Analysis of optic flow in the monkey parietal area 7a. *Cereb. Cortex, 7 (4)*, 327-346.

Snippe, H. (1996). Parameter extraction from population codes: A critical assessment. *Neural Comput., 8*, 511-530.

Snowden, R. J., & Milne, A. B. (1996). The effects of adapting to complex motions: position invariance and tuning to spiral motions. *J. Cognit. Neurosci., 8* (4), 412-429.

Snowden, R. J., & Milne, A. B. (1997). Phantom motion aftereffects - evidence of detectors for the analysis of optic flow. *Curr. Biol., 7*, 717-722.

Softky, W. (1995). Simple codes versus efficient codes. *Curr. Opin. Neurobiol., 5*, 239-247.

Softky, W., & Koch, C. (1993). The highly irregular firing of cortical cells is inconsistent with temporal integration of random EPSPs. *J. Neurosci., 13 (1)*, 334-350.

Stemmler, M., Usher, M., & Niebur, E. (1995). Lateral interactions in primary visual cortex: A model bridging physiology and psychophysics. *Science, 269*, 1877-1880.

Sundareswaran, V., & Vaina, L. M. (1996). Adaptive computational models of fast learning of motion direction discrimination. *Biol. Cybern., 74*, 319-329.

Tanaka, K., Fukada, Y., & Saito, H.-A. (1989). Underlying mechanisms of the response specificity of expansion/contraction and rotation cells in the dorsal part of the medial superior temporal area of the macaque monkey. *J. Neurophysiol., 62 (3),* 642-656.

Tanaka, K., & Saito, H.-A. (1989). Analysis of motion of the visual field by direction, expansion/contraction, and rotation cells clustered in the dorsal part of the medial superior temporal area of the macaque monkey. *J. Neurophysiol., 62 (3),* 626-641.

Taylor, J. G., & Alavi, F. N. (1996). A Basis for Long-Range Inhibition Across Cortex. In: J. Sirosh, R. Miikkulainen, & Y. Choe (Eds.), *Lateral Interactions in the Cortex: Structure and Function,* Electronic Book (http://www.cs.utexas.edu/users/nn/web-pubs/htmlbook96/taylor/). Austin: The UTCS Neural Networks Research Group.

Te Pas, S. F., Kappers, A. M., & Koenderink, J. J. (1996). Detection of first-order structure in optic flow fields. *Vision Res., 36 (2),* 259-270.

Teich, A. F., & Qian, N. (2002). Learning and adaptation in a recurrent model of V1 orientation selectivity. *J. Neurophysiol.*, in press.

Tootell, R. B. H., Reppas, J. B., Kwong, K. K., Malach, R., Born, R. T., Brady, T. J., Rosen, B. R., & Belliveau, J. W. (1995). Functional analysis of human MT and related visual cortical areas using magnetic resonance imaging. *J. Neurosci., 15 (4),* 3215-3230.

Ts'o, D., Gilbert, C. D., & Wiesel, T. N. (1986). Relationships between horizontal interactions and functional architecture in cat striate cortex as revealed by cross-correlation analysis. *J. Neurosci., 6 (4),* 1160-1170.

Vaina, L. M. (1998). Complex motion perception and its deficits. *Curr. Opin. Neurobiol., 8 (4),* 494-502.

Vaina, L. M., Solovyev, S., Kopcik, M., & Chowdhury, S. (2000). Impaired self-motion perception from optic flow: a psychophysical and fMRI study of a patient with a left occipital lobe lesion. *Soc. Neurosci. Abst., 26,* 1065.

Vaina, L. M., Sundareswaran, V., & Harris, J. G. (1995). Learning to ignore: psychophysics and computational modeling of fast learning of direction in noisy motion stimuli. *Cognit. Brain Res, 2* (3), 155-163.

van den Berg, A. V. (2000). Human Ego-Motion Perception. In: M. Lappe (Ed.) *Neuronal Processing of Optic Flow,* 44 (pp. 3-25). New York: Academic Press.

Van Essen, D. C., & Maunsell, J. H. R. (1983). Hierarchical organization and functional streams in the visual cortex. *Trends Neurosci., 6 (9),* 370-375.

Wang, R. (1995). A simple competitive account of some response properties of visual neurons in area MSTd. *Neural Comput., 7,* 290-306.

Wang, R. (1996). A network model for the optic flow computation of the MST neurons. *Neural Netw., 9 (3),* 411-426.

Wiskott, L., & von der Malsburg, C. (1996). Face Recognition by Dynamic Link Matching. In: J. Sirosh, R. Miikkulainen, & Y. Choe (Eds.), *Lateral Interactions in the Cortex: Structure and Function,* Electronic Book (http://www.cs.utexas.edu/users/nn/web-pubs/htmlbook96/wiskott/). Austin: The UTCS Neural Networks Research Group.

Worgotter, F., Niebur, E., & Christof, K. (1991). Isotropic connections generate functional asymmetrical behavior in visual cortical cells. *J. Neurophysiol, 66 (2),* 444-459.

Zemel, R., Dayan, P., & Pouget, A. (1998). Probabilistic interpretation of population codes. *Neural Comput., 10*, 403-430.

Zemel, R. S., & Sejnowski, T. J. (1998). A model for encoding multiple object motions and self-motion in area MST of primate visual cortex. *J. Neurosci., 18 (1),* 531-547.

Zhang, K., Sereno, M. I., & Sereno, M. E. (1993). Emergence of position-independent detectors of sense of rotation and dilation with Hebbian learning: an analysis. *Neural Comput., 5*, 597-612.

Zhao, L., Vaina, L. M., LeMay, M., Kader, B., Chou, I. S., & Kemper, T. (1995). Are there specific anatomical correlates of global motion perception in the human visual cortex? *Invest. Ophthalmol. Vis. Sci., 36 (4),* S56.

Zohary, E. (1992). Population coding of visual stimuli cortical neurons tuned to more than one dimension. *Biol. Cybern., 66*, 265-272.

10. Circular Receptive Field Structures for Flow Analysis and Heading Detection

Jaap A. Beintema[1], Albert V. van den Berg[1], Markus Lappe[2,3]

[1] Functionele Neurobiologie
Universiteit Utrecht, Utrecht, The Netherlands

[2] Allgemeine Psychologie
Westf. Wilhelms-Universität Münster, Münster, Germany

[3] Allgemeine Zoölogie & Neurobiologie
Ruhr-Universität Bochum, Bochum, Germany

1 INTRODUCTION

Recent years have brought forward different models on how the brain might encode heading from optic flow. Neurons in these models can encode heading for a variety of self-motion conditions, while responding to optic flow stimuli similarly as found in electrophysiological studies. Yet, little attention has been given to the receptive field structure of neurons that integrate local motion signals to analyze the optic flow. Intuitively, radial structures might seem suited for the task of heading detection, since pure observer translation causes flow to emanate from the point of heading (Figure

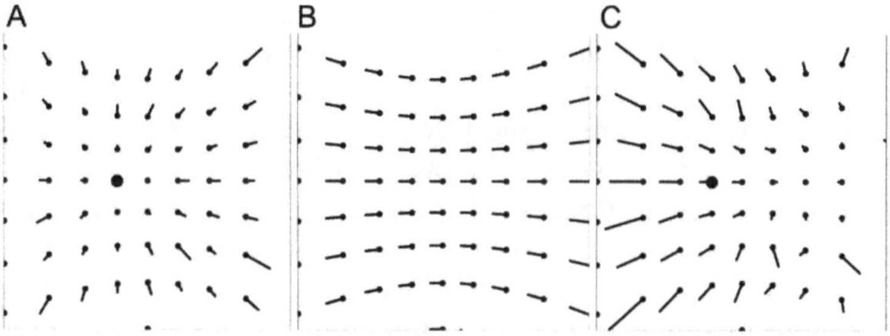

Figure 1. Flow during a) observer translation through a 3-D-cloud of dots, headed 10 degrees towards the left (filled circle), during b) observer rotation about the vertical towards the right, and during c) the combination of both. The dots indicate the start of motion vectors.

L.M. Vaina, S.A. Beardsley and S.K. Rushton (eds.), Optic Flow and Beyond, 223–248.

1a). However, rotational flow (Figure 1b) during simultaneous eye rotation can cause the retinal flow to be shifted away from the heading (Figure 1c). Moreover, variation in point distances with respect to the translating eye results in retinal motion differences, called motion parallax, that can cause the flow during eye rotation to deviate even more from being purely radial. This poses the question what would be the optimal receptive field structure to deal with complex flow fields for retrieving heading.

Research on neural mechanisms has focused on a number of cortical areas that contain cells responsive to optic flow, such as area MST (human V5a) but also areas beyond such as VIP (Colby et al., 1993) or 7a (Read & Siegel, 1997). MST neurons typically have large receptive fields that cover more than a quarter of the visual field, and respond well to complex motion patterns, such as expanding/contracting motion, clockwise/counterclockwise rotation (Tanaka et al., 1989; Tanaka & Saito, 1989), or combinations of rotation and expansion (Graziano et al., 1994). Importantly, MST cells seem systematically tuned to the location of foci of expansion or rotation (Duffy & Wurtz, 1995; Lappe et al., 1996). Convincing evidence for MST's involvement in heading perception has been found from the covariation of MST activity with the heading responses of trained monkeys (Britten & van Wezel, 1998). MST cells acquire their selectivity to optic flow by integrating signals from local motion sensors, found in area MT (human V5). MT neurons have smaller RFs and typically respond to motion in one particular direction at a particular speed (see review van Wezel & Britten, 2002). Still, relatively little is known about the organization of the MT neurons that form the receptive field (RF) of an MST cell tuned for heading.

Several schemes have been proposed to recover heading from optic flow in the presence of rotational flow (see review Lappe, 2000). Intuitively, one might think that detectors for expansion would be ideal for detecting heading. Radial receptive fields are for example assumed by differential motion models that take local motion differences to remove the effect of rotation (Longuet-Higgins & Prazdny 1980; Hildreth 1992; Rieger & Lawton 1985), whereby the remaining motion parallax vector field radiates from the heading direction. Indeed, such approaches find some physiological support as MT cells have been found that respond well to motion differences in their center and surround receptive field (Allman et al., 1985), and a heading detection scheme based on motion-opponent cells seems feasible (Royden, 1997). The analysis of the remaining differential motion vectors could be realized by pure translation detectors with radial receptive fields (Hatsopoulos & Warren, 1991). But, differential motion parallax models can only partly explain human psychophysics on heading perception. Importantly, these models fail to detect heading in the absence of depth differences, whereas humans still correctly can judge heading by relying on extra-retinal rotation signals (Royden et al., 1992), or given a sufficiently large field of view (Grigo & Lappe, 1999).

Yet, the last decade has brought forward a number of physiologically inspired models that detect heading in the presence of eye rotation for a variety of conditions and are able to simulate response properties of MST cells (Beintema & van den Berg, 1998; Lappe & Rauschecker, 1993b; Perrone & Stone, 1994). As these models deal with eye rotation at the level beyond local motion detection, the receptive field of their MST-like neurons must differ from purely radial.

So far, the receptive field structure using small field stimuli has only been roughly probed (Duffy & Wurtz, 1991b). Those results give little support for simple RF structures such as radial motion or uni-circular motion (clockwise or counterclockwise rotation). Another intriguing question is why a large proportion of MST cells respond to clockwise or counterclockwise rotation (Duffy & Wurtz 1991a,b; Lappe et al., 1996; Tanaka & Saito, 1989) at rotation speeds up to 80 deg/s not usually experienced in daily life.

These observations inspired us to examine the receptive fields predicted by three physiologically inspired models. In the following, we first introduce the RF structure assumed in templates of the velocity gain model (Beintema & van den Berg, 1998). Then we present an analysis of the RF structures that emerge under different restrictions of the population model (Lappe & Rauschecker, 1993b). Finally, we consider the effective RF structure of the template model (Perrone & Stone, 1994). Despite the different approaches, a striking similarity is found. The preferred motion driving the neurons turns out to be directed always along circles centered on the neuron's preferred heading, thus perpendicular to the expected radial flow. Moreover, we find similarities and differences in the substructures, such as bi-circularity and motion-opponency, which might reveal different strategies on how the visual system copes with eye rotation.

2 VELOCITY GAIN FIELD MODEL

The velocity gain field model by Beintema and van den Berg (1998) is based on the templates approach (Perrone, 1992; Perrone & Stone, 1994). That approach aims to find the best matching template in a set of motion templates tuned to different ego-motions. Each template evaluates the evidence for a global match with its preferred flow by summating evidence for local matches across different parts in the visual field. How these local comparisons are done is not trivial, but by appropriate selection of the motion sensors at a specific retinal location, templates can be constructed that respond maximally only to flow corresponding to a particular heading in combination with a particular ego-rotation, independent of the distances of points.

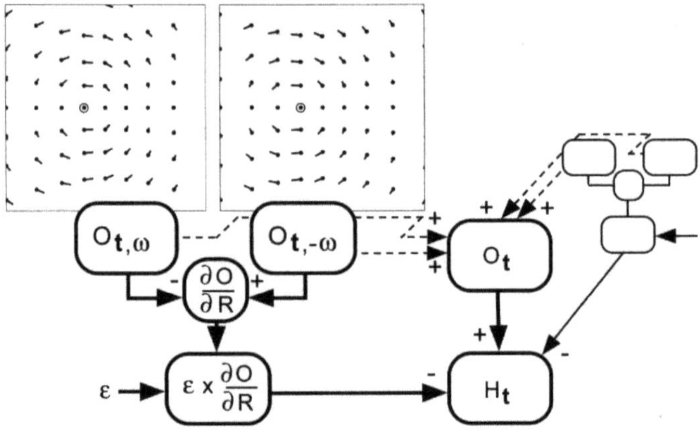

Figure 2. Basic components in the velocity gain field model. The RF structure of typical units, tuned to heading (open circle) and simultaneous left- or rightward rotation about the vertical, is bi-circular. Vectors indicate the preferred direction and velocity of the input motion sensors.

The velocity gain field model differs from the template approach in two important ways. First of all, invariance to rotation is obtained using explicit estimates of rotation, allowing the use of rate-coded extra-retinal signals and reducing the dimensions of templates needed to represent all possible combinations of heading direction and ego-rotations. Secondly, so as to construct templates insensitive to the translational velocity or distances of points, the RF structure of templates in the velocity gain field model was assumed constrained in an explicit non-radial way. These two steps in achieving invariance to parameters other than heading will be explained below.

2.1 Basic Scheme in Velocity Gain Field Model

The basic scheme used by the velocity gain field model is given in Figure 2. Heading is encoded by a collection of templates, each with their own preferred heading **t** and output response H_t. The input to each template H_t is a bundle of templates sharing the same preferred heading **t**, but having different preferred rotations. In principle, three pairs of templates tuned the same heading, each pair having opposite preferred rotation ($\pm\omega$) about one of the cardinal axes of rotation, are sufficient to make the template H_t invariant to rotation magnitudes up to ω (Beintema & van den Berg, 1998).

First of all, invariance for rotation about a given axis is obtained when the template activity H_t equals the activity of a template that shifts its preferred rotation dynamically as much as the actual rotation ε. Generally, a shift of a

Gaussian tuning function can be accomplished by a Taylor series with first and higher order Gaussian derivatives, each term appropriately scaled by first and higher order powers of the desired shift. The response of a template $O_{t,R-\varepsilon}$, shifted in its preferred rotation by an amount ε, is given to first approximation by $O_t - \varepsilon \times \partial O_t / \partial R$ (Beintema & van den Berg, 1998). Note, the signal ε can be an extra-retinal or a visual estimate of rotation velocity.

The idea of dynamically shifting templates can also be understood in terms of compensating activity. The activity H_t needs to be corrected for a change in the activity of the pure translation template O_t in case of rotational flow. Simply subtracting the rate-coded rotation velocity signal ε would not suffice to compensate. For this, ε needs to be multiplied by a visual activity that tells how O_t changes with a change in rotational flow, i.e. the so-called derivative template $\partial O / \partial R$. Since the compensation term $\varepsilon \times \partial O / \partial R$ has the property of being gain-modulated by the extra-retinal velocity signal, this approach has been called the velocity gain field model, analogue to eye position gain fields as reported in area 7a (Andersen et al., 1985).

The model does not require an explicit representation of the subunits as suggested in the scheme. The output of the derivative template $\partial O / \partial R$ can be approximated by the difference activity of two templates tuned to the same heading, but opposite rotation ($O_{t,\pm\omega}$). Alternatively, however, the derivative template could also have been computed directly by summing local difference signals from pairs of motion sensors, each sensor of the pair having equal but oppositely directed preferred motion (Beintema & van den Berg, 1998). Likewise, the pure translation template O_t, tuned to heading but zero rotation, need not be directly computed, but could also be derived from the average response of two rotation-tuned templates with opposite preferred rotation directions, hence the dotted lines feeding the template O_t.

To explain the second step of acquiring insensitivity to the distances of points or the translational velocity, assume a polar coordinate system centered on the direction of heading. Then, the retinal flow can be split into its radial and its circular component (Figure 3). The radial component contains the translational component of the flow, and will vary with translational speed or point distances. The circular component, in contrast, only depends on the rotational component of the flow. This observation leads to the assumption implemented in the velocity gain field model that templates should only measure the flow along circles centered on the point of preferred heading. By assuming such *circular* RF structure, the template strongly reduces sensitivity to variations in depth structure or the translational speed, while preserving its tuning to heading direction and the rotational component of flow (Beintema & van den Berg, 1998).

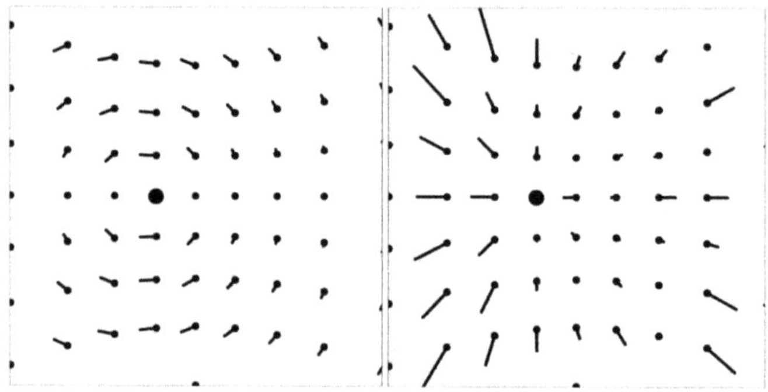

Figure 3. The heading-centered circular (left) and radial (right) component of the flow during combined translation and rotation as in Figure 1c.

2.2 Receptive Fields in Velocity Gain Field Model

Given templates that measure only motion along circles centered on the template's preferred heading, what does the RF structure look like for the units in the velocity gain field model? The RF of a typical pair of templates tuned to heading and opposite rotations about the vertical axis is shown in Figure 2. Note, their preferred structure does not resemble *uni-circular* flow, because the directions of preferred motion vectors reverse across the meridian through the preferred heading and perpendicular to the preferred rotation axis (in this case the horizontal meridian). It is for this reason that the structure has been coined *bi-circular* (van den Berg et al., 2001). Another characteristic of the bi-circular structure is the gradient in motion magnitude along the circle, vanishing at points along the meridian that is perpendicular to the rotation axis. Generally, any unit tuned to rotation about an axis perpendicular to the heading will have a bi-circular structure, modulo a clockwise rotation. For rotation about the horizontal axis, for instance, the bi-circular pattern will be 90 degrees rotated about the heading point.

A similar RF arises if the derivative template is not computed from the difference activity of a pair of rotation-tuned templates, but computed directly. In that case, the RF is expected to be closely related to that of the bi-circular RF. Since the derivative template then is computed from the local difference of the outputs of two motion templates, the RF is represented by the superposition of the two RFs. Interestingly, the receptive field then consists of pairs of motion sensors with opponent-motion directions along circles centered on the preferred heading, and with a magnitude gradient as found for the bi-circular RF structure.

An even more counterintuitive RF structure is expected for the pure translation template O_t, whose output is compensated by the rotation-tuned units. This template is maximally excited when presented with zero motion along its preferred motion directions, i.e. along circles centered on the preferred heading. Such local responses could be accomplished by MT cells that respond maximally to zero-motion while being inhibited by motion in their preferred or anti-preferred direction. Alternatively, as stated before, a pure translation tuned template might also sum the activities of template pairs with opposite preferred rotations but identical heading. Stimulated with flow corresponding to the preferred heading and zero rotation, the output of each rotation-tuned template is balanced by its counterpart with opposite preferred rotation, so that the mean activity will peak at the desired preferred heading. In this case, the predicted RF structure locally consists of motion-opponent pairs as well. But, in contrast to the derivative template, these motion-opponent pairs do not compute the local difference activity of such pairs, but their sum.

3 POPULATION MODEL

The population model by Lappe and Rauschecker (1993b) takes a very different approach to encode heading. The model is an implementation of the subspace algorithm by Heeger and Jepson (1992b) that sets the connection strengths and preferred directions of local motion inputs to heading-specific flow units.

First, we recapitulate the basics of the subspace algorithm, and its neural implementation. Then, we analytically derive the RF structure. Different versions of the model have been proposed (Lappe & Rauschecker, 1995). In the restricted version of the model, the eye movements are assumed to be limited to tracking a stationary point in the scene during the observer translation, without torsion about the line of sight. In the non-restricted version, eye movements are assumed to be independent of the heading, and free to rotate about any axis. The effect of both constraints we shall analyze. Part of the results for restricted eye movements have been published in short form (Beintema et al., 2001). Here, we present the mathematics and the extension to unrestricted eye movements as well.

3.1 Subspace Algorithm

The subspace algorithm resolves the unknown translation direction T from optic flow vectors without requiring knowledge of the three-dimensional rotation vector $\mathbf{\Omega}$ and depths Z_i of m points. In this approach, each flow vector at a given image location (x,y) is written as

$$\boldsymbol{\theta}(x,y) = \frac{1}{Z(x,y)} \mathbf{A}(x,y)\mathbf{T} + \mathbf{B}(x,y)\mathbf{\Omega} \tag{1}$$

with the matrices **A** and **B** depend on image positions (x,y) and focal length f (Heeger & Jepson 1992b):

$$\mathbf{A}(x,y) = \begin{pmatrix} -f & 0 & x \\ 0 & -f & y \end{pmatrix} \text{ and } \mathbf{B}(x,y) = \begin{pmatrix} xy/f & -f - x^2/f & y \\ f + y^2/f & -xy/f & -x \end{pmatrix} \tag{2}$$

By putting all flow components sequentially in a generalized flow vector Θ', the $2m$ flow equations can be written in matrix form by $\Theta' = \mathbf{C}(\mathbf{T}) \cdot \mathbf{q}$. Here, $\mathbf{C}(\mathbf{T})$ is a $2m \times (m+3)$ matrix depending on n image positions and the translation vector $\mathbf{T}$

$$\mathbf{C}(\mathbf{T}) = \begin{pmatrix} \mathbf{A}(x_1,y_1)\mathbf{T} & \cdots & 0 & \mathbf{B}(x_1,y_1) \\ \vdots & \ddots & \vdots & \mathbf{B}(x_2,y_2) \\ 0 & 0 & \mathbf{A}(x_m,y_m)\mathbf{T} & \mathbf{B}(x_m,y_m) \end{pmatrix} \tag{3}$$

and $\mathbf{q}$ is a $(m+3)$-dimensional vector containing the m unknown inverse depths and the three-dimensional rotation vector. The unknown heading $\mathbf{T}_j$ can be solved least squares by minimizing a residual function $R(\mathbf{T}_j) = \left\| \Theta' - \mathbf{C}(\mathbf{T}_j) \cdot \mathbf{q} \right\|^2$. This is equivalent to finding the translation vector $\mathbf{T}_j$ that minimizes

$$R(\mathbf{T}_j) = \left\| \Theta' \mathbf{C}^{\perp}(\mathbf{T}_j) \right\|^2 \tag{4}$$

where $\mathbf{C}^{\perp}(\mathbf{T}_j)$ is the orthogonal complement of $\mathbf{C}(\mathbf{T}_j)$ (Heeger & Jepson, 1992b). For the general case of unrestricted eye movements, the columns of $\mathbf{C}(\mathbf{T}_j)$ form a basis of a $m+3$ subspace of the $\Re^{2m}$, called the range of $\mathbf{C}(\mathbf{T}_j)$. The orthogonal complement is a matrix that spans the remaining $(2m-(m+3))$-dimensional subspace of $\mathbf{C}(\mathbf{T}_j)$. Its columns form the nullspace of the $\mathbf{C}(\mathbf{T}_j)$. Each column is a nullvector and is orthogonal to every row vector of $\mathbf{C}(\mathbf{T}_j)$ (Heeger & Jepson, 1992b).

3.2 Neural Implementation of the Subspace Algorithm

Lappe and Rauschecker (1993b) implemented the subspace algorithm in a neural network to compute *a priori* the connections between MT and MST neurons. Their model approach starts by representing the measured flow vector θ_i by the activity of n local motion sensors s_{ik} with MT-like direction tuning and velocity tuning functions so that $\theta_i = \sum_{k=1}^{n} s_{ik}\mathbf{e}_{ik}$. We represent the motion vector by four mutually perpendicular preferred motion sensors in directions $\mathbf{e}_{ik}$ with cosine-like direction tuning and linear velocity tuning (Lappe & Rauschecker, 1993b, 1995). The measured motion vector could equally well be represented by an extended set of activities based on motion sensors with tuning to specific speeds (Lappe, 1998) without changing our results.

The motion signals from different locations i are collected by second layer neurons that compute the residual function. Their output is a sigmoidal function g of the weighted sum of the inputs s_{ik}

$$u_{jl} = g(\sum_{i=1}^{m}\sum_{k=1}^{n} J_{jl,ik} s_{ik} - \mu) \tag{5}$$

where $J_{jl,ik}$ is the synaptic strength between the l-th output neuron in the second layer population representing heading direction $\mathbf{T}_j$ and the k-th input neuron in the first layer representing the optic flow vector θ_i.

According to the subspace algorithm, the residual function is minimized when all of the measured flow vectors, described by vector Θ', are perpendicular to the l-th column of $\mathbf{C}^{\perp}(\mathbf{T}_j)$. By requiring the input to the neuron to be $\Theta'\mathbf{C}_l^{\perp}(\mathbf{T}_j)$, the synaptic strength for a single second layer neuron will need to be (Eq. 20 in Lappe, 1998):

$$J_{jl,ik} = e_{ik}^{t}\begin{pmatrix} C_{l,2i-1}^{\perp}(\mathbf{T}_j) \\ C_{l,2i}^{\perp}(\mathbf{T}_j) \end{pmatrix} \tag{6}$$

Finally, the population model computes the residual function in a distributed way, partitioning the residual function $R(\mathbf{T}_j)$ into a sum of subresidues, each computed by the l-th neuron. Heading is then represented by the sum of the neurons' responses such that $U_j = \sum_l u_{jl}$. Each l-th neuron, tuned to the correct heading direction, exhibits a sigmoid surface with a rising ridge (not a Gaussian blob) as function of the two-dimensional heading. Although each l-th neuron may have a ridge of different orientation, its position is constrained by the algorithm to pass through the preferred heading $\mathbf{T}_j$. Thus, the population sum activity U_j will peak at the preferred heading. To help ensure that a peak is created, each sigmoid function in Eq. 5 is slightly offset by μ (Lappe & Rauschecker, 1993b).

3.3 Receptive Fields in Population Model

What set of preferred motion directions and connections strengths would be optimal to ensure that the residual function is minimized when presented with the correct heading? According to Eq. 6, the synaptic strengths at each location are given by the subvector in the l-th column of $\mathbf{C}^{\perp}(\mathbf{T}_j)$. We can map the preferred motion $\mathbf{v}_{jl,i}$ by the vector sum of responses to equal stimulation in the $n = 4$ motion sensor directions:

$$\mathbf{v}_{jl,i} = \sum_{k=1}^{n} J_{jl,ik}\mathbf{e}_{ik} = \begin{pmatrix} C^{\perp}_{l,2i-1}(\mathbf{T}_j) \\ C^{\perp}_{l,2i}(\mathbf{T}_j) \end{pmatrix} \qquad (7)$$

Thus, we see that the preferred motion input (motion direction and synaptic strength) are directly given by the two-dimensional subvectors (by the orientation and length, respectively) that make up the orthogonal complement. The population model divides the labor of computing the residual function over different neurons u_{jl} that can have an arbitrary number of inputs. By taking the minimum number of flow inputs required to compute the orthogonal complement, we can reduce the matrix $\mathbf{C}(\mathbf{T}_j)$ so that its orthogonal complement can be solved analytically. Note that the set of motion vectors found this way forms a subset of the receptive field encoding a specific heading. The smallest number of flow inputs for which the orthogonal component can be computed given unrestricted eye movements is $m = 4$. For this case, $\mathbf{C}^{\perp}(\mathbf{T}_j)$ consists of a single column $l = 1$ with 8 elements. For restricted eye movements that are constrained to null the flow at the center of the image, i.e. stabilize gaze on the object at the fovea, without rotation about the line of sight, the rotation $\mathbf{\Omega}$ can be expressed in terms of $\mathbf{T}$ and the fixation distance Z_f. This assumption reduces $\mathbf{C}(\mathbf{T}_j)$ to a

$2m \times (m+1)$-dimensional matrix, and $\mathbf{C}^{\perp}(\mathbf{T}_j)$ to a $(m-1) \times 2m$-dimensional matrix (Lappe & Rauschecker, 1993a). Then, for $m=2$ flow vectors, the orthogonal complement $\mathbf{C}^{\perp}(\mathbf{T}_j)$ consists of only a single column ($l=1$) with length 4 and can be exactly solved. We will first analyze the RF structure for the simplest case of tracking eye movements and two input vectors. Then, we analyze the results for the case of unconstrained eye movements and arbitrary numbers of input vectors.

3.4 Case of Restricted Eye Movements

First, we compute the matrix $\mathbf{C}(\mathbf{T}_j)$ for restricted eye movements. We can choose the preferred heading $\mathbf{T}_j$ to be horizontal ($T_y = 0$) without loss of generality because of symmetry in the xy-plane. Furthermore, we can also set T_z and the focal distance f to unity, so that the heading vector intersects the image plane at the point $f(T_x/T_z, T_y/T_z) = (T_x, 0)$. Then $\mathbf{C}(\mathbf{T}_j)$ becomes

$$\mathbf{C}(\mathbf{T}_j) = \begin{pmatrix} -T_x + x_1 & 0 & T_x(1+x_1^2)/Z_f \\ y_1 & 0 & T_x x_1 y_1/Z_f \\ 0 & -T_x + x_2 & T_x(1+x_2^2)/Z_f \\ 0 & y_2 & T_x x_2 y_2/Z_f \end{pmatrix} \tag{8}$$

The orthogonal complement was computed analytically. For this we used Mathematica to solve the nullspace of the inverted matrix of $\mathbf{C}(\mathbf{T}_j)$. The result presented here is written short by transforming the image coordinates to a system centered on the projection of the translation vector (T_x, T_y) (i.e. $\tilde{x}_i = x_i - T_x$). Then, $\mathbf{C}^{\perp}(\mathbf{T}_j)$ becomes

$$\mathbf{C}^{\perp}(\mathbf{T}_j) = \begin{pmatrix} \tilde{y}_2/\tilde{x}_2 \dfrac{1+T_x(T_x+\tilde{x}_2)}{1+T_x(T_x+\tilde{x}_1)} \\ -\tilde{y}_2\tilde{x}_1/(\tilde{y}_1\tilde{x}_2) \dfrac{1+T_x(T_x+\tilde{x}_2)}{1+T_x(T_x+\tilde{x}_1)} \\ -\tilde{y}_2/\tilde{x}_2 \\ 1 \end{pmatrix} \tag{9}$$

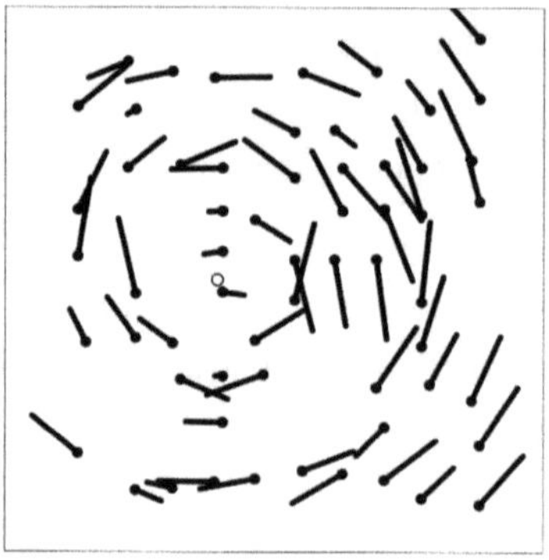

Figure 4. Examples of RF structure of 30 pairs, each pair being input to an u_{jl} neuron with a preferred heading 10 degrees towards the left (open circle).

3.4.1 Circular RF Structures

An example of the RF structure that emerges for arbitrarily chosen pairs of neurons u_{jl}, whose summed responses encode the preferred heading, is given in Figure 4. Most noticeably, the RF structure appears to be organized circularly, and centered on the preferred heading. This would mean that the preferred motion vector $\mathbf{v}_i$ is oriented perpendicular to the vector that radiates from the preferred heading and passes through the location of the motion vector $(\tilde{x}_i, \tilde{y}_i)$. Indeed, as can be seen from Eq. 9, $\mathbf{v}_i \cdot (\tilde{x}_i, \tilde{y}_i) = 0$ for $i = 1$ and $i = 2$.

3.4.2 Motion Opponent RF Structures

We then analyzed in more detail how the preferred motion directions and magnitudes relate to the relative position of motion inputs. First of all, we found that for pairs of motion sensors at nearby locations, the vectors of each pair tend to be opponent (pair A in Figure 5). Indeed, for two motion inputs located infinitely close to each other ($\tilde{x}_2 = \tilde{x}_1$ and $\tilde{y}_2 = \tilde{y}_1$), the orthogonal complement simplifies to $\mathbf{C}^{\perp}(\mathbf{T}_j) = (\tilde{y}_1 / \tilde{x}_1, -1, -\tilde{y}_1 / \tilde{x}_1, 1)'$. Denoting the angular velocity of a vector $\mathbf{v}$ along the circle by scalar v, we find the ratio of preferred angular velocities is ($v_1 / v_2 = -1$). This proves the preferred directions in this case are always opposite and of equal magnitude.

To describe the preferred motion for pairs of motion sensors at different locations, let us define the image location of points in a polar coordinate system centered on the preferred heading. So, points on the same circle have different polar angles but constant eccentricity with respect to the heading.

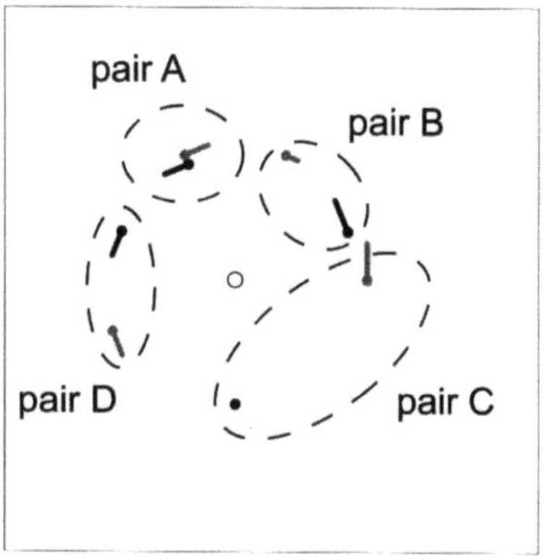

Figure 5. Examples of local direction preferences in a RF from a population of 4 motion pairs (A-D). The MST neuron encodes heading 10 degrees towards the left (open circle).

We exemplify our observations by a few pairs of motion input in Figure 5. First of all, the difference between motion magnitudes increases sinusoidally with the polar angle difference between the two points (compare pair A, B and C). Secondly, consider the meridian through the fixation and heading point. The preferred motion directions are opponent if the pair is located on the same side with respect to that meridian (pair A, B, C).

3.4.3 Bi-circular RF Structures

We also found other interesting substructures under certain constraints on the locations of motion inputs. The preferred motion directions are uni-directional if the pair is split across the meridian through fixation and heading (Figure 5, pair D). Moreover the magnitudes are equal for positions that mirror in the meridian (pair D). Also, eccentricity with respect to the heading had no influence on the preferred motion vectors. In principle, this would allow the construction of a uni-circular RF. Interestingly, we found that if pairs of motion sensors have their partners at image locations 90 degrees rotated about the heading point, a magnitude gradient appears, and a RF structure can be constructed (Figure 6) that has great similarity to the bi-circular RF pattern predicted by the velocity gain field model (Figure 2).

Figure 6. Distribution of pairs of neurons with preferred heading 10 deg left, arranged to make a bi-circular RF.

3.5 Case of Unrestricted Eye Movements

We then computed analytically the orthogonal complement for the minimum number of required motion inputs in the case of unrestricted eye movements ($m = 4$). Again, we find only circular RFs structures. We analyzed the preferred motion magnitudes and directions for various spatial arrangements. In Figure 7 we plotted three possible spatial arrangements. In contrast to results for restricted eye movements, none of these structures appeared to fit a uni-circular structure, nor the bi-directionality and magnitude gradient expected for a bi-circular RF structure. We do find an influence of eccentricity, and a strong degree of motion-opponency. In the special case when pairs of motion vectors were located at exactly the same position, each pair turns out to have equal but opposite preferred motions (Figure 7c).

3.5.1 General Solution

Simulations showed the result of circularity applied to any number of vector inputs. A similar result can actually be found in the proof of the existence and uniqueness of solutions to the subalgorithm (Heeger & Jepson, 1990). From Eq. 3, one can see that any nullvector of $\mathbf{C}(\mathbf{T}_j)$ must be a linear summation of m vectors $\mathbf{\Phi}_i$ of the following form:

$$\mathbf{\Phi}_i = (0,0,0,0,\cdots,\Phi_{ix},\Phi_{iy},\cdots,0,0) \tag{10}$$

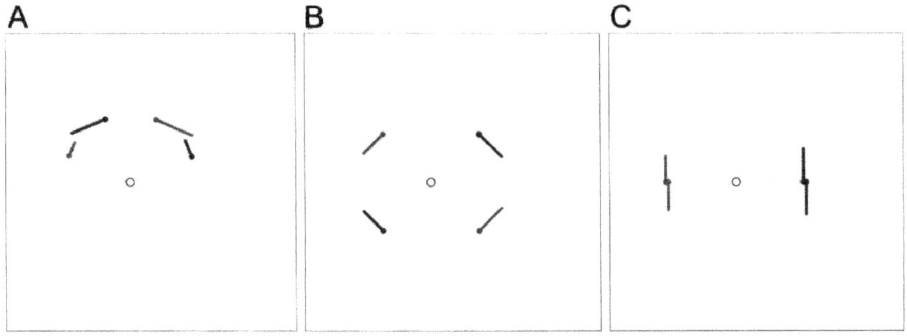

Figure 7. Preferred motion vectors for neuron u_{j1} in the case of unrestricted eye movements given four motion inputs. The motion inputs are chosen on a circle centered on the heading direction (open circle) and separated by a) 45, b) 90, and c) 180 degree.

These vectors are perpendicular to the first m columns of $\mathbf{C}(\mathbf{T}_j)$ and have two non-zero elements given by $\mathbf{\Phi}_i \cdot \mathbf{A}(x_i, y_i)\mathbf{T}$. Substituting ($T_y = 0$, $T_z = 1$, $f = 1$) for $\mathbf{A}$ in Eq. 2, the elements of each vector are related by:

$$\Phi_{iy} / \Phi_{ix} = \frac{x_i - T_x}{-y_i} = \frac{\tilde{x}_i}{-\tilde{y}_i} \tag{11}$$

As each $\mathbf{\Phi}_i$ is oriented along a circle centered on the heading point, any weighted combination $\sum_{i=1}^{m} c_i \mathbf{\Phi}_i$ will be oriented circularly as well. The weights c_i follow from the requirement of rotation invariance, i.e. the weighted vector sum must also be perpendicular to the remaining columns of $\mathbf{C}(\mathbf{T}_j)$ made out of $\mathbf{B}$ matrices (Eq. 2).

When the number of flow vector inputs is larger than the minimum required (2 for tracking eye movements, 4 for unrestricted eye movements), the problem of computing the nullspace of $\mathbf{C}(\mathbf{T}_j)$ becomes underdetermined, such that a continuum of possible solutions exist and the nullvector is just an arbitrary pick among them. Nevertheless, we found that the directions in the case of unrestricted eye movements were always balanced.

4 TEMPLATE MODEL

The template model and the velocity gain field model both rely on templates tuned to heading and a specific component of rotation. But, so as to make each local contribution to a template invariant to the distances of points, the velocity gain field explicitly assumes that only the component of flow

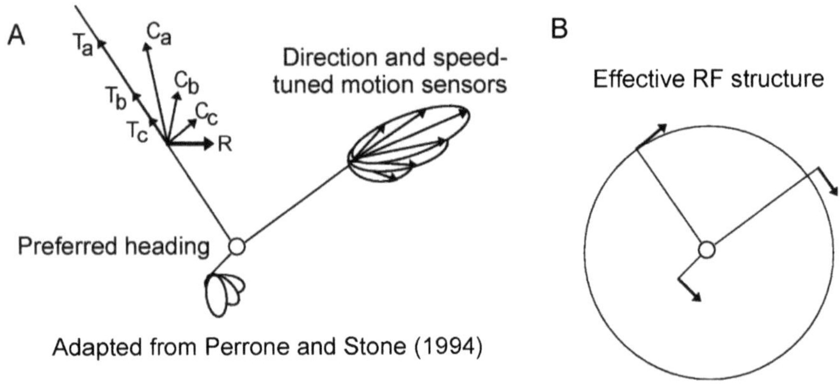

Figure 8. a) Each template sums the responses of the most active sensor at each location. This most active motion sensor is selected from a pool of sensors tuned to different depth planes (C_a, C_b, etc). These vectors are the vector sums of the preferred rotation component R and translational components T_a, T_b, etc. b) At each location the effective motion vector driving the template is tangent to a circle centered on the template's preferred heading.

along the circular structure is measured, while the template approach (Perrone & Stone, 1994) selects the maximum response of various motion sensors. Does the template approach predict circular receptive fields as well?

4.1 Receptive Fields in Template Model

In the template approach, at each location not one motion sensor is read-out, but the most active one in a set of motion sensors tuned to different depth planes but the same preferred heading and rotation. Each motion sensor has a preferred motion that is the vector sum of a radial vector, expected from a translating point on a certain depth plane, combined with a motion vector expected from the template's preferred rotation (Figure 8a). As one can see, the expected vector sums C_i at one specific location are constrained by a line that runs parallel to the radial line. Motion parallel to this constraint line will only change which sensor is maximal active, but not lead to a change in maximal activity (unless the motion lies outside the range of each motion sensor). Only motion perpendicular to the constraint line will change the maximal activity. This motion is directed along the circle centered on the preferred heading, and equals the expected motion from the template's preferred rotation (R), taken along the circle centered on the preferred heading (C_c).

Note that the most active motion sensors need not be directed along heading-centered circles. This is clearly the case for the pure translation

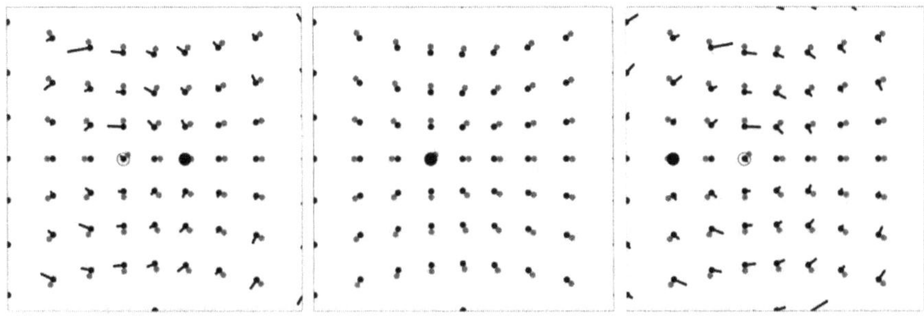

Figure 9. Circular component of flow (solid inner vectors) resulting from pure translation through a 3-D cloud in three different directions (filled circle in each plot), measured along the preferred circular RF structure (gray outer vectors) centered on the preferred heading (open circle).

template. But, also for a template with a preferred component of rotation, the contributing motion sensor need not be oriented along the circle, for instance if the expected rotational motion vector is not tangent to the circle. Then, the local contribution would be determined by the directional tuning rather than the velocity tuning of the most active sensor. Still, the effective motion vector that drives the template's output is oriented along a circle centered on the template's preferred heading (Figure 8b). Whereas the velocity gain field model leaves unspecified how the velocity and direction tuning to motion along the circle might be accomplished at MT level, the selection of maximum activity from a set of constrained sensors, as proposed by the template model, might be a way to realize such tuning.

5 DISCUSSION

5.1 Heading-Centered Circular RF Structures: Intuitive or Counterintuitive?

We looked at the structure of receptive fields for flow analysis in three models for heading detection. The models apply different methods to solve the heading. Nevertheless, we find in all three cases the directions of preferred motion to be constrained along circles centered on the heading.

Circular RFs for heading detection seem rather counterintuitive, for the flow expected from ego-translation is radial. Why would a circular structure be optimal to detect heading? The important point to realize here is that only a deviation from expected heading is of interest. Thus, only the component of translation perpendicular to expected translation really is of importance to sense heading, not the component in the expected heading direction. As the components in the expected heading direction are radial, what is left to be analyzed are components of flow along the circles centered on the expected heading.

How can circular RFs still measure heading? This can be understood by examining the effects of components of translational and rotational flow separately. As shown in Figure 9, a heading-centered circular RF only measures the translational flow caused by translation perpendicular to the preferred heading. By measuring the evidence for zero motion along its preferred circular structure, a neuron can sense a deviation from its preferred heading direction, while not being too sensitive to variations in translational speed along the preferred heading and distances of points.

Circular receptive field structures could explain why area MST is found to be not only selective for expanding motion patterns, but also has a significant proportion of cells are selective to rotation patterns (Duffy & Wurtz, 1991a,b; Lappe et al., 1996; Tanaka et al., 1989; Tanaka & Saito, 1989). The link between selectivity for circular flow structures and heading detection mechanisms also suggests that testing selectivity for expanding motion might be a bad indicator for determining a cell's preferred heading. This point has been noted before, as MST seems to be systematically tuned to the focus of rotation, exactly like model neurons in the population model (Lappe et al., 1996). Another way to test for circular RF structure might be to look at sensitivity to the variations in translational speed or in the speed gradient, when radial flow is centered on the cell's preferred heading. Indeed, early studies suggest that expansion cells are not sensitive to the speed gradient (Tanaka et al., 1989; Orban et al., 1995) or speed itself (Duffy & Wurtz, 1991a). Recent studies do suggest speed sensitivity in MST (Duffy & Wurtz, 1997, Upadhyay et al., 2000) and area 7a (Phinney & Siegel, 2000). However, it would still be critical to know whether the expanding motion was presented exactly centered on the cell's preferred heading.

5.2 RF Structures and Rotation Invariance

Given neurons that measure the evidence for zero-motion along their circular structure, how do they acquire invariance to rotation and what substructures might arise?

In both the velocity gain field model (Beintema & van den Berg, 1998), as well as the template model (Perrone & Stone, 1994), the evidence for zero-motion along the circular structure is measured by a pure translation template. Its activity (O_t in the velocity gain field model) is computed by summing the outputs of motion sensors, where each local motion sensor is oriented along the circular structure and assumed to be tuned to zero-motion. However, because each pure translation template will be influenced by a component of rotation, some trick must be applied to assure the correct template is maximally active.

In the template model, the rotation problem is simply solved by assuming an array of templates that includes all possible combinations of two-dimensional heading and three-dimensional rotations. Such an approach requires many rotation-tuned templates per rotation axis, although one can reduce the dimensions when assuming the eye movements are restricted (Perrone & Stone, 1994). An alternative solution to the rotation problem is used in the velocity gain field approach (Beintema & van den Berg, 1998). Any change in activity of a pure translation template O_t due to a component of rotation is compensated by subtracting an appropriate derivative template activity, multiplied by a measure of the evidence for rotation about that axis ε, i.e. $\varepsilon \times \partial O / \partial R$. The range of rotation velocities is limited by the preferred rotation of the templates, but can be expanded by assuming also pairs tuned to larger rotation velocities.

Thus, both the template and velocity gain field model predict templates specially tuned to heading and zero-preferred rotation (zero-preferred motion along circles). Their receptive field would consist of motion detectors directed along the circular structure, but with zero-preferred motion. But, even more abundantly represented might be templates tuned to heading and a component of rotation about a given axis, especially about an axis perpendicular to the heading, giving rise of *bi-circular* RF structures. Units tuned to rotation about the heading axis would have a *uni-circular* RF structure.

The population model acquires invariance to rotation and depths of points in a different way. Given the locations (image coordinates) of motion inputs to the neuron u, the subspace algorithm is used to generate sets of preferred motion vectors $\mathbf{v}_i$ whose summed inner products with the presented motion vectors must always be zero $\sum_{i=1}^{m} \mathbf{v}_i \cdot \theta_i = 0$ (i.e. the residual function $R(\mathbf{T})$) when the preferred heading is presented. The subspace algorithm splits the flow equations into a part related to translation and a part related to the rotation. The requirement of invariance to translational speed and distances of points constrains all preferred motion vectors to lie along circles centered on the preferred heading. Given that the sum of vector inner products along the circles must also be zero in case of rotation, this additionally constrains the directions (clockwise or counterclockwise) and magnitudes of preferred motion vectors.

Taken the results from our analysis, we can interpret how the RF structure plays its role in obtaining invariance to rotation. The simplest case occurs for arrangements in local pairs, in which case the preferred motion vectors of pairs were found to be opponent and equal of magnitude. Such an arrangement evidently will exclude any contribution from rotational flow to the residual function, because rotation causes two points to move exactly in same direction at same speed. Since one motion vector excites one of the motion sensors, and the same motion inhibits the response of the other motion sensor, their contributions exactly cancel. That the model is able to detect heading for such local motion-opponent pairs, relies on the fact differential motion will be presented as soon as the heading deviates from the preferred heading. Obviously, such local pair arrangements makes heading detection rely on the presence of depth differences along the same visual direction (edges, etc.). For more spatially separated motion inputs, the results for restricted eye movement results are helpful to interpret how invariance to rotation is obtained, for then only one possible axis of rotation was assumed. In that case, we found that the preferred motion magnitude varied sinusoidally as function of the position on the circles (Figure 5), such that it is small when the motion vector expected from rotation is large, and vice versa. This way the sum of two motion responses will cancel during rotation about the vertical axis.

More difficult to analyze is the receptive field structure for unrestricted eye movements, in which case a minimum of four motion vectors is required as input to a u_{jl} neuron. The receptive field structure turned out to be opponent on average. This can be interpreted as a consequence of assumed invariance to rotation about all possible axes. Because the motion directions are balanced in clockwise and counterclockwise direction, flow caused by a rotation about the line of sight does not give a net motion input to the residual function. Also, because the four vectors are balanced vertically and horizontally, the contributions of flow caused by rotation perpendicular to the heading axes cancel.

In summary, we find that the population arrives at rotation-invariance in a way essentially different from template approaches. It relies on zero-motion parallax detection, rather than on zero motion detection as assumed in the velocity gain field model or template model. This difference leads to different expected substructures within the circular RF. Receptive fields meant to measure zero uniform motion require additional RFs that provide estimates of ego-rotation.

5.3 Opponent-Motion RF Structures

Especially, when the motion inputs are arranged in pairs of local neighboring motion inputs, a clear opponent motion RF structure emerges in the population model. This property of the subspace algorithm was also noted by Heeger and Jepson (1992a), who proposed a center-surround motion-opponent structure to measure heading. The population model's reliance on motion-parallax to obtain rotation invariant heading estimates has great analogue to Longuet-Higgins & Prazdny's (1980) original use of differential motion parallax. Their approach exploited the fact that local difference motion vectors are constrained to be oriented radially and intersect at the heading direction. As this approach relies on the presence of motion parallax in one visual direction, it requires scenes with depth edges. The differential motion approach can be improved by computing motion differences within a larger visual area, therefore not requiring the presence of depth edges (Rieger & Lawton, 1985; Hildreth, 1992; Royden, 1997). Indeed, a physiological implementation of this idea assumes the local motion-opponency outputs are summed by a MST-like template which measures the evidence for non-zero differential motion along the radial structure centered on the cell's preferred heading (Royden, 1997). The downside of exploiting motion differences this way, however, is that radial flow detectors are still sensitive to variations in distances of points or to translational velocity.

In contrast, the population model arrangement in local motion-opponent pairs analyses the heading by MST-like cells that detect the evidence for *zero motion-opponency* along their preferred circular structure. This approach is invariant to the translational velocities and distances of points. In that respect, the population model approach is an improvement over traditional motion-parallax theories. Furthermore, the opponent character of pairs in the population model is not restricted to locally near sensors, but is also seen for motion sensors further apart.

Do the receptive fields of templates also exploit motion-parallax by pairs of opponent-motion sensors? The derivative template response, computed by locally subtracting the responses of a pair of templates with opposite preferred rotation, does predict such an opponent structure. However, this local subtraction of motion activities is meant to give a local estimate of the rotation, not an estimate of a local motion difference. Thus, motion parallax is considered noise by these templates.

The velocity gain field model, however, included an extension to allow pure visual solutions (Beintema & van den Berg, 1998) that might be the analogue for the detection of zero-motion parallax. The added feature was the suppression of those H_t templates that use false estimates of the rotation. Such wrong estimates occur for derivative template activities that are contaminated by components of translational flow along their preferred circular structure. As shown in Figure 10 this occurs for non-preferred headings. To get around this problem, the authors theorized that some measure of variance in the presented flow along the circular structure due to motion parallax might be exploited for suppression in case of a non-preferred heading. To this end they used the variance of local estimates of the flow along the circular structure, each scaled with the preferred local flow of a

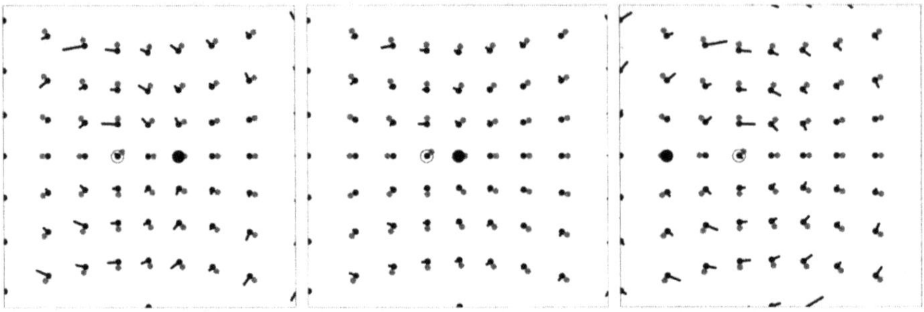

Figure 10. Circular component of flow (solid inner vectors) resulting from rotation combined with observer translation in three different directions (filled circle in each plot), compared with the preferred bi-circular RF structure (gray outer vectors) centered on the preferred heading (open circle).

rotation-tuned template. This measure of variance in local rotational gain is not sensitive to rotation about the preferred axis, but does rise with increased motion parallax along the preferred circular structure. Suppressing the output of a template H_t when the variance is high due to the presence of motion parallax seems similar to reducing the response of neuron u_{jl} in the population model when it measures evidence for non-zero-motion parallax. Therefore, this extension to the velocity gain field model might also be based on a motion-opponent structure that truly measures differences in motion.

In summary, motion opponency seems to emerge in models encoding heading at the level of MST as well, except for the template approach (Perrone & Stone, 1994), which does not exploit the use of motion-parallax cues. So far opponent-motion detectors have only been found at the level of MT cells. Our analysis suggests looking for such opponent-motion structures, at a higher level as well, be it along circular RF structures.

5.4 Bi-circular RF Structures

A typical substructure in circular receptive fields predicted by the velocity gain field model and the template model is bi-circularity. Interestingly, for the restricted version of the population model that assumes tracking eye movements, we were able to construct a *uni-circular* RF and *bi-circular* RF for specific spatial arrangements of the motion inputs. In terms of velocity gain field templates; these two substructures would be optimally tuned to a rotation about the line of sight, and about a fronto-parallel axis parallel to the heading axis, respectively. Such sensitivity to rotation is perfectly in line with the assumption of tracking eye movements, because this only assumes invariance for rotation about the axis perpendicular to the heading axis and the line of sight. However, for unrestricted eye movements, we found no possible arrangement of points that would fit a *bi-circular* or *uni-circular* RF. Therefore, these substructures are unlikely an essential feature of the population model.

Bi-circular receptive fields might play a role as visual rotation estimators. Visual estimates of rotation can only be reliably made at the level of global flow analysis. Such visual rotation estimates could, for instance, be important to gauge extra-retinal signals about rotation of the eye. Bi-circular RFs would also allow the use of extra-retinal signals at a more global level of motion analysis (Beintema & van den Berg, 1998). The population model can also be extended to use extra-retinal signals, but at a more local level (Lappe, 1998). So far, clear evidence for an interaction between visual and extra-retinal signals has only been found at the level of MST responses (Bradley et al., 1996; Shenoy et al., 1999). Moreover, psychophysical experiments on perceived heading suggest that extra-retinal signals must interact beyond the level at which neurons analyze the local flow (Beintema & van den Berg, 2001). A direct comparison of the extra-retinal and visual estimates of rotation allows one model (Beintema & van den Berg, 1998) to explain on one side the benefit of extra-retinal signals during real eye rotations for judging heading when motion parallax cues are absent (Royden et al., 1992), and on the other side the benefit of motion-parallax cues during simulated eye rotation (van den Berg, 1992; Warren & Hannon, 1990).

Thus, bi-circular RF structures seem important substructures of receptive fields to look for. MST responses to uni-circular motion, where the focus of rotation is presented outside the visual field (Duffy & Wurtz, 1995; Graziano et al., 1994), might be in line with bi-circular RFs, because then only part of uni-circular flow matches bi-circular flow. But whether MST cells truly have such substructures still remains to be investigated.

6 CONCLUSION

The analysis of the receptive field structure in neurophysiologically inspired models provides new insights into how the brain might detect heading most effectively. The most important conclusion is that the models predict a circular, not radial, receptive field centered on the neuron's preferred heading. Different strategies to arrive at rotation invariance lead to different predicted substructures. To exploit motion parallax cues, detectors might be organized in motion-opponent pairs and measure the evidence for null motion-opponency along the circular structure. Explicit representations of visual evidence for rotation would predict bi-circular structures. Seeking evidence for circular and more refined structures offer challenging directions for future research.

ACKNOWLEDGEMENTS

ML is supported by the German Science Foundation, the German Federal Ministry of Education and Research, the HFSP and the EC Ecovision Project. AB and JB are supported by NWO and the HFSP.

REFERENCES

Allman, J., Miezin, F., & McGuinness, E. (1985). Stimulus specific responses from beyond the classical receptive field: Neurophysiological mechanisms for local-global comparisons in visual neurons. *Ann. Rev. Neurosci., 8*, 407-430.

Andersen, R. A., Essick, G. K., & Siegel, R. M. (1985). Encoding of spatial location by posterior parietal neurons. *Science, 230* (4724), 456-458.

Beintema, J. A., & van den Berg, A. V. (1998). Heading detection using motion templates and eye velocity gain fields. *Vision Res., 38* (14), 2155-2179.

Beintema, J. A., & van den Berg, A. V. (2001). Pursuit affects precision of perceived heading for small viewing apertures. *Vision Res., 41 (18),* 2375-2391.

Beintema, J. A., van den Berg, A. V., & Lappe, M. (2002). Receptive field structure of flow detectors for heading perception. In: T.G. Dietterich, S. Becker, & Z. Ghahramani (Eds.), *Neural Information Processing Systems, 14*, MIT Press, Cambridge, MA.

Bradley, D. C., Maxwell, M., Andersen, R. A., Banks, M. S., & Shenoy, K. V. (1996). Mechanisms of heading perception in primate visual cortex. *Science, 273* (5281), 1544-1547.

Britten, K. H., & van Wezel, R. J. (1998). Electrical microstimulation of cortical area MST biases heading perception in monkeys. *Nat. Neurosci., 1 (1),* 59-63.

Colby, C. L., Duhamel, J. R., & Goldberg, M. E. (1993). Ventral intraparietal area of the macaque: anatomic location and visual response properties. *J. Neurophysiol., 69 (3),* 902-914.

Duffy, C. J., & Wurtz, R. H. (1991a). Sensitivity of MST neurons to optic flow stimuli. I. A continuum of response selectivity to large-field stimuli. *J. Neurophysiol., 65 (6),* 1329-1345.

Duffy, C. J., & Wurtz, R. H. (1991b). Sensitivity of MST neurons to optic flow stimuli. II. Mechanisms of response selectivity revealed by small-field stimuli. *J. Neurophysiol., 65 (6),* 1346-1359.

Duffy, C. J., & Wurtz, R. H. (1995). Response of monkey MST neurons to optic flow stimuli with shifted centers of motion. *J. Neurosci., 15 (7),* 5192-5208.

Duffy, C. J., & Wurtz, R. H. (1997). Medial superior temporal area neurons respond to speed patterns in optic flow. *J. Neurosci., 17* (8), 2839-2851.

Graziano, M. S., Andersen, R. A., & Snowden, R.J. (1994). Tuning of MST neurons to spiral motions. *J. Neurosci., 14 (1),* 54-67.

Grigo, A., & Lappe, M. (1999). Dynamical use of different sources of information in heading judgments from retinal flow. *J. Opt. Soc. Am. A Opt. Im. Sci. Vis., 16 (9),* 2079-2091.

Hatsopoulos, N. G., & Warren, W. H., Jr. (1991). Visual navigation with a neural network. *Neural Netw., 4 (3),* 303-318.

Heeger, D. J., & Jepson, A. (1990). Subspace methods for recovering rigid motion, II: theory. *Technical Report,* RBCV-TR-90-36 (Dept of Computer Science, University Toronto.

Heeger, D. J., & Jepson, A. (1992a). Recovering observer translation with center-surround motion-opponent mechanisms. *Invest. Ophthalmol. Vis. Sci. Supp., 32,* 823.

Heeger, D. J., & Jepson, A. (1992b). Subspace methods for recovering rigid motion I: algorithm and implementation. *Int. J. Comput. Vis., 7,* 95-117.

Hildreth, E. C. (1992). Recovering heading for visually-guided navigation. *Vision Res., 32 (6),* 1177-1192.

Lappe, M. (1998). A model of the combination of optic flow and extraretinal eye movement signals in primate extrastriate visual cortex. *Neural Netw., 11,* 397-414.

Lappe, M. (2000). Computational mechanisms for optic flow analysis in primate cortex. *Int. Rev. Neurobiol., 44,* 235-268.

Lappe, M., Bremmer, F., Pekel, M., Thiele, A., & Hoffmann, K. P. (1996). Optic flow processing in monkey STS: a theoretical and experimental approach. *J. Neurosci., 16 (19),* 6265-6285.

Lappe, M., & Rauschecker, J. P. (1993a). Computation of heading direction from optic flow in visual cortex. In: C.L. Giles, S.J. Hanson, & J.D. Cowan (Eds.), *Advances in Neural Information Processing Systems, 5* (pp. 433-440). San Mateo, CA: Morgan Kaufmann.

Lappe, M., & Rauschecker, J. P. (1993b). A neural network for the processing of optic flow from ego-motion in man and higher mammals. *Neural Comput., 5,* 374-391.

Lappe, M., & Rauschecker, J. P. (1995). Motion anisotropies and heading detection. *Biol. Cybern., 72* (3), 261-277.

Longuet-Higgins, H. C., & Prazdny, K. (1980). The interpretation of a moving retinal image. *Proc. R. Soc. Lond. B Biol. Sci., 208 (1173),* 385-397.

Orban, G. A., Lagae, L., Raiguel, S., Xiao, D., & Maes, H. (1995). The speed tuning of medial superior temporal (MST) cell responses to optic-flow components. *Perception, 24 (3),* 269-285.

Perrone, J. A. (1992). Model for the computation of self-motion in biological systems. *J. Neurophysiol., 9 (2),* 177-194.

Perrone, J. A., & Stone, L. S. (1994). A model of self-motion estimation within primate extrastriate visual cortex. *Vision Res., 34 (21),* 2917-2938.

Phinney, R. E., & Siegel, R. M. (2000). Speed selectivity for optic flow in area 7a of the behaving macaque. *Cereb. Cortex, 10 (4),* 413-421.

Read, H. L., & Siegel, R. M. (1997). Modulation of responses to optic flow in area 7a by retinotopic and oculomotor cues in monkey. *Cereb. Cortex, 7 (7),* 647-661.

Rieger, J. H., & Lawton, D. T. (1985). Processing differential image motion. *J. Opt. Soc. Am. A, 2 (2),* 354-360.

Royden, C. S. (1997). Mathematical analysis of motion-opponent mechanisms used in the determination of heading and depth. *J. Opt. Soc. Am. A, 14 (9),* 2128-2143.

Royden, C. S., Banks, M. S., & Crowell, J. A. (1992). The perception of heading during eye movements. *Nature, 360 (6404),* 583-585.

Shenoy, K. V., Bradley, D. C., & Andersen, R. A. (1999). Influence of gaze rotation on the visual response of primate MSTd neurons. *J. Neurophysiol., 81 (6),* 2764-2786.

Tanaka, K., Fukada, Y., & Saito, H. A. (1989). Underlying mechanisms of the response specificity of expansion/contraction and rotation cells in the dorsal part of the medial superior temporal area of the macaque monkey. *J. Neurophysiol., 62 (3),* 642-656.

Tanaka, K., & Saito, H. (1989). Analysis of motion of the visual field by direction, expansion/contraction, and rotation cells clustered in the dorsal part of the medial superior temporal area of the macaque monkey. *J. Neurophysiol., 62 (3),* 626-641.

Upadhyay, U. D., Page, W. K., & Duffy, C. J. (2000). MST responses to pursuit across optic flow with motion parallax. *J. Neurophysiol., 84 (2),* 818-826.

van den Berg, A. V. (1992). Robustness of perception of heading from optic flow. *Vision Res., 32 (7),* 1285-1296.

van den Berg, A. V., Beintema, J. A., & Frens, M. A. (2001). Heading and path percepts from visual flow and eye pursuit signals. *Vision Res., 41 (25-26),* 3467-3486.

van Wezel, R. J., & Britten, K. H. (2002). Multiple uses of visual motion. The case for stability in sensory cortex. *Neuroscience, 111 (4),* 739-759.

Warren, W. H., Jr., & Hannon, D. J. (1990). Eye movements and optical flow. *J. Opt. Soc. Am. A, 7 (1),* 160-169.

11. Parametric Measurements of Optic Flow by Humans

José F. Barraza and Norberto M. Grzywacz

Department of Biomedical Engineering and Neuroscience Program
University of Southern California, Los Angeles, CA, USA

1 THEORETICAL FRAMEWORK

Motion pervades the visual world (Marr, 1982). This is so not only because images in nature are unlikely to be static, but also because motion constitutes a rich source of information to understand those images. The strategy that the brain uses to measure motion in images has been extensively studied and many theories have been proposed. The challenge has been to design a biologically plausible theory that will predict all motion phenomena and work on real images. With this approach, Yuille and Grzywacz (1998) have proposed a theoretical framework for visual motion that accounts for most existing psychophysical and physiological experiments. The theory proposes that the visual system fits internal models to the incoming retinal data, selecting the best models and their parameters. The fit begins with a measurement stage that performs local estimates of motion, such as local velocity (Bravo & Watamaniuk, 1995). These local estimates, which are noisy (Shadlen & Newsome, 1998) and sometimes ambiguous (Movshon et al., 1985), are then clustered (in a space of measurement variables) into regions whose boundaries correspond to motion boundaries. This clustering is performed by a number of competitive processes corresponding to different motions. The theory proposes that different types of tests can compete to perform this clustering. For example, the grouping can be done by either parametric or non-parametric tests. The former test will try to detect familiar motions defined by prior statistics-of-natural-scenes models, while the latter will allow the visual system to deal with general types of motion that may never have been seen before.

Although the knowledge of motion statistics in natural images is important to understand what these familiar motions are, some theoretical

L.M. Vaina, S.A. Beardsley and S.K. Rushton (eds.), Optic Flow and Beyond, 249–271.

Figure 1. The pictures are two consecutive frames of a movie showing the road where the viewer is driving. The three vectors plotted on the left picture represent, schematically, the local linear velocities of three arbitrary points on the road. The vectors on the right picture indicate the local velocities of the same three points a unit of time later. Comparing the two pictures, it can be seen that the optic flow presents a combination of two different components: an expansion and a counter-clockwise rotation. This is an example of a complex flow field that can be present in real-life vision.

studies can shed light over this issue. Koenderink and van Doorn (1976) have shown that any complex flow field generated by motions of planar surfaces can be decomposed into several elementary components including translation, expansion, and rotation. Because small patches of natural surfaces are approximately planar, it may be possible to perform this decomposition for small portions of natural images. Therefore, the visual system could take advantage of this decomposition by incorporating these elementary components as models for the analysis of motions in real scenes (Figure 1). Experimental justification for this hypothesis comes from psychophysical experiments showing the existence of looming and rotation detectors (Regan & Beverly, 1985; Freeman & Harris, 1992; Morrone et al., 1995; Snowden & Milne, 1996) and from physiological studies showing that there are cortical neurons sensitive to translation, rotation, expansion, and spiral motion (Maunsell & Van Essen, 1983; Tanaka & Saito, 1989; Tanaka et al., 1989; Duffy & Wurtz, 1991a,b; Graziano, et al., 1994; Lagae et al., 1994). However, to meet the requirements of the framework these specialized neural mechanisms should work parametrically. For example, a parametric model for rotation would have a center of rotation and an angular velocity as parameters to be determined.

It is known that humans can precisely estimate parameters of translational motions such as direction (De Bruyn & Orban, 1988) and speed (McKee, 1981; Johnston et al., 1999). However, little is known about other types of motions. Some pieces of evidence can be found in recent investigations showing that the visual system would not be using just the estimates of local velocity to evaluate the speed of rotation and the rate of expansion (Geesseman & Qian, 1996, 1998). For instance, an expansion appears to move faster than a rotation, even if the local linear velocities at corresponding points of the images are equal. This effect was explained by arguing that the subject does not perform the task by using the local velocities but some form of global velocity (Clifford et al., 1999). The subject could be comparing the rate of expansion with the angular velocity of the rotation. Why a subject sees expansion faster than rotation is a fascinating question, which however, is outside of the scope of this article. For us, what is important is that these motions are seen different, because the rate of expansion, the angular velocity or both of these parameters are not equally perceptually scaled.

Whether the mechanisms that are specialized in the detection of Koenderink and van Doorn's elementary components are metric is the question that we will address in this chapter. Moreover, we will discuss how the visual system can compute these complex motion parameters from real scenes.

2 MEASURING ROTATIONAL MOTION

Werkhoven and Koenderink (1993) suggested that human subjects cannot make precise estimates of angular velocity from rotational motions. These authors found that such estimates were systematically biased towards linear velocity. Unfortunately, Werkhoven and Koenderink's results could not be explained by tangential linear-speed discrimination either. If one tries to explain their results in terms of linear velocity, then the data present a systematic bias towards angular velocity. We wondered whether angular velocity was not well discriminated in their experiment, because the stimuli had few dots and thus provided poor rotational information. We retook this issue and investigated whether there are conditions under which angular velocity can be estimated more precisely.

We used the same strategy as Werkhoven and Koenderink (1993) to dissociate angular velocity from linear velocity. We performed a matching velocity experiment in which the subjects had to evaluate the speed of rotation of a test annulus with respect to a reference annulus stimulus. The perceived angular velocity was measured as a function of the ratio between the test and the reference radii (R_t/R_r). Stimuli consisted of quasi-random-dot-annuli revolving around a fixation mark (Barraza & Grzywacz, 2002a).

Figure 2 shows the angular-velocity-matching results of one of the naïve subjects. The symbols represent the experimental data and the dotted line represents the reference angular velocity. The solid line shows the predicted angular velocity for a matching performed using tangential speeds. The data fit the reference angular velocity prediction, which means that human subjects can measure angular velocity. The systematic bias towards tangential speed

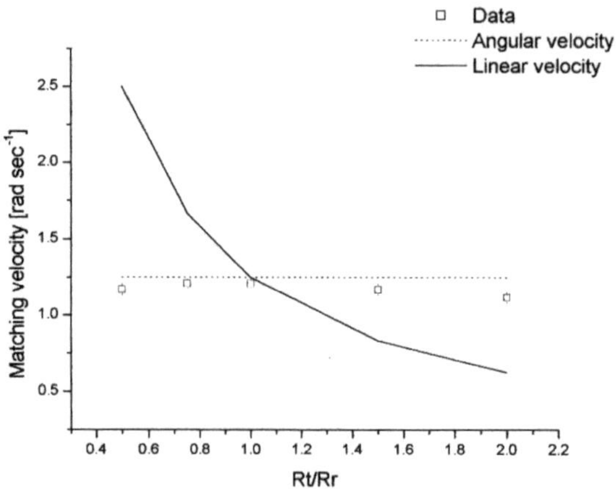

Figure 2. Perceived angular velocity as a function of Rt/Rr. The symbols represent the experimental data, the dotted line represents the angular velocity of the reference, and the solid line represents the angular velocity expected if the task were performed through tangential-speed matching. The results show that the subjects can use angular velocity to match speed of rotation (from Barraza & Grzywacz, 2002a).

found by Werkhoven and Koenderink (1993) did not appear in our data. Perhaps, this was because our stimulus had more dots; the mean distance between dots (d_{nd}) in our stimulus was 0.19°, whereas theirs was approximately 1.35°.

2.1 Effect of Dot Density on the Estimate of Angular Velocity

We wondered whether dot density could account for the differences between the Werkhoven-and-Koenderink's results and ours. To explore this possibility, we measured the bias in the perceived angular velocity for a wide range of d_{nd} (0.12° to 1.51°).

The bias in the perceived angular velocity (matching) is expressed as a percentage of the actual (reference) angular velocity. A bias of 100%, for example, means that the subject perceived the rotation as twice faster than the actual velocity. Because in this experiment the radius of the test was half that of the reference radius, such a bias would indicate that the subject is using the linear-velocity information to perform the task. Figure 3 shows the bias as a function of the inverse of d_{nd}. Results show that, consistent with Werkhoven and Koenderink (1993), there is a bias towards tangential speed for large

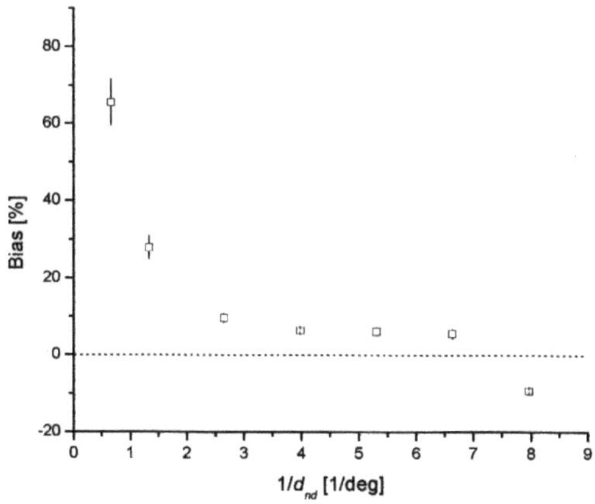

Figure 3. Bias in the perceived angular velocity towards tangential speed as a function of the inverse of the distance between neighboring dots. The results show that when this distance is large the bias is high, but that it falls rapidly as the distance decreases (from Barraza & Grzywacz, 2002a).

values of d_{nd}. However, when the d_{nd} decreases, the bias falls rapidly from approximately 70% to nearby zero. In other words, when the d_{nd} is sufficiently small, subjects use angular velocity to perform the task. In contrast, for large values of d_{nd}, which means poorer rotational signals, subjects use the tangential speed.

This result suggests that the brain can switch between tangential and rotational motion mechanisms depending on the density of the incoming retinal data. Interestingly, this switching is not abrupt, but there are conditions in which the perceived speed of rotation depends on tangential speed, true angular velocity, or both.

In terms of Yuille and Grzywacz's (1998) theoretical framework, this switching can be interpreted as the result of two different mechanisms competing for the parameterization of the visual motion. When the number of dots is small, the brain would not trust the signal for rotational motion and thus, the rotational model would be rejected most of the time. On the other hand, when the number of dots increases, the rotational signal would be strengthened, causing the brain to assume the motion as a rotation.

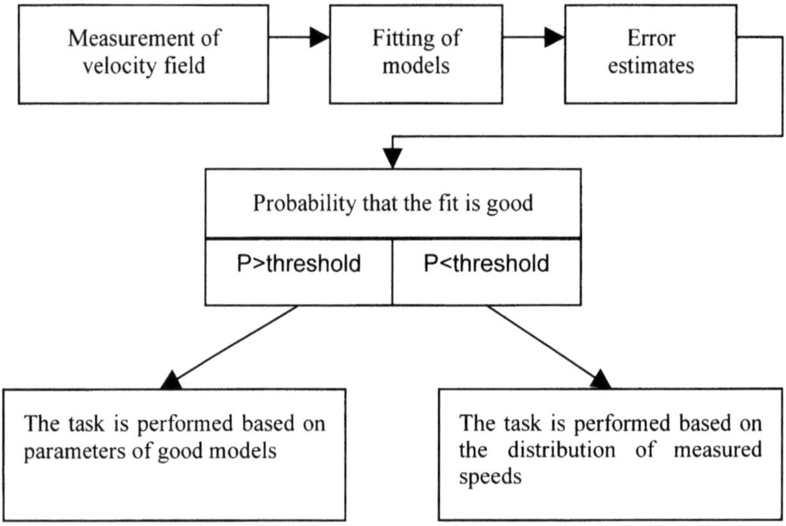

Figure 4. Schematic of the decision process for the estimate of angular velocity. The brain would estimate angular velocity when the stimulus information is good enough for the model to consider the motion as a rotation.

2.1.1 A Bayesian Interpretation

The schematic in Figure 4 shows graphically how the Yuille and Grzywacz framework would account for the density data. The idea is that as the density rises, the probability that a rotational model would provide a good fit would increase. To formalize the idea, we will take advantage of Bayesian Decision theory (Berger, 1985). Let Π_i be the ith motion model, then, according to Bayes' theorem, the probability that a given set of N velocity measurements $\{\vec{v}_m\}$ can be described by the model Π_i is

$$P(\Pi_i|\{\vec{v}_m\},N) = \frac{P(\{\vec{v}_m\}|\Pi_i,N)\cdot P(\Pi_i|N)}{P(\{\vec{v}_m\}|N)} \tag{1}$$

Expanding the denominator, this equation can be written as

$$P(\Pi_i|\{\vec{v}_m\},N) = \frac{P(\{\vec{v}_m\}|\Pi_i,N)\cdot P(\Pi_i|N)}{\sum P(\{\vec{v}_m\}|\Pi_j,N)\cdot P(\Pi_j|N)} \tag{2}$$

We need to express Equation 2 for the particular case of rotation. To do this, we assume that the probabilities of other motion models, such as translation and divergence, are negligible. In other words, only the models for rotation ($\Pi_i = \Omega$) and independent ($\Pi_i = I$) motions will compete for the explanation of the stimuli. Then, Equation 2 can be written as

$$P(\Omega|\{\vec{v}_m\},N) = \frac{P(\{\vec{v}_m\}|\Omega,N)\cdot P(\Omega|N)}{P(\{\vec{v}_m\}|\Omega,N)\cdot P(\Omega|N) + P(\{\vec{v}_m\}|I,N)\cdot P(I|N)} \tag{3}$$

Therefore, the probability of rotation is

$$P(\Omega|\{\vec{v}_m\},N) = \frac{P(\{\vec{v}_m\}|\Omega,N)}{P(\{\vec{v}_m\}|\Omega,N) + Ak^N}, \tag{4}$$

where A is the ratio $P(\mathrm{I}|N)/P(\Omega|N)$ and k^N reflects the independence of the various $\vec{v}_m$ and that $\mathrm{P}(\vec{v}_j) = \mathrm{P}(\vec{v}_k)$ in our experiments, as $|\vec{v}_j| = |\vec{v}_k|$. The ratio A is a parameter of the model that reflects an assumption about the statistics of natural motions. Because in our experiments, the local velocities are exactly consistent with rotation, the only reason $P(\{\vec{v}_m\}\,|\,\Omega,N) \neq 1$ is the noise in the measurement. We assume here for simplicity that this noise is Gaussian and independent across position. Therefore,

$$P(\{\vec{v}_m\}|\Omega,N) = \left(\frac{1}{2\pi\sigma^2}\right)^N \prod_{m=1}^{N} \exp\left(-\frac{(\vec{v}_m - \vec{u}_m(\Omega))^2}{2\sigma^2}\right) \tag{5}$$

where σ is the standard deviation of the noise and $\vec{u}_m(\Omega)$ is the expected measurement if there was no noise. Substituting Equation 5 for $P(\{\vec{v}_m\}\,|\,\Omega,N)$ in Equation 4 gives

$$P(\Omega|\{\vec{v}_m\},N) = \frac{\prod_{m=1}^{N} \exp\left(-\frac{(\vec{v}_m - \vec{u}_m(\Omega))^2}{2\sigma^2}\right)}{\prod_{m=1}^{N} \exp\left(-\frac{(\vec{v}_m - \vec{u}_m(\Omega))^2}{2\sigma^2}\right) + Ah^N} \tag{6}$$

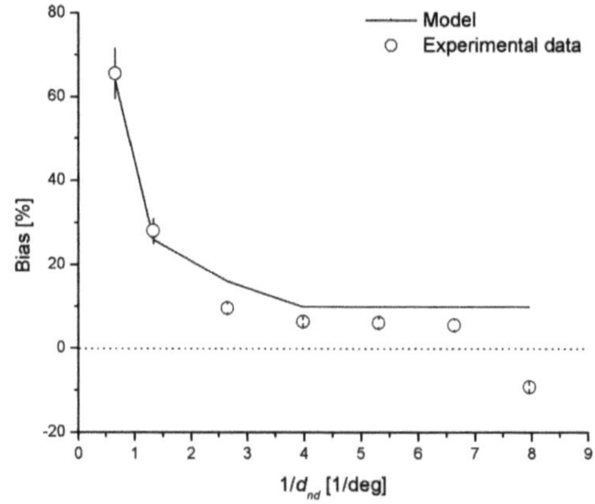

Figure 5. Result of the simulation of the model. Symbols represent the experimental data and the line represents the prediction of the model. Consistent with the psychophysics, the model predicts that for a small number of dots the brain uses linear velocities to describe the motion, but when the number of dots is large the brain estimates angular velocity.

where $h = 2\pi\sigma^2 k$. We propose that the system decides that there is a rotation if $P(\{\vec{v}_m\} \mid \Omega, N)$ is larger than a threshold (Figure 4).

Figure 5 shows the results of a computer simulation of this model based on the Yuille and Grzywacz framework. The line represents the prediction of the model and the symbols are the experimental results of Figure 3. The good fit of the model shows that a decision based on a statistical pooling of local measurements can account for the effect of switching between different motion mechanisms.

2.2 Angular Velocity is Computed Locally

As discussed earlier, the theoretical decomposition into the motion models proposed by Koenderink and Van Doorn (1976) is only valid for planar surfaces. Fortunately, if small portions of an image are considered, then they can most often be assumed to come from planar patches of surfaces, allowing the application of the decomposition. Therefore, to take advantage of this decomposition the visual system would need to compute angular velocity relatively locally. Support for this being true comes from evidence that the visual system makes independent estimates of angular velocity at different portions of an image. For example, if all the dots of a random-dot pattern move along concentric circular trajectories with the same tangential speed, the

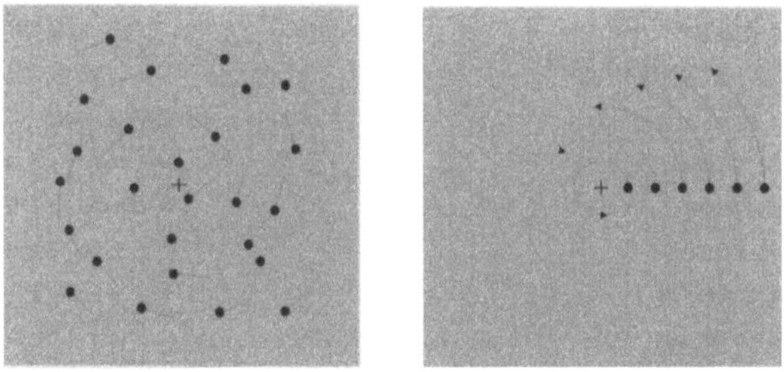

Figure 6. Panel A shows an example of a random-dot pattern undergoing a non-rigid rotation. The line segments indicate the flow of the pattern, with their lengths indicating the speed of each dot. The dots near the center rotate much faster than the outer dots. Panel B shows schematically the non-rigid percept. The curved arrows represent the path along which the dots move in a period of time. The lengths of the arrows are equal, which indicates equal speed, but the angle traversed is larger for the dots near the center of rotation. This indicates a higher angular velocity.

percept is of a non-rigid rotation (Figure 6). The inner portions of the display appear to rotate much faster than the outer portions. This illusion suggests that the visual system is not using the linear velocity to evaluate the speed of rotation in the display. Presumably, the percept is that of local angular velocity, since angular velocity is inversely proportional to the distance from the center of rotation. In this case, non-rigidity would mean that the observer could perceive various angular velocities in the same object without segmenting the image. To ascertain whether this is true, we measured the perceived angular velocity as a function of the distance from the center of rotation (Barraza & Grzywacz, 2002a).

Figure 7 shows that the brain applies the rotational model locally. This figure plots the perceived angular velocity as a function of the radius of the reference disk. Symbols represent the experimental data and the solid lines represent the actual angular velocities in the annuli. Results show that the perceived angular velocity falls hyperbolically with radius and match well the actual angular-velocity prediction. Therefore, the percept of non-rigidity in this stimulus is due to the brain computing angular velocities independently at different radii of rotation. All subjects reported that the stimulus did not appear to be segmented into annuli of different radii. This is important because the interpretation of the results would be different if one thinks that the visual system is segmenting the image and integrating the measurement of angular velocity along annuli. Consequently, if there is no segmentation and the stimulus is perceived as a non-rigid object, then the brain is not computing a single global estimate of angular velocity, but appears to compute it point by point.

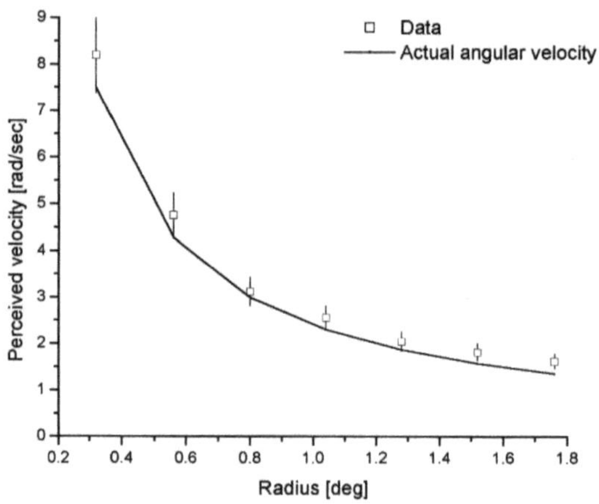

Figure 7. Perceived angular velocity as a function of radius in the non-rigid disk. Symbols represent the experimental data and the solid line represents the actual angular velocity along the disk. The data match well the actual angular velocities (from Barraza & Grzywacz, 2002a).

2.3 The Estimate of the Center of Rotation

So far, we have shown that the human visual system can estimate the angular velocity of rotational motions. Moreover, the perception of non-rigid rotations suggests that this estimate is performed locally. If this is the case, then the perception of angular velocity should depend strongly on the estimate of the center of rotation. In terms of Yuille and Grzywacz's (1998) theoretical framework, the center of rotation is the other parameter of the model to be determined in addition to the angular velocity. We investigated whether the visual system can make precise estimates of the center of rotation. To do this, we conducted an experiment to measure the minimum distance needed for the subject to discriminate the position of the center of rotation, which was deviated from the fixation point along the diagonal of one of four quadrants in a square display. Because the center of rotation is the only point in the display where the speed is zero, we wondered whether the subject uses this particular information for the estimate. To test this hypothesis, the experiment was repeated by masking the central part of the stimulus, thereby hiding the dots falling in this area. The distance threshold was measured as a function of the radius of the mask and was obtained for two different conditions: by keeping constant either the number of dots or the density in the visible part of the display.

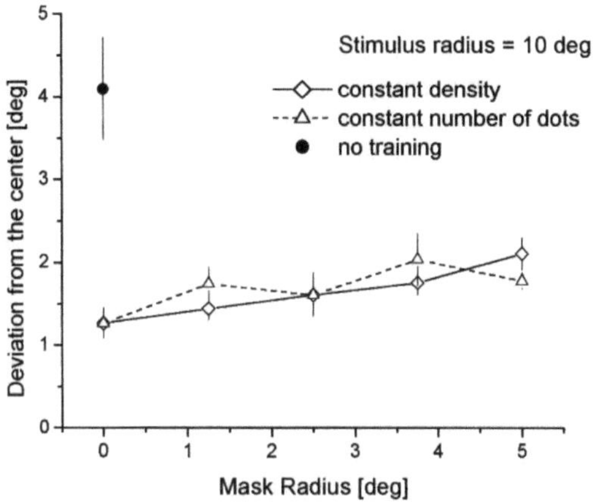

Figure 8. Position estimation of the center of rotation as a function of the mask radius. Even trained subjects cannot estimate the center of rotation precisely. Before training, errors were much higher for both subjects (black circles). Errors do not increase significantly with the radius of the mask.

Figure 8 shows that for both conditions, the subject makes an error of about 1.3° in the estimate of the center of rotation. Before training, the error was even larger (4°), as indicated by the black point in the graph. The flatness of the curves suggests that the visual system does not use the local information of the center of rotation to estimate its position, as otherwise, the error should increase significantly with the size of the mask.

What should be the effect of mislocating the center of rotation on the estimate of angular velocity? We hypothesized that when the center of rotation is deviated from the fixation point, the perceived location of the former is biased towards the latter, causing an error in the estimate of angular velocity. Because angular velocity would be computed locally, then a mislocation of the center of rotation would result in high estimates of angular velocity for those points near the perceived center of rotation. This is because velocity vectors near it would be high rather than the expected near-zero values. Therefore, in the hypothetical case that local measurements need to be integrated to give a global estimate of the motion, these errors would raise the final estimate of angular velocity. Experimental results not detailed here show that when the center of rotation is mislocated, subjects systematically overestimate the angular velocity confirming our hypothesis (Barraza & Grzywacz, 2003). This is further evidence that angular velocity is computed locally.

3 MEASURING RADIAL MOTION

That the visual system computes angular velocity is evidence that the brain uses a parametric model for rotational motion, consistent with the theoretical framework proposed by Yuille and Grzywacz (1998). This model should work in parallel with models for other types of motion such as expansion/contraction, deformation, and translation. In this section, we report that the visual system uses a parametric model for expansion as well.

Previous studies showed that human subjects can estimate the parameter named time-to-collision for simulated objects moving towards the subject or for simulated self-motions (Regan & Hamstra, 1992). One can show that for constant velocity motions, time-to-collision is the inverse of the instantaneous rate of expansion of a flow field

$$\rho = \frac{\left(\frac{d\theta}{dt}\right)}{\theta} \tag{7}$$

where θ is the angular extent of the viewed object. There is no agreement yet on how the time-to-collision is computed by the visual system. One possibility is that it measures the rate of expansion instantaneously and then inverts it to compute the time-to-collision. However, a measurement of the rate of expansion cannot be exactly instantaneous, since it needs some integration time. This temporal limitation could explain the systematic errors in the estimates of time-to-collision showed in several psychophysical studies (Gray & Regan, 1999). We performed an experiment to explore whether human subjects can make precise estimates of the rate of expansion despite misestimating the time-to-collision (Wurfel, 2003). In this experiment, the motion was generated in such a way to produce a constant rate of expansion in the display. As for rotations, we controlled for the use of linear velocity by comparing expansions in disks of different radii.

Figure 9 shows that the brain also appears to use a parametric model for expansions. This figure plots the rate of expansion as a function of the ratio between test and reference radii. The symbols represent the experimental data and the dashed line, the actual rate of expansion. The solid line indicates the predicted result for the task performed by using local linear-velocity information instead of rate of expansion. The procedure used in this expansion experiment is similar to that used to measure angular velocity (Barraza & Grzywacz, 2002a). Results show that subjects can estimate precisely the rate of expansion.

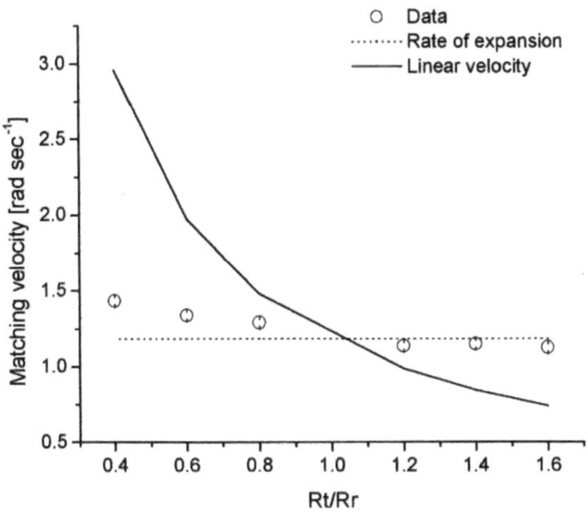

Figure 9. Perceived rate of expansion as a function of Rt/Rr. The symbols represent the experimental data, the dotted line represents the rate of expansion of the reference, and the solid line represents the rate of expansion expected if the task were performed through speed matching. The results show that the subjects can measure rate of expansion in radial motions.

Is the rate of expansion computed locally as the angular velocity? If so, then the effect of non-rigidity should appear in expansions, as it does in rotational motions. In a preliminary experiment, we observed that subjects could discriminate between rigid and non-rigid expansions, reporting non-rigidity when all the dots in the display move along radial trajectories with the same speed. We also tested the effect of mislocating the focus of expansion on the perceived rate of expansion. As in rotation, we found that when large random-dot fields are used, a mislocation of the focus of expansion produces a systematic overestimation of the rate of expansion. Consistently with data obtained for rotation, the overestimation reaches 20% when the deviation of the focus of expansion is about 5°. This is further evidence of a local computation of the rate of expansion as explained after Figure 8. (Interestingly, when the experiment is performed with a small expanding disk the overestimation disappears, perhaps because there is now a strong geometric clue about the position of the center of rotation.)

4 TEMPORAL COHERENCE

If some of the parametric motion models discussed so far can describe an optic flow at an instant, then it is likely that they will continue to do so for a while. This continuity follows from motion inertia in nature. Consequently, an implication of the use of parametric models by the brain is that the visual system may disambiguate, predict, and estimate motion through temporal coherence, that is, by assuming that objects move in consistent trajectories rather than abruptly changing their direction (Barraza & Grzywacz, 2002b; Burgi et al., 2000; Grzywacz et al., 1995; Ramachandran & Anstis, 1983; Verghese & McKee, 2002; Watamaniuk et al., 1995). Other motion phenomena involving temporal coherence include the improvement of velocity estimation over time (McKee et al., 1986), blur removal (Burr et al., 1986; Watamaniuk, 1992), detection of motion-outliers (Watamaniuk et al., 1995), and motion occlusion (Watamaniuk & McKee, 1995).

According to Yuille and Grzywacz's theoretical framework, temporal coherence is possible thanks to the predictions of the parametric models. Therefore, as the brain makes estimates about the future course of objects moving along linear trajectories (Verghese & McKee, 2002; Watamaniuk et al., 1995), if there is a temporal-coherence mechanism for other motion models then the brain could predict the future of the motion based on their parameters.

We tested this temporal-coherence hypothesis for rotation by exploiting the effect of mislocation of moving objects at the moment of their disappearance, such as reported by Mateeff and colleagues (1991a). These authors found a systematic location error towards where the motion was going for translational trajectories. We tested whether if one were to produce such an effect by using rotation instead of translation, then the mislocation would be consistent with the circular trajectory rather than with the instantaneous linear velocity at the time of disappearance or with a model for recent-past translation (Barraza & Grzywacz, 2002b).

We measured the perceptual location of a moving white dot (test) with respect to a static line at the moment of the luminance transition of the dot. Before the luminance transition, this dot was undergoing a clockwise rotation. The test dot was one of thirty dots arranged in an annulus concentric and with the same radius as the circular motion (Figure 10a). All the dots, except for the test, were black. The white dot was set to change its luminance in one of six points on the upper portion of the annulus (Barraza & Grzywacz, 2002b). (One of these points corresponded to the top of the circle $\phi = 0$.) Subjects mislocated this dot according to rotational temporal coherence (Figure 10b)

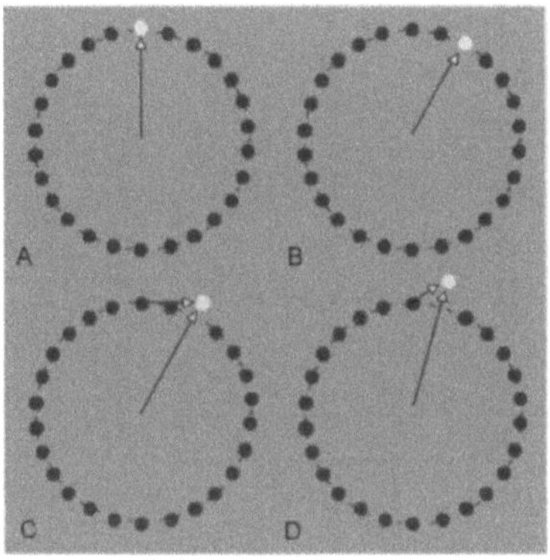

Figure 10. Schematics of visual stimuli and percepts in the rotational temporal coherence experiment. Panel A shows an example of the stimulus in which the test dot (white) "disappears" at the top of the circle ($\phi = 0$). Panel B shows schematically the subject's percept of the location of the dot. Panel C shows the prediction of the hypothesis that the mislocation depends on the instantaneous linear velocity at the time of disappearance. The hypothesis that the mislocation depends on a translational version of temporal coherence makes an even worse prediction (Panel D).

but not according to either the instantaneous velocity (Figure 10c) or a translational temporal-coherence assumption (Figure 10d).

We measured both the X (horizontal) and Y (vertical) perceptual locations of the dot at the time of the luminance transition as a function of the angle of luminance transition. Figure 11 shows the results of these measurements for two subjects. Results in the Y dimension (top-right plots of both panels) show that for negative angles, the test dot is perceived above its actual position (dashed line). In contrast, for zero and positive angles, the perceptual transition takes place below its actual position. These mislocations cause the data curve to be displaced to the left of the actual-positions curve. Furthermore, if one shifts the actual-positions curve leftward (by 0.14 rad), then the resulting curves fit the data well (solid line). This fit shows that subjects see the test dot disappear ahead of its actual position but in a point on its circular trajectory. As a result, the magnitude of the mislocation is not proportional to the instantaneous vertical component of velocity. (For instance, for $\phi = 0$, despite the vertical component of velocity being zero, the dot is seen to disappear below its actual position.) Results obtained in the X dimension (bottom-left plots of both panels) are also consistent with a mislocation based on a rotational assumption. They show that the test is perceived to disappear to the right of its actual position, resulting in a downward shift of the data curve. And again, the data can be fitted well by shifting the actual positions curve downward. Importantly, these fits were obtained for both the Y and X dimensions independently, showing the same

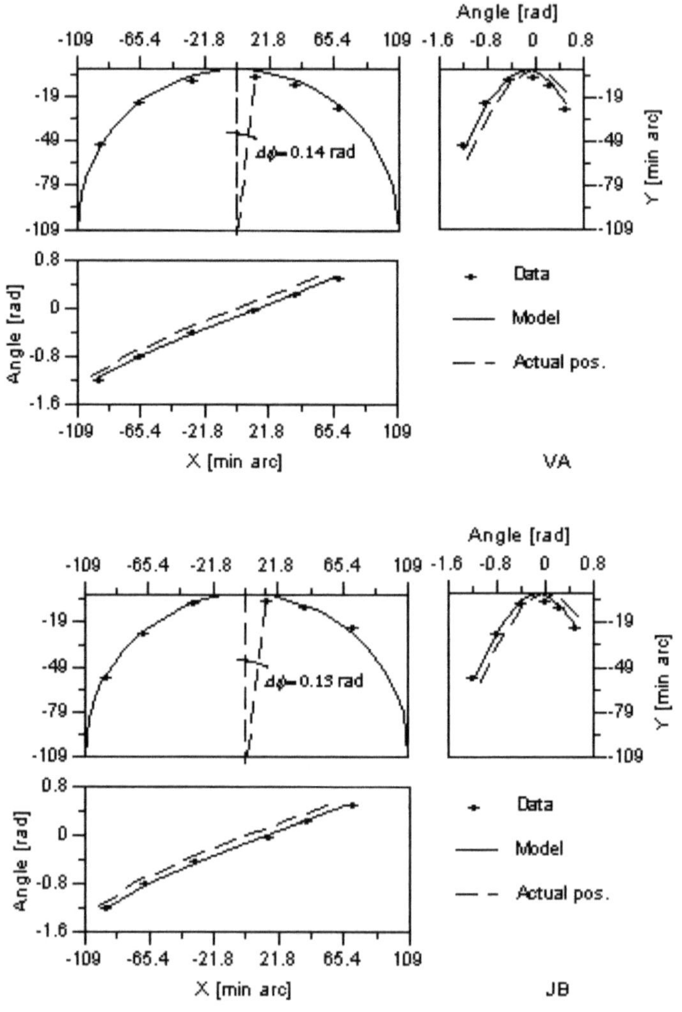

Figure 11. The two panels correspond to two different subjects, VA and JB. In each panel, the perceptual location of disappearance of the test dot as a function of the angle of disappearance for the Y dimension appears on the top-right plot. The same plot for the X dimension appears on the bottom left. Finally, the perceptual location of disappearance of the test dot in the X-Y plane appears in the top-left plot. In the X-Y plot, each perceptual location obtained in Y is plotted as a function of that obtained in X for the same angle of disappearance. Both the top-right and bottom-left plots show the data as solid diamonds, phase-shifted sinusoidal functions as solid lines, and the actual position of disappearance as dashed lines. In the X-Y panel, a circumference arc with the same radius as the annulus and concentric with it appears as a solid line. The $\Delta\phi$ angle shows the measured angular mislocation of the test (from Barraza & Grzywacz, 2002b).

phase shift. The top-left plots in Figure 11 shows a combination of the Y and X panels. In this new panel, the perceptual location of the test dot at the time of the luminance transition in the Y dimension is plotted as a function of the perceptual location in the X dimension. This plot shows how the data produce a "rotated" circle. Therefore, subjects see the test dot change its luminance ahead of its actual position in a point belonging to the circular trajectory.

5 DECOMPOSITION OF COMPLEX MOTION

That the brain appears to use particular parametric models to interpret optic flows raises a vexing problem. Natural optic flows rarely display pure basic motion components such as translation, rotation, or expansions. Can the brain estimate the parameters of the models from motions containing a combination of more than one of these components? To answer this question, we measured the ability of human subjects to estimate either angular velocity or rate of expansion from spiral motions that is a combination of rotation and expansion. The test consisted of a matching-velocity experiment, similar to that described in Barraza & Grzywacz (2002a), but using one of the components as a mask. For instance, when the angular velocity was the parameter to be measured, the rate of expansion was the mask, varying randomly in a range defined by its variance. The distribution of the mask was homogeneous and thus, the variance indicated the maximum value that the mask component could reach. The components were measured in terms of local speeds such that a variance of 1, for example, meant that for each position in the display, the maximum speed of the radial motion was equal to the speed of the rotational motion. With this procedure, we prevented the subject from performing the matching task by using the information of the resultant vector of the spiral.

Figure 12a shows that the brain appears to solve the decomposition problem. This figure plots the ratio between matching and reference (actual) angular velocities of the rotational component of the spiral motion as a function of the variance of the radial motion (mask). (Negative values of variance apply to contraction and positive values apply to expansion.) Results show that the subject measures the correct angular velocity independently of the amount of radial motion in the spiral. This result suggests that the visual system decomposes the spiral motion in rotational and radial components. Interestingly, the slope of the psychometric functions decreases with increase of variance, which implies a rise of the threshold for angular-velocity discrimination. In Figure 12b, we plot the sensitivity for angular-velocity discrimination, the reciprocal of threshold, as a function of the variance. This sensitivity strongly depends on the amount of the mask, suggesting that

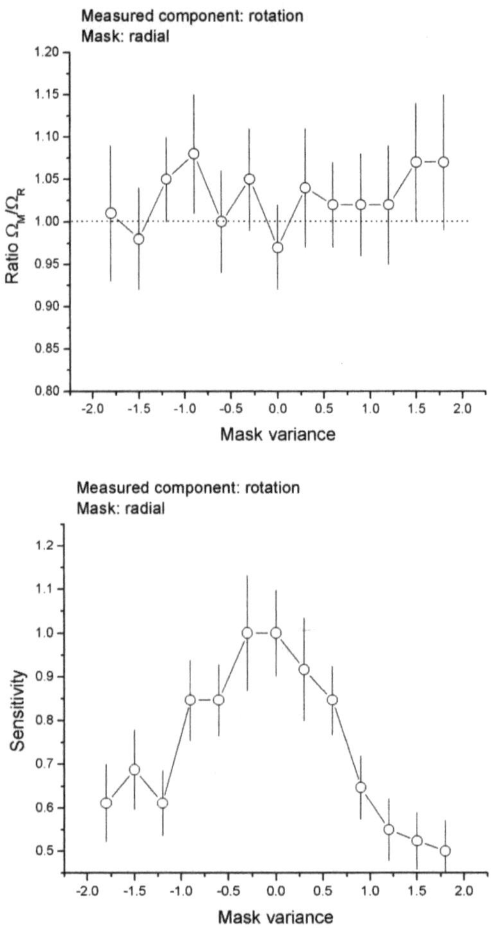

Figure 12. Panel A shows the ratio between the matching and the reference angular velocities as a function of the variance of the mask. The results show that subject can estimate angular velocity when the rotation is masked with a radial motion. Panel B shows the sensitivity for discrimination of angular velocity as a function of the variance of the mask. The figure shows that there is a maximum of sensitivity when the stimulus contains no mask (pure rotation), while it decreases a 50% when the variance is twice the rotational component.

although the visual system decomposes the spiral and computes the correct angular velocity, its precision falls with the increase of the mask.

How can this decomposition be instantiated? Previous studies have shown that the decomposition of spiral motions is probably not performed at the level of a single cell (Orban et al., 1992). The response of a rotation-selective cell decreases substantially when the rotational stimulus is

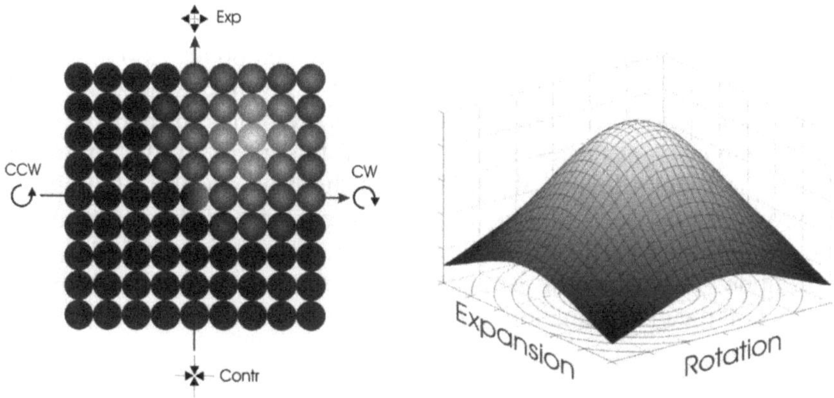

Figure 13. Panel A shows a schematic of the rotational-radial space. Disks represent neurons sensitive to rotational (horizontal axis), radial (vertical axis), and spiral motions. The gray level represents the response of the cells, with white and black indicating strong and weak responses respectively. Panel B shows the distribution of responses in a three-dimensional plot. If this distribution is separable, then the estimates of angular velocity and rate of expansion are given by the distribution of responses along the respective axis.

embedded in a radial motion. The alternate strategy would be a population-cell coding. It was found that there are cells in area MST of primates' brain that are selective to spiral motions (Graziano et al., 1994). We hypothesize that these cells form a rotational-radial Cartesian-like space, in which the angle with respect to one of the axis indicates the proportion of rotational and radial motions in an optic flow. In turn, the distance from the origin in this space would be proportional to a cell's optimal angular velocity and rate of expansion. In support of this notion, Graziano et al. (1994) found that spiral sensitive cells may constitute a continuum in such a space. We further hypothesize that when a particular combination of rotational and radial motions is presented, a sub-population of the cells would fire such that they agglomerate around the coordinate in this space corresponding to the stimulus. The brain may then estimate the parameters of rotational and radial motions by projecting the centers of these agglomerations onto the axes of the space.

Figure 13a shows a schematic of a hypothetic response distribution of cells in this rotation-radial space. The gray levels of the disks represent the amplitudes of the cells responses. Hence, the cell in white indicates the spiral motion to which the system is responding. Figure 13b shows, in a three-dimensional plot, a continuum of the distribution of responses to a spiral motion. The plot is truncated to illustrate the response profiles on the axes. If the response distribution is separable, then these profiles also present peaks that correspond to the component velocities, implementing the aforementioned projection to the axes. Therefore, these profiles may encode the angular velocity and rate of expansion of the component motions of the spiral. Importantly, the height of the profile on a given axis depends on the value of the orthogonal motion component. For example, an increase of one of the components produces a reduction of this peak in the orthogonal component, without affecting the position of the peak. In turn, its reduction would produce a noisier and thus, less precise estimate of the corresponding motion parameter. This would explain why an increase of the radial component does not affect the estimated value of angular velocity, but it has a large effect on the discrimination sensitivity.

6 CONCLUSION

Humans appear to make sense of optic flows by at least in part decomposing them into a few elementary motion components. These components include translational, rotational, and radial motions. For each of the components, the brain seems to estimate their basic parameters, such as linear velocity, angular velocity, and rate of expansion. These estimates may not occur in individual cells, but rather use a population-cell code. However, the brain estimates are not always precise, with errors in the center of rotation and focus of expansion affecting the estimate of other basic parameters. The estimation of these basic parameters appears to be local and to follow a competition between the elementary motion models. If no model is sufficiently good, then the description of the optic flow is in terms of local velocity vectors, otherwise, the winner models do not only describe the instantaneous flow, but may be used to predict its future.

REFERENCES

Barraza, J. F., & Grzywacz, N. M. (2002a). Measurement of angular velocity in the perception of rotation. *Vision Res., 42,* 2457-2462.

Barraza, J. F., & Grzywacz, N. M. (2002b). Temporal coherence in visual rotation. *Vision Res., 42,* 2463-2469.

Barraza, J. F., & Grzywacz, N. M. (2003). Local computation of angular velocity in rotational motion. *J. Opt. Soc. Am. A., (in press).*

Berger, J. O. (1985). *Statistical Decision Theory and Bayesian Analysis.* New York: Springer-Verlag New York Inc.

Bravo, M. J., & Watamaniuk, S. N. (1995). Evidence for two speed signals: a coarse local signal for segregation and a precise global signal for discrimination. *Vision Res., 35,* 1691-1697

Burgi, P. Y., Yuille, A. L., & Grzwacz, N. M. (2000). Probabilistic motion estimation based on temporal coherence. *Neural Comput., 12,* 1839-1867.

Burr, D. C., Ross, J., & Morrone, M. C. (1986). Seeing objects in motion. *Proc. R. Soc. Lond. B,* 227, 249-265.

Clifford, C. W. G, Beardsley, S. A., & Vaina, L. M. (1999). The perception and discrimination of speed in complex motion. *Vision Res., 39,* 2213-2227.

De Bruyn, B., & Orban, G. A. (1988). Human velocity and direction discrimination measured with random dot patterns. *Vision Res., 28,* 1323-1335.

Duffy, C. J., & Wurtz, R. H. (1991a). Sensitivity of MST neurons to optic flow stimuli. I. A continuum of response selectivity to large-field stimuli. *J. Neurophysiol., 65,* 1346-1359.

Duffy, C. J., & Wurtz, R. H. (1991b). Sensitivity of MST neurons to optic flow stimuli. II. Mechanisms of response selectivity revealed by small-field stimuli. *J. Neurophysiol., 65,* 1346-1359.

Freeman, T. C., & Harris, M. G. (1992). Human sensitivity to expanding and rotating motion: effects of complementary masking and directional structure. *Vision Res., 32,* 81-87.

Geesaman, B. J., & Qian, N. (1996). A novel speed illusion involving expansion and rotation patterns. *Vision Res., 36,* 3281-3292.

Geesaman, B. J., & Qian, N. (1998). The effect of complex motion pattern on speed perception. *Vision Res., 38,* 1223-1231.

Gray, R & Regan, D. (1999). Motion in depth: adequate and inadequate simulation. *Percept. Psychophys., 61,* 236-245.

Graziano, M. S. A., Andersen, R. A., & Snowden R. J. (1994). Tuning of MST neurons to spiral motions. *J. Neurosci., 14,* 54-67.

Grzywacz, N. M., Watamaniuk, S. N. J., & McKee, S. P. (1995). Temporal coherence theory for the detection and measurement of visual motion. *Vision Res., 35,* 3183-3203.

Johnston, A., Benton, C. P., & Morgan, N. J. (1999). Concurrent measurement of perceived speed and speed discrimination using the method of single stimuli. *Vision Res., 39,* 3849-3854.

Koenderink, J. J., & van Doorn, A. J. (1976). Local Structure of movement parallax of the plane. *J. Opt. Soc. Am., 66,* 717-723.

Lagae, L., Maes, H., Raiguel, S., Xiao, D., & Orban, G. A. (1994). Response of pacaque STS neurons to optic flow components: a comparison of areas MT and MST. *J. Neurophysiol., 71,* 1597-1626.

Mateeff, S., Yakimoff, N., Hohnsbein, J., Ehrenstein, W. H., Bohdanecky, Z., & Radil, T. (1991). Selective directional sensitivity in visual motion perception. *Vision Res., 31,* 131-138.

Maunsell, J. H. R. and Van Essen, D. C. (1983). Functional properties of neurons in middle temporal visual area of the macaque monkey. I: Selectivity for stimulus direction, speed, and orientation. *J. Neurophysiol., 49,* 1127-1147.

McKee, S. P. (1981). A local mechanism for differential velocity detection. *Vision Res., 21,* 491-500.

McKee, S. P., Silverman, G. H., & Nakayama, K. (1986). Precise velocity discrimination despite random variations in temporal frequency and contrast. *Vision Res., 26,* 609-619.

Morrone, M. C., Burr, D. C., & Vaina L. M. (1995). Two stages of visual processing for radial and circular motion. *Nature, 376,* 507-509.

Marr, D. (1982). *Vision. A Computational Investigation into the Human Representation and Processing of Visual Information.* New York: W H Freeman and Company.

Movshon, J. A., Adelson, E. H., Gizzi, M. S., & Newsome, W. T. (1985). The Analysis of Moving Visual Patterns. In Chagas, C., Gattas, R., & Gross, C. G. (Eds.) *Pattern Recognition Mechanisms* (pp. 117-151). Rome: Vatican Press.

Orban, G. A., Lagae, L.,Verri, A., Raiguel, S., Xiao, D., Maes, H., & Torre, V. (1992). First-order analysis of optical flow in monkey brain. *Proc. Natl.Acad. Sci. USA, 89,* 2595-2599.

Ramachandran, V. S., & Anstis, S. M. (1983). Extrapolation of motion path in human visual perception. *Vision Res., 23,* 83-85.

Regan, D., & Beberley, K. J. (1985). Visual responses to vorticity and the neural analysis of optic flow. *J. Opt. Soc. Am. A, 2,* 280-283.

Regan, D., & Hamstra, S. J. (1992). Dissociation of discrimination thresholds for time to contact and rate of expansion. *Vision Res., 33,* 447-462.

Shadlen, M. N., & Newsome, W.T. (1998). The variable discharge of cortical neurons: implications for connectivity, computation, and information coding. *J. Neurosci., 18,* 3870-3896.

Snowden, R. J., & Milne, A. B. (1995). The effects of adapting to complex motions: position invariance and tuning to spiral motions. *J. Cogn. Neurosci., 8,* 435-452.

Tanaka, A., & Saito, H. (1989). Analysis of motion of the visual field by direction, expansion/contraction, and rotation cells clustered in the dorsal part of the medial superior temporal area of the macaque monkey. *J. Neurophysiol., 62,* 626-641.

Tanaka, A., Fukuda, Y., & Saito, H (1989). Underlying mechanisms of the response specificity of expansion/contraction, and rotation cells, in the dorsal part of the medial superior temporal area of the macaque monkey. *J. Neurophysiol., 62,* 642-656.

Verghese, P., & McKee, S. (2002). Predicting future motion. *J. Vis., 2,* 413-423.

Watamaniuk, S. N. J. (1992). Visible persistence is reduced by fixed-trajectory motion but not by random motion. *Perception, 21,* 791-802.

Watamaniuk, S. N. J., & McKee, S. P. (1995). Seeing motion behind occluders. *Nature, 377,* 729-730.

Watamaniuk, S. N. J., McKee, S. P., & Grzywacz, N. M. (1995). Detecting a trajectory embedded in random-direction motion noise. *Vision Res., 35,* 65-77.

Werkhoven, P., & Koenderink, J. J. (1993). Visual size invariance does not apply to geometric angle and speed of rotation. *Perception, 22,* 177-184.

Wurfel, J., Barraza, J. F., & Grzywacz, N. M. (2003). Measurements of rate of expansion in radial motion. *(Accepted VSS03 abstract).*

Yuille, A. L., & Grzywacz, N. M. (1998). A Theoretical Framework for Visual Motion. In Watanabe, T. (Ed.) *High-Level Motion Processing – Computational, Neurobiological, and Psychophysical Perspectives* (pp. 187-211). Massachusetts: MIT Press.

12. Fast Processing of Image Motion Patterns Arising from 3-D Translational Motion

Venkataraman Sundareswaran[1,2], Scott A. Beardsley[1] and Lucia M. Vaina[1,3]

[1] Brain and Vision Research Laboratory, Department of Biomedical Engineering
Boston University, Boston, MA, USA

[2] Rockwell Scientific Company
Thousand Oaks, CA, USA

[3] Department of Neurology
Harvard Medical School, Boston, MA, USA

1 INTRODUCTION

As we drive on the highway, walk down the street, or move around in the house, we make constant use of a complex perceptual mechanism: vision. In these situations, we rely on the visual system's ability to process the perceived motion of the visual scene across the retina, termed optic flow, for our perception and estimates of self-motion. When self-motion is known, it is possible to identify obstacles and moving objects, determine *time to collision*, and the three-dimensional structure of the environment.

For most people, the perception of motion and its application in everyday tasks are seemingly effortless. In some cases, however, visual impairments seriously degrade motion perception to the point of reduced functionality in real world tasks such as walking and navigation, in which estimate of self-motion is crucial. For these impairments, mobility-assistive devices capable of processing visual motion in real-time and calculating perceptually relevant information would be invaluable. However, theoretical and computational limitations significantly restrict our ability to develop devices capable of processing the full visual motion field in real-time.

Several methods have been proposed in the literature to compute the motion of the observer from optic flow; for reviews see Barron, et al., 1994, Heeger & Jepson, 1992, or Hummel & Sundareswaran, 1993. In these

L.M. Vaina, S.A. Beardsley and S.K. Rushton (eds.), Optic Flow and Beyond, 273–287.

methods, however, the *aperture problem* restricts the recovery of the full optical flow field. Specifically, only the component of optical flow in the direction of the local image intensity gradient can be reliably computed without the application of computationally expensive local and/or global smoothness constraints (Adelson & Bergen, 1986, Horn & Schunck, 1981, Nagel & Enkelmann, 1986). For estimates of self-motion based on full optic flow, the application of such constraints via iterative or least-squares techniques significantly increases the amount of pre-processing necessary to recover self-motion estimates, limiting their real-time utility.

Computationally efficient algorithms based on the local projection of optical flow, termed "normal flow", have been developed to recover estimates of *time to contact* and heading direction in real-time (Aloimonos & Duric, 1992, Alok & Aditya, 1995, Camus, 1995, Coombs, et al., 1998, Fermuller & Aloimonos, 1995, Herwig, et al., 1998, Horn & Weldon, 1988, Negahdaripour & Horn, 1989, Sinclair, et al., 1994). In a robotic navigation task, Herwig et al. (1998) used normal flow in conjunction with a cost function based on the half-plane constraint (Aloimonos, et al., 1993) to estimate heading direction. They found that robust heading performance could be achieved in the presence of statistical noise and for small amounts of rotation (<0.05 rad/s). In a real-time wandering task, Coombs et al. (1998) used normal peripheral flow in conjunction with flow field divergence estimates to compute *time to contact* for obstacle avoidance. Together with a direction centering mechanism based on the maximal peripheral flows, they demonstrated that robust image motion cues such as *time to contact* could be extracted from normal flow and used in real-time to safely navigate complex visual environments for extended periods of time.

In this chapter, we present a method for the *rapid* calculation of the observer's translational velocity based on normal flow estimates and we demonstrate its applicability in two tasks: in the first a camera is actively controlled to orient itself in the direction of translation, and in the second, the method is used for fast collision detection. The goal of this work is to develop methods for performing tasks using visual information rather than building a representation based on visual processing. Together with the continued reductions in the size and increasing power of computer systems we expect that such methods could play a central role in the real-time vision processing applications necessary for the development of autonomous robotic systems and mobility assistive devices for the visually impaired.

2 BACKGROUND

The relative motion of an observer with respect to a rigid, unknown environment can be represented in a coordinate system centered in the observer's center of projection. We will use perspective projection on a plane as the camera model. The two-dimensional motion of the image intensity pattern on the image plane, namely the *optic flow*, depends on the three-dimensional motion (translation and rotation) of the observer. For instance, an optic flow pattern radially expanding from an image location corresponds to translational motion towards the 3-D location projected at that image location. Rotational motion produces elliptical patterns, and a combination of the two (translation and rotation) produces complex optic flow patterns that are the summation of the individual patterns. In the work outlined below we are interested in the *inverse* problem of determining the motion of the observer from optic flow patterns.

Any method to compute optic flow must deal with the aperture problem which restricts measurement of the local motion component to the direction of the local intensity gradient (Horn, 1985). This direction is normal to the local *edge*; hence the 2-D motion along this direction is referred to as "normal flow." Typically, a regularization approach is taken to compensate for the lack of motion information along the edge (Anandan, 1989, Heeger, 1988, Hildreth, 1984, Horn, 1985). However, regularization involves a function minimization process that can be time-consuming; for an exception, see Hildreth (1984). Also, regularization assumes smoothness across the optic flow field and often introduces deviations from the actual motion field, in turn reducing the accuracy of 3-D motion and structure calculations based on the optic flow. Due to these reasons, we are interested in developing a method that computes relative 3-D motion directly from the normal flow. The goal is to develop a method that is approximate but fast enough to be useful in real-time or near- real-time applications.

3 PROPOSED METHOD

The proposed method is based on the well-known observation that for *translational* motion of the camera, image motion everywhere is directed away from a singular point corresponding to the projection of the translation vector on the image plane. This point, also called the Focus of Expansion (FOE), corresponds to the intersection of the optic flow vectors depicting the image motion. The FOE can be determined by a simple analysis of the direction of the components of the optic flow field vectors, as described below. In the current approach we focus on translational motion under gaze-stabilized viewing conditions.

3.1 Approach to Determine the FOE from Optic Flow

In a rigid environment the sign of the horizontal component of an optic flow vector (e.g. positive: rightward, negative: leftward) is determined by the location of the FOE. At a qualitative level, the distribution of the component signs in an image region is determined by the position of the FOE relative to the region. For example, if the FOE is in the middle of the region, motion vectors in one half of the region will have negative sign, and those in the other half will have positive sign. This observation suggests the possibility to use the distribution of the signs to determine the horizontal location of the FOE. In a similar fashion, by using the signs of the vertical components, the vertical coordinate of the FOE may be determined.

We are interested in examining the applicability of this simple approach to determine the FOE based on the normal flow field instead of the full optic flow field. In the proof of the following theorem we demonstrate that the use of the normal flow field in this manner is likely to yield usable FOE estimates under the previously prescribed conditions.

Theorem: *With probability greater than chance, the signs of the components of a normal flow vector agree with the signs of the components of the corresponding optic flow vector, providing the local intensity gradient direction is distributed uniformly*

Proof: Let the local gradient subtend an angle α, and the optic flow vector subtend an angle θ, with respect to the x-axis. Without loss of generality, we can assume that both α and θ span the range $[0, \pi/2]$, where the local intensity gradient and the optic flow vector lie along the unit vectors $[\cos(\alpha), \sin(\alpha)]$ and $[\cos(\theta), \sin(\theta)]$ respectively. Since, by definition, the normal flow vector corresponds to the component of the optic flow that is parallel to the local intensity gradient (i.e. perpendicular to the local edge) the normal flow direction is given by $\cos(\theta-\alpha)[\cos(\alpha), \sin(\alpha)]$.

For the x-component of this (normal) vector to have the same sign as the x-component of the optic flow vector,

$$\cos(\theta-\alpha)\cos(\alpha)\cos(\theta) > 0 \tag{1}$$

We consider two possibilities:

$$\begin{aligned}&1.\ \cos(\theta-\alpha)>0 \quad and \quad \cos(\alpha)\cos(\theta)>0\\&2.\ \cos(\theta-\alpha)<0 \quad and \quad \cos(\alpha)\cos(\theta)<0\end{aligned} \tag{2}$$

The first situation occurs if $|\theta-\alpha| < \pi/2$, and if α and θ are in the same vertical set of quadrants (first and fourth, or second and third). For a given θ, α has a range of π-θ for which an agreement of signs results. For the second situation, $|\pi + \theta - \alpha| < \pi/2$, and α and θ must be in the same horizontal set of quadrants (first and second, or third and fourth). This corresponds to a range of π-θ for α. Since the total possible range for α is 0 to 2π, assuming a uniform distribution for α, the probability that the signs of the x components agree is $(\pi - \theta)/\pi$. Using a similar argument, it can be easily shown that the probability that the y components agree is $(\pi/2+\theta)/\pi$. In passing, we note that for $\alpha = 0$ or π (vertical edges), the signs of the x components always agree!

In light of this theorem, we conclude that it is possible to determine FOE based on normal flow vectors. Below, we present results based on the application of this method to compute the FOE in real image sequences.

4 EXPERIMENTS

The procedure described above has been implemented on three different platforms: in the image processing package HIPS on a Sun SPARC IPX to determine FOE in off-line image sequences, in a robot control program on a Sun SPARC 10 for controlling the motion of a six degree-of-freedom (DOF) robot in real-time, and in the public-domain software NIH Image for a real-time collision warning system.

4.1 FOE Calculation

Following the theorem in Section 3.1, the focus of expansion (FOE) is computed based on the principle that flow vectors are oriented in specific directions relative to the FOE. Specifically, the horizontal component h_L of an optic flow vector L to the left of the FOE points leftward while the horizontal component h_R of an optic flow vector R to the right of the FOE points rightward, as shown in Figure 1. In a full optic flow field the horizontal location of the FOE, corresponding to the position about which a majority of horizontal components diverge, can be estimated using a simple counting method to tally the signs of the horizontal components centered on each image location. At the point where the divergence is maximized, the difference between the number of h_L components to the left of the FOE and

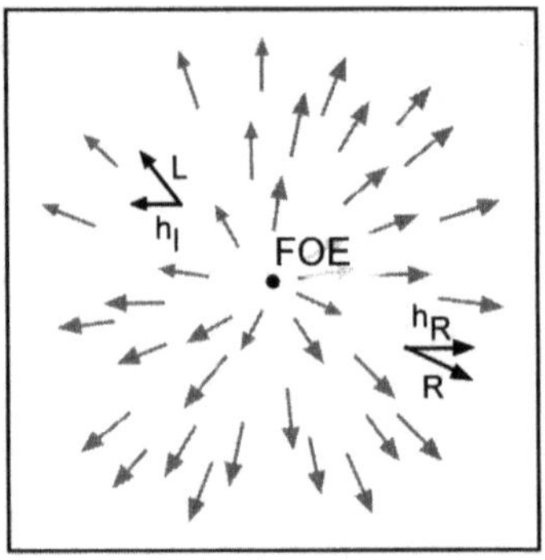

Figure 1. Schematic representation of two local optic flow components resulting from forward translation. For motion vectors located to the left (L) and right (R) of the Focus of Expansion (FOE) the horizontal components of motion point to the left and right respectively.

the number of h_R components to the right of the FOE will be minimized. Similarly we can estimate the vertical location of the FOE by identifying the position about which a majority of vertical components (v_U and v_D) diverge.

Here we extend this principle to the field of normal flow vectors. As outlined in the theorem in Section 3.1, we can reliably estimate the FOE from a normal flow vector field, providing that the intensity gradient (edge) distribution in the image is uniform (i.e., the probability that a certain orientation is seen in an image location is constant across orientations).

We implemented a version of the proposed method in HIPS to determine FOEs in standard image sequences. To calculate the normal flow vector fields, the images were first convolved with one-dimensional Gaussian derivative kernels to estimate the spatial derivatives of the intensity gradient. The standard deviation of the Gaussians was 0.75 pixels. The results of applying this kernel on an image are shown in Figure 2. The image, and its x- and y- spatial derivatives (I_x, I_y) are shown respectively from left to right. The temporal derivative (I_t) was computed as the pixel-wise difference between sequential pairs of image frames. Normal flow was then computed using the image flow constraint equation (Horn, 1985)

$$I_x u + I_y v + I_t = 0 \tag{3}$$

where I_x, I_y, and I_t correspond to the spatial (x,y) and temporal gradients in the local image intensity respectively and (u,v) correspond to the x- and y-components of motion.

The FOE coordinates were estimated from the fraction of rightward horizontal components and the fraction of downward vertical components of the normal flow. This method is admittedly crude because it does not consider

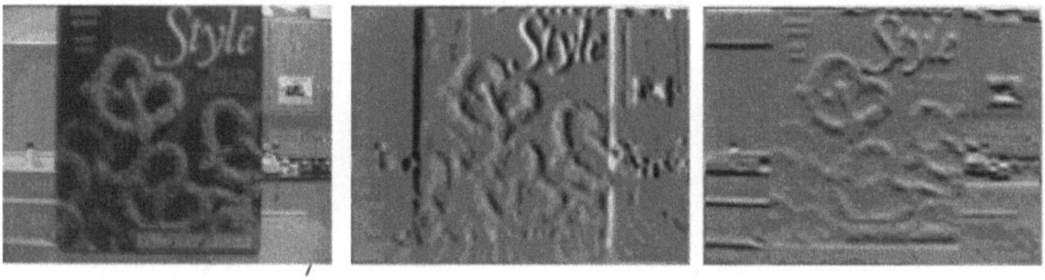

Figure 2. A sample image (left) together with the corresponding spatial gradients, I_x (middle) and I_y (right), formed by convolving the difference kernels with the image.

the detailed distribution of the signs (we used a more elaborate implementation in the collision detection application described later) and the temporal difference calculation does not incorporate temporal smoothing (which can improve the results for typical 3-D observer motion where transitions in the FOE are generally smooth). Using this straightforward procedure, we calculated the FOEs on four different standard image sequences, and found the FOEs to be qualitatively correct in each. The results are illustrated in Figure 3 where the FOE is marked as a dark dot within a white square (only typical results across the images sequences are shown).

Naturally, this method is limited because it is useful only for motion with pure translational velocity. The applicability of the method will be enhanced if we can demonstrate its utility in applications requiring rapid computation of the FOE such as control loops or fast collision detection. Accordingly, we describe below applications where we have used this fast approach to compute the FOE from the normal flow vector field.

4.2 Closed-Loop Control

The goal of this application was to control a camera so as to align its optical axis with the unknown direction of translation. The direction of translation (FOE) was computed using the method outlined above, and a standard visual servoing method (Espiau, et al., 1992) was used to calculate the rotational velocity control to accomplish the desired alignment. The overall control procedure, shown schematically in Figure 4 and explained in detail elsewhere (Bouthemy & Sundareswaran, 1993, Sundareswaran, et al., 1996, Sundareswaran, et al., 1994), resulted in correct alignment of the camera optical axis direction with the unknown direction of translation. A graph illustrating the convergence of these two directions in a typical experiment is shown in Figure 5.

Figure 3. Frames from standard image sequences showing the results of the FOE computations using the method proposed here. The FOE is shown in each case as a black dot embedded within a white square.

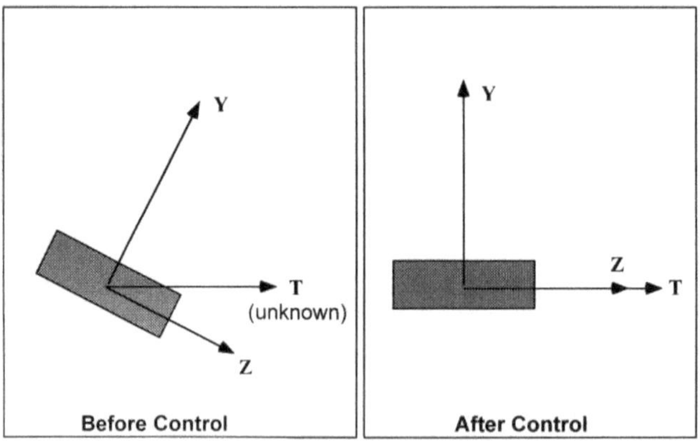

Figure 4. Schematic of the control system. Before control, the relative orientation between the camera optic axis (Z) and the translational direction is arbitrary. After control, they are aligned. A two-dimensional projection is shown for simplicity.

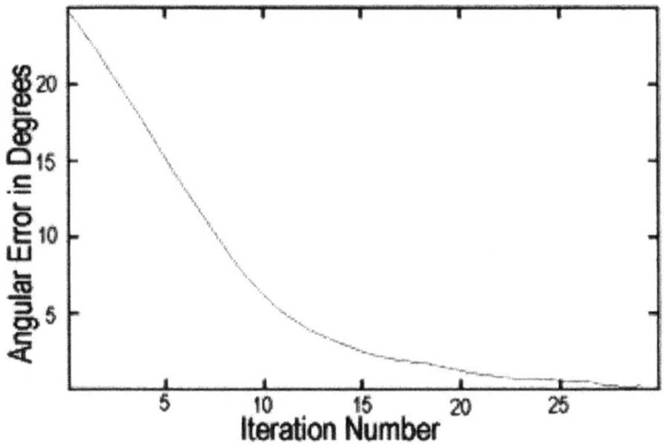

Figure 5. The graph shows the convergence in the angular difference between the camera optical axis direction and the direction of translation for an experiment performed with a camera mounted on a 6-DOF robot arm

4.3 Collision Detection

An important capability for mobile systems is to avoid collisions with static and mobile objects. Knowledge of self-motion is very useful to determine the *time to collision* (TTC) with static objects, and to segment independently moving objects. Using the method outlined above we calculated the FOE and estimated TTC by determining if the area surrounding the FOE was "zooming in" at a sufficiently fast rate to detect possible collision with that area.

The TTC can be estimated from the parameters of the first-order approximation to the optic flow field. The first-order (affine) approximation is given by

$$\begin{aligned} u &= a_1 + a_2 x + a_3 y \\ v &= a_4 + a_5 x + a_6 y \end{aligned} \tag{4}$$

where the parameters a_2 and a_6 encode the *divergence* of the flow. It can be shown that for surfaces that give rise to a flow that is nearly affine, the TTC is given by the expression

$$\tau = \frac{2}{a_2 + a_6} \tag{5}$$

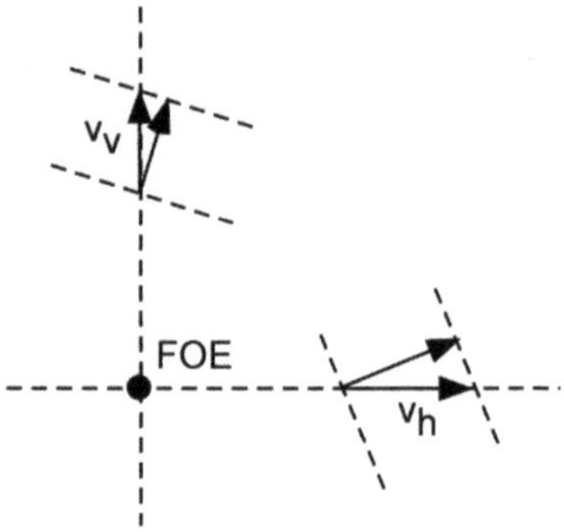

Figure 6. The affine parameters a_2 and a_6 can be estimated from points along the horizontal and vertical lines passing through the FOE as the distance normalized horizontal (v_v) and vertical (v_h) components respectively.

The parameters a_2 and a_6 can be computed from the normal flow over a 2-D image region. However, to simplify the computation, we consider only the horizontal line and the vertical line passing through the FOE. Figure 6 illustrates the simplification achieved by this choice of normal flow components.

Arbitrary local edge directions have been chosen to show the normal flow of two optic flow vectors, one along the vertical line and the other along the horizontal line passing through the FOE. Note that these optic flow vectors must lie along these lines because of the structure of translational flow fields. It can be easily shown that the magnitude of the horizontal optic flow vector is

$$v_h = \frac{I_t}{I_y} \tag{6}$$

and that of the vertical optic flow vector is

$$v_v = \frac{I_t}{I_x} \tag{7}$$

where (I) is the spatio-temporal intensity function. Using several measurements along the horizontal and vertical lines, we can calculate the desired affine parameters as follows

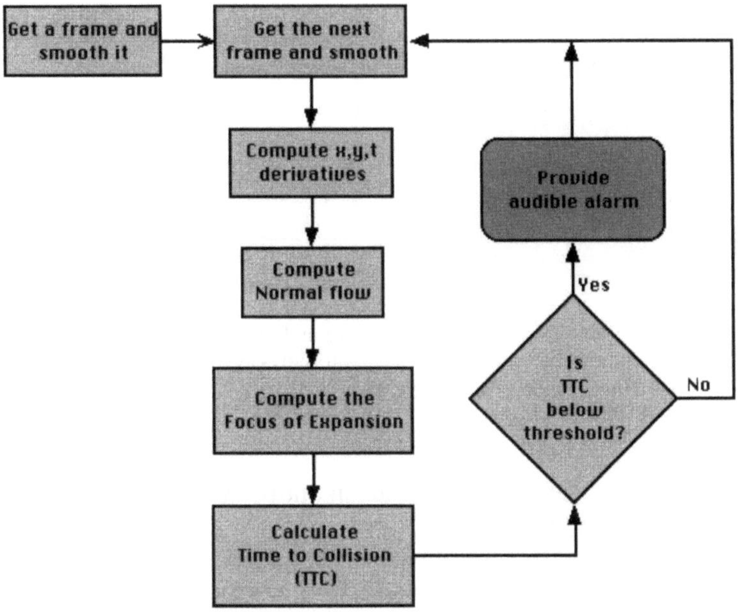

Figure 7. Flow chart of the collision warning method.

$$a_2 = \frac{1}{m}\sum_h \frac{v_h}{x_h}$$
$$a_6 = \frac{1}{n}\sum_v \frac{v_v}{y_v} \qquad (8)$$

where x_h and y_v correspond to the transformed image coordinates of the horizontal and vertical normal flow components with the FOE as the origin, and *m* and *n* correspond to the number of horizontal and vertical motion components computed across the image.

We have implemented a demonstration in NIH Image, based on computation of the TTC on a PowerMacintosh 8500 with a camcorder connected to the built-in video port. The system generates warning beeps whenever an object approaches faster than a pre-set threshold to indicate a "low" time to collision value.

The difference between this system and a general motion sensor is that this system provides a warning only if there is an impending collision, but not if there is any other type of motion. The schematic in Figure 7 illustrates the operation of the system. Image sequences showing results of the collision

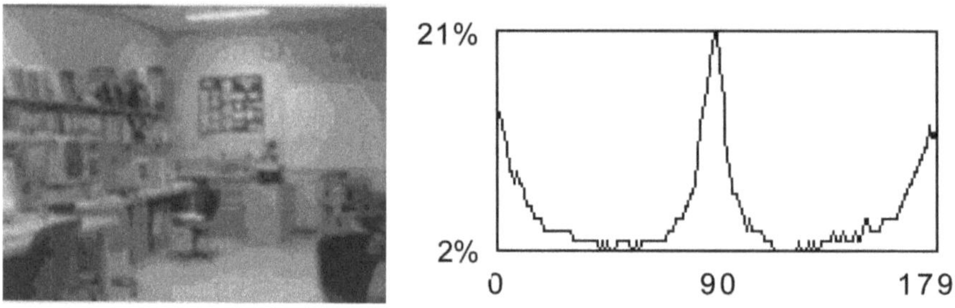

Figure 8. A sample frame from an indoor sequence and the corresponding distribution of edge orientations (across the whole sequence). The graph shows average percentage of pixels in each frame with a certain edge orientation (0 deg corresponds to horizontal edges, and 90 deg corresponds to vertical edges).

detection method can be accessed at http://www.bu.edu/bravi/research/MagicHat/MagicHat.html.

Incoming frames were smoothed using a built-in (NIH Image) weighted averaging filter, and spatial derivatives were computed using 3X3 difference filters. Temporal derivatives were computed by differencing (two) smoothed frames. The normal flow was computed as before using the optic flow constraint equation. The FOE was computed as the coordinate with the least number of "wrongly oriented" components on either side of it (e.g., for the horizontal coordinate of the FOE was chosen so that the sum of the number of rightward components to the left of it and the number of leftward components to the right of it was minimal).

The linear coefficients of the first-order approximation to the optic flow provide effective measures of the rate of expansion, and hence the time to collision. Thus the time to collision can be directly related to the coefficients of x and y in the first order expansion. The linear coefficients in a region around the FOE were estimated by considering the normal flow on the vertical and the horizontal lines passing through the FOE. The time to collision based on the estimated coefficients was thresholded, i.e., if it was below a threshold a warning beep was sounded.

To verify our assumption that the edge distribution is uniform, we implemented a method to histogram the orientation of moving edges in typical image sequences (using bins of 1 degree width), assuming that the edge distribution is a stationary process (probability distribution at a pixel location is the same as the spatial distribution). We observed empirically that for indoor image sequences the distribution of edge orientations was *not* uniform. We found a preponderance of vertical and horizontal edges, and a relatively uniform distribution across the other orientations (Figure 8).

This result, however, does not invalidate the method proposed here. Following the reasoning in the theorem in Section 3.1, it can be shown that for vertical edges, the sign of the x components of the optic flow and normal flow *always* agree; likewise for the horizontal edges, the sign of the y components *always* agree. Thus, the effect of the non-uniform edge distribution for indoor scenes only improves the performance of our implementation, as long as the camera pose is such that the vertical and horizontal edges of the scene are imaged as vertical or horizontal edges (i.e., near-zero camera tilt angle).

5 DISCUSSION

In this research we proposed the development of specific methods for performing tasks using visual information rather than building a representation based on visual processing. Toward this goal we have presented an approximate but fast method to locate the focus of expansion (FOE) based on the normal optic flow in an image sequence. Together with the use of a simple divergence estimator that does not require significant post-processing beyond the normal flow calculations, we have shown that the FOE in conventional image sequences can be reliably estimated in real-time and we have demonstrated its use in gaze control and collision detection.

The algorithm described here has several advantages that make it appealing as a method for extracting real-time FOE estimates for use in visually-guided tasks. First and foremost, the system can be implemented using off-the-shelf frame grabbers and cameras and the algorithm's performance can be easily scaled up to real-time levels using commercially available computer systems.

Second, no camera calibration information is required. While it is true that the inter-pixel distance is required for the TTC computation, it can be specified in arbitrary units (e.g., pixel count, as in our implementation) since the "collision sensitivity" or threshold of the system can be adjusted to compensate for the unknown scale factor.

Due to the simplicity and reliability of the computations, the approach can also be extended for use in low-light situations or infrared image sequences. Similarly, the approach can be easily adapted for tracking and motion encoding applications.

However, before the proposed method can be applied to more complex environments involving multiple motion sources it will be important to incorporate algorithms to segment the image motion and identify observer motion. This is a complex problem in itself and we do not propose a solution here. Nevertheless, we note from empirical observations that for the camera alignment task, small object movement did not significantly affect the computation of the FOE and the subsequent convergence (Bouthemy & Sundareswaran, 1993).

In future work we will investigate these issues with the goal of generalizing the proposed algorithm to systematically more complex visual scenes. Together with the increase in computer power and portability, we expect that the ongoing development of robust and computationally efficient real-time methods will continue to facilitate application-driven research into autonomous robotic systems and mobility assistive devices for the visually impaired.

ACKNOWLEDGMENTS

This work was supported in part by NIH grant EY-2R01-07861-13 to LMV and NSF-SGER CDA-9528079 to LMV.

REFERENCES

Adelson, E.H., & Bergen, J.R. (1986). The extraction of spatio-temporal energy in human and machine vision. *Proceedings of the IEEE Workshop on Motion: Representation and Analysis* (pp. 151-155). Charleston, South Carolina.

Aloimonos, Y., & Duric, Z. (1992). Active egomotion estimation: a qualitative approach. *Second European Conference on Computer Vision* (pp. 497-510). Santa Margherita Ligure, Italy: Springer.

Aloimonos, Y., Rivlin, E., & Huang, L. (1993). Designing Visual Systems: Purposive Navigation. In: Y. Aloimonos (Ed.) *Active Vision* (pp. 47-102): Lawrence Erlbaum.

Alok, M., & Aditya, V. (1995). Real time vision system for collision detection. *J. Comput. Sci. Inform., Special Issue on "Robotics and Automation", 25 (1),* 174-208.

Anandan, P. (1989). A Computational Framework and an Algorithm for the Measurement of Visual Motion. *Int. J. Comput. Vision, 2*, 283-310.

Barron, J.L., Fleet, D.J., & Beauchemin, S.S. (1994). Performance of optical flow techniques. *Int. J. Comput. Vis*ion, 43-77.

Bouthemy, P., & Sundareswaran, V. (1993). Qualitative motion detection with a mobile and active camera. *Proceedings of the International Conference on Digital Signal Processing* (pp. 444-449). Cyprus.

Camus, T.A. (1995). Calculating time-to-contact using real-time quantized optical flow. (pp. 1-12). Tübingen, Germany: Max Planck Institute for Biological Cybernetics.

Coombs, D., Herman, M., Hong, T., & Nashman, M. (1998). Real-time obstacle avoidance using central flow divergence and peripheral flow. *IEEE Trans. Rob. Autom., 14 (1),* 49-59.

Espiau, B., Chaumette, F., & Rives, P. (1992). A new approach to visual servoing in robotics. *IEEE Trans. Rob. Autom., 8 (3),* 313-326.

Fermuller, C., & Aloimonos, Y. (1995). Direct perception of three-dimensional motion from patterns of visual motion. *Science, 270*, 1973-1976.

Heeger, D.J. (1988). Optical flow using spatiotemporal filters. *Int. J. Comput. Vision, 1*, 279-302.

Heeger, D.J., & Jepson, A.D. (1992). Subspace methods for recovering rigid motion I: Algorithm and implementation. *Int. J. Comput. Vision, 7 (2),* 95-117.

Herwig, C., Carmesin, H.-O., Hamker, F., & Wandtke, D. (1998). Real-time estimation of heading direction. *J. Real-Time Im., Special Issue on "Real Time Motion Analysis", 4 (1).*

Hildreth, E.C. (1984). *Computations Underlying the Measurement of Visual Motion.* (Cambridge, MA: MIT Press).

Horn, B. (1985). *Robot Vision.* (Cambridge MA: MIT Press).

Horn, B.K.P., & Schunck, B.G. (1981). Determining optical flow. *Artif. Intell., 17*, 185-203.

Horn, B.K.P., & Weldon, E.J. (1988). Direct methods for recovering motion. *Int. J. Comput. Vision, 2 (1),* 51-76.

Hummel, R., & Sundareswaran, V. (1993). Motion Parameter Estimation from Global Flow Field Data. *IEEE Trans. Pattern Anal. Mach. Intell., 15 (5),* 459-476.

Nagel, H.H., & Enkelmann, W. (1986). An estimation of smoothness constraints for the estimation of displacement vector fields from from image sequences. *IEEE Trans. Pattern Anal. Mach. Intell., PAMI-8*, 565-593.

Negahdaripour, S., & Horn, B.K.P. (1989). Direct method for locating the focus of expansion. *Comput. Vis. Graph. Im. Process., 46 (3),* 303-326.

Sinclair, D., Blake, A., & Murray, D. (1994). Robust estimating of egomotion from normal flow. *Int. J. Comput. Vision, 13*, 57-69.

Sundareswaran, V., Bouthemy, P., & Chaumette, F. (1996). Exploiting image motion for active vision in a visual servoing framework. *Int. J. Rob.. Res., 15 (6),* 629-645.

Sundareswaran, V., Chaumette, F., & Bouthemy, P. (1994). Visual servoing using image motion information. *Proceedings of the IAPR/IEEE Workshop on Visual Behavior* (pp. 102-106). Seattle.

13. On the Computation of Image Motion and Heading in a 3-D Cluttered Scene

Michael S. Langer[1] and Richard Mann[2]

[1] School of Computer Science
McGill University, Montreal, Quebec, Canada

[2] School of Computer Science
University of Waterloo, Waterloo, Ontario, Canada

1 INTRODUCTION

When an observer moves through a static 3-D scene, the retinal image deforms in a way that depends both on the scene geometry and on the observer's motion. These deformations thus provide information about the scene geometry and the observer's motion (Gibson, 1950). Human observers can in many cases perceive the direction of 3-D heading from retinal image motion, even under passive viewing such as watching TV. Performance depends on several factors, such as whether there is sufficient depth variation in the scene and whether the image motion is due to real or simulated eye movements. A variety of scene geometries have been used ranging from a single ground plane to a "3-D cloud of dots". A recent review of human heading perception can be found in (Warren, 1998).

Several computational models of heading perception have also been proposed. These models typically assume that accurate estimates of retinal image velocities can be pre-computed and they take these velocities as their starting point. Such an assumption is often reasonable. Many computer vision algorithms have been developed for computing image velocities, and these algorithms often perform well enough to be used reliably as input to a heading computation. A review of computational models of heading which assume pre-computed image velocities can be found in (Hildreth & Royden, 1998).

Unfortunately, there are many types of scenes for which one cannot simply assume that image velocities are computable. An example is densely cluttered 3-D scenes such as a forest or grassland. When an observer moves through a densely cluttered 3-D scene, the occlusion relationships between the

L.M. Vaina, S.A. Beardsley and S.K. Rushton (eds.), Optic Flow and Beyond, 289–304.

objects present are so complex that it is unlikely that the visual system can recover an accurate velocity field. Think of the many branches and leaves visible in a bush. When an observer moves past the bush, motion parallax occurs and the image projection of the various leaves and branches slide past each other. To our knowledge, there is no evidence – either computational or psychophysical – that an accurate velocity field can be computed from such complex motions.

Despite the difficulty of recovering an accurate velocity field in a cluttered 3-D scene, animals such as cats, monkeys, and birds can easily navigate through such scenes. Humans appear to have this ability as well, as cited above. These observations raise the question of how this remarkable performance is achieved. How can an observer compute heading in a 3-D cluttered scene while by-passing the step of computing an accurate image velocity field? Our goal in this paper is to suggest an answer to this question.

We argue that the goal of estimating accurate image velocities in a cluttered 3-D scene is not merely unattainable. The goal is not even well-defined. To estimate the image velocity at a point, the visual system must analyze the time varying intensities in a local neighborhood of the point. But in a cluttered 3-D scene, multiple surfaces are present in most local image neighborhoods because of dense occlusions. Since each surface present typically lies at a different depth, each surface defines a different velocity because of motion parallax. Thus any single velocity vector that is computed really represents a weighted average velocity for the neighborhood, where the weighting depends in a complex manner on the luminance and relative areas of surfaces visible in the neighborhood. It is quite unclear what information this average velocity vector carries about scene depth.

This argument leads us to abandon the traditional goal of pre-computing an image velocity field prior to the heading computation. Instead we introduce a different goal, which is to estimate a family of velocities that is present in a local image region. We show that for an observer moving through a 3-D cluttered scene, the family of velocities in a local image region has a linear structure in image velocity space. This linear structure represents the local motion parallax. We then briefly review our recent algorithm for recovering this family of velocities in a local image region (Langer & Mann, 2001, 2002; Mann & Langer, 2002). Finally, we suggest how these local velocities could in principle be used to compute heading.

2 THE MOTION FIELD SEEN BY A MOVING OBSERVER

The general equations of retinal image velocities field seen by an observer moving through a static scene were derived in (Longuet-Higgins & Prazdny, 1980). Let the observer's translation velocity be (T_x, T_y, T_z) and rotational velocity be $(\Omega_x, \Omega_y, \Omega_z)$. Let $Z(x, y)$ be the depth of the surface point visible at image position (x, y). Assuming the image plane is at $Z = 1$, the velocity field can be written as the sum of a translation field and a rotation field as follows

$$\begin{bmatrix} v_x \\ v_y \end{bmatrix} = \frac{T_z}{Z(x,y)} \begin{bmatrix} x - x_T \\ y - y_T \end{bmatrix} + \begin{bmatrix} xy & -(1+x^2) & y \\ 1+y^2 & -xy & -x \end{bmatrix} \begin{bmatrix} \Omega_x \\ \Omega_y \\ \Omega_z \end{bmatrix} \tag{1}$$

The first term is the translation field. It depends on inverse depth. The second term is the rotation component. It does not depend on depth. The special image position

$$(x_T, y_T) = \frac{1}{T_z}(T_x, T_y) \tag{2}$$

is called *the axis of translation* (AOT). When no camera rotation is present, this position is called the *focus of expansion* (FOE). Because we are allowing for camera rotation in this paper, we will use the term AOT rather than FOE. We also define the special image position

$$(x_\Omega, y_\Omega) = \frac{1}{\Omega_z}(\Omega_x, \Omega_y) \tag{3}$$

to be the *axis of rotation* (AOR). In general, the AOR can point in any direction. Specific examples are when the camera rotates about the optical axis (roll), the vertical (yaw), or the horizontal axis (pitch).

There are three standard observations about the translation and rotation fields. First, each vector in the translation field points away from the AOT. Second, the speed of the translation vectors depends on depth. The translation field is smooth over image regions in which depth is smooth. At a depth discontinuity, there is a discontinuity in the magnitude (speed) of the field but not the direction. Third, the rotation field is smooth across the image and, in particular, does not depend on depth.

These observations are the basis for several algorithms for recovering the T and Ω vectors, including the algorithm we will introduce later in the chapter. The algorithms stem from the theory of (Longuet-Higgins & Prazdny, 1980). The first implemented algorithm was described in (Rieger & Lawton, 1985). The idea of their algorithm is to subtract the velocities of neighboring image points that straddle a depth discontinuity. In taking the difference of the two nearby velocity vectors, the rotation components of the two vectors approximately cancel because the rotation field is smooth. Thus, the difference of the two velocities is approximately equal to the difference of the translation components only. Since translation components point toward the AOT, the difference of two neighboring vectors should also point toward the AOT. As analyzed in (Rieger & Lawton, 1985) this last result assumes that the depth difference is large, the points are nearby in the image, and that the velocities are computed accurately.

Rieger-Lawton's experiments address the case of two continuous surfaces separated by a single occlusion boundary. Provided that the depth variation is sufficiently large and accurate flow vectors can be computed near the boundary, heading can be accurately computed. Royden extended this analysis to smooth depth variations, such as an inclined plane (Royden, 1997). Others methods, such as that proposed by Heeger and Jepson (1992), avoid taking local differences of the velocity field and instead minimize a global least squares function.

All these methods assume that the image velocity vector is well-defined in a local image region and that this vector can be computed prior to applying the method. We argued earlier in the chapter, however, that such an assumption is not suitable for a 3-D cluttered scene since many depth discontinuities occur even in local patches. In the next section we develop a model of motion extraction that is suitable for 3-D cluttered scenes and we argue that it captures the information necessary to determine heading direction.

3 MODEL OF LOCAL MOTION PARALLAX

Although a cluttered 3-D scene such as a forest produces a dense set of depth continuities, the motion field is still rich in information. Two specific observations can be made from Eq. (1). First, *the direction of the translation field is smooth and hence is locally constant*. The only exception is within an image region that is near the AOT since the direction of the translation field has a singularity at the AOT. Note that the speed of the translation field is not locally constant since it depends on depth, which may have many discontinuities. Rather, the direction of the translation field is locally constant.

a) Translation Fields (Normalized)

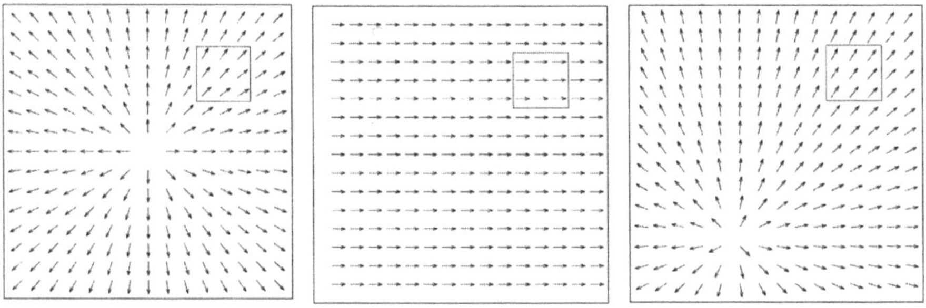

T = (0, 0, 3) *T* = (-1, 0, 0) *T* = (-1, -2, 3)

b) Rotation Fields

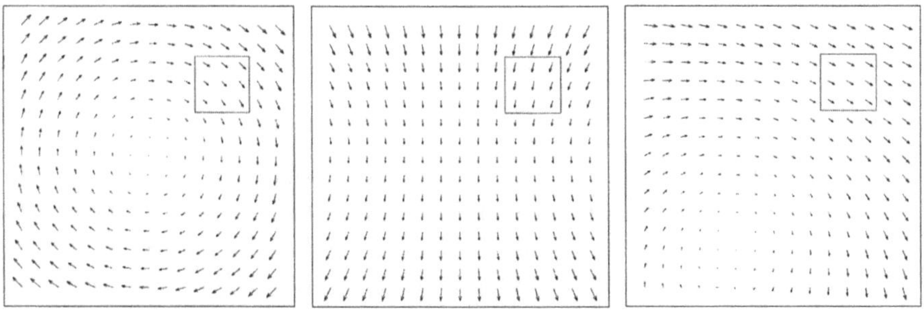

Ω = (0, 0, 3) *Ω* = (-1, 0, 0) *Ω* =(-1, -2, 3)

Figure 1. Examples of image motion for a 90° field of view centered on optical axis. a) Normalized translation fields, i.e. direction vectors only. b) Rotational fields. Each vector is approximately 6° from its neighbor. A large square region containing 3-by-3 vectors is shown that is approximately 30° from the optical axis. For each example, the fields are approximately constant within this square region.

Second, *the rotation field is smooth and hence locally constant.* Note that even though the rotation field vanishes near the axis of rotation (AOR), the field is still smooth.

Examples of the translation and rotation fields are given in Figure 1. These examples support the above two observations. Each vector field shows a 90 degree field of view centered at the optical axis. A 15 × 15 sampling of the vectors fields at roughly 6 degree intervals is shown. Figure 1a shows examples of the translation fields, with velocities normalized to unit length so as to indicate direction only. Except for points near the AOT, the direction of the fields is locally constant. Figure 1b shows examples of the rotation fields. These fields are locally constant everywhere as well.

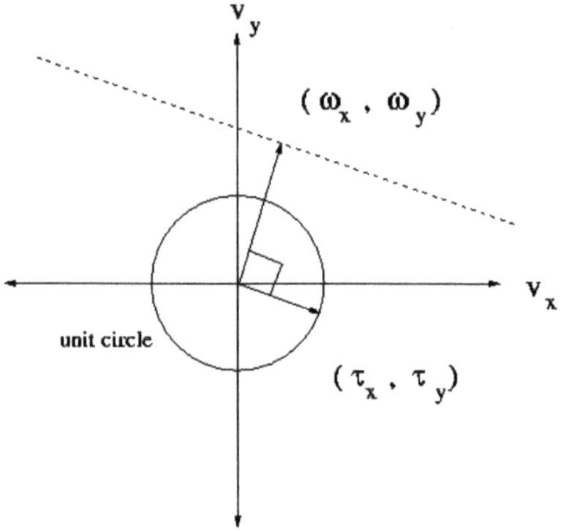

Figure 2. Model of local motion parallax. The set of velocities in a local image region lie on the dotted line of Eq. (4). The τ vector is of unit length and the ω vector is perpendicular to the dotted line.

The two observations above imply a linear model of the image velocities in a local region that does not contain the AOT: *the set of velocities in this region lies on a line in velocity space*. The direction of the line is the direction of the translation component of the velocities, which points to the AOT. The line may be parameterized as follows (see Figure 2)

$$(v_x, v_y) = \alpha(x, y)(\tau_x, \tau_y) + (\omega_x, \omega_y) \tag{4}$$

The velocity at (x, y) depends on the position x, y and on the depth $Z(x, y)$. This dependence is captured by the function $\alpha(x, y)$. The vectors (ω_x, ω_y) and (τ_x, τ_y) are constant for each local region.

When depth is continuous in an image region, the $\alpha(x, y)$ function is also continuous and thus the velocity vectors form a continuous field. An example is the shear field produced by lateral motion relative to a slanted plane. When the 3-D scene is densely cluttered, however, a dense set of depth discontinuities is present and so the velocity field is also discontinuous. When an observer moves past a tree or bush, many leaves are present, even in a 6° field such as the distance between vectors in Figure 1. Since the leaves lie at different depths the velocity field has many discontinuities, yet the model of Eq. (4) still holds.

The vector (τ_x, τ_y) is defined to be the unit vector parallel to the line of Eq. (4). This vector is uniquely defined up to a sign change, provided that $\alpha(x, y)$

indeed takes more than one value over the image region so that the line itself is well-defined. The vector (ω_x, ω_y) is defined to be the perpendicular vector from the origin to the line of Eq. (4). This vector is typically not of unit length. For our model of local motion parallax, (τ_x, τ_y) is the unit vector pointing in the direction of translation. (ω_x, ω_y) is the component of the rotation vector that is perpendicular to (τ_x, τ_y) .

The model of Eq. (4) is similar to the model of (Longuet-Higgins & Prazdny, 1980) and (Rieger & Lawton, 1985). The image velocities within a small region of the image are assumed to be the sum of a rotation component, which is approximately constant, and a translation component, which depends on depth. By treating the constant and depth-dependent vectors abstractly in our model, we may assume without loss of generality that these vectors are perpendicular.

Given local motion parallax, how do we compute heading direction? We suggest an idea similar to (Longuet-Higgins & Prazdny, 1980) and (Rieger & Lawton, 1985). For several image regions, we estimate the line of Eq. (4), in particular, the (τ_x, τ_y) vector. This vector points toward the axis of translation (AOT) since the difference of any two vectors on this line is due to the difference in translation field only, and hence points toward the AOT. Once we compute the τ vector for several regions of the image, one can recover the AOT by taking the intersection of these lines using standard methods.

The problem that remains, of course, is how to estimate the τ vector from the retinal image of a moving observer. This estimation problem is non-trivial because depth discontinuities occur in abundance even in local regions and so classical methods for computing image velocities cannot be relied on. In the next section, we present a recent method we have developed for solving this problem.

4 MOTION PARALLAX IN THE FREQUENCY DOMAIN

Our idea for estimating the τ vector of the translation field in a local image region uses a frequency-based motion estimation technique, which we have introduced previously (Langer & Mann, 2001), that extends previous frequency-based techniques for measuring image motion (Heeger, 1987). The frequency-based techniques use the following *motion plane* property (Watson & Ahumada, 1985). A 2-D image region that is moving with uniform velocity (v_x, v_y) produces a plane of energy in the 3-D spatiotemporal frequency domain. This plane is given by

$$v_x f_x + v_y f_y + f_t = 0 \tag{5}$$

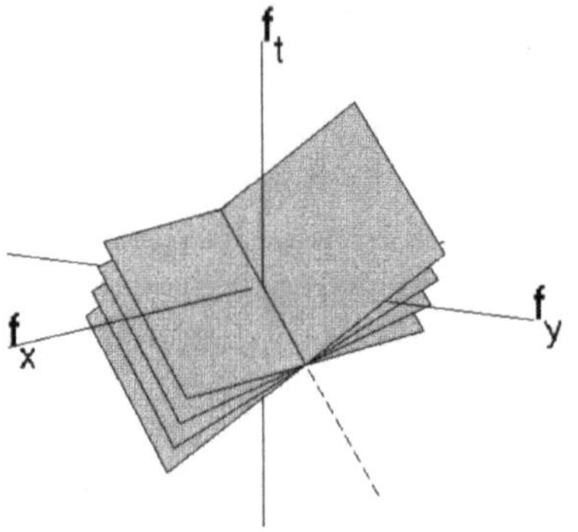

Figure 3. The motion described by Eq. (6) results in a family of planes in the frequency domain. The planes intersect at a line that passes through the origin. We refer to this as a "bowtie" pattern.

where f_x and f_y are the two spatial frequencies and f_t is the temporal frequency. In Langer & Mann (2001, 2002), we extended this motion plane property to motions that satisfy Eq. (4) by substituting Eq. (4) into Eq. (5). This substitution defined a one-parameter family of motion planes

$$(\omega_x + \alpha\tau_x) f_x + (\omega_y + \alpha\tau_y) f_y + f_t = 0 . \tag{6}$$

As in Eq. (4), the ω and τ constants here are fixed for the local image region and the parameter α takes on a range of values that depends on image position and depth.

The motion planes in Eq. (6) all intersect at a common line that passes through the origin, as illustrated in Figure 3. For this reason, we say that the family of planes has a *bowtie pattern.* We refer to the line of intersection of the planes as the *axis of the bowtie.* We show in Langer and Mann (2002) that the direction of the axis of the bowtie in the 3-D frequency domain is

$$(-\tau_y, \tau_x, \sqrt{\omega_x^2 + \omega_y^2}) \tag{7}$$

4.1 Parallel Motion Parallax

An interesting special case occurs when the line of Eq. (4) passes through the origin so that the ω vector is (0,0). An example is an observer moving through a static 3-D scene and tracking a particular visible surface point in the

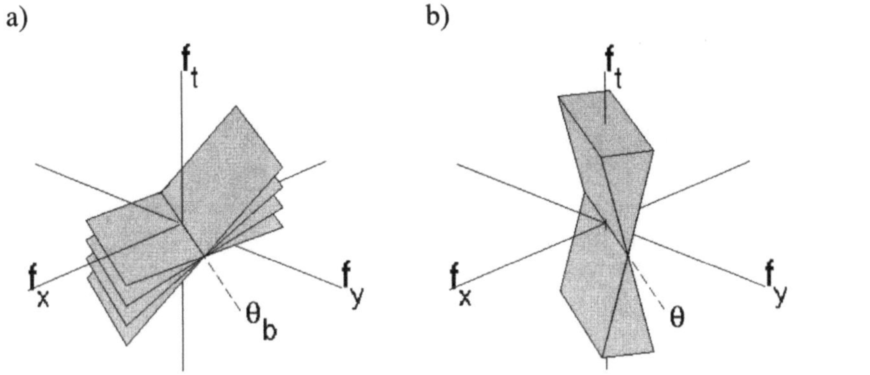

Figure 4. Parallel motion parallax (optical snow). a) Bowtie pattern in the frequency domain. The axis of the bowtie is in the f_x-f_y plane. The angle θ_b is measured from the f_x axis. b) Wedge used to estimate the orientation of the bowtie in Section 4.2. The power within the wedge reaches a minimum when the wedge is aligned with the bowtie in (a), that is, when $\theta = \theta_b$.

scene. To track the point, the observer rotates the eye so that the image velocity of the tracked point is reduced to zero (Lappe & Rauschecker, 1993).

Since the tracked point has zero velocity, the line of Eq. (4) that represents the image velocities near that tracked point must itself pass through the origin. The reason is that this line includes the velocity of the tracked point, which is zero. If the line of velocities passes through the origin, then the ω vector is zero. The bowtie model of Eq. (6) reduces to

$$\alpha\tau_x f_x + \alpha\tau_y f_y + f_t = 0 \tag{8}$$

We refer to local motions for which the ω vector is zero as *parallel motion parallax*. The reason is that all local motions are in the direction of the τ vector, and thus are parallel to each other. In earlier papers (Langer & Mann, 2001; Mann & Langer, 2002) we called such motion *optical snow*. The reason for this term is that the motion is similar to what a static observer sees during snowfall. Snowflakes that are near the observer fall with a greater image velocity than snowflakes that are far from the observer because of the inverse depth effect (parallax). Yet each local image region may contain snowflakes at a wide range of depths. (Note that this example is slightly different since we have in mind for this example a static observer and a translating 3-D scene, namely a rigid cloud of snowflakes falling vertically.)

For the case of optical snow, the ω vector is zero and Eq. (8) holds such that the axis of the bowtie lies in the (f_x, f_y) plane. We parameterize the direction of this axis with an angle θ_b such that $(\cos\theta_b, \sin\theta_b)$ is the direction of the axis of the bowtie (Figure 4a). Since the axis of the bowtie is in

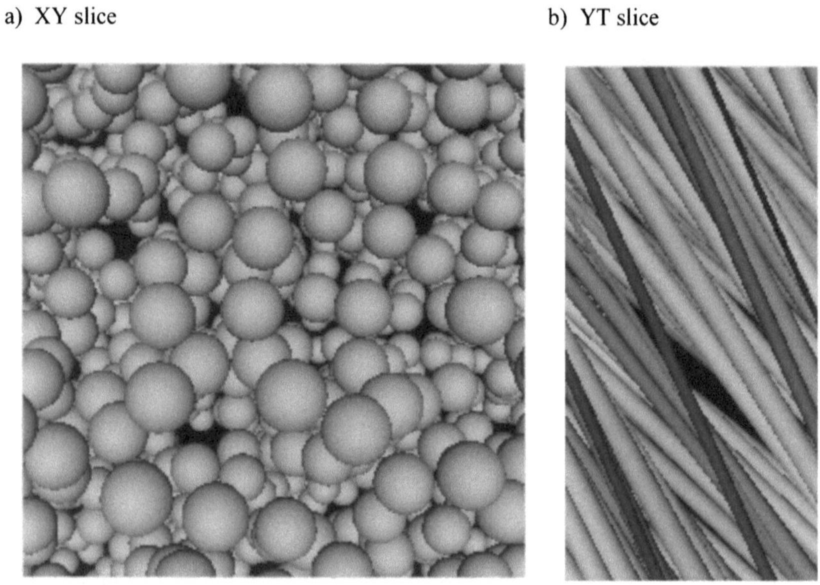

Figure 5. Synthetic "falling spheres" sequence. a) XY slice of the sequence at the first frame. b) YT slice of the sequence taken at rightmost pixel column of (a). The data consists of 128 frames, each of which is a 256×256 image.

direction $(-\tau_y, \tau_x, 0)$, it follows that

$$(\tau_x, \tau_y) = (-\sin\theta_b, \cos\theta_b) \tag{9}$$

As an example, consider a computer generated image sequence of falling snow as discussed above. The scene is a set of static spheres seen by an observer moving in the vertical direction and was generated using OpenGL (Figure 5). The spheres have constant 3-D size and are placed at random positions within a view volume prior to rendering.

Figure 5a shows one frame of the sequence (XY slice) and Figure 5b shows a YT slice (Adelson & Bergen, 1985). Because different spheres lie at different depths, different spheres have different image speeds. In Figure 5b, each sphere appears as a space-time bar. Different bar orientations indicate different speeds (Bolles et al., 1987).

To visualize the bowtie of power in the 3-D frequency domain for this sequence, we compute the 3-D power spectrum and project it orthographically onto a set of planes,

$$\cos\theta \; f_x + \sin\theta \; f_y = 0 \tag{10}$$

where $\theta \in [0,\pi)$. For any θ, we compute the projection by summing the power along lines parallel to $(\cos\theta, \sin\theta)$. Figure 6 shows the projected power for several angles θ. A bowtie indeed appears at $\theta = \theta_b = 0$ as expected. Since the image velocities are in the vertical direction, these velocities are of the form $(0, \alpha)$ and so the axis of the bowtie has $\theta_b = 0$. That is, all the motion planes pass through the f_x axis. Moreover, as θ deviates from $\theta_b = 0$ the bowtie gradually diminishes (Langer & Mann, 2001).

One aside worth noting is the aliasing effects in Figure 6 at $\theta = 0$, where the bowtie wraps around the boundaries of the plot. Aliasing is due to the edges of the spheres, which are represented with floating point accuracy by OpenGL. For image sequences taken by a real moving camera, aliasing effects are reduced because of optical blurring at the sensor level prior to spatial sampling. An example is given in (Langer & Mann, 2001).

In the example of Figure 6, the slopes of the planes making up the bowtie were all positive. The reason is that the camera is not rotating and so all the y speeds are negative. A second example, outlined below, has both positive and negative speeds.

Take the same 3-D scene of spheres and the same moving observer but now rotate the observer about the x-axis direction as the camera translates in the y direction. The rotation is such that the optical axis passes through a 3-D point near the center of the view volume in all frames, that is, the observer tracks this 3-D point, reducing its image velocity to zero. Objects on the near side of the tracked 3-D point have downward image velocity and objects on the far side of the tracked point have upward image velocity. Figure 7 shows the XY and YT plots as well as the projected spectrum where the projection is along the bowtie axis. In the YT plot, the slopes are positive for the background objects and negative for the foreground objects. In the projected bowtie, the speeds are both positive and negative.

4.2 Estimating the Translation Direction

The visualization of the bowtie pattern presented in the last section suggests a method for estimating the direction of the motion, that is, the τ vector. The method assumes the ω vector of Eq. (4) is zero such that the axis of the bowtie lies in the (f_x, f_y) plane (Langer & Mann, 2001). At the end of this section, we discuss how to generalize the method to the case where the ω vector is non-zero. Results for this case will be presented in a future paper.

For the case where ω is zero we need to estimate the angle θ_b, which is equivalent to estimating the τ vector. We consider all possible directions θ for the axis of the bowtie. For each direction, we sum the power in a wedge of

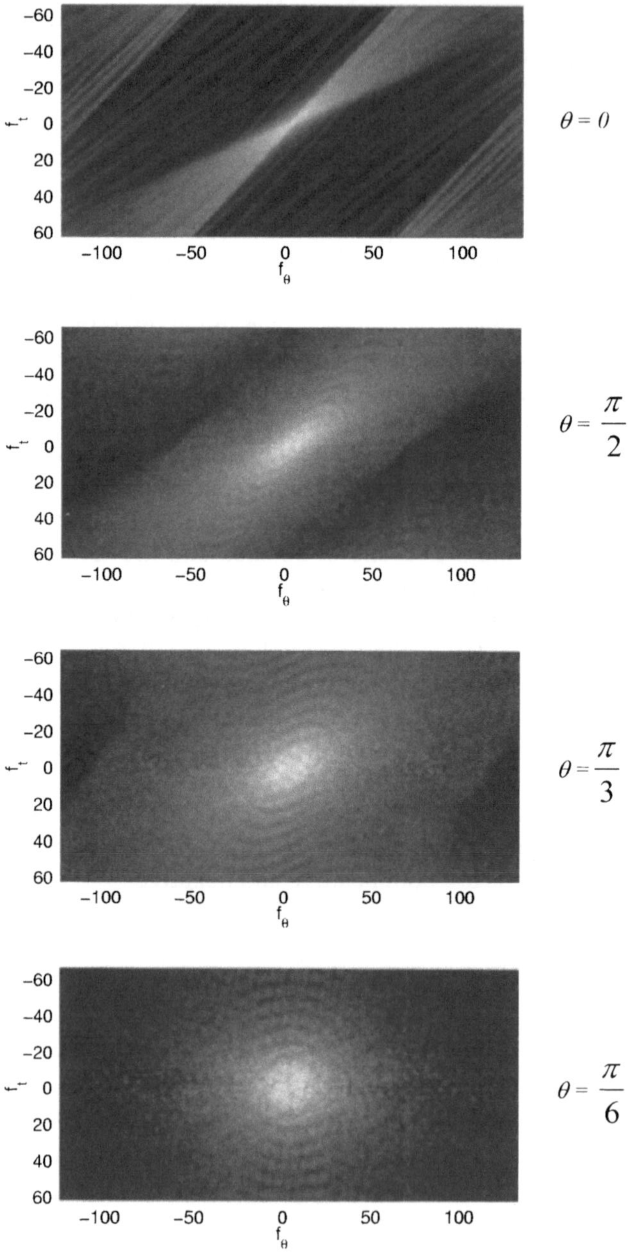

Figure 6. Projected power spectrum for the falling sphere sequence in Figure 5. The bowtie is evident at $\theta = 0$ where the direction of projection is identical to the axis of the bowtie.

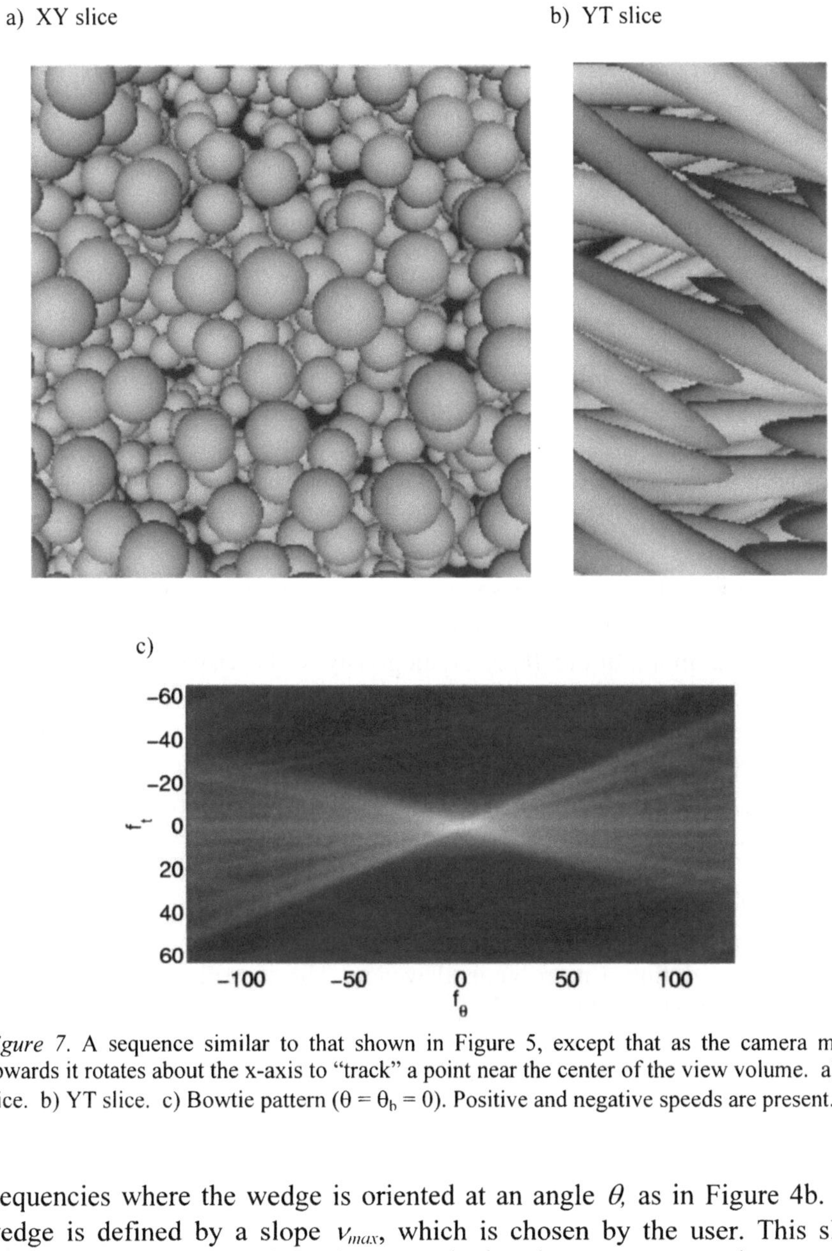

Figure 7. A sequence similar to that shown in Figure 5, except that as the camera moves upwards it rotates about the x-axis to "track" a point near the center of the view volume. a) XY slice. b) YT slice. c) Bowtie pattern ($\theta = \theta_h = 0$). Positive and negative speeds are present.

frequencies where the wedge is oriented at an angle θ, as in Figure 4b. The wedge is defined by a slope v_{max}, which is chosen by the user. This slope corresponds to an upper bound on speeds that the user assumes is present in the image sequence.

The direction θ_h is found by looking for a minimum in power within the wedge over all θ. The minimum in power should occur at θ_h because at this

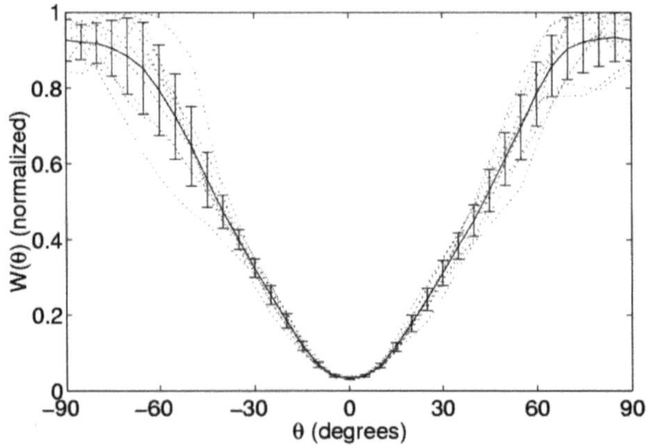

Figure 8. W(θ) for the synthetic sequence in Example 1. The minimum occurs at θ = 0° corresponding to vertical motion. The results of ten runs are shown (dotted lines), along with the mean response (solid lines) and error bars (one standard deviation).

angle the bowtie and the wedge should have little overlap. Formally, let $W(\theta)$ be the power in the wedge when the wedge is oriented in direction θ. We are looking for the minimum of $W(\theta)$. Analogously to the visualization of Figure 6, we compute $W(\theta)$ for a given image sequence by rotating through angles θ at fixed angular increments.

Figure 8 shows $W(\theta)$ for the sequence in Example 1. In this experiment, we chose v_{max} = 8 pixels/frame. We show $W(\theta)$ for ten different sequences, as well as the mean curve and the error bars at one standard deviation. Note that while there is some variability in the response, $W(\theta)$ has a well-defined minimum at $\theta = 0°$ which corresponds to vertical motion.

How could this algorithm be generalized to the case where ω is non-zero? Our idea is to define a more general wedge. The two planes of the wedge again have high slope but now the planes intersect at the axis of the bowtie, which in the general case does not lie in the (f_x, f_y) plane; recall Figure 3. The same principle applies as in the parallel case, namely that the wedge should contain a minimum of power when the axis of the wedge is identical to the true axis of the bowtie. The idea of the method is thus to search for the general bowtie axis. The search would be over both the azimuth angle θ as before, and also over an elevation angle as well. The elevation angle would determine the length of the ω vector. It remains to be seen how well such a method can recover the ω and τ parameters for motion in a small image region. This is a topic of our present research.

5 SUMMARY

In this paper we have considered the problem of computing image motion and heading in a static 3-D cluttered scene. The starting point for our analysis was our observation that the motion fields in such scenes contain a rich set of depth discontinuities, and as a result it is unlikely that accurate image velocities can be computed. Thus, models of heading perception that are based on pre-computed image velocities may not be feasible in 3-D cluttered scenes. We have proposed that instead of trying to recover accurate velocities, an alternative goal may be to recover parameterized families of velocities that are present in local image regions. We have argued that this family of velocities has a simple linear form, which gives rise to a bowtie pattern in the 3-D power spectrum. We then outlined a method that could be used to estimate the family of image velocities in a patch, and subsequently to compute direction of heading.

ACKNOWLEDGEMENTS

This research was supported by the Natural Sciences and Engineering Research Council of Canada (NSERC).

REFERENCES

Adelson, E. & Bergen, J. (1985). Spatiotemporal energy models for the perception of motion. *J. Opt. Soc. Am. A, 2(2)*, 284-299.

Bolles, R. C., Baker, H. H., & Marimont, D. H. (1987). Epipolar-plane image analysis: An approach to determining structure from motion. *Int. J. Comput. Vision, 1,* 7-55.

Gibson, J. J. (1950). *The Perception of the Visual World.* Houghton Mifflin, Boston.

Heeger, D. J. (1987). Optical flow from spatiotemporal filters, *First International Conference on Computer Vision,* p. 181-190.

Heeger, D. J. & Jepson, A. D. (1992). Subspace methods for recovering rigid motion. I: algorithm and implementation. *Int. J. Comput. Vision, 7,* 95-117.

Hildreth, E. C. & Royden, C. S. (1998). Computing Observer Motion from Optical Flow. In Watanabe, T. (Ed.) *High-Level Motion Processing – Computational, Neurobiological, and Psychophysical Perspectives* (p. 269-294), Massachussets: MIT Press.

Langer, M. S. & Mann, R. (2001). Dimensional analysis of image motion, *IEEE International Conference on Computer Vision,* Vancouver, Canada, p. 155-162.

Langer, M. S. & Mann, R. (2002). Tracking through optical snow, In: Bulthoff, H. H., Lee, S. W., Poggio, T. A., Wallraven, C. (Eds.), *2nd Workshop on Biologically Motivated Computer Vision: Lecture Notes in Computer Science, 2525* (p. 181-188), Springer-Verlag.

Lappe, M. & Rauschecker, J. P. (1993). A neural network for processing of optical flow from egomotion in man and higher mammals. *Neural Comput., 5,* 374-391.

Longuet-Higgins, H. & Prazdny, K. (1980). The interpretation of a moving retinal image. *Proc. R. Soc. Lond. B, 208,* 385-397.

Mann, R. & Langer, M. S. (2002). Optical snow and the aperture problem, *International Conference on Pattern Recognition,* Quebec City, Canada, *4*, 264-267.

Rieger, J. H. & Lawton, D. T. (1985). Processing differential image motion. *J. Opt. Soc. Am. A, 2,* 254-260

Royden, C. S. (1997). Mathematical analysis of motion-opponent mechanisms used in the determination of heading and depth. *J. Opt. Soc. Am. A, 14(9),* 2128-2143.

Warren, W. H. (1998). The State of Flow. In Watanabe, T. (Ed.) *High-Level Motion Processing – Computational, Neurobiological, and Psychophysical Perspectives* (p. 315-358), Massachussets: MIT Press.

Watson, A. & Ahumada, A. (1985). Model of human visual-motion sensing. *J. Opt. Soc. Am. A, 2(2),* 322-342.

Section 3

Visual Locomotion and Beyond

14. From Optic Flow to Laws of Control

William H. Warren[1] and Brett R. Fajen[2]

[1]Department of Cognitive & Linguistic Sciences
Brown University, Providence, RI, USA

[2]Department of Cognitive Science
Rensselaer Polytechnic Institute, Troy, NY, USA

1 INTRODUCTION

It is now established beyond a reasonable doubt that people can perceive self-motion from optic flow with sufficient accuracy to guide their locomotion. In particular, the direction of self-motion, or *heading*, can be judged quite accurately under a variety of conditions. However, as pointed out by Nakayama (1994), it remains controversial as to whether optic flow is actually used to control locomotor behavior. The aim of this chapter is to move beyond the perception of optic flow per se to the question of how a variety of information is used to control human locomotion on foot. We offer an interim report on an ongoing research program that seeks to determine the laws of control for steering and obstacle avoidance in a complex, dynamic environment.

2 PERCEPTION OF HEADING FROM OPTIC FLOW

Cutting and colleagues (1992) estimated that an accuracy of 1° to 3° of visual angle is needed to guide ordinary locomotor behavior such as running and skiing. In a series of psychophysical experiments over the last 15 years, it has been shown that one's current heading or future path can be judged with this level of accuracy under a wide range of environmental and viewing conditions (for recent reviews see Lappe et al., 1999; Warren, in press). For example, heading accuracy is on the order of 1° in a variety of environments (ground planes, frontal planes, 3-D clouds, realistic 3-D worlds); with dense,

L.M. Vaina, S.A. Beardsley and S.K. Rushton (eds.), Optic Flow and Beyond, 307–337.

sparse, discontinuous, or noisy flow fields; and on straight or circular paths of self-motion.

For purely translational movements of the observer, there is consistent evidence that the visual system determines heading from the radial pattern of optic flow, in which the *focus of expansion* corresponds to one's current heading direction. When the eye is simultaneously rotating – during a pursuit eye movement, for example – one can also recover the instantaneous heading or the path over time. After more than a decade of controversy over how the visual system handles this *rotation problem* (Banks et al., 1996; Royden et al., 1992, 1994; Stone & Perrone, 1997; van den Berg, 1992, 1996; Warren & Hannon, 1988, 1990), it now appears that both information in the retinal flow pattern and extra-retinal signals about eye rotation contribute to heading and path perception; for a detailed account see Warren (in press). From the retinal flow, the observer's translation in a retinal reference frame is specified by the pattern of *differential motion* (motion parallax) between points at different depths (Rieger & Lawton, 1985); in particular, differential motion goes to zero in the direction of heading. The observer's rotation is specified by the common *lamellar motion* (parallel flow) across the visual field (Perrone, 1992). Consequently, one's *object-relative heading* – the direction of heading with respect to objects that are also given in a retinal reference frame – is specified by the retinal flow pattern. The path through the environment may thus be determined from the sequence of such headings over time (Li & Warren, 2000, 2002). In contrast, one's *absolute heading* in space would seem to require extra-retinal information about eye and head position. As we shall see below, object-relative heading is precisely the sort of information that would be useful to control locomotion with respect to goals and obstacles.

3 CONTROL OF LOCOMOTION FROM OPTIC FLOW

The fact that people can accurately perceive heading from optic flow, and that specialized neural pathways exist to extract this information (Duffy, in press), would seem to imply that optic flow must be good for something in everyday behavior. But it is not a foregone conclusion that optic flow in general, or perceived heading in particular, is actually used to guide human locomotion (Wann & Land, 2000). Gibson (1958/1998; Warren, 1998) originally proposed a set of "formulae" by which optic flow could be used to steer toward goals, avoid obstacles, and chase or escape moving objects. But for any such locomotor task, a number of alternative strategies are also available. The challenge is to formally model and experimentally test the laws of control that actually govern human locomotion. We argue that such control laws must take into account not only visual information, but also the organization of the action system and the physics of the environment.

3.1 Laws of Control

A control law is generally considered to be a mapping from task-specific informational variable(s) to action variable(s) that describe observed behavior.

$$a = f(i) \tag{1}$$

If regularities in behavior can be identified at this level of abstraction, it suggests that there are systematic dependencies of action on information, presumably attributable to the laws of ecological optics. In some instances, these control principles may be quite general, spanning species from insects to humans (Duchon & Warren, 2002; Lee, van der Weel et al., 1992; Srinivasan, 1998; Wang & Frost, 1992). But how, exactly, are we to write such laws of control? There is little agreement in the literature, so let us consider several possible formulations.

First, control laws may be written in a *kinematic form*. This is a function that relates an informational variable directly to a kinematic movement variable such as limb trajectory, velocity, or timing

$$\dot{a} = f(i) \tag{2}$$

In this vein, Lee (1980, 1985) proposed that the onset of a movement to avoid an approaching object might be triggered at a critical value of the optical variable *tau*, which specifies the first-order time-to-contact. More recently, Lee (1998) suggested that the trajectory of a movement could be determined by the continuous coupling of two *tau* functions relating, say, the rate of closure of the distance to the target and the rate of closure of the angle of approach. Such a formulation provides a summary description of the relation between information and behavior, with the advantage that its terms are directly observable. However, it assumes that the organization of the action system doesn't contribute to the form of the behavior, and leaves out of account how the required movement is generated.

Second, control laws might take a *kinetic form*, a mapping from information to the effector forces that produce movement (Warren, 1988). Specifically, this would be a function that relates an informational variable to a force-related variable

$$F = f(i) \tag{3}$$

For example, Warren, Young, & Lee (1986) proposed that step length during running may be controlled by using the difference in time-to-contact between the next two footholds to regulate the vertical impulse of the push-off, given a constant body mass. However, this description still leaves out the action system and how it generates such forces.

We suggest that control laws be written in a *dynamic form*. The way that information can influence movement is by means of modulating the dynamics of the action system, which in turn generates effector forces. On this view, behavior is a function of the current state of the action system together with information that regulates the control variables of the system (Warren, 2003). This can be expressed functionally as a dynamical system

$$\dot{a} = \Psi(a, i) \tag{4}$$

The control law does not specify the kinematics of movement per se, but rather specifies an *attractor* for the action system. Such a fixed point or stable orbit corresponds to the goal of the intended action, and is converted into joint torques and limb movements given the properties of the musculo-skeletal system. The net result is a force exerted by effectors in the environment.

The difficulty here is that control relations between informational variables and control variables are not directly observable, but must be inferred from behavior. Thus, we will begin at a higher level of analysis with a description of the time-evolution of observable behavior, which we will call the *behavioral dynamics*. Then we may be able to infer *control laws* at a lower level that generate this behavioral outcome. Control laws thus characterize how information about the environment acts to modulate the control variables of a dynamical system, leading to adaptive behavior. In what follows, we will develop these concepts beginning with the information that is used to control locomotion, followed by our recent research on the behavioral dynamics of locomotion, and finally considering how we might derive control laws from these results.

3.2 Strategies for Steering to a Goal

A primary question is whether optic flow is actually used to guide human locomotion. Consider the most fundamental case of steering to a stationary goal. An optic flow strategy supposes that one would walk so as to create a flow pattern corresponding to self-motion toward the target. But there is a simple alternative: one could also walk in the perceived egocentric direction of the target, without relying on optic flow at all.

Normally these two strategies are redundant and would yield similar, successful behavior. One might expect that biology takes advantage of such

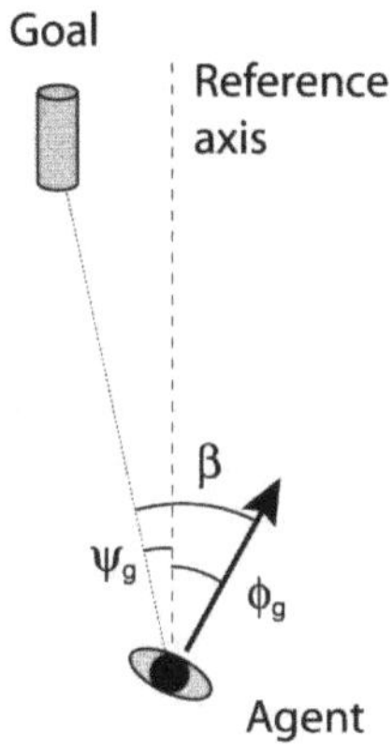

Figure 1. Definition of variables

redundancy to achieve robust locomotor control under a range of conditions. For example, in a visually structured environment, optic flow allows for heading judgments that are an order of magnitude more precise than those based on proprioceptive information about the direction of walking (Telford et al., 1995). This advantage could be due to the fact that object-relative heading is defined within a retinal reference frame, avoiding coordinate transformations that may accumulate error between eye, head, body, and effector frames. But when traveling at night or in fog, optic flow is unavailable and the system can fall back on egocentric direction.

The optic flow and egocentric direction strategies are actually two broad classes that can be broken down into more specific hypotheses. For convenience of analysis, let us define the physical heading by the angle ϕ between the current direction of locomotion and an arbitrary reference axis (see Figure 1). We also define the direction of a goal by the angle ψ_g with respect to the same reference axis. The object-relative heading, or heading error, is thus $\beta = \phi - \psi_g$, and the simplest definition of steering to the goal is to bring β to zero. Consider some possible flow strategies for adopting a straight path to a target.

- *Heading hypothesis*. Gibson (1958/1998, 1979) originally proposed that to aim locomotion at a goal, one should keep the focus of expansion near the goal. This formulation applies to the case of observer translation, but as we have seen it is complicated by the rotation problem, as well as by moving objects (Saunders & Warren, 1996). Thus, a more general version of the same principle is to *keep one's perceived heading near the goal* (Warren, 1998). Note that the heading may be specified by the focus of expansion in the case of pure observer translation, or more generally by the direction in which motion parallax goes to zero. The required turning angle is specified by the object-relative heading (β); the turning rate has been discussed by Lee (1998) and Fajen (2001).

- *Raw retinal flow hypothesis.* Other steering strategies do not require that heading be explicitly determined, but are based on the "raw retinal flow" when fixating the goal. In this hypothesis, the observer would *fixate the goal and steer so as to make the retinal flow pattern radial or symmetrical.* If the goal is fixated at eye level, the required steering adjustments could be determined from the curvature of the flow in the ground plane: if the flow curves to the right, one is heading to the left of fixation, so steer rightward, and vice versa. Alternatively, steering adjustments might be derived from Cutting, et al.'s (1992) concept of differential motion parallax. Environmental objects that are closer than the fixated goal appear to move across the line of sight opposite the heading direction. If such motion is rightward, one is heading to the left of fixation, so one should steer rightward, and vice versa.

One can also formulate specific versions of an egocentric direction strategy, based on information about the visual direction of the target with respect to the body.

- *Locomotor axis hypothesis.* The first alternative is directly analogous to the heading hypothesis above, except that the current direction of locomotion is specified by proprioception, which we will call the body's locomotor axis. Specifically, to walk to a goal, *keep the felt locomotor axis pointing in the direction of the goal.* The heading error β is specified by the angle between the current locomotor axis and the visual direction to the goal. Due to their parallel form, the heading and locomotor axis strategies provide straightforward redundancy in locomotor control. However, this hypothesis relies on a unique relation between effector proprioception and the direction of movement in terrestrial locomotion, which does not hold for aerial or aquatic environments (Gibson, 1966).

- *Thrust hypothesis.* A closely related strategy is to *perceive the egocentric direction of the goal and apply thrust force in the opposite direction.* Whereas the preceding hypothesis was based on the relation between the goal and proprioception, this hypothesis is based on the relation between the goal and motor commands. Thus, it can be used to guide the initiation of walking toward the goal from a standstill. Note that the axis of thrust need not be aligned with the anterior-posterior (AP) axis of the body, for the observer can "crab" sideways. This hypothesis relies on a unique relation between the direction of force application and the resulting body displacement in terrestrial locomotion.

- *Centering hypothesis.* A special case of the locomotor axis hypothesis assumes that the eyes, head, and AP axis tend to align with the locomotor axis during walking. To walk to a goal, the observer can thus *fixate the goal, center it at the midline of the body and walk forward.* If one's gaze and head initially deviate from the AP and locomotor axes, they tend to come into alignment, analogous to the uncoiling of a twisted spring. This hypothesis is consistent with the folk wisdom that you should look in the direction you want to go and not at obstacles you want to avoid.

- *Target drift hypothesis.* Finally, one could walk so as to *cancel target drift, keeping the goal in a constant egocentric direction.* This strategy actually takes advantage of a local feature of the optic flow; that the only fixed point in the flow field ahead is at the focus of expansion. Thus, if one is heading toward the target, its optical drift is zero and hence its egocentric direction remains constant. Otherwise, it will drift away from the current heading point, providing a basis for steering adjustments.

A number of other hypotheses have been suggested that yield a curved path to the goal (Kim & Turvey, 1999; Lee, 1998; Wann & Land 2000; Wann & Swapp, 2000). However, we will show that during actual walking people do not adopt continuously curved paths to a goal, but rather turn onto a straight path. There may be situations in which a curved path is called for, such as driving around a bend, but such lane-following tasks appear to involve special strategies (Beal & Loomis, 1996; Land, 1998; Land & Lee, 1994). Thus, we will not pursue the curved path hypotheses here.

In experiments on joystick steering, with simulated rotation corresponding to fixation of the target, participants are able to steer accurately as long as motion parallax is present in the display (Frey & Owen, 1999; Rushton et al., 1999). By itself, this finding does not differentiate the heading and raw retinal flow hypotheses, because steering could be achieved either by aligning the perceived heading on the target or by making the flow symmetric around the target. But participants can steer straight paths toward a target even with simulated fixation elsewhere in the display (Li & Warren, 2002), an ability that cannot be accounted for by the raw retinal flow hypothesis. The pattern of results for active steering is quite similar to that for heading judgments (Li & Warren, 2000), consistent with the heading hypothesis. Moreover, a moving object biases both heading judgments and joystick steering in precisely the same way (Royden & Hildreth, 1996; Warren & Saunders, 1995), consistent with the idea that locomotor control is based on perceived heading. The preliminary evidence thus seems more in line with a heading strategy than one based on raw retinal flow with a fixated target.

However, experiments on joystick steering cannot test egocentric direction strategies. With a joystick and a computer monitor, the locomotor axis and direction of thrust are not specified, and the mapping between joystick movement and the direction of locomotion in the display must be learned. To test locomotor strategies, therefore, we must turn to experiments on the hoof.

3.3 Is Optic Flow Used to Walk to a Goal?

The first cut through the panoply of hypotheses is to compare the two broad classes of optic flow and egocentric direction strategies. However, because these two strategies are normally redundant and predict the same behavior, they must be dissociated by varying the optic flow independently of the locomotor axis. We originally attempted this in a "virtual treadmill" apparatus, by manipulating the optic flow on a projection screen for a participant walking on a treadmill (Warren & Kay, 1997). However, the presence of the screen and the confines of the treadmill raised the possibility of artifactual strategies. We subsequently transferred the experiments to the large-scale Virtual Environment Navigation Lab (VENLab), a 12 m by 12 m room equipped with a sonic/inertial head-tracking system in the ceiling. Participants can walk freely while wearing a head-mounted display (60° H x 40° V) and be immersed in a virtual environment.

At the same time, Rushton et al. (1998) arrived at a low-tech solution – wedge prisms. Suppose that a participant wearing displacing prisms is walking directly toward a target, which is 10 m away on a grass lawn. The prisms displace both the target and the optic flow by 16° from the direction of walking (to the right, say), such that the focus of expansion remains on the target. Thus, if participants rely on the optic flow, they should keep the flow centered on the target (virtual heading error = 0°) and continue walking on a straight path. However, if they rely on the egocentric direction of the target, they should walk 16° to the right, toward the displaced image of the target. This shifts the focus of expansion to the right of the target (virtual heading error = 16°), causing the target to drift slowly to the left in the visual field. As the participant continues to turn toward the drifting image of the target, they will trace out a curved path. And this is exactly what Rushton, et al. (1998) reported: participants followed curved paths to the target, with virtual heading errors close to 16°. These results are wholly consistent with an egocentric direction strategy, showing no influence of optic flow on walking.

However, the optic flow available in this experiment was rather minimal, defined only by the fine texture of the grass on the ground plane. In addition, prisms introduce blur and optical distortion that warps the flow pattern. These effects may have reduced reliance on the optic flow, resulting in the dominance of an egocentric direction strategy.

In the VENLab, we manipulated the area and magnitude of optic flow in the display by varying the visual structure of the environment (Warren et al., 2001). We created a similar displacement by offsetting the focus of expansion in the HMD 10° to the right or left (randomly) of the participant's actual direction of walking. The predictions are the same: if participants rely on optic flow, they should keep the focus of expansion aligned with the goal (a virtual heading error of 0°), resulting in a straight path; whereas if they rely on egocentric direction, they should keep the locomotor axis aligned with the goal (a virtual heading error of 10°), resulting in a curved path.

With an isolated target line on a black background, there was little optic flow, and participants closely followed the curved path predicted by the egocentric direction strategy (Figure 2a). Informal observations suggest that people tended to align their head and AP axis with the target and walk forward, as suggested by the centering hypothesis. However, as a textured ground plane (Figure 2b) and a textured ceiling and frontal wall (Figure 2c) were added, paths became significantly straighter and heading error significantly decreased. Finally, when an array of textured posts was added, creating salient motion parallax, heading errors decreased to about 2° (Figure 2d). Participants started walking in the egocentric direction of the target, but once the optic flow became available their paths quickly straightened out and heading error dropped. In this case, observations suggest that they tended to align their head and AP axis – but not their locomotor axis – in the visual direction of the target, so that they "crabbed" slightly sideways. These results strongly indicate that both egocentric direction and optic flow information contribute to the control of walking, but the latter increasingly dominates as more flow becomes available.

We modeled this as an additive relation in a simple dynamical control law. Specifically, the rate of change in heading ($\dot{\phi}$) is a function of the current heading error, which is given by a linear combination of egocentric direction and optic flow

$$\dot{\phi} = -k\left[\left(\phi_{ego} - \psi_g\right) + wv\left(\phi_{flow} - \psi_g\right)\right] \tag{5}$$

The coefficient k is a turning rate constant. The current heading error is redundantly specified by the egocentric direction of the goal with respect to

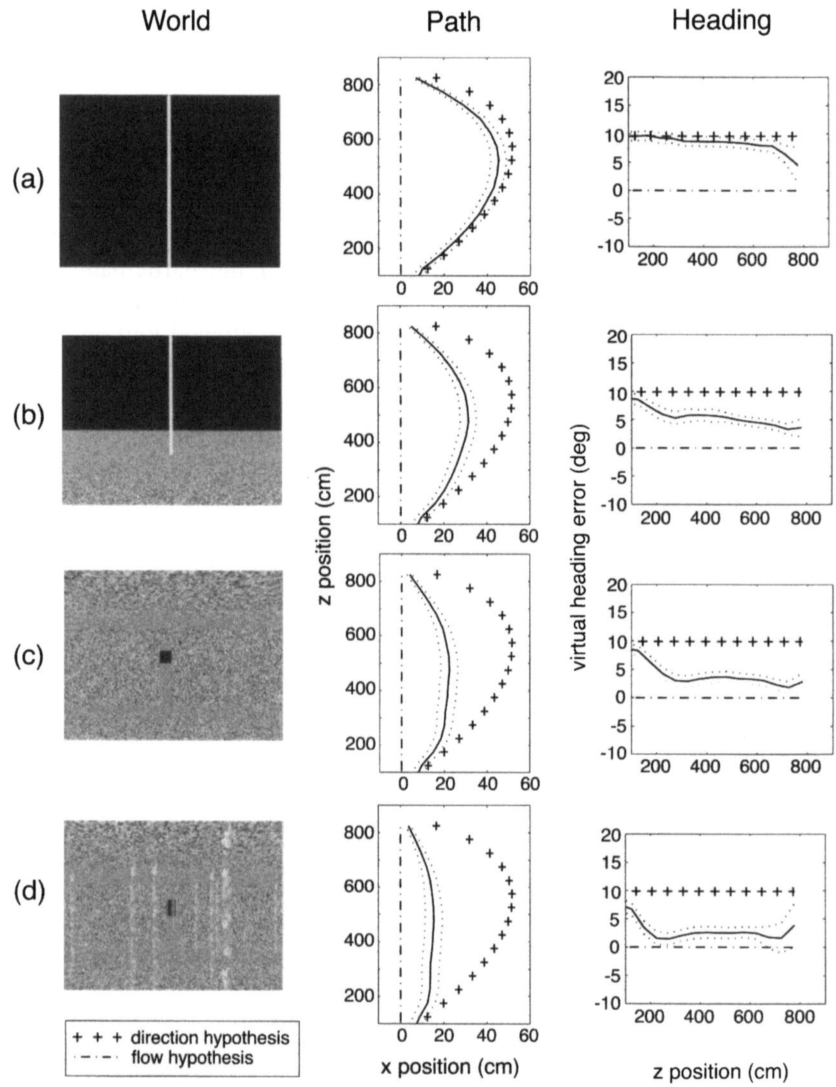

Figure 2. Walking to a target in four virtual environments that vary the amount of optic flow. Center column: mean path in the virtual world. Right column: mean virtual heading error as a function of distance from the start. Curves indicate between-subject SE (...), egocentric direction prediction (+++), optic flow prediction (_ . _). [From Warren, et al., 2001; used by permission.]

the locomotor axis ($\phi_{loco} - \psi_g$), and by the visual angle between the goal and the heading given by optic flow ($\phi_{flow} - \psi_g$). Finally, the flow contribution is weighted by the observer's velocity *v*, which influences the flow rate, and by *w*, a measure of the visual area and magnitude of optic flow due to environmental structure. Thus, if the observer is walking slowly or there is little visual structure, the flow contribution will be minimal. We simulated the initiation of walking by increasing *v* from 0 to 1 m/s as a logistic function of time over the first 2 sec.; the initial direction of walking was toward the goal. Simulations in which *w* ranged from 0 to 6 show a pattern of results similar to the human time series of heading error in Figure 2.

Why do our results differ from those of Rushton, et al. (1998)? Minimal flow from the grass may be part of the answer. To test the role of the prisms, we had participants wear wedge prisms inside our HMD while they walked in the same four environments. The prisms displaced the optic flow by 10°, always to the right, which we compared with a computed offset of the optic flow to the right. The influence of the flow was significantly reduced by viewing it through the prisms. As visual structure was added, the drop in heading error was significantly less with the prisms than with the computed offset. This suggests that Rushton, et al.'s (1998) null effect of optic flow may have been due to a combination of prisms and minimal flow.

This interpretation is supported by other open-field prism experiments, in which the influence of optic flow increases with visual structure. For example, Wood et al. (2000) found that walking paths became much straighter when random markings were placed on the grass, and almost completely straight when an array of small squares was arranged in a grid pattern. Similarly, Harris and Carré (2001) reported that paths are significantly straighter when subjects crawl rather than walk, which increases the flow rate and the visual coarseness of floor texture by lowering the eye height.

Equation 5 predicts an influence of flow rate on the contribution of optic flow, consistent with Harris and Carré's (2001) finding. We directly tested flow rate in the VENLab by manipulating the gain of the visual display (Fink & Warren, 2002). Participants walked at a normal speed to a target in a coarsely textured environment, while the visual gain was varied between 0.5, 1.0, 2.0, or 4.0. With a gain of 1.0, the flow speed in the display matched the participant's walking speed, whereas with a gain of 4.0, it was four times the walking speed. As before, the optic flow was randomly offset by 10° to the left or right of the walking direction. We observed a direct relation between flow rate and path straightness: as gain increased, the walking paths became significantly straighter and the virtual heading error significantly smaller. At the highest flow rate (gain=4.0), the lateral deviation of the path decreased by 33% and the heading error was reduced to 2°. This confirms an increasing contribution of optic flow with speed, as well as a residual influence of egocentric direction, as predicted by the additive model.

Following Rogers and Dalton (1999), we also examined adaptation to the visual offset to determine whether it was influenced by the available optic flow (Zosh et al. 2000). Participants received 38 adaptation trials either in the fully textured world (floor, ceiling, wall, posts) or with a single target line (little flow). The optic flow was offset by 10°, always to the right, inducing walking paths that curved leftward. They were then transferred to the same or opposite environment for 10 trials, with normal flow. An aftereffect of adaptation would thus be a path that curved rightward. Two important results stand out in the data. First, the textured world produced greater adaptation than did the target line, as revealed by a larger aftereffect when participants were transferred to the line environment. Second, when they were transferred to the textured environment, the aftereffect was abolished within one or two trials. These results indicate that participants depend upon the optic flow as a reliable "teaching signal", so that they rapidly adapt to a mismatch between optic flow and egocentric direction. This again confirms the dominance of optic flow in the visual guidance of walking to a stationary goal.

In contrast, somewhat to our surprise, we recently found no contribution of optic flow to intercepting a moving target (Fajen & Warren, 2002). In the VENLab, participants walked to a target that moved on a linear path at 0.6 m/s, but randomly varied in its initial position and direction of motion. They successfully intercepted the target by turning onto a straight path that led the target by a β angle that approached 15°-20°. However, the paths were unaffected by manipulations of the global flow from the background, which specifies the heading in the environment, or the local flow from the target itself, which specifies the heading relative to the target. This strongly suggests that steering toward a moving target is based on the egocentric direction of the target alone.

Why might this be the case? With a stationary target, the observer can bring β to zero by placing the global FOE on the target, or by nulling the motion parallax between the target and the surrounding environment. But if the target is moving the FOE is eliminated as the observer turns to track the target, and the motion parallax with the target cannot be nulled. Thus, optic flow may dominate with a stationary target when it is particularly effective for bringing β to zero, but not with a moving target to guide turning to a constant β ahead of the target. In this case egocentric direction dominates instead.

There is some disagreement as to whether the results for a stationary goal should be attributed to optic flow per se, or might be due to "alignment cues" such as local motion parallax. For example, keeping the target aligned with a nearer or farther feature on the ground by nulling the motion parallax between them would result in a straight path. Such motion parallax is in fact a local property of the optic flow, and zeroing parallax with the target is just one species of an optic flow strategy, so there is no inconsistency here. However, Harris and Carré (2001) reported little influence of local parallax during walking, when they manipulated the distance between the target and a background wall; but, parallax with the ground plane was still available. Similarly, when Li and Warren (2000) removed local target parallax, there was no effect on heading judgments during simulated rotation, implying that the global optic flow was sufficient to perceive heading. On the other hand, an alignment cue such as a grid or checkerboard that defines a line (or lane) to the target presents a special case. One could follow a straight path to the target simply by maintaining one's position with respect to the line, without relying on optic flow at all. However, to the extent that optical rotation of the line is used to guide walking (Beal & Loomis, 1996), this case is also related to motion parallax.

Although the existing evidence is consistent with a heading strategy, there is as yet no direct evidence that the perception of heading per se is involved in the control of locomotion. The next item on the agenda for this line of research is thus to identify the specific properties of optic flow that are used to guide walking, and tease apart the particular hypotheses for steering to a goal.

4 BEHAVIORAL DYNAMICS OF LOCOMOTION

If our aim is to formulate laws of control that characterize how information modulates action, we need to have a good description of the behavioral outcome. By *behavioral dynamics*, we simply mean a description of the time-evolution of observed behavior. This way of formulating the problem allows us to formalize behavior in terms of systems of differential equations and to use methods from nonlinear dynamics to analyze perception and action. Our approach is inspired by that of Schöner et al. (1995), who developed a dynamical control system for mobile robots. In the present case, we wish to develop an empirical dynamical model of human locomotor behavior.

The behavioral dynamics of locomotion must cover the tasks of steering to a stationary goal, avoiding stationary obstacles, intercepting moving targets, and avoiding moving obstacles. Our current research program seeks to experimentally measure human walking behavior for each of these tasks, and use the data to specify a dynamical model of heading control. Once these components are modeled individually, we attempt to combine them to predict the routes that people adopt in more complex environments. A common approach to this problem is to explicitly plan a route based on a detailed world model, an internal representation of the positions and motions of all objects in the scene. But in the present approach, steering is based on current information about one's heading with respect to nearby objects, so the path emerges on-line from the interaction between agent and environment. If we can formalize the locomotor "rules" for an individual agent, this may ultimately allow us to model interactions among multiple agents in a complex environment.

4.1 Behavioral Dynamics of Steering to a Goal

Let us assume that goal-directed behavior can be described by a small number of *behavioral variables* which express aspects of action that are relevant to the goal. These define the dimensions of a *state space* for the system, and an instance of behavior can be represented as a *trajectory* in state space. Goals correspond to regions in state space to which trajectories converge, known as *attractors*, whereas states to be avoided correspond to regions from which trajectories diverge, known as *repellors*. Sudden changes in the number or type of such fixed points are known as *bifurcations*, which correspond to qualitative transitions in behavior. These trajectories may be formally expressed as solutions to a system of differential equations, and thus the problem is to formalize such a *dynamical system* whose solutions capture the observed behavior in question.

For a terrestrial agent, we take the current heading direction ϕ and turning rate $\dot{\phi}$ in the horizontal plane as behavioral variables, assuming travel at a constant speed v. From the agent's current (x, z) position, a goal lies in the direction ψ_g at a distance d_g, and an obstacle lies in the direction ψ_o at a distance d_o. The simplest description of steering toward a stationary goal is for the agent to bring the heading error or *goal angle* between the current heading direction and the goal direction to zero, such that the heading is stabilized on the goal $\beta = \phi - \psi_g = 0$. In this basic case, the goal direction would behave like an attractor in state space at $(\phi, \dot{\phi}) = (\psi_g, 0)$. On the other hand, the simplest description of obstacle avoidance is to increase the *obstacle angle* between the current heading and the obstacle direction, $\phi - \psi_o > 0$. The obstacle direction would thus act like a repellor, or unstable fixed point, at $(\phi, \dot{\phi}) = (\psi_o, 0)$. In addition, one might suspect that object distance (or equivalently, time-to-contact) also influences behavior, for nearby obstacles should be avoided before more distant obstacles.

However, the form of the turning functions is an empirical question. Given that a physical body with inertia must undergo angular acceleration to change heading direction, it is reasonable to assume that a description of steering behavior requires at least a second-order system. To get an intuition, imagine that the agent's current heading direction is attached to the goal direction by a damped spring. Angular acceleration toward the goal direction would thus depend on the stiffness of the spring and be resisted by the damping. In addition, the distance of the goal would modulate the spring stiffness. At the same time, imagine that the heading direction is repelled from the direction of each obstacle by another spring, whose stiffness is modulated by the distance of the obstacle. Thus, at any moment, the heading is determined by the resultant of all spring forces acting on the agent; specifically, the current attractor direction is determined by the sum of all components. As the agent moves through the environment to the next (x,z) position, the goal angle (ψ_g) and obstacle angles (ψ_o) change and the directions of attractors and repellors shift, influencing the next heading direction. Locomotion in an environment is thus a four-dimensional system, for to predict the agent's future position we need to know it's current position (x, z), heading (ϕ), and turning rate ($\dot{\phi}$), assuming that speed is constant.

To determine the forms of these functions, we embarked on a series of studies in the VENLab to measure how a walker's heading direction is influenced by the angles and distances of goals and obstacles (Fajen & Warren, 2003). Our first experiments investigated how participants steered toward a stationary goal, as we varied the initial goal angle (0° to 25°) and goal distance (2 m to 8 m). On a given trial, the participant walked toward a marker on a textured ground plane to establish an initial direction and speed, and then a blue goal post appeared.

The results demonstrate that participants turn onto a straight path to the goal (e.g. Figure 3a), but do so more rapidly when the goal is at a greater angle or a closer distance. The time series of heading error show that β converges to zero from all initial conditions (Figure 3b), with an angular acceleration that increases linearly with goal angle and decreases exponentially with goal distance. The goal direction thus behaves like an attractor of heading.

We modeled this behavior with an angular "mass-spring" equation, in which angular acceleration $\ddot{\phi}$ is a function of both goal angle ($\beta=\phi-\psi_g$) and goal distance (d_g),

$$\ddot{\phi} = -b\dot{\phi} - k_g(\phi - \psi_g)(e^{-c_1 d_g} + c_2) \tag{6}$$

The "damping" term indicates that the resistance to turning is proportional to turning rate; the b parameter determines the slope of this function, expressing the ratio of damping to the body's moment of inertia (in units of 1/s). The "stiffness" term reflects the finding that angular acceleration increases linearly with goal angle, at least over the tested range of –25° to +25°. The k_g

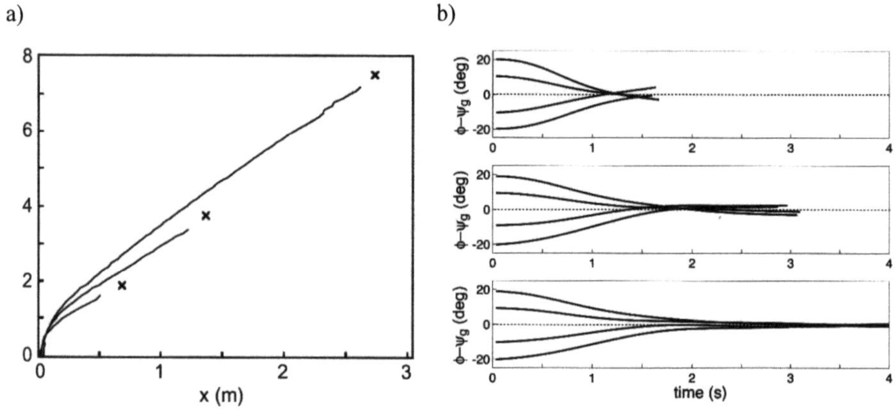

Figure 3. Walking to a goal at a distance of 2, 4 or 8 m. a) Mean paths with an initial goal angle of 20°. b) Mean time series of heading error with initial goal angles of ±10° or 20°. [From Fajen & Warren, in press; used by permission.]

parameter determines the slope of this function and hence the attractiveness of the goal, expressing the ratio of stiffness to the moment of inertia (in units of $1/s^2$). Finally, the attractiveness of the goal decreases exponentially with distance to some minimum value (to ensure that the agent steers toward distant goals), so the "stiffness" is modulated by an exponential function in which c_1 determines the rate of decay and c_2 the minimum angular acceleration. Least-squares fits to the mean time series of β yielded parameter values of $b = 3.25$, $k_g = 7.50$, $c_1 = 0.40$, and $c_2 = 0.40$.

Simulations of our experimental conditions generate locomotor paths that are very close to the human data (Figure 4a), and β time series that converge to zero in a similar manner (Figure 4b). The fits to the mean time series averaged r^2=0.98 over all conditions, indicating that model behavior is virtually identical to the mean human behavior. Thus, the model successfully captures the behavioral dynamics of walking to a goal, in which the goal direction behaves like an attractor of heading, whose strength increases with angle and decreases with distance.

4.2 Behavioral Dynamics of Obstacle Avoidance

Now consider how people avoid a stationary obstacle. In these experiments, we recorded detours taken around an obstacle en route to a goal, which presumably depend on the obstacle's position. On each trial, the participant began walking toward a green goal post, and after 1 m a blue obstacle post appeared slightly to the left or right of their path. We

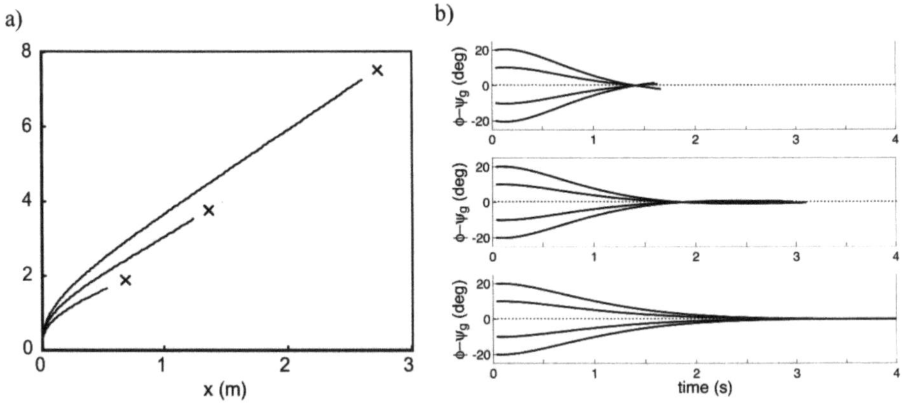

Figure 4. Model simulations of walking to a goal at 2, 4, and 8 m. a) Paths with an initial goal angle of 20°. b) Time series of heading error with initial goal angles of ±10° or 20°. [From Fajen & Warren, in press; used by permission.]

manipulated the initial angle between the obstacle and the path (1° to 8°) and the initial obstacle distance (3 m to 5 m), and observed their effects on the participant's heading direction. Once again, both the angle and distance of the obstacle influenced the locomotor path (Figure 5a). The time series of the obstacle angle β_o show that the heading was repelled from the obstacle direction, such that the curves diverge from zero in all conditions (Figure 5b). In this case, the angular acceleration decreased exponentially with both obstacle angle and obstacle distance.

To incorporate this behavior in the model, we simply added an obstacle component to the previous goal component. The net angular acceleration is thus also a function of the obstacle angle ($\beta_o = \phi - \psi_o$) and distance (d_o),

$$\ddot{\phi} = -b\dot{\phi} - k_g(\phi - \psi_g)(e^{-c_1 d_g} + c_2) + k_o(\phi - \psi_o)(e^{-c_3|\phi - \psi_o|})(e^{-c_4 d_o}) \qquad (7)$$

The obstacle "stiffness" term reflects the finding that angular acceleration decreases exponentially with a positive (right) or negative (left) obstacle angle; the amplitude of this function is determined by the parameter k_o, its decay rate by c_3 (in units of 1/rad), and it asymptotes to zero. The stiffness also decreases exponentially to zero with obstacle distance, where parameter c_4 is the decay rate (in units of 1/m). Keeping the previous parameter values for the goal component fixed, we fit the extended model to the mean β time series for the obstacle data, yielding obstacle parameter values of k_o = 198.0, c_3 = 6.5, and c_4 = 0.8.

Simulations for the initial conditions of the obstacle experiments reproduce the human paths (e.g. Figure 6a) and the β_o time series (Figure 6b),

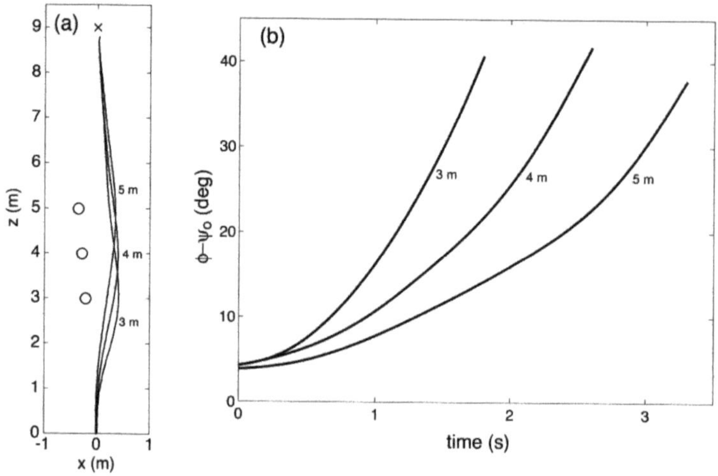

Figure 5. Obstacle avoidance with an initial distance of 3, 4, or 5 m and obstacle angle of –4°. a) Mean paths. b) Mean time series of heading error (obstacle angle). [From Fajen & Warren, in press; used by permission.]

with a mean r^2 = .975. Thus, the extended model captures the behavioral dynamics of obstacle avoidance, such that the obstacle direction behaves like a repellor of heading, and angular acceleration decreases with both obstacle angle and distance.

It is important to note that the fitted model only relies on information about the environment within a limited window, not a full world model. In particular, the influence of obstacles asymptotes to zero at a distance of around 4 m and an angle of ±60° about the heading direction, and the influence of the goal asymptotes at a distance of around 8 m. This implies that a limited sample of the environment is sufficient to account for human locomotor behavior. Moreover, because the distance functions are gradually decreasing exponentials, the model can tolerate a fair amount of error in perceived distance (or time-to-contact), particularly at larger distances, without greatly affecting the steering behavior. Adding 10% Gaussian noise into the perceived distance variables only induces a standard deviation of a few centimeters in the lateral position of the path around an obstacle.

4.3 Routes as Emergent Behavior

In the steering dynamics model, a route emerges from the agent's interaction with a structured environment, rather than being explicitly planned

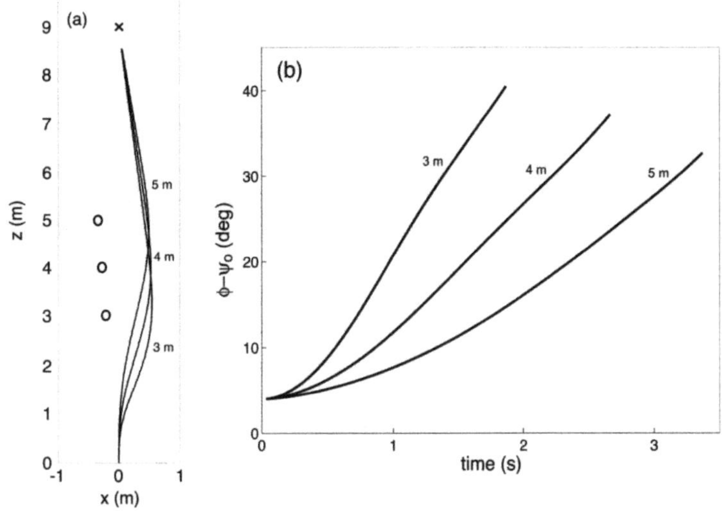

Figure 6. Model simulations of obstacle avoidance, with an initial distance of 3, 4, or 5 m and obstacle angle of –4°. a) Paths. b) Time series of heading error (obstacle angle). [From Fajen & Warren, in press; used by permission.]

in advance. Now that we have formulated basic components for a goal and an obstacle, the question arises as to whether the model can be used predict more complex behavior. The simplest case involves selecting one of two possible routes around an obstacle en route to a goal – the most direct route on an "inside" path, or the long way around on an "outside" path. Such a choice appears as a bifurcation in the model dynamics, and the branch that is taken depends on the agent's initial conditions.

The aim of our third study was to record the routes that people adopt around an obstacle under different initial conditions and to test whether the parameterized model can account for them (Fajen & Warren, in press). On each trial, the participant walked toward a marker for 1 m, and then both a blue goal post and a red obstacle post appeared. The obstacle lay between the heading direction and the goal direction at a distance of 4 m, such that the participant was initially heading on an outside path. The position of the obstacle was manipulated so that the goal-obstacle angle varied from 1° to 8°, while the attractiveness of the goal was manipulated by varying its initial distance from 5 to 7 m. This allowed us to test the conditions under which participants switch from an outside to an inside route.

Participants switched to an inside path when the initial goal-obstacle angle increased to 2°-4°, and as the goal got closer. When we tested the model with the previous parameter values, it exhibited the same pattern of switching, although the switch occurred at a somewhat higher angle, indicating that the model was biased toward outside paths. This may be because our first experiments did not sample cases in which the participant had to cross in front of an obstacle. However, adjusting a single obstacle parameter, from c_4=0.8 to 1.6, was sufficient to induce the switch in the human range. Parameter c_4 might be thought of as a "risk" parameter, for increasing it makes the repulsion decay more rapidly with distance, allowing a closer approach to obstacles. Thus, the agent's risk level and body size are implicitly represented in the model by this parameter. These parameter values were held fixed for the remaining experiments.

Such route switching behavior results from competition between the attractiveness of the goal, which increases with the angle and nearness of the goal, and the repulsion of the obstacle, which decreases with angle. If the obstacle is positioned between the agent and the goal, the model is bistable, such that both outside and inside heading directions are attractive; the one that is selected depends on the agent's initial conditions. As the agent moves around the obstacle, the model shifts to only one stable heading, exhibiting a *tangent bifurcation* (see Fajen & Warren, 2003). Thus, switching behavior and route selection can be understood as a consequence of bifurcations and attractors in the underlying dynamics of the system.

Whereas the model produces a unique path for each set of initial conditions, human routes are more variable. We wanted to see whether simply adding noise to the model would be sufficient to simulate the relative frequency of inside and outside paths. We thus added error to each perceptual variable and parameter independently at the onset of a trial, drawn from a Gaussian distribution with a standard deviation of 10%. The agent's initial x position and heading direction were also randomly varied, matching the human standard deviations. The resulting simulations effectively reproduced the distribution of inside and outside paths across environmental conditions in the human data (Fajen & Warren, 2003).

These results demonstrate that simple route selection can be accounted for as a consequence of the on-line steering dynamics for goals and obstacles. But what about routes through more complex environments? One advantage of the present model is that it scales linearly with the complexity of the scene, simply adding one term for each object in the immediate environment. Thus, in principle, we could predict locomotor paths by continuing to add a term for each new obstacle, while holding parameter values fixed.

We first tested this prediction with a configuration of two obstacles en route to a goal straight ahead (Fajen & Warren, 2001). The nearer obstacle was in a fixed position slightly to one side of the goal direction, whereas the farther obstacle was manipulated so its initial goal-obstacle angle varied from 0° to 10°. As this angle increases, the model predicts a particular sequence of route switching: from the outside of the far obstacle, to the outside of the near obstacle, and finally to a route between them. Participants demonstrated

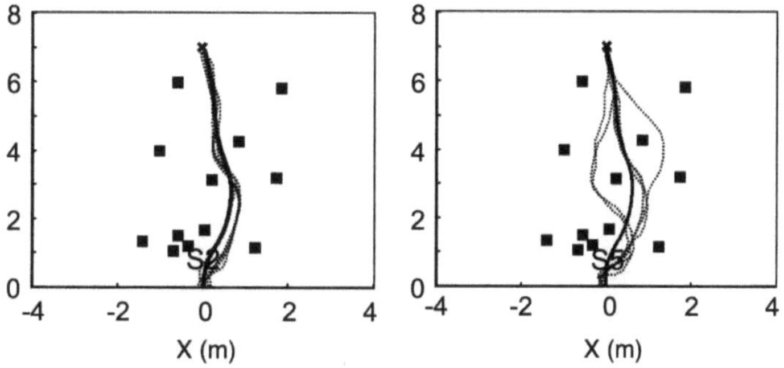

Figure 7. Routes to a goal (X) through an array of 12 obstacles (O), for two participants (S2 and S5). Dotted curves represent 6 trials from each participant, solid curves represent the model simulation. Starting point is at (0,0).

exactly this sequence of switching. Once again there was some variability in human route selection, but the distribution of paths was closely reproduced by adding 10% Gaussian noise to the model's perceptual variables and parameters at the onset of a simulated trial (Fajen et al., 2002).

A strong test is whether the model can predict human routes through a complex field of obstacles, simply by adding terms with fixed parameters to the equation. To examine this possibility, we recorded participants walking through random arrays of 12 yellow posts to get to a blue goal post (Warren et al., 2001). The model did a reasonable job of reproducing the human paths (e.g. Figure 7). One measure of model performance is the number of obstacles by which the model differed from the most frequent human path. On half of the eight arrays they were identical, on two arrays they differed by only one obstacle, and on the remaining two arrays they differed by two and four obstacles. Of course, there was once again some variability in human paths across trials and individuals (see Figure 7). Given the number of bifurcation points in such an environment, behavior in this case is particularly susceptible to influences at multiple time scales that could affect the current state of the participant and send them down a different path. These might include variations in the initial walking direction, the foot one is currently on, the obstacles to which one is attending, and the adaptive history of one's postural state. We are currently trying to simulate the distribution of human paths by adding noise to perceptual variables and parameters at each step en route. But the deeper point is that the model captures the qualitative structure of locomotor paths and route switching in terms of the dynamics of attractors, repellors, and bifurcations.

One limitation of the model is that all obstacles are currently treated as points. This may be adequate for posts, but is unrealistic for environments that contain large obstacles or extended surfaces such as walls. One solution may

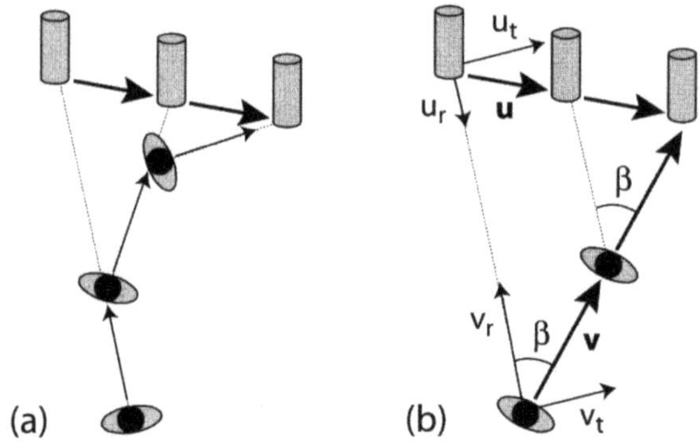

Figure 8. Walking to a moving target. a) Pursuit strategy. b) Interception strategy.

be to adjust the decay rate of the repulsion function for each obstacle (parameter c_3) based on its visual angle, or to treat a fat obstacle as a set of points at finite intervals and sum their influence.

In sum, human route selection can be understood as a form of emergent behavior, resulting from an agent with certain steering dynamics interacting with a structured environment. Somewhat surprisingly, the influences of objects in the environment can be treated as independent and additive, so the model scales linearly with the complexity of the scene. Yet nonlinear behavior such as route switching emerges from the interactions of attractors and repellors. The results demonstrate that the on-line steering dynamics are empirically sufficient to account for human locomotor paths, even in fairly complex environments, rendering explicit path planning and an internal world model unnecessary.

4.4 Behavioral Dynamics of Intercepting a Moving Target

Thus far we have modeled locomotion with increasingly complex configurations of stationary goals and obstacles. But people typically operate in a dynamic environment in which some goals and obstacles are moving – for example, walking through a busy train terminal to catch a friend. We have begun to model such dynamic situations by investigating the case of a moving target (Fajen & Warren, 2002).

If people walk to a moving target the same way they do to a stationary one, they would simply head in the current direction of the target. Such a *pursuit strategy* would result in a curved path of locomotion (Figure 8a), as the walker continually changes direction to track the target. Alternatively, the walker might try to intercept a moving target by walking ahead of it, which

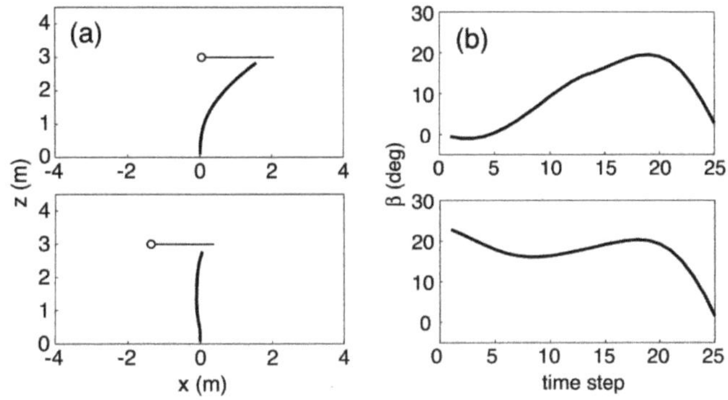

Figure 9. Intercepting a moving target on foot. a) Mean path for target initially straight ahead (top) or 25° to the left (bottom). The target (O) moves laterally. b) Mean time series of obstacle angle β for the same initial conditions.

we call an *interception strategy* (Figure 8b). A good example is the open-field tackle in American football, in which the defenseman tries to cut off the ball-carrier by running on a short, straight interception path.

The interception path may be determined as follows. The ball-carrier's velocity (**u**) has two components, a radial component toward the defenseman (u_r), and a transverse component in the perpendicular direction (u_t). A defenseman moving with velocity (**v**) could match the ball-carrier's transverse component ($v_t = u_t$), reducing it to a one-dimensional problem. Then he or she can approach the ball-carrier ($v_r > u_r$). If both players are traveling at constant velocities, this yields a straight path to the interception point with a constant angle β between the heading direction and the target direction. But even if the ball-carrier's velocity changes, maintaining these conditions in a closed-loop manner will lead to successful interception. One way to implement the interception strategy is to perceive the distal velocity of the target and compute the required β ($\hat{\beta} = \sin^{-1}(u_t/v)$). Alternatively, one could try to adopt a straight path that keeps β constant, effectively nulling $\dot{\beta}$. Sailors are familiar with this *constant bearing* strategy, for if another boat approaches with a constant β, it is a clear indicator that you are on a collision course.

To investigate this question, we asked participants to walk to moving targets in a textured environment the VENLab. The target was a blue post that moved with a constant velocity. On each trial, the participant walked forward for 1 m, whereupon the target appeared at a distance of 3 m; its initial angle from the heading and its direction of motion were varied across trials. It turns out that participants do not simply adopt straight paths with constant interception angles, but rather exhibit transient dynamics (Figure 9): they gradually turn onto a straight path and decelerate as they arrive at the target, while β approaches the expected angle and then falls to zero.

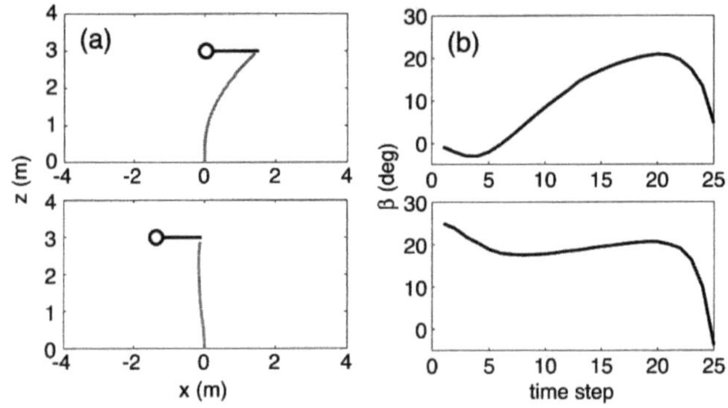

Figure 10. Model simulations of intercepting a moving target, for the same initial conditions as Figure 9. a) Locomotor path. b) Time series of obstacle angle β.

We modeled this interception behavior with a simple modification of the goal component. In order to null $\dot{\beta}$, we substituted it for β in the stiffness term:

$$\ddot{\phi} = -b\dot{\phi} - k_g(\dot{\phi} - \dot{\psi}_g)(e^{-c_1 d_g} + c_2) \tag{8}$$

Consequently, angular acceleration goes to zero as both the turning rate ($\dot{\phi}$) and the change in target-heading angle ($\dot{\beta} = \dot{\phi} - \dot{\psi}_t$) go to zero. The "stiffness" term thus yields a constant β, while the "damping" term tends to produce a straight path, thereby avoiding the infinitely many equi-angular spiral paths that also hold β constant. We modeled the detection of target motion at the onset of a trial with a sigmoidal function having a latency of 0.5 s. Based on the human data, we held walking speed constant (1.42 m/s) with a final deceleration in the last half-second before contact with the target. Fitting this model to the mean time series of β resulting in parameter values of $b = 7.00$, $k_g = 0.17$, $c_1 = 0.013$, and $c_2 = 0.45$.

Model simulations yielded interception paths that were very similar to the walking data (Figure 10). The time series of β also closely matched the evolution of mean target-heading angle, with $r^2 = 0.92$. We also determined that the final drop in β is due to the deceleration near contact, which might be controlled by the optical expansion of the target (Lee, 1976; Yilmaz & Warren, 1995).

Thus, the null-$\dot{\beta}$ strategy is sufficient for controlling interception with a moving target. It is interesting to note that simulations in which the required constant β is explicitly computed from distal target velocity produced nearly identical behavior, even when the parameters were the same as those for a stationary goal. This version has the advantage of reducing to the stationary goal model when target speed is zero, yielding a smooth transition between stationary and moving targets – but at the price of explicit computation from additional informational variables. This observation raises an important question about the organization of behavior: is there a switch between two distinct strategies for stationary and moving targets, or a single continuous strategy? In either case, a perceptual threshold for the detection of target motion could yield what appears to be nonlinear switching behavior.

The next step in this research program is to study the avoidance of moving obstacles, such as other pedestrians. Most simply, the model might be extended to a moving obstacle by flipping the sign of the "stiffness" term, turning the interception point from an attractor into a repellor of heading. Successfully modeling these four basic locomotor components would then permit us to investigate their interactions, such as walking to a moving target while avoiding stationary or moving obstacles. Once the locomotor "rules" for an individual agent are worked out, this may allow us to simulate interactions among multiple agents and structured environments, such as pedestrian traffic flow and crowd behavior.

5 CONTROL LAWS FOR LOCOMOTION

Let us return briefly to our original question, how perceptual information is used to control locomotor behavior. Now that we have a formal description of the behavioral dynamics of locomotion, we can see how information plays a role by contributing to the dynamics, rather than directly determining behavior. In particular, it is possible to consider whether specific control laws can give rise to the observed behavior.

Schöner, et al. (1995) developed a control system for a mobile robot based on a first-order dynamical system, which is always in an attractor state and thus always stable. The advantage of such a system is that it can assure a stable solution under multiple constraints, such as goals and obstacles in arbitrary positions. In contrast, observed behavior is a consequence of such control laws interacting with the physics of the agent and its environment. For example, an inertial body must angularly accelerate and decelerate, such that the actual heading lags behind the intended heading. Our model of the behavioral dynamics is thus a higher-order system that treats steering adjustments as transient behavior toward the current attractor. The question is whether a first-order control law could give rise to such higher-order behavior.

To test this idea, our colleague Philip Fink simulated a first-order control law that is embedded within a second-order system representing the physical agent. Given that our model captures the influence of goals and obstacles, we used the same form for the control law,

$$\dot{\phi}' = -k_g(\phi - \psi_g)(e^{-c_1 d_g} + c_2) + k_o(\phi - \psi_o)(e^{-c_3|\phi - \psi_o|})(e^{-c_4 d_o}) \quad (9)$$

The control law is thus a first-order system that immediately relaxes to an attractor for the intended heading in the direction ϕ*, which is determined by the current configuration of goals and obstacles. The angular acceleration of the body toward this intended heading is then determined by a second-order system with fixed parameters,

$$\ddot{\phi} = -b_b\dot{\phi} - k_b(\phi - \phi^*) \quad (10)$$

Due to the body's inertia, the actual heading lags behind the intended heading, so the observed behavior is transient. When this model was tested with one obstacle en route to a goal (Section 4.2), the paths were nearly identical to those of our original model and the time series of β fit the mean human data with an $r^2 = 0.991$ (for parameter values $k_g = 59.1625$, $c_1 = .0555$, $c_2 = .01125$, $k_o = 842$, $c_3 = 2.74063$, $c_4 = .04653$, $b_b = .0375$, $k_b = 592$). Thus, the behavioral dynamics can be accounted for by a 1st-order control law driving a 2nd-order body.

Finally, the control law incorporates certain perceptual variables, including the angle between the current heading direction (ϕ) and an object's current direction (ψ), as well as the object's current distance (d). As reviewed in Section 2, the direction of heading is redundantly specified by optic flow and the proprioceptive locomotor axis, whereas the direction of a goal or obstacle is given by its visual direction. We previously determined that both optic flow and proprioception contribute to walking to a goal (Eq. 5), (Warren et al., 2001). Thus, we may expand the informational term in the goal component of Eq. 9 as

$$\phi - \psi_g = (\phi_{loco} - \psi_g) + wv(\phi_{flow} - \psi_g) \quad (11)$$

We have yet to empirically test whether the same relation holds for the obstacle component. Note that the distance of an object may be specified either by static distance information such as its angle of elevation on the ground plane or stereoscopic depth. Alternatively, equivalent information over short distances may be provided by the first-order time-to-contact with the object.

Such dynamical control laws are quite different from a simple mapping between an optical variable and a movement variable. Rather than directly determining the kinematics of the movement, the control law determines an attractor for the intended action, thereby modulating the dynamics of the system. This is converted into a force and thence an angular acceleration, resulting in the observed behavior.

6 CONCLUSION

In this chapter we have sought to present an integrated account of perceptually guided locomotion. On our view, such an account must include the multi-sensory information for self-motion, the control laws by which that information modulates the action system, and the behavioral dynamics to which they give rise. Locomotor paths and choices about routes can then be understood as emergent behavior, which unfolds as an agent interacts with a structured environment. Locomotion offers a relatively simple case study in how adaptive behavior can emerge from information and dynamics. It is our belief that it provides a model for the way in which such processes of pattern formation give rise to more complex forms of human behavior.

ACKNOWLEDGMENTS

This research was supported by grants from the National Institutes of Health (EY10923, KO2 MH01353) and the National Science Foundation (LIS IRI-9720327). Permission to reproduce Figures 1, 3-6, and 8 was obtained from the American Psychological Association, and Figure 2 from the Nature Publishing Group.

REFERENCES

Banks, M. S., Ehrlich, S. M., Backus, B. T., & Crowell, J. A. (1996). Estimating heading during real and simulated eye movements. *Vision Res., 36,* 431-443.

Beal, A. C., & Loomis, J. M. (1996). Visual control of steering without course information. *Perception, 25,* 481-494.

Cutting, J. E., Springer, K., Braren, P. A., & Johnson, S. H. (1992). Wayfinding on foot from information in retinal, not optical, flow. *J. Exp. Psychol. Gen., 121,* 41-72.

Duchon, A. P., & Warren, W. H. (2002). A visual equalization strategy for locomotor control: Of honeybees, robots, and humans. *Psychol. Sci., 13,* 272-278.

Duffy, C. J. (in press). The Cortical Analysis of Optic Flow. In: L. M. Chalupa & J. S. Werner (Eds.), *The Visual Neurosciences.* Cambridge, MA: MIT Press.

Fajen, B. R. (2001). Steering toward a goal by equalizing taus, *J. Exp. Psych. Hum. Percept. Perform., 27*, 953-968.

Fajen, B. R., & Warren, W. H. (2002). Behavioral dynamics of intercepting a moving target on foot, *(submitted).*

Fajen, B. R., & Warren, W. H. (2003). Behavioral dynamics of steering, obstacle avoidance, and route selection. *J. Exp. Psychol. Hum. Percept. Perform, 29,* 343-362.

Fink, P. W., & Warren, W. H. (2002). Velocity dependence of optic flow strategy for steering and obstacle avoidance. Paper presented at the Vision Science Society, Sarasota, FL.

Frey, B. F., & Owen, D. H. (1999). The utility of motion parallax information for the perception and control of heading. *J. Exp. Psychol. Hum. Percept. Perform.*, 25, 445-460.

Gibson, J. J. (1958/1998). Visually controlled locomotion and visual orientation in animals. *Br. J. Psychol., 49,* 182-194., Reprinted in *Ecol. Psychol., 10,* 161-176.

Gibson, J. J. (1966). *The Senses Considered as Perceptual Systems.* Boston, MA: Houghton-Mifflin.

Gibson, J. J. (1979). *The Ecological Approach to Visual Perception.* Boston: Houghton Mifflin.

Harris, M. G., & Carre, G. (2001). Is optic flow used to guide walking while wearing a displacing prism? *Perception, 30,* 811-818.

Kim, N. –G., & Turvey, M. T. (1999). Eye movements and a rule for perceiving direction of heading, *Ecol. Psych., 11*, 233-248.

Land, M. (1998). The Visual Control of Steering. In: L. R. Harris & H. Jenkins (Eds.), *Vision and Action.* Cambridge University Press.

Land, M. E., & Lee, D. N. (1994). Where we look when we steer. *Nature, 369,* 742-744.

Lappe, M., Bremmer, F., & van den Berg, A. V. (1999). Perception of self-motion from visual flow. *Trends Cogn. Sci., 3,* 329-336.

Lee, D. N. (1976). A theory of visual control of braking based on information about time-to-collision. *Perception, 5,* 437-459.

Lee, D. N. (1980). Visuo-motor Coordination in Space-Time. In G. E. Stelmach & J. Requin (Eds.), *Tutorials in Motor Behavior* (pp. 281-295). Amsterdam: North-Holland.

Lee, D. N. (1998). Guiding movement by coupling taus. *Ecol. Psychol., 10,* 221-250.

Lee, D. N., van der Weel, F. R., Hitchcock, T., Matejowsky, E., & Pettigrew, J. D. (1992). Common principles of guidance by echolocation and vision. *J. Comp. Physiol. A, 171,* 563-571.

Lee, D. N., & Young, D. S. (1985). Visual Timing of Interceptive Action. In D. Ingle & M. Jeannerod & D. N. Lee (Eds.), *Brain Mechanisms and Spatial Vision* (pp. 1-30). Dordrecht, The Netherlands: Martinus Nijhoff.

Li, L., & Warren, W. H. (2000). Perception of heading during rotation: sufficiency of dense motion parallax and reference objects. *Vision Res. 40,* 3873-3894.

Li, L., & Warren, W. H. (2002). Retinal flow is sufficient for steering during simulated rotation. *Psychol. Sci,, 13,* 485-491.

Nakayama, K. (1994). James Gibson -- An appreciation. *Psychol. Rev., 101,* 329-335.

Perrone, J. A. (1992). Model for the computation of self-motion in biological systems. *J. Opt. Soc. Am. A, 9,* 177-194.

Rieger, J. H., & Lawton, D. T. (1985). Processing differential image motion. *J. Opt. Soc. Am. A, 2,* 354-360.

Rogers, B. J., & Allison, R. S. (1999). When do we use optic flow and when do we use perceived direction to control locomotion? *Perception, 28 (Suppl),* 2.

Royden, C. S., Banks, M. S., & Crowell, J. A. (1992). The perception of heading during eye movements. *Nature, 360,* 583-585.

Royden, C. S., Crowell, J. A., & Banks, M. S. (1994). Estimating heading during eye movements. *Vision Res., 34,* 3197-3214.

Royden, C. S., & Hildreth, E. C. (1996). Human heading judgments in the presence of moving objects. *Percept. Psychophys., 58,* 836-856.

Rushton, S. K., Harris, J. M., Lloyd, M., & Wann, J. P. (1998). Guidance of locomotion on foot uses perceived target location rather than optic flow. *Curr. Biol., 8,* 1191-1194.

Rushton, S. K., Harris, J. M., & Wann, J. P. (1999). Steering, optic flow, and the respective importance of depth and retinal motion distribution. *Perception, 28,* 255-266.

Saunders, J. A., & Warren, W. H. (1996). Perceived heading biased by a moving object: effects of disparity and object position. *Invest. Ophthalmol. Vis. Sci., 37,* S454.

Schöner, G., Dose, M., & Engels, C. (1995). Dynamics of behavior: Theory and applications for autonomous robot architectures. *Rob. Auton. Sys.,* 16, 213-245.

Srinivasan, M. V. (1998). Insects as Gibsonian animals. *Ecol. Psychol., 10.*

Stone, L. S., & Perrone, J. A. (1997). Human heading estimation during visually simulated curvilinear motion. Vision Research, 37, 573-590.

Telford, L., Howard, I. P., & Ohmi, M. (1995). Heading judgments during active and passive self-motion. *Exp. Brain Res., 104,* 502-510.

van den Berg, A. V. (1992). Robustness of perception of heading from optic flow. *Vision Res., 32,* 1285-1296.

van den Berg, A. V. (1996). Judgements of heading. *Vision Res., 36,* 2337-2350.

Wang, Y., & Frost, B. J. (1992). Time to collision is signalled by neurons in the nucleum rotundus of pigeons. *Nature, 356,* 236-238.

Wann, J. P., & Land, M. (2000). Steering with or without the flow: is the retrieval of heading necessary?, *Trends Cogn. Sci., 4,* 319-324.

Wann, J. P., & Swapp, D. K. (2000). Why you should look where you are going, *Nat. Neurosci., 3,* 647-648.

Warren, W. H. (1988). Action Modes and Laws of Control for the Visual Guidance of Action. In O. Meijer & K. Roth (Eds.), *Movement Behavior: The Motor-Action Controversy* (pp. 339-380). Amsterdam: North Holland.

Warren, W. H. (1998). Visually controlled locomotion: 40 years later. *Ecol. Psychol., 10,* 177-219.

Warren, W. H. (2003). The dynamics of perception and action. *(Unpublished manuscript).*

Warren, W. H. (in press). Optic Flow. In: L. M. Chalupa & J. S. Werner (Eds.), *The Visual Neurosciences.* Cambridge, MA: MIT Press.

Warren, W. H., & Hannon, D. J. (1988). Direction of self-motion is perceived from optical flow. *Nature, 336(6195),* 162-163.

Warren, W. H., & Hannon, D. J. (1990). Eye movements and optical flow. *J. Opt. Soc. Am. A, 7(1),* 160-169.

Warren, W. H., & Kay, B. A. (1997). Control law switching during visually guided walking. *Abs. Psychonomic Soc., 2,* 52.

Warren, W. H., Kay, B. A., Zosh, W. D., Duchon, A. P., & Sahuc, S. (2001). Optic flow is used to control human walking. *Nat. Neurosci., 4,* 213-216.

Warren, W. H., & Saunders, J. A. (1995). Perception of heading in the presence of moving objects. *Perception, 24,* 315-331.

Warren, W. H., & Yaffe, D. M. (1989). Dynamics of step length adjustment during running. *J. Exp. Psychol. Hum. Percept. Perform., 15,* 618-623.

Warren, W. H., Young, D. S., & Lee, D. N. (1986). Visual control of step length during running over irregular terrain. *J. Exp. Psychol. Hum. Percept. Perform.*, *12,* 259-266.

Wood, R. M., Harvey, M. A., Young, C. E., Beedie, A., & Wilson, T. (2000). Weighting to go with the flow? *Curr. Biol., 10,* R545-R546.

Yilmaz, E. H., & Warren, W. H. (1995). Visual control of braking: a test of the tau-dot hypothesis. *J. Exp. Psychol. Hum. Percept. Perform., 21,* 996-1014.

Zosh, W. D., Duchon, A. P., & Warren, W. H. (2000). The role of optic flow in adaptation to visual displacements during walking, *Abs. Psychonomic Soc., 5*, 40.

15. Egocentric Direction and Locomotion

Simon K. Rushton

Centre for Vision Research & Department of Psychology
York University, Toronto, Ontario, Canada.

1 INTRODUCTION

For 50 years it was assumed that humans rely on optic flow for the visual guidance of locomotion. It has been suggested that the flow theory originated with Grindley (see Mollon, 1997), but its popularization was undoubtedly due to Gibson (e.g. Gibson, 1958). Gibson's publications motivated psychophysical studies, neurophysiology, imaging and computational modelling (see Lappe et al., 1999, for a review), however, recently the theory has been challenged.

Rushton et al. (1998) reported an experimental result seemingly at odds with the use of optic flow and proposed instead a simple heuristic that better described the behavior they observed. The proposal is that visual guidance of locomotion to a target is achieved by keeping the target at a fixed direction, or *eccentricity,* relative to the body, rather than regulating behavior so as to maintain a certain pattern of flow on the retina (the optic flow solution). In short, if the current direction of a target object is known, and the observer walks so as to keep the direction constant then they will reach the target. If the target is kept straight-ahead then a straight-line course to the target will result. If the target is maintained at some other direction then the path will be an equi-angular spiral.

The Rushton et al. study has now been replicated by many others (Rogers & Dalton, 1999; Wood et al., 2000; Harris & Carre, 2001; Warren et al., 2001; Harris & Bonas, 2002). A concise summary of the original study is provided below. Following this, in the second section there is a selective review of the factors that influence the perception of egocentric direction. Section three discusses subsequent studies on the visual guidance of locomotion and the role of optic flow and egocentric direction, and section four outlines new questions that are prompted by a direction account.

L.M. Vaina, S.A. Beardsley and S.K. Rushton (eds.), Optic Flow and Beyond, 339–362.

1.1 Prism Study

The Rushton et al. (1998) study involved observers wearing prism glasses. Observers were asked to walk briskly towards a target held out by an experimenter positioned about 10m to 15m away. The glasses contained either paired base-left or base-right wedge prisms. Prisms deflect the image and so shifted the perceived location of objects (relative to the body) approximately 15° to the right or left. Wearing prism glasses had a dramatic effect on the trajectory taken by observers when asked to walk towards the target. Observers veered while attempting to walk 'straight towards' the target. A typical veering trajectory is shown in the bottom left panel of Figure 1.

1.2 A Flow Explanation?

Can use of optic flow account for such a trajectory? Flow based strategies rely on keeping the flow-specified direction of heading (DoH) and the target coincident. More generally, they are concerned with relative positions or patterns within the flow-field. As can be seen from the top left panels of Figure 1, although prisms displace the scene and change the perceived location of objects, the critical relations in the flow field are not perturbed. Specifically, the relative positions of the DoH and the target remain unchanged. Therefore, perception of direction of locomotion should remain unchanged and veridical if flow is used, and observers should end up on a straight trajectory towards to the target. The bottom left panel of Figure 1 shows a markedly curved trajectory indicating that the observer did not use a DoH-target strategy. A model based on using the flow field specified DoH is therefore incompatible with the experimental results.

1.3 Direction Hypothesis

A simple model, the *perceived direction* model (Rushton et al., 1998), is compatible with the data. The model predicts that observers take a curved path because they attempt to keep the target perceptually straight-ahead of them. When wearing prisms, the perceived position of the whole scene, relative to the observer's body, is changed by the angular deflection of the prism. For example, if the prisms shifts the scene by 15° to the left, an object at 0° relative to the trunk will be seen at approximately 15° to the left. Thus, keeping the target

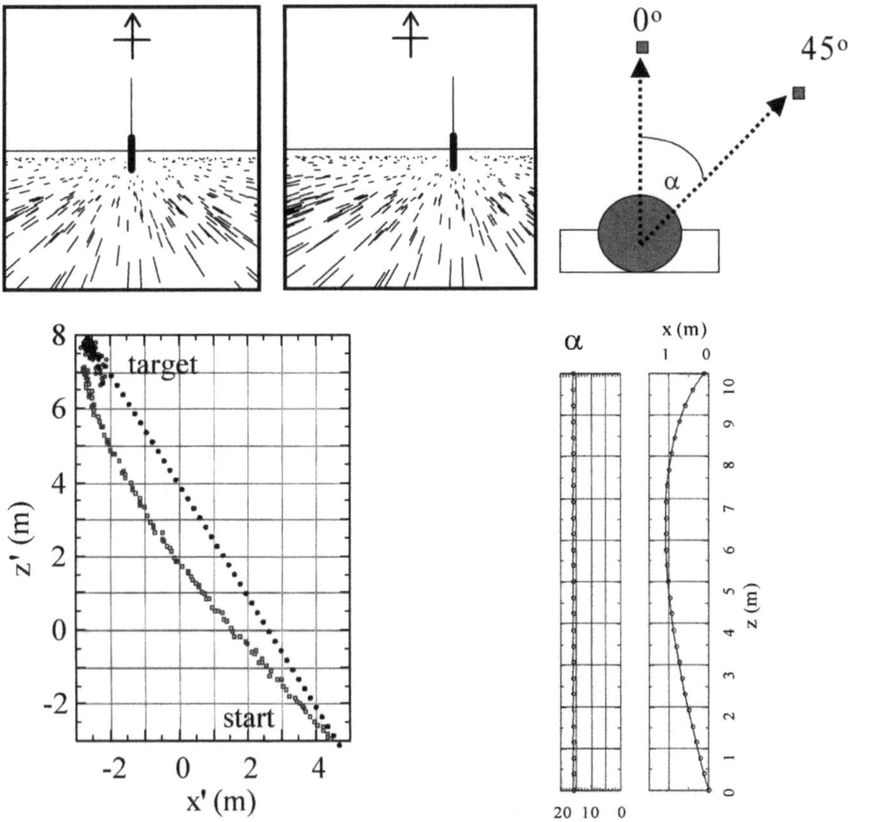

Figure 1. Top Left Panel – Flow-field during forward translation (magnitude indicates image speed) toward a target tower (solid black rectangle) at 16m. The thin vertical line indicates the direction of travel. The arrow indicates egocentric straight-ahead; 'focus of expansion' (FoE) is coincident with the tower, which indicates the observer is travelling directly towards tower. The arrow above tower indicates the perceived 'straight-ahead' direction; note it coincides with the tower. *Top Center Panel* –_Same as the left panel but with displacement of the image by a prism. Note the FoE is still directly over tower, thus flow indicates the observer is travelling directly towards tower. However, the perceived straight-ahead (denoted by the arrow above) no longer coincides with the tower. *Top Right Panel* – Egocentric directions, 'eccentricity', α, measured angle in cardinal plane. *Bottom Left Panel* – A representative trajectory of an observer approaching a target wearing a pair of wedge prisms that deflect right. The plot shows raw digitized data with axes x' and z' showing distances in world co-ordinates. *Bottom Right Panel* – Simulation of the trajectory and direction error when wearing prisms by a simple model using target direction. *Left* – angle (α) between the instantaneous direction of the target and the direction of locomotion (tangent to the curve), which remains constant throughout the trajectory. *Right* – plan view of the predicted trajectory of a prism-wearing participant walking in the perceived direction of the target (perceived direction is offset from actual position by the angular deflection of the prism glasses). The simulation plot is for a 16° deflection, which is the approximate angular deflection of the wedge prisms used in the walking experiment. X and z are distances parallel and perpendicular, respectively, to the starting position of the participant (facing along the z-axis). [Reprinted from Current Biology, 8, Rushton et al, "Guidance of locomotion on foot.", pp11191-1194, 1998 with permission from Elsevier.]

perceptually straight-ahead requires the observer to keep the target at a fixed eccentricity (relative to the body) of 15° to the right of the trunk mid-line. If this strategy is used then it should lead to a veering trajectory to the target. The trajectories walked by observers were very similar to those predicted by this simple *perceived-direction* model (compare the bottom left and bottom right panels of Figure 1).

1.4 Direction vs. Flow and Prism Distortion

It had been suggested that use of optical prisms might invalidate any conclusions reached because prisms "distort" the optic flow field (Warren & Kay, 1999; Warren et al., 2001). If this were correct then it would have important consequences, calling into question not only the prism-locomotion study (Rushton et al., 1998), but also an extensive body of work on optic flow and perceptuo-motor calibration based upon prism studies (see Redding & Wallace, 1997 for the most recent treatise on this subject). However, there is good reason to believe that the distortion objection can be dismissed. First, if we consider observers in the locomotion guidance and adaptation studies, they walked with no hesitation towards the targets; there was no obvious indication that they had to navigate using impoverished information. Second, there is empirical data that shows that observers can make veridical judgements of direction of heading with severe distortions of the flow field. Kim et al. (2000) used dramatic perturbations of flow vectors and examined judgement of heading. In an important condition, equivalent to a prism displacement that varies across the field, they found that heading was still perceived correctly. This is fortuitous because if perception of heading from flow was impaired by prisms then it would it would lead to problems for any model that involves optic flow because a significant proportion of the population wears spectacles with prescriptions that produce worse optical distortions than wedge prisms. Also head-mounted displays, which have been used as an electronic alternative to prism glasses in locomotion experiments (e.g. Warren et al., 2001), typically introduce considerably worse optical distortions than prisms.

1.5 Moving Targets

The experiment reported above involved locomotion towards a static target. Rushton et al. (1998) also reported a second experiment that was concerned with the interception of moving targets. They concluded that, as with static targets, the interception of moving targets is based upon egocentric direction.

1.6 Strong vs. Weak Direction Accounts

Rushton et al. (1998) concluded "that perceived location, rather than optic or retinal flow, is the predominant cue that guides locomotion on foot" (p1191). This statement can be interpreted in a strong form as excluding all other cues apart from egocentric direction, or in a weaker form that egocentric direction is the primary cue (the one that is always used and which generally dominates), but that secondary cues (such as optic flow) might contribute under some circumstances. Parsimony suggests that the stronger (simpler) account be championed until it is found to be severely challenged. In the rest of this chapter we will consider the strong version of the egocentric direction theory.

2 PERCEPTION OF EGOCENTRIC DIRECTION

As noted, an observer will reach their target if they keep the target at its current egocentric direction and ensure that they are moving towards the target rather than away from it (Llewellyn, 1972). The trajectories that result are a family of equi-angular spirals. Straight-ahead is the most important egocentric direction as it produces a direct course from the current position to the target. The straight-ahead case of the egocentric direction strategy was highlighted in Rushton et al. (1998) and is assumed in most discussions of the use of egocentric direction (but see the companion chapter by Rushton & Harris).

The question that arises is how does the observer judge that a target is straight-ahead when the body and the target cannot be viewed simultaneously? The geometric answer is that the target is straight-ahead when the angular lateral distance, from the fovea, of the target on the retina is equal to the gaze angle (head-body orientation + eye-head orientation). The simplest case is when the target is fixated and the gaze direction is zero. If this information is used then it follows that if there is any error in perception of eye or head orientation, or retinal location, then straight-ahead will be misjudged.

A number of factors have been found to effect perception of egocentric direction. Some of these factors can be directly attributed to a misperception of eye or head orientation or head-centric direction. Below is a selective review of some of the factors.

2.1 Measuring a Shift in Perceived Direction

Straight-ahead is the most commonly measured egocentric direction. The normal motivation for choosing straight-ahead, rather than any other direction, is that it falls on one of the primary axes of egocentric space and it is very easy to measure.

Perceived straight-ahead may be assessed in a number of ways, the simplest is to stand an observer in front of a wall, give them a laser pointer and ask them to indicate the point that is straight-ahead. Alternative ways include having an observer move a marker indirectly using a joystick, or presenting points and asking the observer to judge if they are to the left or right of their perceived straight-ahead.

2.2 Demonstrating the Role of Gaze Direction Information

Egocentric direction is given as a function of eye-head orientation, head-trunk orientation and retinal target location. The role of gaze orientation can be demonstrated by optically displacing the scene relative to the body, allowing the observer to re-fixate, and measuring the angular shift in perceived straight-ahead that results from changing the gaze signal. Optical displacement can be achieved by interposing prisms, mirrors, or a rotated camera and display, between the eyes and the scene. Any of these displacements will lead to a shift in perceived straight-ahead.

Perceived head-trunk orientation can be perturbed by vibration of the posterior neck muscles: Bigeur et al. (1988) and Karnath et al. (1994) demonstrated a shift in perceived straight-ahead as a result of such a manipulation. Karnath et al. (1994) additionally investigated the effect of caloric stimulation – stimulation of the vestibular system by the injection of cold or warm water into the ear – on perception of straight-ahead. The shift that resulted from caloric stimulation was found to combine additively with a shift from neck vibration when the two were performed simultaneously. It should be noted that in the Bigeur et al. and Karneth et al. studies the magnitude of the shift varied between subjects. Also the effect was nulled or attenuated when a structured background was visible.

A similar manipulation of perceived eye-in-head orientation is also possible: Velay et al. (1997) demonstrated that vibration of the eye-muscles could shift perceived egocentric position of a point light source in a dark room. It would be expected that direct measurement of perceived straight-ahead would also reveal a shift with eye-muscle vibration.

Perceived eye-orientation can also be manipulated by adaptation. If an observer maintains eccentric fixation for 20 sec. or longer then perceived straight-ahead will shift in the direction of fixation. The shift is proportional to the period and eccentricity of fixation. When gaze is returned to straight-ahead an error will still be apparent in eye-orientation – away from true straight-ahead – and the judgement of straight-ahead (Paap & Ebenholtz, 1976).

The manipulations described above directly perturb real or perceived gaze direction. In the section that follows 'scene' or 'retinal' influences on perceived direction are described.

2.3 Retinal Influences on Perceived Egocentric Direction

Rock et al. (1966) reported an 'immediate correction' effect with prisms. They found that when observers viewed through prisms, the shift in perceived straight-ahead that was observed was not as great as would be expected from the magnitude of the optical displacement of the prism. It appeared observers 'immediately corrected' for the prism. As the effect did not occur when observers were tested with a single luminous point in an unseen room, it was deduced that structure, or information, in the optic array contributes to the perception of straight-ahead. Recent data on this matter is reported later in this chapter.

Roelofs' effect is well known and describes the behavior of an observer who is given the task of indicating straight-ahead when viewing an illuminated panel, in an otherwise dark field. When the panel is directly ahead of them they will point towards the center of the panel. This is as expected as true straight-ahead coincides with the center of the panel. However when the panel is presented off-center (see Figure 2), the observer's response is drawn away from their true straight-ahead and towards the center of the panel (Roelofs, 1935). How strongly the center of the panel attracts the observer is a function of the size of the display. Interestingly a difference in the magnitude of the shift is reported between displacement of the frame to the left and displacement to the right (Bruell & Ablee, 1955; Werner et al., 1953).

Bridgeman & Graziano (1989) conducted a study using an eye-press technique to perturb perceived egocentric direction and asked subjects to indicate straight-ahead. They varied the background against which judgements were performed, a blank field, a projected picture or a projected texture. The shift in straight-ahead was reduced when viewing a regular texture and further reduced when viewing a picture. The extensive work of Matin and colleagues (e.g. Matin et al., 1980), bears on the same issue.

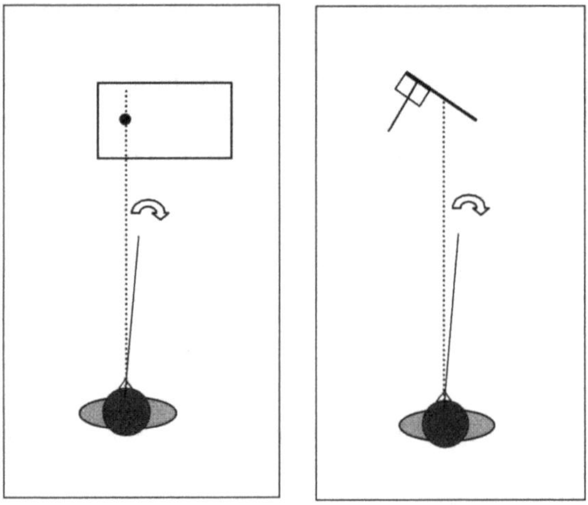

Figure 2. Left Panel – Perceived straight-ahead should be biased towards the center of the frame. *Right Panel* – Perceived straight-ahead should be biased towards a closer part of the surface, or the normal to the surface.

Kleinhans (1970, reported in Perrone, 1982) found that perceived straight-ahead is biased when viewing a slanted surface (see Figure 1, right panel). Straight-ahead was drawn in the direction of the surface normal. Interestingly, Perrone (1982) proposed that misperception of straight-ahead may contribute to misperception of surface slant. His proposal was that perceived straight-ahead is drawn towards the portion of the surface that is closest to him or her (Figure 2, right panel).

Additionally, retinal information about the position of body parts (i.e., being able to view parts of the body) has also been claimed to reduce the influence of prisms, for example Wallach et al. (1963). An authoritative review of the literature on prisms can be found in Welch (1978).

The mechanisms responsible for the above influences of the structure of the optic array on judgement of straight-ahead are not clear (see Bridgeman & Graziano, 1989, for pointers to the relevant literature). Nonetheless, the critical point is that there appears to be an abundance of evidence that the structure of the optic array influences perceptual judgement of straight-ahead. In light of the contemporary 'two visual system' hypothesis (Milner & Goodale, 1995) it should be noted that most of the studies cited above involve perceptual judgements, rather than actions. Also the interested reader may wish to consult Harris (1974) for a critical appraisal of methods and measures used in prism adaptation studies.

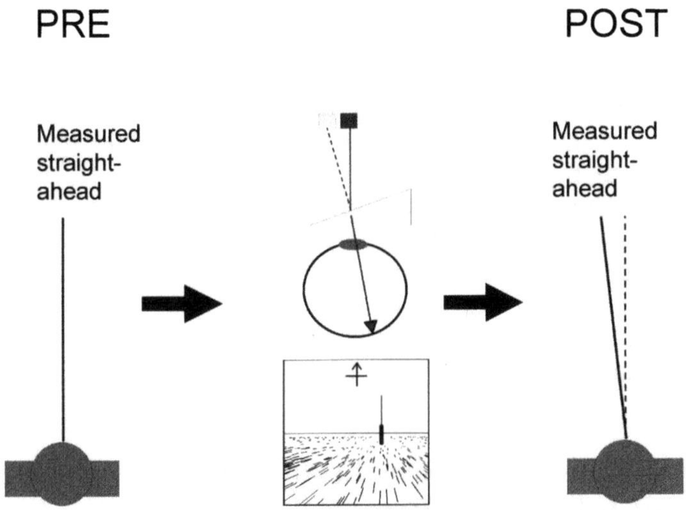

Figure 3. Change in perceived straight-ahead that occurs after locomotion in Redding & Wallace studies.

2.4 A Dynamic Retinal Specification of Straight-Ahead and Calibration

During natural walking observers normally walk straight-forward or on a curved path, they seldom translate diagonally, therefore the focus of expansion of a radial flow field is a fairly reliable dynamic specification of straight-ahead (especially when averaged over time).

There is a long history to the idea that optic flow might be used in the calibration of egocentric directions (see Rock, 1966). More recently, Redding & Wallace have conducted extensive work on prism adaptation (see Redding & Wallace, 1997 for a review). In their studies observers are typically given prism glasses and asked to walk up and down a corridor for a period of approximately 10 minutes. Perceived straight-ahead is measured before and after the walking. After walking the direction of perceived straight-ahead shifts so as to compensate in part for the prism displacement (Figure 3).

The stimulus for the adaptation is hypothesised to be the discrepancy between the retinal position of the focus of expansion and the perceived straight-ahead. Redding & Wallace's work suggests a slow interaction between retinal flow and egocentric direction, with the flow leading to a 'recalibration' of egocentric directions. Rogers and colleagues (e.g. Rogers,

2001) have also reported compatible adaptation and after-effects in their locomotion studies.

Interestingly Redding & Wallace found a significant influence of attention during adaptation, suggesting that use of optic flow information is attentionally demanding (see their book for a summary of the findings). In an unpublished study by Rushton & Rosenthal, prism-wearing observers walked towards a target through a scene with very rich motion parallax. The influence of the motion-parallax – to compensate for the prism displacement – was found to be dependent upon attention. Also Wann et al. (2000) demonstrated the role of attention in judgement of heading from optic flow. Therefore, as an aside, the role of attention is an important point to be considered when interpreting data on walking and optic flow.

2.5 A Dynamic Retinal Specification of a Gaze Movement and its Influence on Perceived Direction

A rotation of gaze to the left produces rightward translational flow on the retina and vice-versa. All translational vectors within the retinal image will be the same size because the magnitude of a flow vector is independent of the distance of the element in the scene. The magnitude of the flow vectors will be a function of the change of gaze direction. Therefore, the angular magnitude of the translational vectors directly specifies the magnitude of the gaze rotation and thus is a retinal indicator of *gaze rotation*. Such a signal could be used as a complement to any extra-retinal signals about gaze-rotation.

The suggestion that retinal flow could be used as a source of information about gaze rotation is a somewhat strange suggestion. Traditionally the retinal flow that results from eye-movement is considered a *problem*. It has been described as 'retinal slip', motion that must be cancelled by use of other eye-movement information so that object and self-motion can be discriminated and perceptual stability maintained. Recently in the locomotion literature, it has been described as a source of noise that contaminates the optic flow field resulting from the translation of the observer through space. In the heading literature the concern is whether this 'noise' can be discriminated from the 'signal' within the flow field or whether it is necessary to use 'extra-ocular' information to help remove it – retinal slip cancellation again. Consequently, the potential utility of retinal flow as a source of gaze movement information appears to have received little attention, (although see Perrone, 1992 and note the suggestion by Duffy & Wurtz, 1993, that their 'optic flow illusion' could be explained by assuming that flow provides 're-afferent gaze movement'

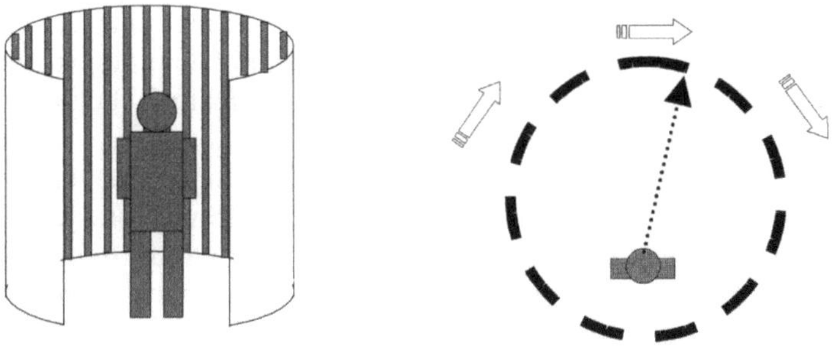

Figure 4. Left Panel – Observer in a vection cylinder. *Right Panel* – Perceived straight-ahead.

information. Also note that Rushton et al. (1999) highlighted the potential utility of retinal information about gaze rotation in natural locomotion).

It was noted earlier that perturbation of the vestibular system produces a change in perceived egocentric direction. Presumably the vestibular signal is integrated twice to give the change in head orientation. Therefore, if a vestibular motion signal is combined with positional information in the perception of egocentric direction then it is reasonable to assume that a retinal motion signal might also be utilized. Studies reported in the following section bear on this matter.

2.5.1 Studies on the Perceived Egocentric Direction and Translational Retinal Flow

Brecher et al. (1972) described the influence of translational flow on perception of straight-ahead. Observers stood in a rotating vection cylinder (see Figure 4), that produces translatory visual motion and were asked to indicate straight-ahead. Straight-ahead was found to shift in the direction of the flow. This finding has been replicated many times, for example by Post & Heckmann (1986) and Karnath (1994).

An earlier study by Brecher & Brecher (1969) is particularly interesting. They placed observers in a vection cylinder and asked them to shuffle very slowly in an 'absolutely straight-line'. The distance moved was about 70cm and the movement lasted about 6 seconds. A pen taped to their foot recorded trajectory. They were found to not take a straight path, but to veer. The amount of veering increased as a function of the cylinder speed. If the observer identifies the point furthest from them and orients themselves to place it 'straight-ahead' and walks towards it then they would produce a trajectory similar to that recorded. However, it is not known whether observers adopt such a strategy so caution should be exercised before attempting to read too much into their data. The critical point to take from their study is that the shift in perceived egocentric direction produced by translation flow is not a purely abstract 'visual illusion', but also has an impact on perceptuo-motor control.

From the hypothesis that retinal flow is used as an eye-movement signal, simple predictions can be derived: If translational flow produces a shift in perceived straight-ahead because it provides a gaze rotation signal then there should be a simple relationship between the speed of translational flow and shift of straight-ahead, and also between the time viewing translational flow and shift. There should be a linear relationship between shift and viewing time and between shift and speed.

Post & Heckmann (1986) conducted a study that provides data on the former. They found that the shift of straight-ahead increases with viewing time and that the relationship between shift and viewing time in their data is reasonably linear for at least the first 40 sec. of viewing a 5 deg/sec translational flow field. Brecher et al. (1972) looked at the latter and found that the shift in perceived straight-ahead increased with speed. Inspection of their data revealed the relationship to be approximately linear up to 80 deg/sec. Both of these findings would be predicted if observers integrate translational flow over a period of time to get an estimate of gaze rotation that is used in the calculation of gaze direction.

There are outstanding questions that need to be addressed regarding the use of translational retinal flow as a gaze-rotation signal; such as how are eye, head and body rotations discriminated? In the former cases it is appropriate to update perceived straight-ahead, in the latter it is not.

2.5.2 An Aside: The Human Fly Effect

Interestingly, this fast acting interaction between retinal flow and perceived direction could provide an alternative explanation of the 'human fly' effect that has been reported: when steering down a corridor or tunnel, unequal flow presented in the left and right hemifields leads to a bias in steering away from the field with the highest amount of flow (Duchon & Warren, 2002). The unequal flow corresponds to a net translational flow across the field and so should shift the perceived straight ahead towards the side with the higher amount of flow. If the observer steered by trying to keep the middle of the corridor perceptually straight ahead then this would lead to the corrective action that has been reported. Thus the 'human fly' effect can be accounted for with recourse to an additional flow-equalization mechanism.

2.5.3 Retinal Flow and Updating Egocentric Positions.

It is suggested above that retinal flow provides information about movement of the gaze system. This is distinct from the assumption that flow provides abstract information about where in the environment you will get to at some point in the future if you do not change you current trajectory (and possibly what your path will be).

Further retinal flow also provides information about gaze translation (radial flow) and so could be used in the updating of egocentric positions of objects in close space during or after observer translation. This would help in the maintenance of perceptual stability and a veridical map of egocentric object positions to support action. The work of Lepecq et al. (1993) on vection and spatial memory, and Miles and colleagues (e.g. Miles, 1998) on parsing the optic flow field to aid gaze stabilization may be relevant here. Also Rushton & Bremmer (2000) reported that the perceived egocentric position of objects could be changed by background optic flow.

2.5.4 Target Drift

If an observer is not heading straight towards a target then the target will drift on each step. If the target is to the left of straight-ahead then the target will drift left and vice-versa. If the target is close to straight ahead then the drift will be small, if the target is far from straight-ahead then the drift will be large (drift also varies as a function of step size and distance from the target). Therefore, target drift, or the local motion of the target against the global scene could provide dynamic information about egocentric target direction. This issue is addressed in the companion chapter by Rushton & Harris.

2.5.5 Dissociating Optic Flow and Perceived Egocentric Direction

It should now be apparent that the claim of a dissociation of flow and direction as set out in the original papers on locomotion, flow and direction (Rushton et al., 1998; Warren et al., 1999) was not strictly correct. A review of the literature reveals that optic flow would be expected to influence perceived egocentric direction and hence should indirectly influence locomotion (see Rushton & Salvucci, 2001).

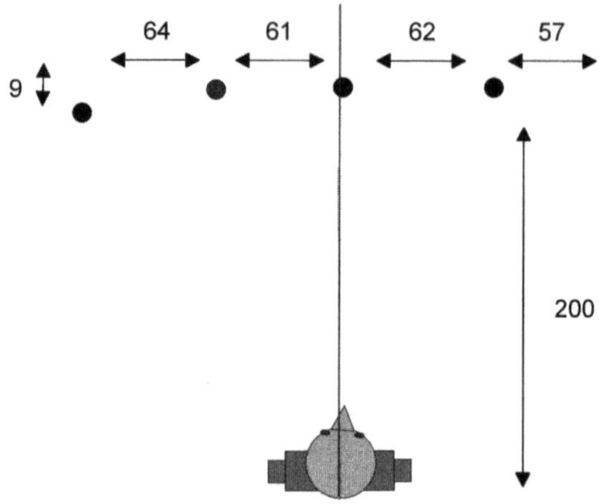

Figure 5. Observer and targets. Targets were miniature lights mounted at eye level on lengths of doweling. The illumination of the lights was computer controlled. A light would be illuminated for approximately eight seconds and the observer would turn and face it. After eight seconds had elapsed the observer's orientation would be recorded and then the next light illuminated. Six data points were collected for each target light. For each prism a baseline was first recorded, followed by the experimental data. Changes in orientation were calculated by subtracting the baseline orientations from the prism orientations. In the dark condition, the room was pitch black and only the target light could be seen. In the later replication with distant targets the experiment was conducted in an open space with targets (coloured pieces of paper) beyond 30m and in approximately the same directions as in the original experiment.

2.6 Some Recent Empirical Data on Perception of Egocentric Direction

As noted above, Rock et al. (1966) reported in their experiment that observers wearing prism glasses did not experience as large a shift in perceived egocentric direction as would be expected from the optical displacement of the prisms. Recently I have re-examined this matter, and some illustrative data are presented below.

Rather than use a judgement or pointing task, an alignment task was selected. Observers were instructed to turn and face a target object, the specific instructions being to "turn and face the target so that if you closed your eyes and walked straight forward you would collide with the target". Five targets were used and the observer's head and trunk position was measured with electromagnetic Ascension (6 degrees of freedom) trackers. Their alignment was recorded with and without prism glasses (paired base left and paired base right of approximately 8 prism dioptres). The setup for the first experiment is show in Figure 5.

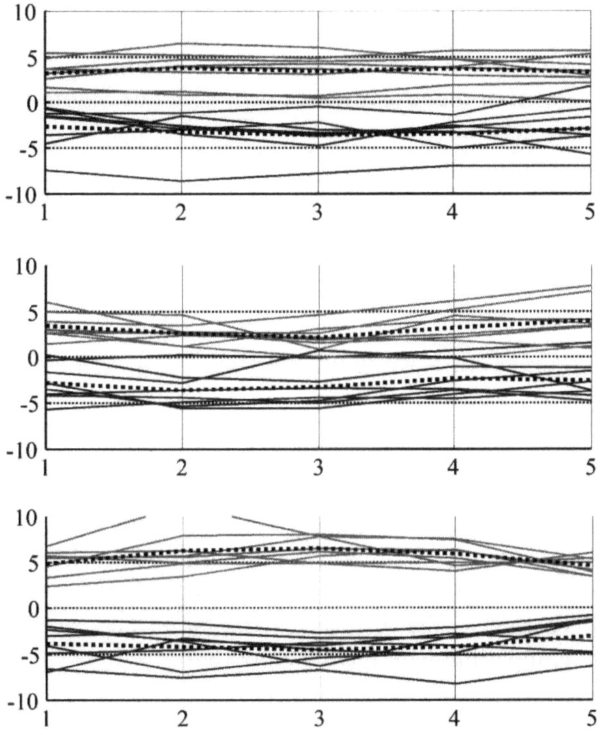

Figure 6. Change in body (trunk) alignment with base left and base right prisms. Fine lines indicate individual observer means (from six data points per target). The thick dotted line is the group mean and the optical displacement of the prisms was approximately 4.5°. *Top Panel* – Dark room. Overall mean = 3.321° (head = 3.174°). *Middle Panel* – Replication in the light. Overall mean = 2.971° (head = 1.314°). *Bottom Panel* – Replication outside with distant targets. Overall mean = 4.798° (head =4.512°).

The changes in orientation are shown in the top and middle panels of Figure 6. The prism-induced shift is significantly less than the prism deflection (in line with Rock et al.'s finding). The change in orientation is approximately the same in both the light and dark conditions. This finding is compatible with a recent paper by Harris & Bonas (2002) who reported that observers walking down a corridor towards a light while wearing prisms took a trajectory that was less curved than would be expected from the prism displacement, and was approximately the same in the light or dark.

The finding that the prism-induced change in orientation was less than the optical displacement of the prisms prompted the question, what would happen if the experiment were replicated outdoors with more distant targets? The assumption was that the prism effect would be further attenuated because outdoors there is an extended scene with a broader range of disparities. In fact the opposite was found. Outdoors the full effect of the prism was found, the change in orientation was approximately what would be expected from the optical displacement of the prisms (see lower panel of Figure 6). The findings briefly described here have subsequently been replicated and extended and details will appear elsewhere.

So what might be happening; what visual information could the observer be using indoors to attenuate the effect of the prism? In the following section a couple of potential sources of information are briefly discussed.

2.6.1 Consequences of Results for the Visual Guidance of Locomotion

As an aside, we can make three coarse predictions about locomotion studies conducted with observers walking to targets while wearing prisms from the data above. First, studies run with a close target indoors are likely to show less veering trajectories than those run outdoors with far targets. Second, trajectories associated with running the studies in the light and the dark are likely to be very similar. Third, there is predicted to be a reduction in the amount of veering as the distance of the target decreases. Most of these predictions are compatible with the data reported.

2.6.2 Retinal Information About Eye Orientation or Head-Centric Direction

It is possible to identify many sources of retinal or visual information that could be used in the perception of egocentric direction. There is potentially a lot of information available about head-centric direction or eye-orientation. In the former case the direction of an object relative to the head could be determined from local information, in the latter case global information from across the visual field could be used to estimate eye-orientation. Below we describe a few potential sources of information for illustration.

2.6.3 Retinal Markers of Eye Orientation

Gibson (1966) pointed out that view of the nose could be used to determine eye orientation as the retinal coordinates of the nose vary as a function of eye-orientation. However, we can discount this source of information in the experiment reported above as a similar attenuation of the prism-effect was found in the light and the dark, and in the latter case the nose and other features would not be visible. Also target distance influences the prism-effect and visibility of the nose does not vary as a function of fixation distance.

2.6.4 Binocular Disparities, Eye Orientation and Head-Centric Direction

Another potential source of information is binocular disparities, that is the small differences between the two eyes' images.

2.6.4.1 Vertical Disparity Field

Vertical disparities can be most simply thought of in terms of size ratios. If an object is 40cm in front of the nose then the image of the object will be the same size in the left and right eye. If the object is shifted to the right, the distance of the object from the right eye will now be less than the distance from the left eye. Therefore the size of the image of the object will be larger in the right eye than the left eye. It follows that if the relative size of the left and right eye images of an object is known then it is possible to determine if the object is to the left or right of the head. If the (Cyclopean) distance of the object is known the exact direction can be determined.

The usefulness of the size ratio varies as a function of the distance of the object. When an object is close there will be a significant size difference. As the object becomes more distant so the ratio tends to unity, independent of direction. The size ratio for an object at a range of distances is plotted as a function of direction in Figure 7. The use of vertical disparity information in the perception of direction has recently been explored experimentally by Banks et al. (2002) and Berends et al. (2002).

2.6.4.2 Horizontal Disparity Field

Horizontal disparity, or the difference in relative position of an object at the left and right eye, is not especially useful by itself as a cue to direction, although a probabilistic relationship relates horizontal disparity to direction (as can be deduced from Figure 8). However, the range of disparities across the visual field is more informative. Consider two extremes. Two objects both in front of the nose, one close and one far, will have a relative disparity – they will appear at different places in the two eyes. If the objects are rotated about the Cyclopean eye so that they are now both at 60°, then the relative disparity will be decreased. This is because the "effective" inter-ocular separation, or

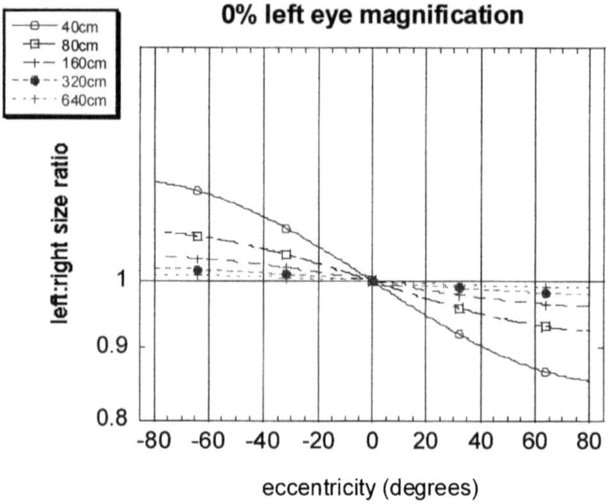

Figure 7. A 1 mm point at eye-height: range of eccentricities (-80° to 80°) and a range of distances (40 cm to 640 cm).

more correctly, the distance between the ocular axes, will have reduced. When the objects are at 90° to the head then the disparities will be zero – an object at 90° would of course in practice only be visible to one eye.

Therefore if an observer can identify the part of the visual field with the largest range of horizontal disparities then that will be straight-ahead. Of course if the scene is uneven, such as when the observer is stood close to a side-wall, then the direction with the largest range of disparities may not in fact be true straight-ahead. However this could be confirmed through eye-movements, or through considering the gradient across the field. Figure 8 plots the range of disparities across the visual field as a function of head-centric direction.

Two potential sources of retinal information about direction have been identified. There is no suggestion here that either of these sources of information accounts for the attenuated prism effect reported above. But the information identified serves to illustrate how much information is potentially available.

3 SUBSEQUENT AND FUTURE LOCOMOTION STUDIES

Since the original study a number of replications have been published (Rogers & Dalton, 1999; Wood et al., 2000; Harris & Carre, 2001; Warren et

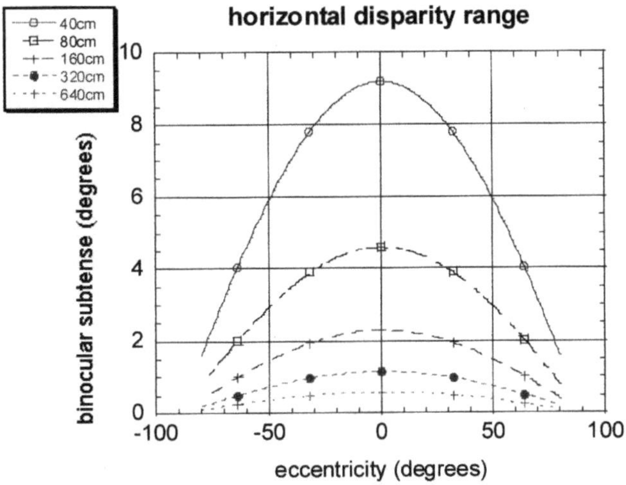

Figure 8. The range of horizontal disparities is illustrated by plotting the disparity of objects at a range of distances from the Cyclopean eye, or nose, as a function of head-centric direction. Distant objects have a disparity of zero in all directions, so the disparity of close objects determines the range.

al., 2001; Harris & Bonas, 2002; if the reader is familiar with any of these studies they may wish to consider them in light of the data on the perception of egocentric direction that is described in the previous section). Although there was initially a large gap between the views of researchers (e.g., Warren & Kay, 1997 and Rushton et al., 1998), the points of disagreement have reduced considerably despite the apparent difference of opinion (such as the exchange of letters in *Trends in Cognitive Sciences*).

Points of consensus are that egocentric direction is the sole cue in the following situations: (i) when a target is moving (Rushton et al., 1998; Fajen & Warren, 2002); (ii) when moving through a sparse environment; (iii) at the beginning of a trajectory.

The condition that continues to generate dispute is what happens when an observer is already in motion, moving through a cluttered and lit environment towards a static target. From the literature reviewed above it is clear that the structure, and changing structure (optic flow), of the visual environment will influence perceived egocentric direction. Therefore it should also influence walking trajectory. The outstanding question is does the changing visual structure of the environment (optic flow) also have a direct locomotion-specific influence or role? We can try to clarify this point with a thought experiment. Let us identify a situation that would allow a locomotion-specific role to be identified.

Consider an observer walking towards a target with a ball in their hand that can be thrown at the target at any time: If a vibrator was attached so as to stimulate the observer's neck muscle and the observer threw the ball while in motion, the instantaneous error in the direction of locomotion towards the target should be the same as the error in the direction of the throw. This is because the vibrator would be expected to bias perceived egocentric direction that is a source of information used commonly in both actions (locomotion and throwing). A similar prediction would apply to caloric stimulation, or any of other manipulation reviewed in the previous section, that has an influence on perceived egocentric direction.

If optic flow has a locomotion-specific use, then in contrast, it would be expected that the error in locomotion direction and throwing direction would be different when the optic flow is manipulated. Recall from the above review we already expect that optic flow would influence throwing because it influences perceived egocentric direction. Therefore, it would be necessary to show that the error in throwing and walking is different.

If indeed there is a locomotion-specific role of optic flow that differentiates its influence from all the other manipulations that change perceived direction, then it might in practice be difficult to demonstrate. An extensive parametric study that varied walking speed, environmental clutter and so on might be necessary to allow the locomotion-specific influence to be identified. Until such data has been collected it seems safer to assume there is not a locomotion specific role for optic flow.

4 NEW QUESTIONS

How is direction perceived? This chapter provided a brief review of some the literature that is relevant when attempting to answer this question. However, as will be apparent there is not very much literature on this issue. It seems peculiar that so much is known about the perception of distance (see the twin volumes on the perception of distance by Howard & Rogers, 2002), and so little about the perception of direction. To interact with an object it is necessary to know both components of the object's position. Will the research from distance perception and cue contribution, combination, and dominance translate to the perception of direction?

If optic flow processing is not primarily to support locomotion, then what is it there for? Can we re-interpret all the flow-locomotion data in terms of perception of direction? One idea is that optic flow does not support control of movement but rather, maintenance of perceptual stability.

How might an observer get around an environment using egocentric direction? In a companion chapter by Rushton & Harris heuristics are described that could be used to guide an observer around an environment, including intercepting moving and static targets, avoiding obstacles and following paths. Recent empirical work by Warren & Fajen and Wann & Wilkie is beginning to address this question (see their chapters in this book).

It is important to clarify that the egocentric direction account of the visual guidance of locomotion is not, as sometimes perceived, simply a critique of the classic flow theories. The direction account cannot be dismissed as soon as 'an influence of flow' on locomotion is reported. The direction account is a fully-fledged alternative. Conceiving of locomotion as being guided by perceived egocentric direction brings locomotion into line with reaching, throwing, catching and other actions. This chapter, and the companion one by Rushton & Harris, simply sketch out some of the issues associated with the direction account.

ACKNOWLEDGEMENTS

The research reported in this chapter was supported in part by Nissan Technical Center, North America Inc. Thanks to Andrew Welchman for comments on the penultimate draft. Thanks to John Wann for use of the flow pictures in Figure 1.

REFERENCES

Banks, M. S., Backus, B.T. & Banks, R.S. (2002). Is vertical disparity used to determine azimuth? *Vision Res., 42,* 801-807.

Biguer, B., Donaldson, I. M. L., Hein, A. & Jeannerod, M. (1988). Neck muscle vibration modifies the representation of visual motion and direction in man, *Brain, 111,* 1405-1424.

Berends, E. M., van Ee, R. & Erkelens, C. J. (2002). Vertical disparity can alter perceived direction, *Perception, 31,* 1323-1333.

Brecher, M. H. & Brecher, G. A. (1969). Motor Effects from visually induced disorientation in man, *Office of Aviation Medicine. AD708425, National Technical Information Service, U.S. Department of Commerce.*

Brecher, G. A., Brecher, M. H, Kommerell, G., Sauter, F. A. & Sellerbeck, J. (1972). Relation of optical and labyrinthean orientation. *Optica Acta, 19,* 467-471.

Bridgeman, B. & Graziano, J. A. (1989). Effect of context and efference copy on visual straight ahead, *Vision Res., 29,* 1729-1736.

Bruell, J. H. & Ablee, G. W. (1965). Effect of asymmetrical retinal stimulation on the perception of the median plane, *Percept. Motor Skills, 5,* 133-139.

Duchon, A. P. & Warren, W. H. (2002). A visual equalization strategy for locomotor control: of honeybees, robots, and humans, *Psychol. Sci., 13,* 272-278

Duffy, C. J. & Wurtz, R. H. (1993). An illusory transformation of optic flow fields, *Vision Res., 33,* 1481-1490.

Fajen, B. R. & Warren, W. H. (2001). Interception of moving objects on foot [Abstract], *J. Vis. 1(3),* 187a, http://journalofvision.org/1/3/187, DOI 10.1167/1.3.187.

Gibson, J. J. (1958). Visually controlled locomotion and visual orientation in animals, *Br. J. Psychol., 49,* 182-194.

Gibson, J. J. (1966). *The Senses Considered as Perceptual Systems*. Boston: Houghton Mifflin.

Harris, C. S. (1974). Beware of straight-ahead shift - nonperceptual change in experiments on adaptation to displaced vision, *Perception, 3,* 461-476.

Harris, J. M. & Bonas, W. (2002). Optic flow and scene structure do not always contribute to the control of human walking, *Vision Res., 42,* 1619-1626.

Harris, M. G. & Carre, G. (2001). Is optic flow used to guide walking while wearing a displacing prism? *Perception*, 30, 811-818.

Howard, I. P. & Rogers, B. J. (2002). *Depth Perception.* Toronto: I. Porteous.

Karnath, H. O., Sievering, D., Fetter, M. (1994). The interactive contribution of neck muscle proprioception and vestibular stimulation to subjective "straight-ahead" orientation in man, *Exp. Brain Res., 101,* 140-146.

Kim, N. G., Fajen, B. R. & Turvey, M. T. (2000). Perceiving circular heading in noncanonical flow fields, *J. Exp. Psychol. Hum. Percept. Perform., 26,* 31-56.

Kleinhans, J. L. (1970). *Perception of Spatial Orientation in Sloped, Slanted, and Titled Visual Fields,* PhD thesis, Rutgers University, Brunswick, NJ.

Lappe, M., Bremmer, F. & van den Berg, A. V. (1999). Perception of self-motion from visual flow. *Trends Cogn. Sci., 3,* 329-336.

Lepecq, J. C., Jouen, F. & Dubon, D. (1993). The effect of linear vection on manual aiming at memorized directions of stationary targets, *Perception, 22,* 49-60.

Llewellyn K. R. (1971). Visual guidance of locomotion, *J. Exp.Psychol., 91,* 245-261.

Matin, L., Picolut, E., Stevens, J. K., Edwards Jr, M. W., Young, E. & MacArthur, R. (1982). Ocuoloparalytic illusion: visual-field dependent spatial mislocalisations by humans partially paralyzed by curare, *Science, 216,* 198-201.

Miles, F. A. (1998). The neural processing of 3-D visual information: evidence from eye movements. *Eur. J. Neurosci., 10,* 811-822.

Milner D. A. & Goodale, M. A. (1995). *The Visual Brain in Action.* New York: Oxford University Press.

Mollon, J. (1997). "....on the basis of velocity cues alone": some perceptual themes, *Q. J. Exp. Psychol., 50A,* 859-878.

Paap, K. R. & Ebenholtz, S. M. (1976). Perceptual consequences of potentiation in extraocular-muscles - alternative explanation for adaptation to wedge prisms, *J. Exp. Psychol. Hum. Percept. Perform., 2,* 457-468.

Perrone, J. A. (1982). Visual slant underestimation: a general model, *Perception, 11,* 641-654.

Perrone, J. A. (1992). Model for the computation of self-motion in biological systems, *J. Opt. Soc. Am. A, 9,* 177-194.

Post, R. B. & Heckman, T. (1986). Induced motion and apparent straight-ahead during prolonged stimulation, *Percept. Psychophys., 40,* 263-270.

Redding, G. M. & Wallace, B. (1997). *Adaptive Spatial Alignment.* New Jersey: Lawrence Erlbaum Associates.

Rock, I., Goldbery, J. & Mack, A. (1966). Immediate correction and adaptation based on viewing a prismatically displaced scene, *Percept. Psychophys., 1,* 351-354.

Rock, I. (1966). *The Nature of Perceptual Adaptation.* New York: Basic Books.

Roelofs, C. O. (1935). Die optische localisation. *Archive für Augenheilkunde, 109,* 395–415.

Rogers, B. J. & Dalton, C. (1999). The role of (I) perceived direction and (II) optic flow in the control of locomotion and for estimating the point of impact, *Invest. Ophthalmol. Vis. Sci., 40,* s4036.

Rogers, B. J. (2001). Heading in the wrong direction? *Invest. Ophthalmol. Vis. Sci., 42,* S4981.

Rushton, S. K. & Brenner, E. (2001). The use of retinal flow in the perception of position, *Invest. Ophthalmol. Vis. Sci., 42,* s3316.

Rushton, S. K., Harris, J. M., Lloyd, M. R. and Wann, J. P. (1998). Guidance of locomotion on foot uses perceived target location rather than optic flow, *Curr. Biol., 8,* 1191-1194.

Rushton, S. K., Harris, J. M. & Wann J. P. (1999). Steering, optic flow, and the respective importance of depth and retinal motion distribution, *Perception, 28,* 255-266.

Rushton, S. K. & Harris J. M. (2003). The Utility of Not Changing Direction and the Visual Guidance of Locomotion. In: Vaina, L. M., Beardsley, S.A., & Rushton, S.K. (Eds). *Optic Flow and Beyond.* Kluwer Academic Publishers.

Rushton, S. K. & Salvucci, D. D. (2001). An egocentric account of the visual guidance of locomotion, *Trends Cogn. Sci., 5,* 6-7.

Velay, J. L., Allin, F. & Bouquerel, A. (1997). Motor and perceptual responses to horizontal and vertical eye vibration in humans, *Vision Res., 37,* 2631-2638.

Wallach, H., Kravitz, J. H, & Lindauer, J. (1963). A passive condition for rapid adaptation to displaced visual direction, *Am. J. Psychol., 76,* 568-578.

Wann, J. P., Swapp, D. & Rushton, S. K. (2000). Heading perception and the allocation of attention, *Vision Res., 40,* 2533-2543.

Warren, W. H. & Kay, B. A. (1999). The Focus of Expansion is used to Control Walking. In: Schmuckler, M. A & Kennedy, J. M. (Eds). *Studies in Perception and Action IV / Ninth International Conference on Perception and Action.* New Jersey: Erlbaum.

Warren, W. H., Kay, B. A., Zosh, W. D., Duchon, A. P. & Sahuc, S. (2001). Optic flow is used to control human walking, *Nat. Neurosci., 4,* 213-216.

Werner, H., Wapner, S. & Bruell, J. H. (1953). Experiments on sensory-tonic field theory of perception: VI. The effect of position of head, eyes, and other object on the position of apparent median plane, *J. Exp. Psychol., 46,* 293-299.

Welch, B. (1978). *Perceptual Modification: Adapting to Altered Sensory Environments.* New York: Academic Press.

Wood, R. M., Harvey, M. A., Young, C. E., Beedie, A. & Wilson, T. (2000). Weighting to go with the flow? *Curr. Biol., 10,* 545-546.

16. The Utility of not Changing Direction and the Visual Guidance of Locomotion

Simon K. Rushton[1] and Julie M. Harris[2]

[1]Centre for Vision Research & Department of Psychology
York University, Toronto, Ontario, Canada

[2]School of Biology (Psychology)
University of Newcastle-upon-Tyne, Newcastle-upon-Tyne , UK

1 INTRODUCTION

In this chapter we describe how the simple heuristic *maintain egocentric direction of the target* and the associated heuristic *cancel target drift* can, in theory, be used to successfully visually guide an observer around an environment. The motivation for this chapter is the proposal by Rushton et al. (1998) that observers visually guide locomotion using perceived egocentric target direction rather than optic flow. We concentrate on the visual control aspects of self-motion. The other approach is to develop models that capture the walking behavior of an observer attempting to reach a target. These models describe the final avoidance or closure behavior, rather than just the visual control laws themselves. Therefore they include terms such as a "damping" parameter and take into account perceptuo-motor processing delays. These terms reflect the physical embodiment of the observer, specifically, the physical limitations of the observer. An observer can only process information so fast, and the observer's body has a mass whose velocity must be regulated.

Although we appreciate the value of modeling the observer as a complete system, here we are interested in highlighting the use of simple visual information and basic heuristics. We also acknowledge that a skilled observer may go beyond these heuristics and use an inverse model to optimize behavior taking into account the physical characteristics of the whole perceptuo-motor system, i.e. factoring in understeer or compensating for processing delays through use of a predictive procedure, but we leave such

L.M. Vaina, S.A. Beardsley and S.K. Rushton (eds.), Optic Flow and Beyond, 363–381.

matters for treatment elsewhere. Last, we note that although we concentrate on the visual information available, rather than its implementation, the heuristics we describe here do indeed work "in practice" if perceptuo-motor delays, noise, hysteresis etc. are present. One example of this is given towards the end of this chapter where we refer to a robotic implementation of some of the heuristics described.

Some of the ideas described here are not novel. We would in particular draw the reader's attention to Llewellyn (1972) for an earlier drift cancellation proposal, Thompson (1966) on the occurrence of equi-angular spirals in nature, the various articles of Land and Collett on insects (e.g. Land & Collett, 1974) and Lee (1998) on general laws for the visual guidance of action. This chapter simply aims to highlight the potential utility of maintaining target direction during locomotion.

2 CHECKLIST OF FUNDAMENTAL LOCOMOTOR BEHAVIORS

Before we begin, it is worth identifying a list of locomotor behaviors that humans routinely exhibit. Numerous behaviors have been enumerated, and we note three lists, those of Gibson (1958), Loomis & Beall (1998) and Lee (1998). We synthesise these lists to produce the following short summary:

- Steering with respect to a static or moving target to be reached
- Steering with respect to static or moving obstacles to be avoided
- Steering with respect to a straight or curving path

In this chapter we will attempt to identify heuristics that could be used in all of the above cases.

3 TARGET DIRECTION OR "ECCENTRICITY"

In the rest of the chapter we will talk about *eccentricity*, or α, which we define as the egocentric direction of the target, more specifically the direction of the target measured relative to the mid-line of the trunk (see Figure 1). In the section after this we will talk of *target drift*, this being defined as simply a change in eccentricity, $\dot{\alpha}$.

We now consider some possible trajectories that could be used to approach a target. The trajectories that follow are generated by regulating the direction of locomotion so as to keep the target at a fixed eccentricity. In many cases the target eccentricity will be 0°, i.e. straight-ahead, but it is not

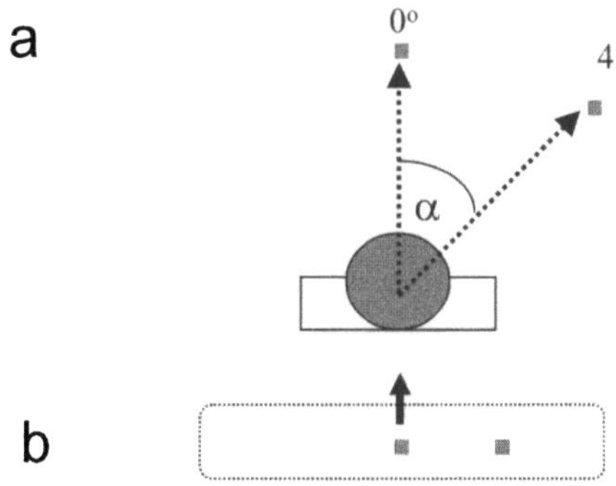

Figure 1. a) Ego-centric directions, 'eccentricity', α, measured as an angle in cardinal plane. b) The 'Cyclopean view': the view as if looking through a single eye at the center of the head.

necessary to maintain a target at 0° to intercept it, any direction greater than -90° and less than 90° will suffice provided the observer gets closer to the target on every step. The consequence of maintaining an eccentricity other than 0° is that the observer will approach the target on a spiral trajectory. If the eccentricity is low then this will only be identifiable by an increase in the rate of change of curvature of the trajectory as the target is approached, if the eccentricity is high then the observer may revolve around the target.

3.1 Travel with Respect to a Target

First we illustrate a family of movement trajectories towards a target that are generated by maintaining the target at a series of different, fixed, eccentricities during approach. The left panel of Figure 2 shows the trajectories towards a static target, the middle panel for a target moving at a constant velocity, and the right panel for a target that is accelerating.

An observer's trajectory can be described by a pair of simple iterative equations. If the observer is at position (x_{obs},z_{obs}) on the ground plane and the target is at (x_{tar},z_{tar}), the eccentricity of the target from the observer, ϕ, is given by:

$$\phi = \tan^{-1}\left(\frac{x_{tar} - x_{obs}}{z_{tar} - z_{obs}}\right) \quad (1)$$

or if we define the observer's position as (0,0), if the position of the target is (x'_{tar}, z'_{tar}), then

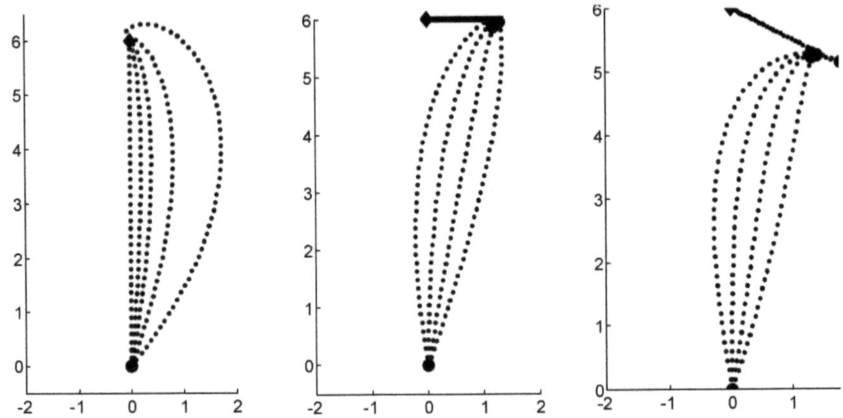

Figure 2. Left – Plan view of trajectories that would result from holding a target at a fixed eccentricity, α, of 0°, 5°, 10°, 20° and 40° (from left to right). Observer starts at (0,0), target is at (or starts at) (0, 6). Holding the target 'straight ahead', i.e. at 0°, would produce a straight trajectory leading directly to the target. Any other trajectory based upon holding the target at an eccentricity other than zero results in the observer 'veering' to one side before finally reaching the target. *Middle* – Intercepting a moving target. Target starts at (0, 6), and moves rightwards, observer starts at (0,0). Four fixed eccentricity trajectories shown, -10°, 0°, 10°, 20°. *Right* – Intercepting an accelerating target. Target starts at (0,6), and moves rightwards and downwards with increasing speed (constant acceleration), observer starts at (0,0). Fixed eccentricity trajectories shown are -10°, 0°, 10°, 20°.

$$\phi = \tan^{-1}\left(\frac{x'_{tar}}{z'_{tar}}\right) \tag{2}$$

Now, if the observer rotates so as to place the target at a chosen eccentricity of α and then takes a stride of length d, after the next step they will be at a new position

$$x_{n+1} = x_n + d.\sin(\phi + \alpha) \tag{3}$$

$$z_{n+1} = z_n + d.\cos(\phi + \alpha) \tag{4}$$

If we keep the observer as the origin of the coordinate frame then we can say that, following a step of length d, the x coordinate of all the objects in the scene will have decreased by $d.\sin(\phi + \alpha)$ and the z coordinate by $d.\cos(\phi + \alpha)$.

Using these equations for a range of different fixed eccentricities, α, generates the trajectories illustrated in Figure 2. As already noted, for many purposes the best choice of path will be to take α = 0°, but there are circumstances when a different choice could be useful. For example, in the

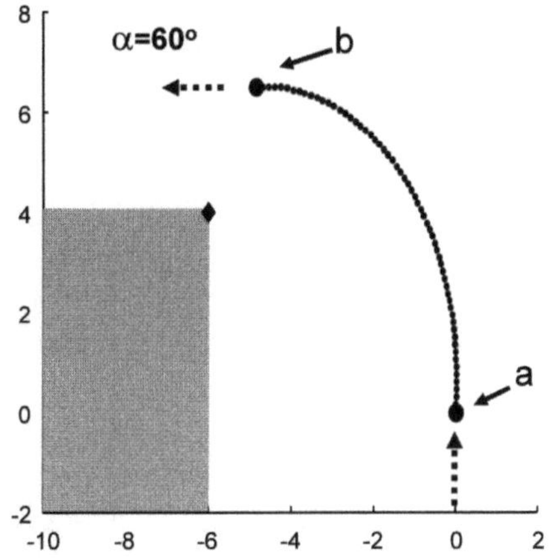

Figure 3. A plan view showing the rounding of a corner using a 'pivot point', fixed eccentricity strategy. The trajectory results from walking to point (at point **a**) and then holding the corner at its current eccentricity (60°) until the corner is rounded (at point **b**), at which time a new aim-point can be chosen to return to a straight path.

next section, a point held at a relatively wide eccentricity is used as a pivot point.

3.2 A Static Target as a Pivot-Point

Instead of wanting to walk towards an object, an observer may wish to turn ('pivot') around it. Consider walking down a corridor that changes direction at 90° (Figure 3).

The following strategy, based on using the eccentricity of just two or three scene points, allows an observer to pass around such a corner. Initially the observer can walk towards the end of the current section of the corridor, keeping a straight-path by maintaining the middle of the corridor straight-ahead ($\alpha = 0$). As the corner approaches, a smooth turn into the next section of corridor can be achieved by using the corner, or nearest edge, as a pivot point. To start this pivoting motion, the current eccentricity of the edge should be noted and then the observer changes their direction of locomotion so as to keep the eccentricity constant at that value. Use of this heuristic will smoothly turn them round the corner. Once the corner is almost rounded a new distant aim-point can be identified and used to once more maintain a straight path ($\alpha = 0$).

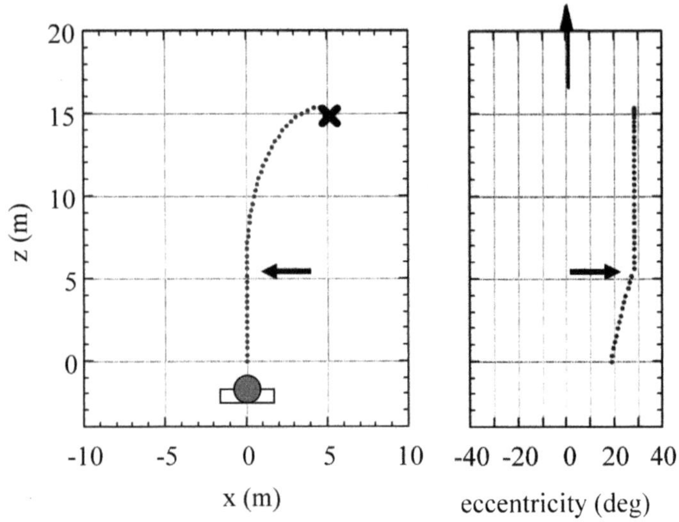

Figure 4. Approach to an object off to one side using a fixed eccentricity strategy. *Left Panel* – plan view of the trajectory. *Right Panel* – eccentricity, α, as a function of z. When the observer decides to approach an object of interest (decision point indicated by bold arrow), they simply hold the target at its current eccentricity, α (as shown in the right panel).

3.3 Selection of Approach Eccentricity

We have just shown one case where it might be useful to use a non-zero eccentricity approach to pivot around a point. Where else might a non-zero approach be useful, and what are the relative advantages and disadvantages of zero and non-zero approaches?

3.3.1 Change of Target: Approaching a New Target Without Explicitly Turning

If an observer is walking and then selects a new target off to one side there is no need to first turn in the direction of the target and then proceed to walk keeping the target straight-ahead. The observer may instead just walk to maintain the current egocentric direction of the new target (see Figure 4) .

3.3.2 Orientation, Position or Path

If an observer wished to arrive at or travel with a particular orientation relative to the target they could choose a trajectory appropriately. For example, consider if the target is a person that is standing ahead facing to the right. Taking an eccentric trajectory (say $\alpha = 40^o$; see left panel of Figure 2) it would be possible to maintain eye contact while approaching, or (with $\alpha = -40^o$) to sneak up behind the person.

Of course another reason to take a non-zero trajectory rather than a $\alpha = 0^o$ trajectory might be to avoid an obstacle. Return again to the left panel of Figure 2. If there was an obstacle between the observer and the target then taking a $\alpha = 40^o$ trajectory may avoid the obstacle while still taking the observer to the target. This principle of switching between trajectories in the family of trajectories available forms the basis for the obstacle avoidance ideas in Section 5.

3.3.3 Advantages and Disadvantages of Zero and Non-Zero Trajectories

Let us consider two trajectories (0° and 40°) in some detail, look at the differences between the two, and try to quantify the relative advantages or disadvantages associated with following them. We will use these examples to define a parameter we call *difficulty*, a measure of the costs associated with using any particular fixed eccentricity.

3.3.3.1 'Straight-Ahead' Trajectory ($\alpha = 0^O$)

Initially the target is straight-ahead. To maintain this path it is necessary to monitor for any error in direction of the target and correct when necessary. If the observer glances away for a few seconds while maintaining their direction of motion, they are still likely to be approximately on the correct path (the target is 'stable' and not prone to rapid drift).

3.3.3.2 $\alpha = 40^o$ Trajectory

As shown in Figure 2, keeping a target at 40° will result in the observer reaching it. However, because people typically walk forwards (perpendicular to a line joining their shoulders) on each step, to maintain this trajectory it is

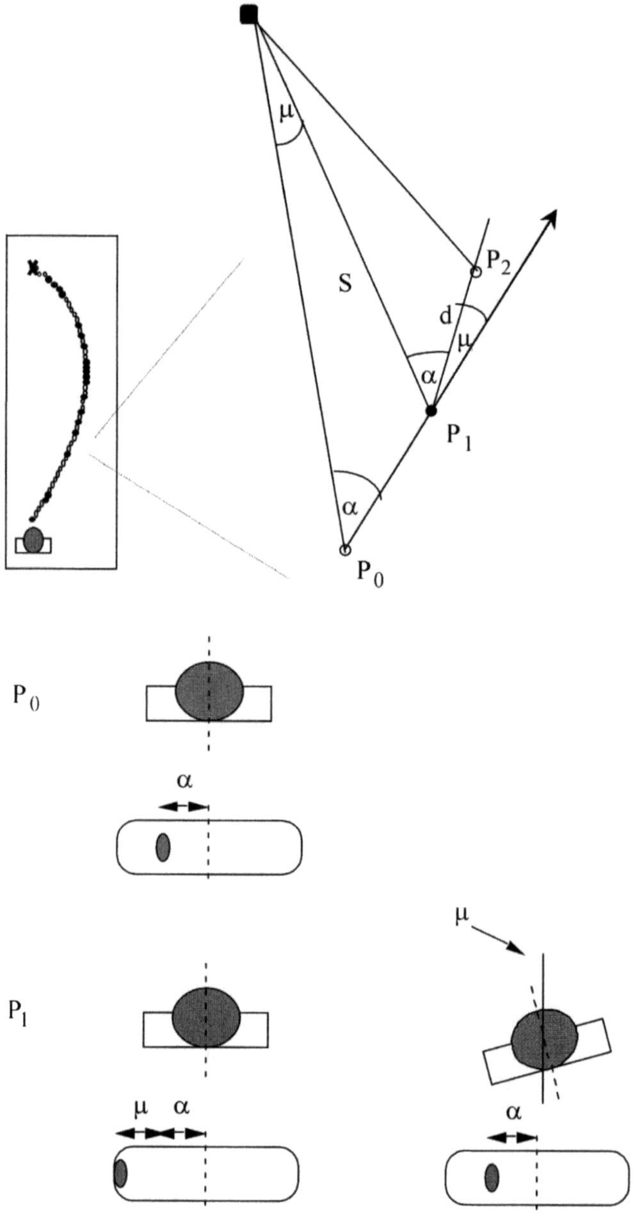

Figure 5. Top panel shows an observer walking towards a target while maintaining a fixed eccentricity, α. Inset shows an expanded view of the position along the trajectory. At P1 the observer may either continue in the same direction, or regulate their trajectory so as to reach P2. Lower panel shows these two alternatives. At P1 the target has drifted by μ degrees to a new eccentricity (left). The observer can cancel the drift by rotating through μ degrees and putting the target back to an eccentricity of α (right).

necessary to change direction slightly after every step (see Figure 5). If the observer glances away for a few seconds and does not continuously regulate their direction of locomotion then they will quickly find themselves a long way off course. Correcting for such a lapse will require a large compensatory change. The position of the target is thus 'unstable'; it will readily drift if the change of direction is not correct and continuously maintained from step to step.

Another potential problem is the act of turning after each step to maintain the chosen eccentricity. The larger magnitude of rotation required on each step to maintain a high eccentricity trajectory can be problematic as it may challenge the observer's stability. This would be particularly relevant when walking on a slippery surface such as ice, or in awkward footwear such as high heels. Finally, a high eccentricity trajectory is longer than a low eccentricity one, thus it takes longer to travel and requires more energy. In summary, a large eccentricity trajectory is longer and more '*difficult*', i.e., the direction of locomotion needs to be continuously regulated and there is a greater consequence if there is a lapse in regulation.

Can we define a simple way of quantifying the 'difficulty' of following a given trajectory, or the difficulty at a given point in a trajectory? As noted previously, a critical parameter is the rate at which the target drifts on each step. This defines the magnitude and speed of the rotation the observer needs to perform to maintain the trajectory, it is also correlated with the 'stability' of the target and the cost of missing a regulatory turn. Thus, the rate of target drift gives a measure of difficulty of maintaining a particular eccentricity. This is easily determined. On each step, the amount of drift, μ, where μ is the angle one must turn through on a step to correct the drift in eccentricity, will be dependent upon the chosen eccentricity, α, the stride length, d, and the distance to the target, S (Figure 5).

If the observer is moving at a constant velocity V, the instantaneous speed in a direction perpendicular to a line joining the target and observer is given by $V.\sin(\alpha)$ and $\dot{\alpha}$ the angular velocity is given by:

$$\dot{\alpha} = sin(\alpha)\frac{V}{S} \qquad (5)$$

Therefore, if walking a given trajectory becomes too problematic, the difficulty can be decreased by reducing the foot speed or by changing to a lower eccentricity approach.

Note, it also follows that if the rate of change of drift, $\dot{\alpha}$, could be measured on the fly, an observer could in principle use this information in the perception or calibration of eccentricity, α. In particular, the direction that is least *difficult* and most visually stable is that which corresponds to taking a 0°

straight-line trajectory to the target. Thus if an observer is moving on the most stable path, they must be on a 0° trajectory.

There are obvious advantages to taking a 0° eccentricity approach to a target. In the next section we will look further at target drift and how it could be used to straighten a path into a 0° approach through systematically reducing the eccentricity of the approach. First though we briefly examine some data that appears compatible with the influence of *difficulty* on trajectory choice.

3.3.3.3 *Difficulty vs. Straight-Ahead*

The prism manipulation employed by Rushton et al. (1998) pits straight-ahead against *difficulty*. Because participants are wearing deflecting prisms, the trajectory that results from keeping the target at the *perceived* straight-ahead is a physically curved path.

In the prism study, this trajectory is not especially challenging for the young participants walking on the slabbed surface. However, in the second experiment reported in Rushton et al. (1998) with a moving target, in some trials the target was moved into the path of the observer as they approached and in some trials it was moved away from the path. When the target is moved into the path then keeping the target straight-ahead is "easy" and when the target is moved away from the path then keeping the target straight-ahead is "hard". Therefore comparison of the mean eccentricity in the "into" and "away" trials may indicate an influence of difficulty on trajectory choice.

When a prism-wearing observer starts walking they initially walk away from the target. If the target is moved into their path then the trajectory that results from keeping the target perceptually 'straight-ahead' is almost physically straight-ahead and so 'easy' to walk. Therefore, it is likely that the angle between the instantaneous direction of travel and the direction of the target, α, will be approximately the angular deflection of the prism.

However, if the target is moved in the direction opposite the observer's current path then the trajectory required to keep the target perceptually 'straight-ahead' is a lot more sharply curved. Therefore, it is possible that the trajectory will become too *difficult* (require the observer too turn too sharply) and so the observer will change to a different constant eccentricity trajectory.

Figure 6 shows the mean value of α across the time-course for trials in which the target was moving into the observer's path and trials in which it was moving away from the path. α is close to the deflection of the prism for the 'into' trails but notably reduced for the more *difficult* 'away' trials. Looking at the graph it appears that the difficulty of the $\alpha = 0^{\circ}$ approach leads

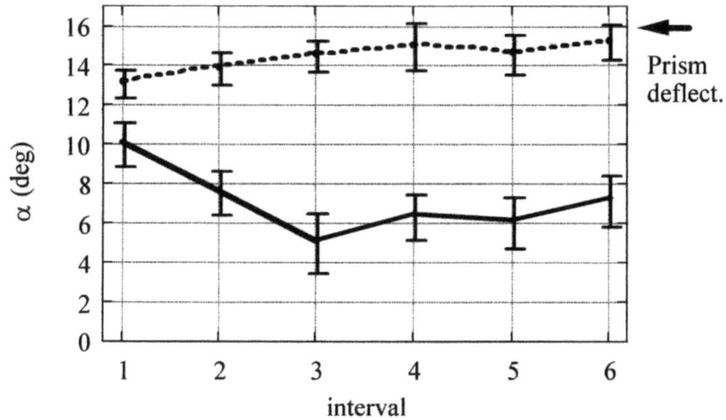

Figure 6. Mean target eccentricity from five trajectories (from five observers) with the target moving into path (dotted line) and moving away from path (solid line). Error bars indicate standard deviation. Length of trajectories varied so results were normalised by dividing each trial into six intervals. Note trajectories were initially smoothed by a Gaussian ($\sigma = 8$ data points or 1/3 sec., window = 16 data points or 2/3 sec.), as a consequence the first 2/3 sec. and last 2/3 sec. is lost.

the observers to switch to an easier trajectory, settling on $\alpha = 6^{\circ}$, which corresponds to an egocentric direction of 9°.

This data appears compatible with the hypothesis that observer trajectory is constrained by perceptuo-motor factors. It should be possible to reduce the eccentricity of an observer's trajectory similarly by requiring them to walk faster (though there is a danger of 'overshoot') or having the observer walk in high-heeled shoes or over ice.

3.3.4 Path Following

We have discussed the use of a maintain-eccentricity heuristic to guide locomotion to targets. The other common behavior that humans demonstrate is path following. Below we show how a maintain-eccentricity approach might be useful, and indeed how non-zero eccentricity can provide a simple range of behaviors for path following.

3.3.4.1 Travelling a Straight Path

An observer may stay on a straight path by maintaining a distant point at the end of the path at an eccentricity of 0°. This should keep them on the path and stop them straying towards the edges. Evidence in support of the use of a distant aim-point can be found in a study by Strelow & Brabyn (1981). They had observers walk parallel to a 'shoreline' of fluorescent poles in the dark (which would provide splay and splay-rate information). Their performance was found to improve when there was a visible object beyond the end of the shoreline that they could use as an aim-point. In principle, the distant targets need not be at an eccentricity of 0°. For example, the observer could regulate their direction of travel so as to keep constant the eccentricity of a path or road edge a fixed distance ahead (see Figure 6a; Lee, 1998).

3.3.4.2 Travelling Round Curving Paths

When an observer is walking down a corridor or along a curving path then a number of simple strategies can be employed to regulate their motion. Figure 7 shows three strategies based upon regulation of the egocentric direction of a portion of a path edge that are sufficient to guide an observer along a curving path (c.f. Lee, 1998). The first panel (A) shows the trajectory that results from keeping a portion of the inside path edge a given distance ahead at a fixed egocentric direction – this strategy therefore requires knowledge of distance in addition to direction. The second panel (B) shows the same strategy as (A) but using the outside path edge rather than the inside edge. If there was a center-line down the path then that could be used instead of a path edge. The third panel (C) shows the trajectory that results from keeping the tangent point of the bend (see Land & Lee, 1994) at a fixed direction. It might aid the reader to think of the first two strategies by comparing them to running a cane along the edge of a corridor wall. The first and second strategy require distance perception, but when walking on a flat surface, a simple visual equivalent of a cane for identifying a point a fixed distance ahead, would be the egocentric direction measured in the vertical plane; the vertical gaze angle.

If we consider the three strategies, one disadvantage of using a fixed distance ahead strategy is that it only works if the observer does not use a portion of the road too far ahead. The maximum distance ahead that could be used is proportional to the radius of curvature of the bend. A strategy that automatically compensates for the curvature of the bend is to use the tangent point. The result of such a strategy is shown in Figure 7c. Land (1998) and Wann & Land (2000) provide useful reviews of behavioral data on driving around bends and models and provide some alternative perceptual control solutions.

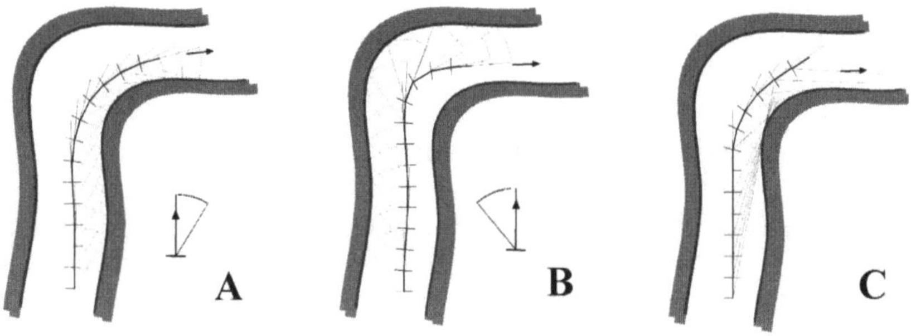

Figure 7. Travelling along a path by maintaining a road edge at a fixed eccentricity. a) Fixed distance, fixed direction inside of bend. b) Fixed distance, fixed direction outside of bend. c) Tangent point, fixed direction.

3.3.4.3 Turning into a Bend

Once in a bend an observer can use one of the fixed direction heuristics described in Section 3.3.4.1 above. How might an observer make the transition into a bend and out of a bend? Consider turning into a bend when currently on a straight. Assume that the current "control point" – the point whose eccentricity the observer is trying to maintain by regulating direction of travel – is the vanishing point or most distant part of the path. When approaching a bend an observer could select a new potential control point on the bend and simply make a transition from the current control point to the next. Figure 4 showed the transition from one target to another. In this case the new target would be a moving target, e.g., the tangent point or a path edge a fixed distance ahead. At what time should the observer choose to make the transition from one control point to another? They could monitor either the target eccentricity or the rate of target drift, waiting until a threshold level is reached.

4 TARGET DRIFT

If an observer is on a trajectory towards a target they can maintain the eccentricity of their approach using either a positional error signal or a motion one. In the former, the observer would compare the current eccentricity of the target with the desired eccentricity, in the latter the observer could simply note the change in eccentricity directly, i.e., pick it up as a motion signal.

Once on a trajectory, use of either error signal will lead to the same resultant trajectory. However it is possible to make additional use of target drift.

4.1 Target Drift and Calibration

In the previous section (3.3.3) it was noted that there are circumstances when an observer may wish to select one eccentricity approach over another eccentricity approach. For example, they may wish to select a 0° trajectory so as to take the shortest path to the target. To be able to do this requires that the observer is able to determine directions, i.e., that his or her perception of direction, and most critically straight-ahead, is calibrated. How might they do this?

First, an observer could use optic flow. The focus of expansion of the flow field sampled over a couple of minutes is a very reliable dynamic specification of straight-ahead. Rushton discusses the use of optic flow in calibration of straight-ahead in the companion chapter. Another possibility is to use target drift (see Llewellyn, 1971). It was noted in Section 3.3.3.2 that when a target is held at an eccentricity other than 0° it will drift. If the target eccentricity corresponds to left of the mid-line then when the observer takes a step forward the target will drift left, when the target is to the right of straight-ahead the target will drift right. Therefore the observer could simply continue to change their direction of travel until they find the travel direction associated with the highest level of target stability and then recalibrate their system with this direction as 0°. A more systematic approach is to use target drift in a feedback loop.

The easiest way to use target drift is to "overcompensate". If after having taken a step the target has drifted 1° to the left (μ in Figure 5), then to maintain a constant eccentricity approach it is necessary for the observer to rotate by 1° to the left (100% compensation). However, if the observer rotated 2° left (200% compensation) then they would reduce the eccentricity of the approach they are taking. If this were done repeatedly on every step then they would decrease the eccentricity down to a limit of zero. Figure 8 shows trajectories with over-compensation. As can be seen the trajectories rapidly straighten. Once the target was noted to have stopped drifting noticeably then the observer could simply average the position of the target over the next few steps to get a running estimate of straight-ahead.

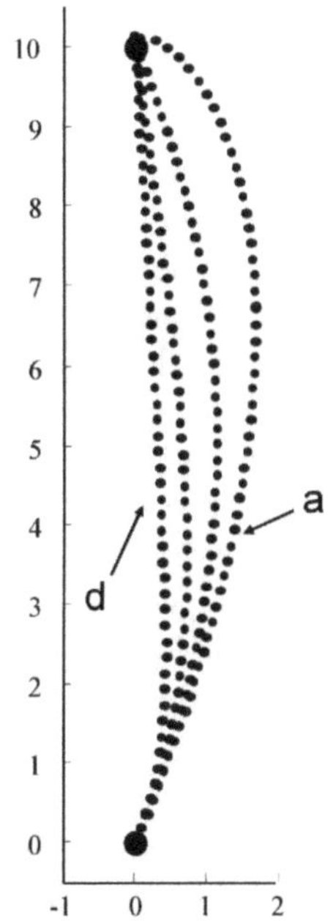

Figure 8. Four approach trajectories (from right to left, 'a' to' d'). Observer heads towards the target (0, 10). Initial target-heading direction is 25°. Trajectory (a) shows the course taken by cancelling target drift on each step (100% compensation) resulting in a constant direction trajectory. Trajectory (b) shows the course taken when the observer "over-compensates" for the target drift by a factor of 2 (200% compensation). Trajectory (c) "over-compensation" is 400%, and trajectory (d) is 800%.

4.2 Target Drift and Path Straightening

It should be apparent that the strategy described in the last section (4.1) could also be used to smoothly close down the eccentricity of the trajectory so as to optimise the path to the target. It would be sensible to only attempt to use such a strategy when the observer is in the clear and not concerned with avoiding obstacles.

If the observer is proceeding through a clear open space then we have shown how they could use target drift to optimise their approach – by closing down the eccentricity of the approach to zero – or use target drift in calibration. However, in other circumstances the observer might better use their attention to identify and avoid obstacles. That problem is tackled in the next section.

5 OBSTACLE AVOIDANCE

Here we outline several ideas regarding how an observer could detect an obstacle and then produce an appropriate change in their approach trajectory. The fixed eccentricity strategy can be simply reversed to test whether an object is on a collision course (i.e. whether an object is an obstacle). If an object remains at a fixed eccentricity during approach then the observer will collide with it. Therefore, should an impending collision be detected it is necessary to change the approach to the target so as to avoid the obstacle.

To move around the obstacle, but not hit it, the observer could simply change the eccentricity of their constant target angle approach (for example shift from a $\alpha = 0^{\circ}$ to $\alpha = 15^{\circ}$ approach). A refinement to this simple rule is required to avoid brushing collisions. The rule above would be sufficient to avoid collisions if both the observer and the obstacle were points. However, in the natural world objects have an extent. We consider two solutions to the problem of how to avoid collision with the edge of an extended target.

First, the observer could use the 'point assumption' until or unless the obstacle got into near space (as defined by distance or temporal proximity). Corrective steps could then be taken to move the obstacle out of near space, e.g., move laterally until clear of the obstacle, and a trajectory towards the goal could then be resumed.

Second, the observer could rely on predictive information, such as the distance from the Cyclopean eye at which the obstacle will cross the fronto-parallel plane that contains the eyes, (see Bootsma 1991), to detect potential obstacles. Once an obstacle is detected, then the imminence of collision can be determined from the time before the obstacle will cross the fronto-parallel plane containing the eyes, (see Lee, 1976). An appropriate change can then be made to the eccentricity of the approach to the target.

5.1 Approaching Obstacles and Avoiding Obstacles, an Implementation

Rushton et al. (2002) implemented a system for guiding a mobile robot based upon these principles. The robot functions in two modes, when it detects that it is in clear space with no obstacles close by then it decays the eccentricity of the approach to the target so straightening the trajectory (using overcompensation for target drift in the manner described in the previous section). When there are obstacles close by, then the robot switches between fixed eccentricity trajectories within the family of available trajectories to

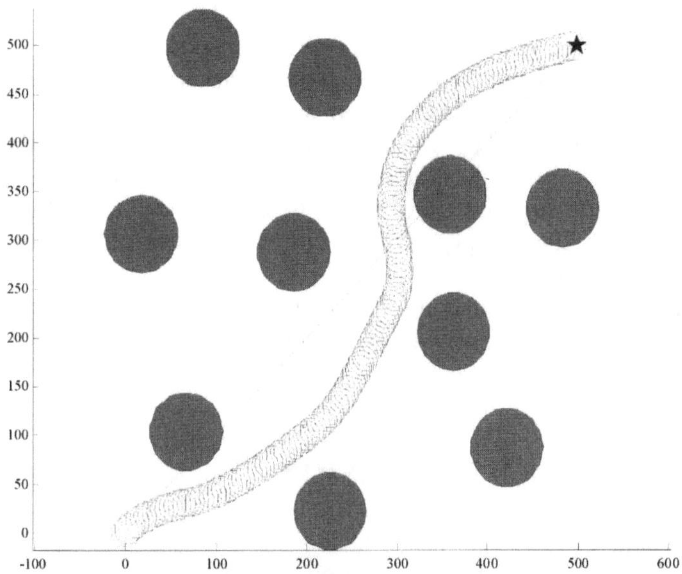

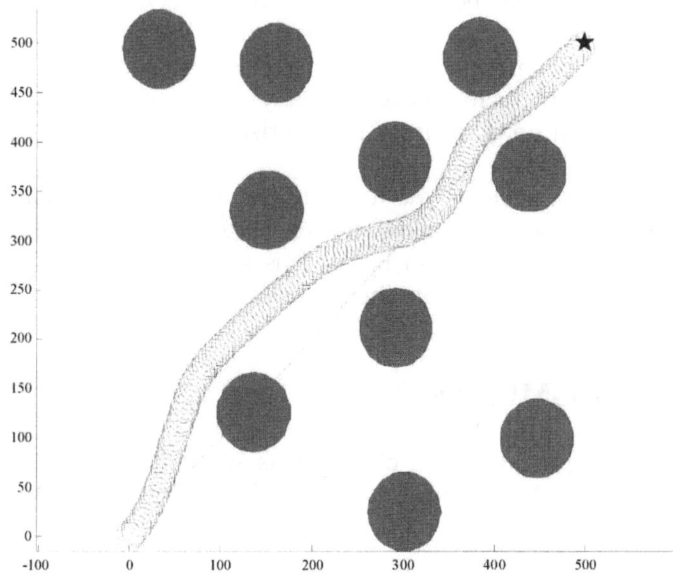

Figure 9. Results of simulations from Rushton et al. (2002). Obstacles are reliably avoided and may be arbitrary sizes or shapes. Trajectory starts at (0,0) and proceeds to target at (500,500). Example movies can be found at http://www.cs.yorku.ca/~percept/robot.html

avoid obstacles. The robot can be conceived of as being "knocked" from one trajectory within a family of trajectories to another by the obstacles.

Obstacle detection and avoidance is based upon crossing distance and time-to-contact – the latter based upon a variant of the *dipole* model of Rushton & Wann (1998). The crossing distance is calculated for the nearest edge of the obstacle, this makes the solution obstacle size and shape invariant. The control law for deciding how much to change the eccentricity of the approach is based upon Peper et al.'s (1994) continuous control law for projectile interception.

The value of this implementation is that it demonstrates that very simple heuristics can produce robust behavior. Examples of the trajectories that result are shown in figure 9 above.

6 SUMMARY AND DISCUSSION

In this chapter we have illustrated how use of a very simple heuristic, maintenance of egocentric target direction, is sufficient to allow successful locomotion around an environment. We gave several examples of how a fixed eccentricity strategy can be used to control behaviors as diverse as approaching an aim point and avoiding obstacles. We then considered the factors involved in choosing a particular fixed eccentricity strategy and the costs associated with such choices. Finally we cited a recent robotic implementation based upon many of the heuristics described in this chapter (see Rushton et al., 2002).

The intent of this chapter was to elaborate the potential utility of the eccentricity heuristic so as to allow and encourage testing of these simple strategies. The strategies can also be used as a reference against which to compare the power or necessity of other more complex models.

ACKNOWLEDGMENTS

The research reported in this chapter was supported in part by Nissan Technical Center, North America Inc and UK EPSRC grants to the first and second authors.

REFERENCES

Bootsma, R. J. (1991). Predictive information and the control of action: what you see is what you get, *Int. J. Sports Psychol., 22,* 271–278.

Gibson J. J. (1958). Visually controlled locomotion and visual orientation in animals, *Br. J. Psychol., 19,* 182-194.

Land, M. F. (1998). The Visual Control of Steering. In: Harris, L. R. & Jenkin M. (Eds). *Vision and Action.* Cambridge University Press.

Land, M. F. & Collett, T. S. (1974). Chasing behavior of houseflies (Fannia canicularis). A description and analysis, *J. Comp. Physiol., 89,* 331-357.

Land, M. F. & Lee, D. N. (1994). Where we look when we steer, *Nature, 369,* 742-744.

Lee, D. N. (1976). A theory of the visual control of braking based on information about time-to-collision, *Perception, 5,* 437-459.

Lee, D. N. (1998). Guiding movement by coupling taus, *Ecol. Psychol., 10,* 221-250.

Llewellyn K. R. (1971). Visual guidance of locomotion, *J. Exp. Psychol., 91,* 245-261.

Loomis, J. M. & Beall, A. C. (1998). Visually controlled locomotion: its dependence on optic flow, three-dimensional space perception and cognition, *Ecol. Psychol., 10,* 271-285.

Peper, L., Bootsma, R. J., Mestre, D. R. & Bakker, F. C. (1994). Catching balls: how to get the hand to the right place at the right time, *J. Exp. Psychol. Hum. Percept. Perform., 20,* 591-612.

Rushton, S. K., Wen, J. & Allison, R. S. (2002). Egocentric Direction and the Visual Guidance of Robot Locomotion: Background, Theory and Implementation. In: Bülthoff, H. H., Lee, S. W., Poggio,T. A. & Wallraven, C. (Eds). *Biologically Motivated Computer Vision: Lecture Notes in Computer Science* (pp. 576-591), Springer.

Rushton, S. K., Harris, J. M., Lloyd, M. L. & Wann, J. P. (1998). Guidance of locomotion on foot uses perceived target location rather than optic flow, *Curr. Biol., 8,* 1191-1194.

Rushton, S. K. & Wann, J. P. (1999). Weighted combination of size and disparity: a computational model for timing a ball catch, *Nat. Neurosci.*, 2, 186-190.

Thompson, D. A. (1966). *On Growth and Form.* Cambridge,England: Cambridge University Press.

Strelow, E. R. & Brabyn, J. A. (1981). Use of foreground and background information in visually guided locomotion, *Perception, 10,* 191-198.

Wann, J. P. & Land, M. F. (2000). Steering with or without the flow: is the retrieval of heading necessary? *Trends Cogn. Sci., 4,* 319-324.

17. Gaze Behaviors During Adaptive Human Locomotion: Insights into How Vision is Used to Regulate Locomotion

Aftab E. Patla

Gait & Posture Lab, Department of Kinesiology
University of Waterloo, Waterloo, Ontario, Canada

1 INTRODUCTION

"Locomotion is guided chiefly by vision", (Gibson & Crooks, 1938). This statement is understandable since vision is the key sensory system that makes adaptive legged locomotion possible by providing accurate distant environmental information about both animate and inanimate objects. The haptic system – a combination of cutaneous and kinesthetic mechano-receptors in the skin surface, muscles, joints and tendons – can be an adequate substitute, but not a one-to-one replacement, when the visual system is compromised. Olfactory and auditory systems are also able to gather information at a distance but both these systems are generally unable to detect inanimate objects. However, species such as bats have adapted their auditory system to detect objects using echolocation.

Advance information available from the visual system makes it possible to use isolated footholds, to step on surfaces of different geometry, to step over or under obstacles and to steer. Vision is the only modality that allows information to be maintained about one item (peripheral vision) while another one is being foveated and examined. Although Braille readers do use the two hands in a similar manner, one hand is used to read ahead while the other hand trails behind and fills in the details later. We have shown that information about limb posture and movements and global self-motion information (called visual kinaesthesis) is also superior to similar information provided by other modalities (Patla, 1999).

There is no debate that vision is critical for the regulation of human locomotion; decoding the nature of the visual information most useful for locomotion control, however, has proved to be a challenge. Psychophysical

L.M. Vaina, S.A. Beardsley and S.K. Rushton (eds.), Optic Flow and Beyond, 383–399.

studies examining exclusively perceptual responses to visual inputs (abstraction of naturally occurring stimuli patterns during locomotion) focus on the sensory side without examining how the relevant information is used to guide action. Recording of neural activity in animals in response to similar stimuli or functional neuro-imaging studies in humans, while fruitful, also does not provide insights into the actual information and strategies used during adaptive locomotion (Warren, 1999). Warren and his colleagues have addressed this limitation using virtual reality technology (that allows them to manipulate specific features of optic flow, including some that are not ecological) coupled with whole body responses to test various theories on visual strategies for control of locomotion (Warren et al., 2001). Others have examined patients with specific neurological impairments (see Vaina & Rushton, 2000).

Our movements through the world result in changes to the images on the retina; this optic flow (Gibson, 1950) provides a rich source of information to guide movements such as locomotion (Gibson, 1958). Determining what optic flow pattern is actually seen by the locomoting observer is essential for understanding the neural basis of visual guidance. Therefore we need to measure gaze (eyes in space) patterns during locomotion. While researchers have postulated, for example, the type of gaze behavior best suited for maintaining or changing direction of locomotion (Wann & Swapp, 2000), experimental data has been lacking until recently (Hollands et al., 2002).

2 SOME INSIGHTS FROM EARLIER GAZE STUDIES DURING PERCEPTION OF AN IMAGE AND DRIVING

"Nearly every muscle of the body can be used to change the stimulation of the eyes and pick up visual information", (Trevarthen, 1995). Retinal mobility for the acquisition of information can be achieved by whole body movement, head movement relative to the body, and/or eye movement relative to the head (Trevarthen, 1995). Monitoring gaze during any movement but in particular during whole body movement such as locomotion is therefore a technically challenging problem.

Technical limitations involved in gaze measurements initially limited scientific efforts to studying visual perception of static two-dimensional images while the individual's head and body were constrained. The work by Yarbus (1967) exemplifies this type of work. He was able to develop a unique cap system that allowed him to monitor eye movements, with minimal slippage over the eyeball, on a moving film of a photokymograph. Besides the technical achievement, the experiments that he carried out took full advantage

Figure 1. Forest picture and eye movement during 10 min of viewing the picture (from Yarbus, 1967).

of the technique to explore gaze patterns during the perception of complex objects (see for example the numerous figures in the last chapter of the book by Yarbus (1967) that show the spatial gaze patterns superimposed on the pictures and objects viewed). He was interested in what features attract visual attention and which do not. Visual palpation of the objects was shown to be influenced, for example, by the type of information the individual was supposed to extract from the picture. Of particular relevance is the forest picture and gaze patterns shown in Figure 118 in the book, replicated above (Figure 1). He made the astute observation that individuals fixated on the horizon more and were interested in the gap between the trees. These gaze patterns indicate that the information that was acquired by the individual could be relevant to locomotion through the forest; detecting goals on the horizon and determining if passage between trees is possible would help chart a travel path.

Research on visual search patterns while driving a vehicle provided the next advances in gaze research: head referenced view of the visual scene of the road ahead (a projected movie) along with visual image of the eye within the head was possible (see Shinar, 1978). Researchers talk about monitoring the beam of a searchlight (fovea), where it lands and the pattern of its movement. The analogy is apropos. The earliest work showed both the spatial

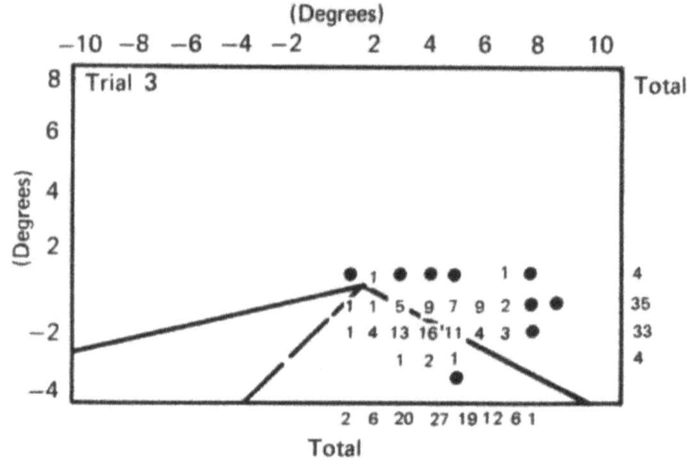

Figure 2. Distribution of percentage of time spent by driver fixating on different areas of the visual field during driving with a car following and the driver not required to read any signs (from Mourant et al., 1969).

and temporal patterns of visual fixation on different aspects of the visual scene while a driver was driving along a straight road (see Mourant et al., 1969). These patterns were affected by the instructions to the driver about what aspects of the road to attend to, by the geometry of the roadway, and whether or not there were other cars around. The figure from this earlier study, replicated above, illustrates the unique insights about where and for how long individuals fixated on different areas of the visual scene. The predominant area of fixation represents where the car is headed: a reasonable driving strategy.

Next, advances made it possible to monitor gaze behavior during actual driving on real roads (Land, 1992, Land & Lee, 1994). Land & Lee (1994) were the first to couple the visual fixation with the motor behavior during steering over an undulating road; the temporal coupling between the direction of gaze and turning of the steering wheel was about 0.75s. These temporal values are critical to our understanding of the sensori-motor transformations.

3 GAZE PATTERNS DURING ADAPTIVE HUMAN LOCOMOTION

Researchers have used various techniques to study where and when people look when walking. Wagner et al. (1981) relied on subjects' reporting

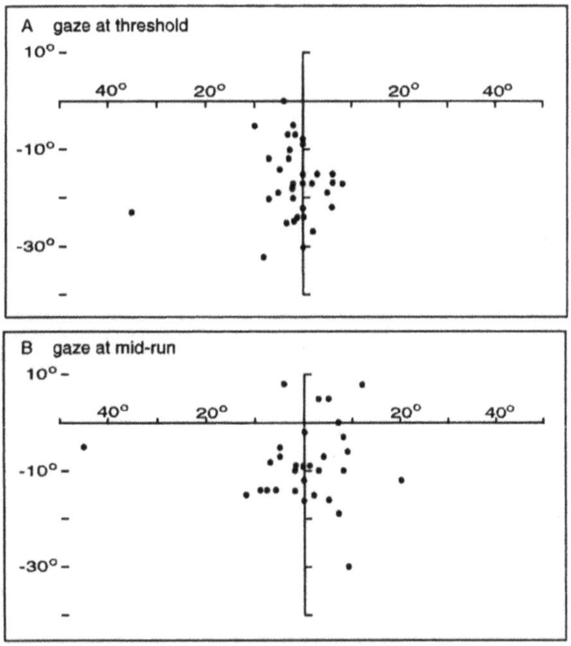

Figure 3. Cat's direction of gaze during locomotion in a cluttered alley (from Sherk & Fowler, 2000).

of what they were looking at when an audio signal (distributed throughout the walk) sounded. They reported that subjects most often tended to look 1.5 to 3m ahead during the walk.

As discussed before, accurate recording of gaze requires simultaneous recording of eye movement relative to the head, and head and whole body movement. Recording of all this information is a daunting task, and often requires constraining and tethering the individual on a specific limited path (see Solomon & Cohen, 1992a,b; Imai et al., 2001).

Others have recorded horizontal eye movements using EOG measurements during a stepping-stone task (Hollands et al., 1995, 1996), and have showed that eye movement (to the landing target) precedes the initiation of the limb lift-off towards the landing target. Despite the many assumptions involved in using horizontal eye movement as the surrogate for gaze, the study by Hollands et al. (1995) did show the feed-forward control of limb trajectory based on visual information. Recently Sherk & Fowler (2000, 2001) estimated gaze from eye level video recording while a cat was required to walk down a cluttered alley. The direction of gaze at an entrance of the alley showed clearly that gaze was within 10 deg of the mid-line and downwards as seen in Figure 3.

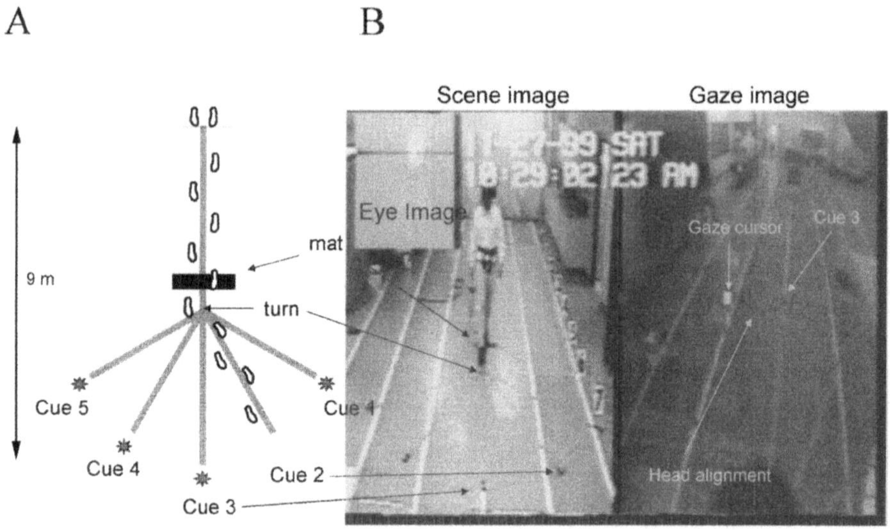

Figure 4. Schematic diagram of the experimental set-up on the left and a frame of the composite video showing the gaze cursor and the subject during a steering task (from Hollands et al., 2002).

Turano et al. (2001) monitored gaze while the subjects viewed the environment in a head mounted display and showed that gaze was directed primarily ahead or at the goal when virtually traversing an unfamiliar but obstacle-free route. They also demonstrated that gaze patterns are affected by visual impairment

We have used techniques similar to those used by Land & Lee (1994) adapted for use by a free-moving person (Hollands et al., 2002; Patla & Vickers, 1997, 2003; Vickers & Patla, 1999). The basic measurement technique involves simultaneous recording of the eyes along with the head referenced scene ahead; all the equipment is mounted on a light helmet and connected to the computer by a long umbilical cord. This makes it possible to monitor gaze while the person moves around freely in different environments simulating different challenges. Figure 4 shows one frame of the video showing the gaze superimposed on the visual scene. The task studied was steering control in response to a light cue. Also shown in the same figure is the schematic diagram of the experimental set-up.

The major findings from our studies (Hollands et al., 2002; Patla & Vickers, 1997, 2003; Vickers & Patla, 1999) are discussed below. The types of gaze behaviors and their spatial and temporal patterns are described. We have not measured limb or whole body movement in detail, therefore the

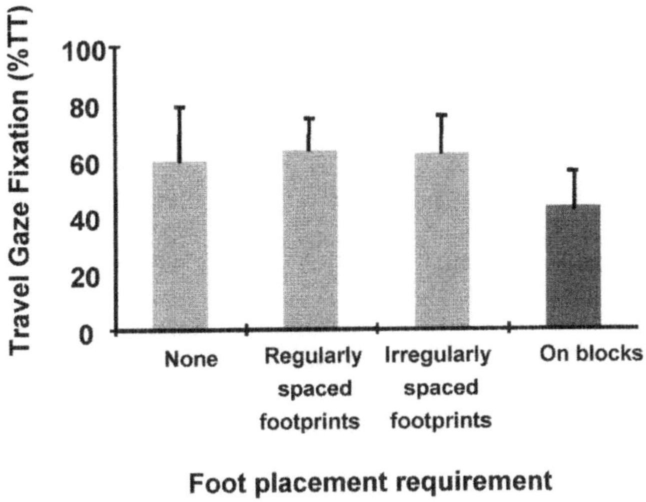

Figure 5. Travel gaze fixation as a percenateg of travel time (TT) duration during a variety of adaptive locomotion tasks.

movement patterns and temporal coupling values are described in terms of step cycle metric.

3.1 Travel Gaze Fixation is the Dominant Gaze Behavior During Adaptive Locomotion

One type of gaze behavior that was commonly observed in all of our locomotion studies, involved the eyes parked in the front of the moving observer and being carried along by the observer. It is akin to carrying a video camera focussed on a target in front. We were not able to determine the most common distance where the gaze was anchored, but if we take the results from earlier study by Wagner et al. (1981), it is likely that gaze was anchored about a two step distance ahead a majority of the time. We call these episodes of gaze behavior "travel gaze fixation". Gaze is fixated relative to the moving person and is travelling with the individual. The fixation episode duration was generally 300 ms or less. Recently, Fowler et al. (2000) showed similar gaze behavior in cats walking in a cluttered alley; the gaze angle was held constant during these episodes, which were 500 ms, or less in duration. What is most surprising is that this type of gaze behavior occupies the majority of the travel time irrespective of the environmental challenges, as seen in Figure 5. Fowler

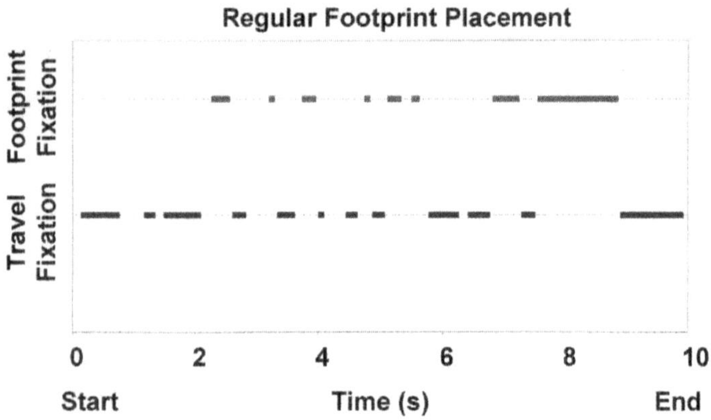

Figure 6. Sequence of travel and object (footprint to be stepped on) gaze fixation patterns during an adaptive locomotor task.

et al. (2002) also showed that the total duration of constant angle of gaze episodes was over 50% of the total time.

We have argued that this type of gaze behavior is attractive since only reflexive eye movements would be necessary to compensate for the head accelerations experienced during locomotion (Patla & Vickers, 1997, 2003). The retinal slip of images during these short duration episodes has been shown to be tolerated by the oculomotor system without substantial degradation in visual acuity (Demer & Amjadi, 1993). Depending on where the gaze is anchored relative to the person and the duration of the travel gaze fixation episode, it is possible that a relevant object such as landing spot or obstacle to be stepped over can appear within the fovea. This obviates the need to fixate on the object.

3.2 Object and Travel Gaze Fixation Behaviors are Interspersed During Locomotion

While the dominant gaze behavior is travel gaze fixation, individuals did also fixate on objects relevant to the task of locomotion. While approaching an obstacle that is to be stepped over, individuals fixate on the obstacle (Patla & Vickers, 1997). When subjects are required to step on specific landing spots, they fixate on these targets (Patla & Vickers, 2002). During steering control they fixate on objects that provide information about the path to be taken: cue lights that direct individuals to a specific path or the mats that act as the trigger for turning on the appropriate cue light (Hollands et al., 2002). The duration of gaze fixation on an object varied: for the footprints most gaze

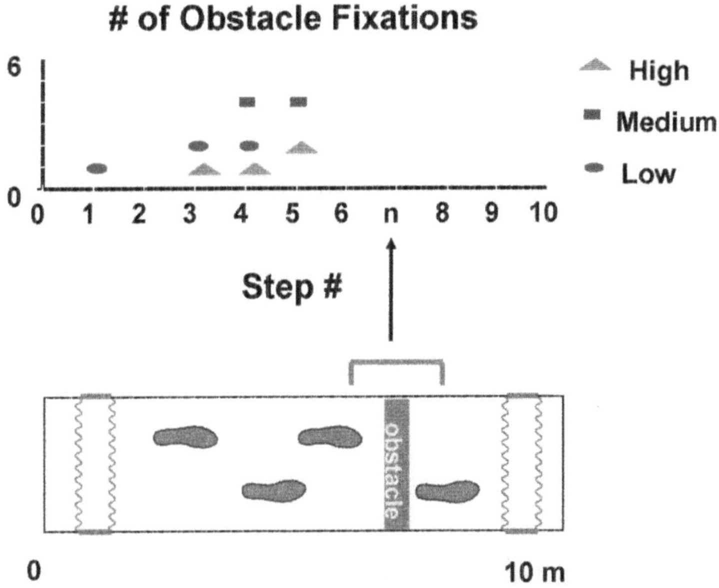

Figure 7. Obstacle fixation patterns during an obstacle avoidance task (of three different heights with five trials each), (from Patla & Vickers, 1997).

fixations were 300 ms or less (Patla & Vickers, 2003). The episodes of gaze fixation on the object were interspersed with episodes of travel gaze fixation (see Figure 6). The transition from travel gaze fixation to fixation on object was always achieved by saccadic eye movements; although slow head movements do also assist as shown by Land (1992) during a driving task. Figure 6 shows the sequence of obstacle and footprint gaze fixation during a trial where the subject was required to step on footprints spaced at regular distance.

3.3 Object Fixation – Whether it is to be Avoided or Accommodated – Precedes Initiation of Limb Movement

Gaze patterns, particularly fixation on objects relevant to the locomotor task, clearly demonstrate that environmental information is acquired in advance; an obstacle to be stepped over or the target to be stepped on is fixated before the limb trajectory is modified to achieve the task. Therefore the planning of gait changes can be carried out in advance. Consistently, information is acquired at least two steps ahead, which is sufficient to

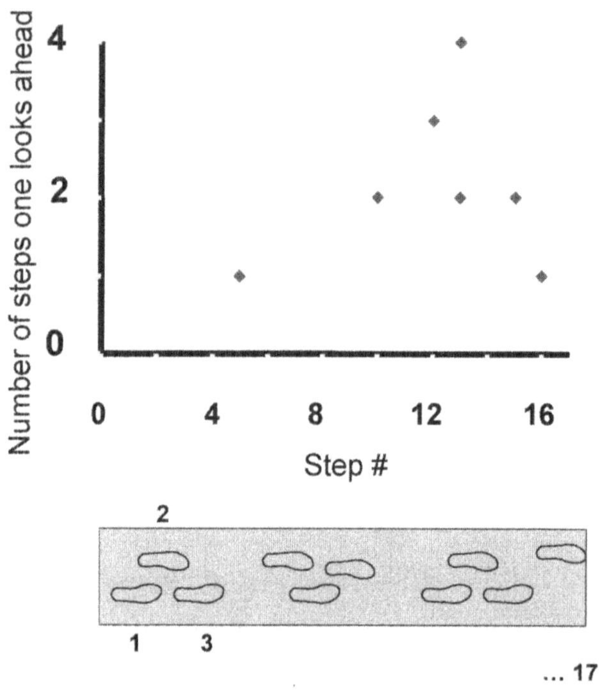

Figure 8. Gaze patterns on landing targets during visually guided stepping (adapted from Patla & Vickers, 2003).

implement all avoidance strategies (Patla, 1997; Patla & Vickers, 2003). Figures 7 and 8 show object fixation pattern during obstacle avoidance and stepping target fixation during locomotion where the landing areas are prescribed. Both figures show that fixation on an object, whether it is to be avoided or stepped on, is accomplished in the preceding steps. Thus visual information about the object can be used to plan and modify the limb trajectory appropriately.

3.4 Scanning of the Travel Path Prior to Gait Initiation is not Observed

In our studies we asked the individual to stand at the beginning of the travel path with their eyes closed, while the travel path was being set. Upon a verbal cue, subjects were to open their eyes and begin walking. There was no specific instruction about beginning immediately upon opening their eyes. Subjects never visually scanned the pathway ahead any longer than 300 ms

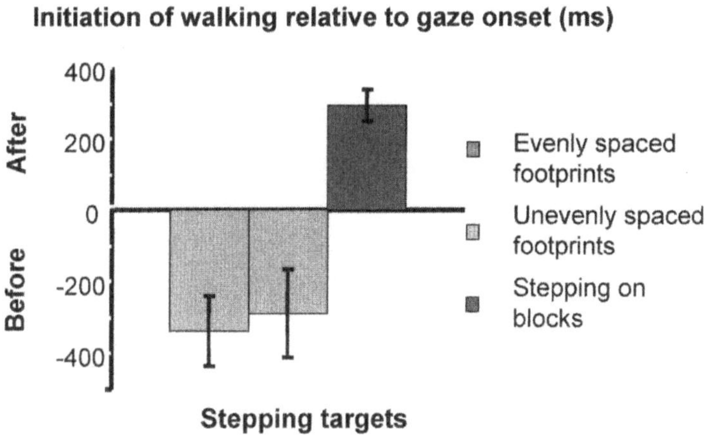

Figure 9. Initiation of walking relative to gaze onset for three different environmental challenges.

before beginning to walk even when, for example, stepping on the first stone of a series with the wrong foot would require them to use a cross stepping pattern midway (Vickers & Patla, 1999). In Figure 9, the time between the appearance of gaze cursor and initiation of walking for three different travel paths are shown. The data clearly show that walk is initiated before or after a brief time following onset of gaze. Thus visual input about the environment was acquired during locomotion rather than before beginning the walk. Visual input acquired while the body is moving leads to superior performance compared to static visual sampling before walking (Patla, 1999).

3.5 Gaze Scanning Patterns During an Adaptive Locomotor Task are Different from Gaze Patterns During a Perception Task

When viewing a picture that includes a person the most expressive elements of the face, eyes and lips, are the primary targets of fixation (Yarbus, 1967). Eye movements appear to visually trace the outline of the object or a face (Yarbus, 1967). Gaze patterns when viewing a picture also show cyclical patterns; the same features are fixated on during each cycle (Yarbus, 1967). During locomotion, gaze fixations on objects appear to show no such predictable patterns as seen in the Figures 10 and 11.

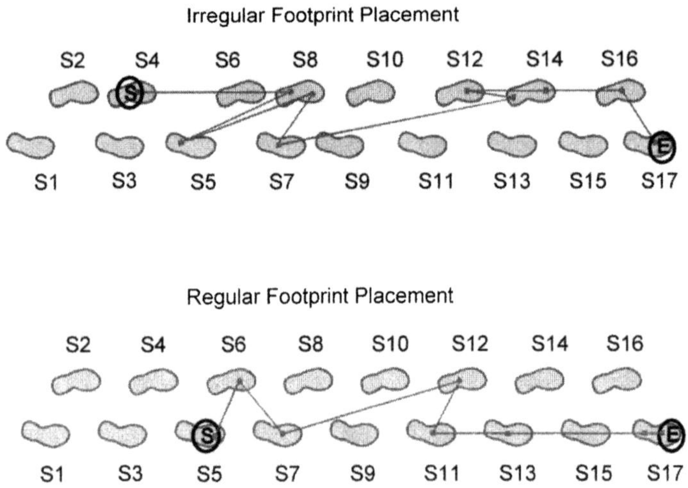

Figure 10. Spatial pattern of object gaze fixation (red line) during a stepping task (S – start of object gaze fixation; E – end of object gaze fixation).

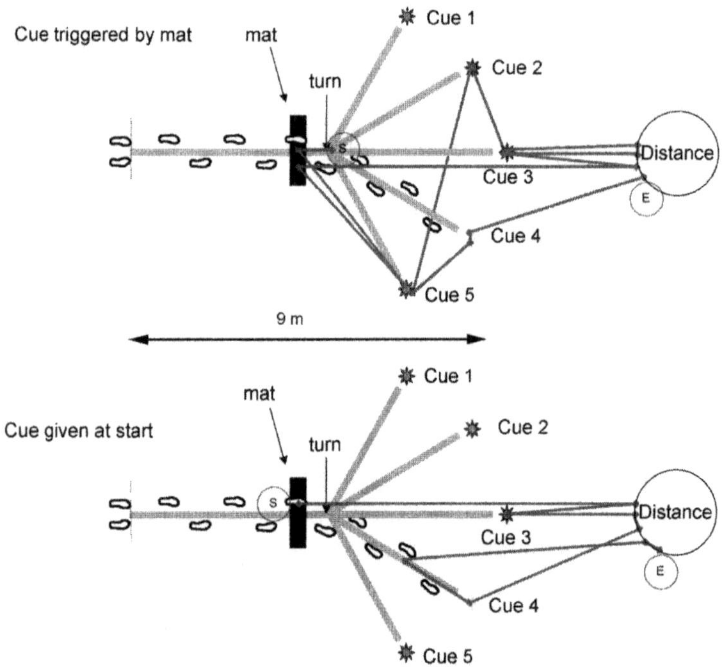

Figure 11. Spatial pattern of object gaze fixation (red line) during a steering task (S – start of object gaze fixation; E – end of object gaze fixation).

3.6 Consequences of Errors in Foot Placement Influence the Gaze Behaviors.

The relative duration of object and travel gaze fixation can be influenced by errors during the locomotor task. For example, in a stepping-stone task subjects often stepped on the ground while executing a cross over step when they started with the wrong foot on the first step. This error led to a reduction in duration of travel gaze fixation and an increase in landing target (blocks) fixation in the next trial. Once they had figured out the correct foot to begin their walk for a successful trial, the duration of gaze fixation returned to the original values.

Figure 12 summarizes this data of relative travel gaze fixation and fixation on the blocks. These results highlight two important points. First, while travel gaze fixation is a simpler gaze pattern and is preferable, locomotion relevant object fixation is needed for more precise walking over cluttered terrain. Second, the transition from travel gaze fixation to object fixation can be voluntarily controlled. The trigger for the transition between the two types of gaze behavior is not known.

When there were no errors in foot placement or accidental contact with the obstacle, successive trials showed no changes in gaze behaviors. One would have expected that prior experience with the footprint laden travel path would result in changes in the number of steps one looks ahead. Rather the number of steps one looks ahead remained similar across successive trials. Researchers have argued that vision for action is used 'denovo' rather than relying on memory in this case from previous travel over the same path (Milner & Goodale, 1995). This is understandable since paths can present unexpected challenges and reliance on memory can lead to errors, compromising safety.

4 WHAT HAVE WE LEARNED FROM GAZE PATTERNS ABOUT HOW VISION IS USED TO REGULATE LOCOMOTION

These studies have provided unique insights into how vision is used to regulate human locomotion. Gaze patterns during steering for example have allowed us to suggest which proposed strategies for steering control are not necessary and which strategies can be used (Hollands et al., 2002). A major problem addressed in many psychophysical studies of heading judgements from optic flow is how the nervous system isolates a translational flow component from the retinal flow which also contains flow effects due to eye

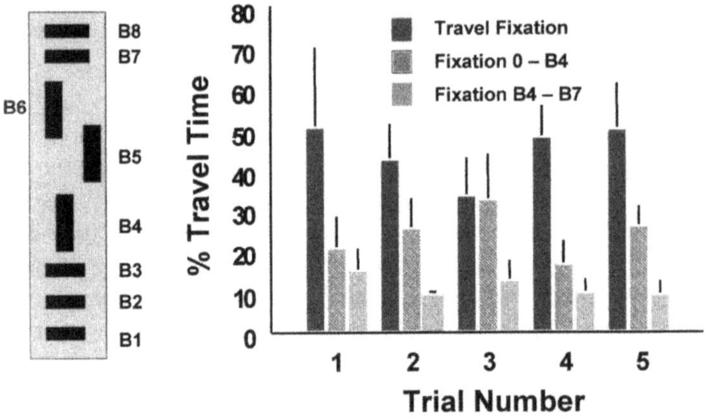

Figure 12. Changes in duration of travel gaze fixation and fixation on blocks following an error in foot placement (see trial 3), (from Vickers & Patla, 1999).

rotation (Lappe & Hoffman, 2000; Warren & Hannon, 1988). Since gaze patterns clearly showed an individual's eye and head are aligned with features lying in the direction of travel, no decomposition of the translational and rotational component (horizontal) is necessary. Fixation on objects other than the end target in the direction of travel will produce vertical eye rotations. This will shift the focus of expansion vertically; since the focus of expansion and heading direction remain in the same sagittal plane, no steering errors should result. Cats walking in an alley also show similar gaze behavior (see Figure 3), (Sherk & Fowler, 2000). Also, during driving similar gaze patterns are seen (see Figure 2), (Mourant et al., 1969). While optic flow can be used to guide steering control, by keeping the difference between focus of expansion and body orientation to zero, it is also possible that individuals were using egocentric control of walking as shown by Rushton et al. (1998) since they were fixating on the end-goal before changing their direction.

Fixation patterns on the obstacle during obstacle avoidance tasks have provided insights into the information necessary for these tasks. Successful obstacle avoidance requires information about obstacle height and shape to plan appropriate limb elevation and reach, and continuous information about the position of the obstacle relative to the person to plan step length adjustments for proper foot placement before the obstacle (Patla, 1999). It is likely that obstacle fixation earlier in the approach phase allows the person to detect relative spatial information. One possibility is that estimation of time-to-contact judgement from the retinal image expansion of the obstacle, coupled with speed of travel from self-motion information, can be used to determine the location of the obstacle relative to the person. Because the obstacle is fixated from a distance, the vertical rotation of the image would be small. Obstacle features could be extracted from optic flow patterns during both obstacle and travel gaze fixation.

The study investigating gaze patterns during visually guided stepping (Patla & Vickers, 2003) showed infrequent and inconsistent fixation on landing targets whether or not they are placed at regular distance or irregular distance. When participants fixate on the landing targets, they did so two steps ahead most often. This would give them sufficient time to plan and modify limb trajectory to land on the target. Hollands et al. (1995) in contrast showed a more consistent eye movement to the stepping area. There are two possible explanations for the different results. First, in their study subjects were required to step on raised stepping-stones. This places greater demand on the balance control system, and an error in foot placement can result in injury or a fall. As discussed before, gaze patterns are influenced by the relative risk during the locomotor task (Figure 12). Fixation on the landing target allows the individual to use egocentric control of limb trajectory, but more often they are using optic flow generated from travel gaze fixation episodes to guide their limb accurately to the landing target. The second reason for the differences between our work and the work by Hollands et al. (1995) could be attributed to the visual environment; in their study the environment was not as visually rich. Warren et al. (2001) have shown that in a visually impoverished environment, individuals are unable to use optic flow to control steering but when other features are added to the environment optic flow is used. Our work on gaze patterns during visually guided stepping shows the use of a combination of visuo-motor strategies to control locomotion.

5 CONCLUDING REMARKS

The study of gaze patterns during adaptive human locomotion is in its infancy. We need to explore greater variations in environmental challenges in travel path, both static and dynamic, and monitor gait movements in greater detail.

ACKNOWLEDGEMENTS

The work was supported by a grant from NSERC, Canada. The author acknowledges the contributions by Drs. Vickers and Hollands to the work discussed in this chapter.

REFERENCES

Demer J. L., & Amjadi F. (1993). Dynamic visual acuity of normal subjects during vertical optotype and head motion, Invest. *Opthalmol. Vis. Sci., 34(6),* 1894-1906.

Fowler, G. A., Soden, C., Brush, S., & Sherk, H. (2002). Gaze during visually guided locomotion, *Soc. Neurosci. Abs., 26.*

Gibson, J. J., & Crooks, L. E. (1938). A theoretical field analysis of automobile driving, *Am. J. Psychol., 51,* 453-471.

Gibson, J. J. (1950*). The Perception of the Visual World.* Boston: Houghton-Miflin.

Gibson, J. J. (1958). Visually controlled locomotion and visual orientation in animals, *Br. J. Psychol., 49,* 182-194.

Hollands, M., & Marple-Horvat, D. (1996). Visually guided stepping under conditions of step cycle-related denial of visual information, *Exp. Brain. Res., 109,* 343-356.

Hollands, M., Marple-Horvat, D., Henkes, S., & Rowan, A. K. (1995). Human eye movements during visually guided stepping, *J. Mot. Behav., 27,* 155-163.

Hollands, M. A., Patla, A. E., & Vickers, J. N. (2002). "Look where you're going!": Gaze behavior associated with maintaining and changing the direction of locomotion, *Exp. Brain Res., 143,* 221-230.

Imai, T., Moore, S. T., Raphan, T., & Cohen, B. (2001). Interaction of the body, head, and eyes during walking and turning, *Exp. Brain Res., 136(1),* 1-18.

Imai, T., Moore, S. T., Raphan, T., & Cohen, B. (2001). Posture and gaze during circular locomotion, *Ann. N. Y. Acad. Sci., 942,* 470-1.

Land, M. F. (1992). Predictable eye-head coordination during driving, *Nature, 359,* 318-320.

Land, M. F., & Lee, D. N. (1994). Where we look when we steer, *Nature, 369,* 42-744.

Lappe, M. & Hoffman, K-P. (2000). Optic Flow and Eye Movements. In: Lappe, M. (ed.), *Neuronal Processing of Optic Flow* (pp. 29-47), Academic Press.

Milner, A. D., & Goodale, M. A. (1995). *The Visual Brain in Action.* New York: Oxford.

Mourant, R. R., Rockwell, T. H., & Rackoff, N. J. (1969). Driver's eye movements and visual workload, *Highway Res. Rec., 292,* 1-10.

Patla, A. E. (1997). Understanding the roles of vision in the control of human locomotion, *Gait Post., 5,* 54-69.

Patla, A. E., & Vickers, J. N. (1997). Where and when do we look as we approach and step over an obstacle in the travel path? *NeuroReport, 8,* 3661-3665.

Patla, A. E. (1999). How is human gait controlled by vision? *Ecol. Psychol., 10 (3-4),* 287-302

Patla, A. E., & Vickers, J. (2003). How far ahead do we look when required to step on specific locations in the travel path during locomotion, *Exp. Brain Res., 148,* 133-138.

Rushton, S. K., Harris, J. M., Lloyd, M. R., & Wann, J. P. (1988). Guidance of locomotion on foot uses perceived target location rather than optic flow, *Curr. Biol., 8,* 1191-1194

Sherk, H. & Fowler, G. A. (2000). Optic Flow and the Visual Guidance of Locomotion in the Cat. In: Lappe, M. (ed.), *Neuronal Processing of Optic Flow* (pp. 141-167), Academic Press.

Sherk, H., & Fowler, G.A. (2001). Neural analysis of visual information during locomotion, *Prog. Brain Res., 134,* 247-264

Shinar, B. (1978). *Psychology on the Road.* New York: Wiley.

Solomon, D., & Cohen, B. (1992a). Stabilization of gaze during circular locomotion in darkness: II. Contribution of velocity storage to compensatory head and eye nystagmus in the running monkey, *J. Neurophysiol., 67(5),* 1158-1170.

Solomon, D, & Cohen, B. (1992b). Stabilization of gaze during circular locomotion in light: I. Compensatory head and eye nystagmus in the running monkey, *J. Neurophysiol., 67(5),* 1146-1157.

Trevarthen, C. (1995). Mother and Baby - Seeing Artfully Eye to Eye. In: Gregory, R., Harris, J., Heard, P. & Rose, D. (Eds.), *The Artful Eye.* New York: Oxford University Press.

Turano, K. A., Geruschat, D. R., Baker, F. H., Stahl, J. W., & Shapiro, M. D. (2001). Direction of gaze while walking a simple route: persons with normal vision and persons with retinitis pigmentosa, *Optom. Vis. Sci., 78(9),* 667-75.

Vaina, L. M., & Rushton, S. K. (2000). What Neurological Patients Tell Us About the Use of Optic Flow. In: Lappe, M. (ed.), *Neuronal Processing of Optic Flow* (pp. 292-309), Academic Press.

Vickers, J., & Patla, A. E. (1999). Object and travel gaze fixation during successful and unsuccessful stepping stone task, *Gait Post., 9(1),* S3.

Wagner, M., Baird, J. C., & Barbaresi, W. (1981). The locus of environmental attention, *J. Environ. Psychol. 1,* 195-206.

Wann, J. P., & Swapp, D. K. (2000). Why you should look where you are going, *Nat. Neurosci. 3(7),* 647-648.

Warren, W. H. & Hannon, D. J. (1988). Direction of self-motion is perceived from optical flow, *Nature, 336,* 162-163

Warren, W. H. (1999). Visually controlled locomotion: 40 years later, *Ecol. Psychol. 10(3-4),* 177-219.

Warren, W. H., Kay, B. A., Zosh, W. D., Duchon, A. P., & Sahuc, S. (2001). Optic flow is used to control human walking, *Nat. Neurosci., 4(2),* 213-216.

Yarbus, A. L. (1967). *Eye Movements and Vision.* New York: Plenun Press.

18. How Do We Control High Speed Steering?

John P. Wann and Richard M. Wilkie

School of Psychology
University of Reading, Reading, UK

1 INTRODUCTION

We routinely travel at high speed, in a car or on a bicycle, and also steer complex paths with relatively little conscious processing or explicit procedures as to how we 'judge a bend'. The consequences of an error, however, could be considerable. Within this chapter we tackle the issue of how optic flow and other sources of information can enable locomotor animals to steer effectively. Although there has been a strong body of research into how we might judge locomotor *heading* we will argue that this does not equate to active locomotor control and that there is relatively little research into effective steering, despite the latter being the ecological skill that all locomotor animals need to achieve. We have written this chapter in a tutorial style to try and make a difficult field accessible to undergraduate and postgraduate students. In this respect we do not attempt to cite every contribution on the use of optic flow or other information sources, but provide some basic background and then concentrate on the components that we believe can be linked into a coherent account of high speed steering.

1.1 Optic Flow, Retinal Flow and Heading

In a seminal paper in the British Journal of Psychology, Gibson (1958) tabled the agenda that vision research should be tackling ecological tasks such as steering & aiming; approaching without collision; and steering amongst obstacles. He proposed that the task of steering could be accomplished by using properties of the optic flow field. A particular assertion was that "*the center of the flow pattern during forward movement of the animal is the direction of movement*" and "*to aim locomotion at an object is to keep the*

L.M. Vaina, S.A. Beardsley and S.K. Rushton (eds.), Optic Flow and Beyond, 401–419.

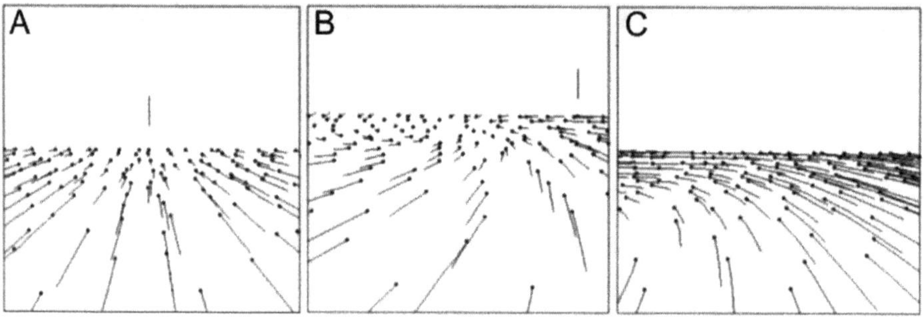

Figure 1. Retinal flow patterns for a) Radial Flow with gaze fixed forward, actual destination indicated by the vertical line above the horizon. b) The same straight line trajectory as in (a), but with gaze fixed on a ground feature to the left of the path, so on the retinal the destination is displaced to the right and there is a local focus of expansion at the point of gaze. c) A curved path with fixed gaze demonstrating locomotor flow lines.

center of flow of the optic array as close as possible to the form which the object projects", (p. 187). These statements are true for *optic flow* when the observer is on a straight path, in which case the flow field will be radial with a clear focus of expansion (FoE) in the direction of heading (Figure 1a). These principles are suitable for *maintaining* heading towards a particular goal, when the observer has a stable gaze position. The problem that arises in ecological tasks is that humans, and many other locomotor animals, have fast, mobile gaze systems that are used to sample from the environment and fixate future targets during locomotion. For a mobile eye the fundamental problem is that *optic flow* (as discussed by Gibson) is not equivalent to the *retinal flow* that will be received at the back of the observer's eye. The classic description of the problem is that if the observer is traveling on a straight-line path so the optic flow is radial, but fixates a feature to the side of locomotor path, then the flow field on the retina is not radial (Figure 1b). It is of particular note that the FoE at the point of heading is now masked and a new local expansion point occurs at the point of fixation (Regan & Beverley, 1982). To recover optic flow from the retinal pattern, processing would be required to subtract the gaze rotation component, render an unbiased flow field with an identifiable FoE and thereby provide an estimate of *heading.* The premise is that once optic flow is recovered the control strategy would be one of keeping the focus of expansion on the target object in order to maintain your heading towards it.

The task of recovering heading from retinal flow has been the main focus of research in this area during the last 15 years. The approaches may be crudely divided into those that propose the use of decomposition algorithms or template methods (Warren & Hannon, 1988; 1990; van den Berg, 1993; Stone & Perrone, 1997; Li & Warren, 2000) and those that propose the extraction of the gaze rotation component using extra-retinal signals specifying the head and eye motion (Royden et al., 1994; Banks et al., 1996; Crowell et al., 1998). The main difference between the two approaches is the extent to which the eye-movements are incorporated. Decomposition algorithms work solely upon the retinal image, relying upon contrasting properties of the rotation and translation components of the flow field to recover optic flow. These methods are limited by the quality of the retinal information and have problems with some potentially ambiguous flow fields. In contrast extra-retinal signals provide a direct measure of the magnitude of motion, but this would require muscle motor commands to be combined with visual information.

The majority of heading research has been applied to the case of a straight-line path. Similar issues do arise for curved trajectories (Figure 1c), and while there is no FoE for optic flow it may be possible to discern 'locomotor flow lines' (Lee & Lishman, 1977). A simple generalization of Gibson (1958) would be that you could steer by keeping the target object centered within the locomotor flow lines, irrespective of whether they are radial (linear path) or curved. But once again if the observer fixates an object in the direction of travel the locomotor flow lines will be biased by gaze rotation. In this case the use of decomposition algorithms is more problematic because there is a potential confusion between the rotation in the flow field that is due to the curvature of the path and the component produced by gaze rotation.

In summary if a driver or cyclist moves his/her eyes during locomotion then this masks any clear specification of the direction of travel in the retinal flow field. This has led to a prolonged debate as to how human participants can recover their direction of heading from retinal flow. The problem with this debate is that it has taken a tangent to Gibson's original agenda. Rather than addressing the issue of active steering using optic flow, it has focused on the recovery of instantaneous heading. Even if the observer is able to recover their instantaneous heading from retinal flow it is still the case that to *steer* they need to implement a procedure to change that heading and perceive whether the rate of change of heading is sufficient to achieve their goal. In this respect perceiving heading can at best be an intermediate stage in the control of locomotion. Our argument will be that it may not be necessary at all, and that active steering can progress without the need to recover heading per se.

In reviewing the previous research on optic/retinal flow Lappe et al. (1999) posited that the questions for future research were: How is path information obtained and how is the path predicted? How is optic flow combined with other navigational strategies? How are eye movements actively used? This chapter outlines the retinal or extra-retinal information that can specify the future path, and how gaze fixation can be used in combination with retinal flow and extra-retinal information to provide a robust steering mechanism.

2 IS FLOW NECESSARY FOR STEERING?

Although there is a strong body of research into the perception of heading from flow, considerably less research has been directed toward exploring the use of retinal flow and/or extra-retinal information in steering. The role of flow in active steering has been questioned: Rushton et al. (1998) demonstrated that participants asked to walk to a target, which was viewed through a prism, took a distinctive curved path. The prism caused the target to be shifted in relation to the observer's head, but the pattern of flow was unchanged by the prism since all scene elements were shifted equally. In this situation, it is predictable that participants set off in the wrong direction. Before they started moving the angle between their body axis and direction of gaze (Visual Direction: VD) was the only information they had regarding the placement of the target, and this was displaced by the prism. Once they were in motion, however, the retinal flow emanating from the fixated target would have been weakly curved, due to gaze rotation, indicating that they were not walking towards the target. If participants were able to use flow information they should have been able to fully compensate for the error and then walk straight towards the target. Instead, Rushton et al. (1998) observed participants fixating the target and then closing down the angle between their body axis and direction of gaze, hence VD seemed to be the primary source of information in walking directly towards a goal. Based on this and a number of follow-on studies (Rogers & Dalton, 1999; Wood et al., 2000; Harris & Bonas, 2002), Harris & Rogers (1999) stated, "*We challenge flow researchers to provide some compelling evidence for a significant role of optic flow in the control of the direction of locomotion on foot*", (p. 449). In response Warren et al. (2001) conducted equivalent walking experiments in a large scale virtual environment, where they could manipulate the quality of optic flow, and proposed that VD and optic flow are both used, but the reliance on either depends upon the strength of optic flow. Warren et al. responded to the challenge by stating, "*contrary to previous claims humans indeed make use of optic flow to walk to a goal*", (p. 216).

In evaluating this debate it is useful to note that in the prism experiment the curved trajectory results because the participant is iteratively adjusting walking direction so that the displaced image appears directly ahead, so from the participants viewpoint the trajectory was 'straight towards the target'. In an everyday steering task, with no prism displacement, the equivalent strategy would be to register the visual angle of the target and then pivot around to cancel that angle so a straight-line trajectory can be taken. This is a very simple solution to some locomotor steering tasks, but it is not a general solution. Most vehicles have a wheelbase that restricts the turning arc, and even in running or skating, momentum makes a curved trajectory necessary and precludes an instantaneous target alignment. In other cases (e.g. a roadway) there may be a course that precludes a direct line approach that has to be followed to avoid obstacles and reach the target. The VD strategy and heading-from-flow strategies that have been debated only provide solutions for the rather simplistic task of walking in a straight line towards a target at low speeds (1-2 m/s). For a more general solution we consider how VD or optic flow might be used to plan and execute curved trajectories, including those of minimal curvature (i.e. almost straight) during higher speed approaches.

The VD strategy can be used for curved trajectories. Land & Lee (1994) and Land (1998) proposed that car drivers fixate the tangent point of a bend (apex of the inside curb) and use the angle of the tangent point to estimate the degree of steering required

$$Curvature = 1/R = \theta^2/2x \tag{1}$$

where R is the radius, θ is the angle of gaze (visual direction relative to the locomotor axis) and x is the distance of the inside of the curve from the intended path. The tangent point can then be used to maintain a curved trajectory that will keep them equi-distant from the curb, and hence within their traffic lane, by keeping the tangent point at a constant VD angle. The gaze velocity (change in gaze angle $d\theta$ divided by the change in time dt) therefore would be zero

$$d\theta/dt = 0 \tag{2}$$

If the gaze velocity is greater than zero this would indicate an understeer whereas $d\theta/dt < 0$ would indicate an oversteer. This technique uses the visual direction of a single road feature and does not require any contribution from global optic flow. This strategy, however, does not provide a solution to the general case where locomotion is in a car park, field or forest where there is no explicit guide to your path such as a curb or white line. Wann & Land (2000) proposed a variant of the tangent point strategy that would be applicable for roads that do not have a clear tangent point, or if the driver wishes to "cut-the corner", or for open field steering. If the observer fixates a point at a distance ahead (d_f) that they wish to pass through then a constant curvature path that will take them to that point is specified by

$$Curvature = 2 \sin \theta/d_f \tag{3}$$

If this trajectory is maintained then as the observer progresses toward the target then

$$d\theta/dt = V/2R \tag{4}$$

where V is the locomotor speed (tangential velocity) and R the radius of curvature. Hence if locomotor speed is constant then $d\theta/dt$ is constant for an appropriate trajectory. The observer does not need to know V or R, they simply need to recognize that for an appropriate 'safe' trajectory the fixated point will sweep from its initial offset, to directly in front of the locomotor axis, at a constant rate. If it accelerates towards their midline then they are oversteering, if it slows, or moves away, then they are understeering. In summary equations 1-4 provide solutions to steering a range of curved paths, using fixation of a single environmental feature, without the use of optic flow. It is therefore appropriate to ask what optic flow can provide in addition to egocentric visual direction.

2.1 Retinal Flow, Gaze Fixation and Steering

Although there has been a considerable amount of research into the perception of heading, there appears to have been an assumption that the recovery of heading equates to locomotor control. But as discussed earlier there have been very few proposals as to how perceived heading is used to control steering. In addition, as highlighted by Lappe et al. (1999), there is a dearth of research investigating how eye-movements and gaze fixation can play an active part in steering judgments. The classic heading studies have treated gaze fixation as a confounding variable rather than an active component of control. We propose that "where you look" is an essential component of visually guided steering. A natural gaze response when steering a bend is to look ahead to your future path and advanced driving manuals recommend this strategy. Fixation of a ground feature that you wish to pass through that is eccentric to your current trajectory will introduce a gaze rotation component into the retinal flow field (Figure 2a). Rather than subscribe to the theme that the observer has to recover heading from this flow field[1], we propose that a simpler solution is to consider the raw retinal flow pattern that arises if the trajectory is appropriate for the steering task (Kim & Turvey, 1999; Wann & Swapp, 2000). If the observer fixates a point on the

[1] The retinal flow field could be decomposed into rotation and translation components, to recover heading. The recovery of the translation component, however, cannot indicate whether the current rate of change of heading is sufficient to achieve the intended trajectory, whereas the rotation component includes both the locomotor rotation and the gaze counter-rotation. So it is not clear that flow decomposition is particularly useful for the task of steering.

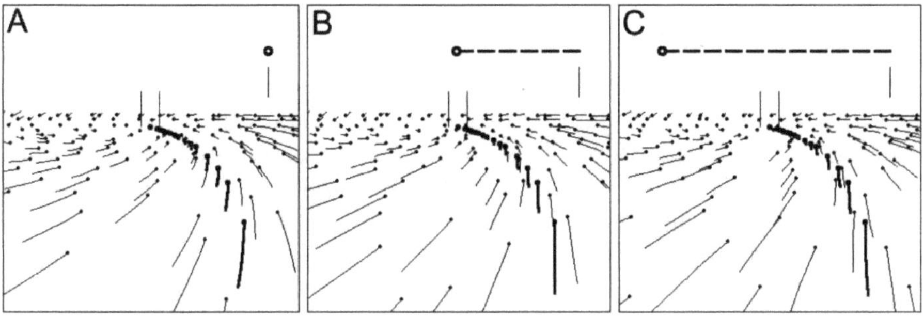

Figure 2. Retinal flow patterns for different steering responses. Simulations display the flow arising over 0.1s while traveling at 30 mph and fixating between the pair of vertical posts, which represent a gate that the observer wishes to steer towards. Initial heading is indicated as a vertical line above the horizon with a horizontal dashed line indicating changing heading due to steering. a) If the observer understeers and is on an eccentric heading, but fixating the target gate then the pattern of curvature is centered around the point of fixation (as in Figure 1b). b) If steering is set to a circular course that will pass through the gate, and fixation is maintained on the target, the flow lines are straight but asymmetrically displaced. c) If the observer oversteers, such that the future trajectory will pass in front of the target then the flow is still centered around the fixation point, but displays curvature in the opposite direction to the understeer and the curvature is correlated with the degree of steering error. See Wann & Swapp (2000) for further details.

intended path and adjusts steering to a curved path that will pass through that point (e.g. using Eq. 3) then the flow-lines for ground elements will remain straight, but will move outwards asymmetrically from the observer's future path (Figure 2b). Points that lie on the observer's future path move vertically in the projected field. If the observer is under-steering, the flow-lines curve away from the direction of steering error and over-steer is reflected in an opposite change in flow curvature (Figure 2a,c).

Wann & Land (2000) proposed that if the observer fixates a point on the future path then a set of principles hold that extend Gibson's original assertions about the use of optic flow to the use of retinal flow (RF):

RF1: If the retinal flow-lines for ground elements curve then you are not on a path to where you are looking and your steering error is in the direction opposite to the flow curvature

RF2: Once you are on a curved path to the point of gaze the flow-lines will be straight, but will be distributed non-radially.

RF3: If you are on a linear path to the point of gaze the flow-lines will be straight and will be distributed radially.

RF4: In both RF2 and RF3, points, which move with pure vertical motion in retinal flow, indicate the future path for the current steering angle (Figure 2b).

Irrespective of whether the trajectory is straight or curved to the point of gaze, RF4 suggests that the observer may be able to detect the future path by elements that move with pure vertical motion, or equivalently by the zero-crossings for horizontal flow[2]. This proposal has been questioned by van den Berg et al. (2001), who commented that in passive tasks observers have perceived changes in path when transformations that do not disrupt the vertical motion properties have been applied to the flow field. There needs to be further research to investigate, in an active control setting, whether there is sensitivity to flow-line shape, over and above the global speed and motion of retinal flow.

We do not propose that the everyday driver explicitly knows principles RF1, RF2, RF3 & RF4, but they do reflect simple principles that any skilled driver or cyclist will recognize. Consider some distinct components of ground flow, such as the near-side curb and the white line: During a safe trajectory around a bend, principle RF2 specifies that their projection on the retina should move outwards to either side of the drivers intended trajectory with no change in their direction of motion. If either of them starts to curve and break inwards we propose that any driver with a modicum of experience would recognize a steering error (principle RF1) that will result in the vehicle crossing the white line or curb in the near future.

2.2 Using Retinal Flow, Visual Direction & Gaze Motion Signals

There is a notable similarity between the strategy we have proposed for the use of visual direction (VD) and the strategy for the use of retinal flow. With VD we proposed that if the observer fixated a point on their intended path, the gaze angle can specify the degree of steering required (Eq. 3) while the rate of change of that angle can confirm if steering is appropriate or if there is under/over-steer (Eq. 4). For the same situation, however, we have argued that retinal flow can also provide some indication of the steering required and can confirm if steering is appropriate or if there is under/over-steer. There is redundancy in the information available through VD and retinal flow and the use of both sources of information should ensure a robust control system across a range of steering tasks and environmental conditions.

---

[2] The zero-crossing is the point where left/right motion changes sign and vertical motion would occur. In principle it is not essential to have texture elements that move with pure vertical motion. The notion of "zero-crossings" is that the observer may perceive the path as being midway between a rightward flow vector and a leftward flow vector.

Figure 3. A car supplies strong visual references for the locomotor axis in the form of the mirror, windscreen and dash.

We propose that VD information can be considered as having two main components. When there is a clear visual indication of the locomotor axis, such as your body or vehicle, this can provide a retinal reference for the target angle (RRD: Retinally Referenced Direction). If there is no visible reference for your body/vehicle, but the observer fixates the target, then the target angle could be recovered from somato-sensory information about head & eye direction relative to the body (GAD: Gaze angle direction).

Consider the differences in the information available when steering a car, motorbike, bicycle or roller-blades. In a car there is a very strong visual reference for the locomotor axis. The VD of a steering target can be coded with respect to the windscreen surround or the front of the car (Figure 3). In such a vehicle the steering task can be reduced to a pure VD strategy of bringing the target to the center of the windscreen. A curious situation occurs with a motorbike: The use of a full-face helmet provides a peripheral visual frame, similar to the car windscreen, but the frame moves with the head, providing a false reference (Figure 4). On larger bikes, however, there are still relatively strong references to the direction of locomotion provided by the mirrors, dash and screen of the bike. In contrast to this, when riding a bicycle, all visual references (such as the handlebars) will be very low in the peripheral field (Figure 5). It is debatable as to whether a horizontal feature in the peripheral field can provide a strong retinal reference for visual direction (RRD). A front bicycle light illuminating the path could provide limited information, but this information is not consistently yoked to the locomotor axis. An alternative strategy for the cyclist is to use proprioception to sense their head angle relative to their body and the angle of the handlebars relative

Figure 4. Left Panel - A motorcycle approaching a bend. Note the driver's head is offset from his locomotor axis supplying GAD information. *Right Panel* – The view from the motorcycle with the point of gaze marked as a black disc. In addition to the visual reference supplied by the windshield and mirror, the helmet both limits the field of view and supplies a visual reference that is fixed to the head angle not the locomotor axis.

to their body and thereby recover an estimate of target angle from somatosensory (gaze-angle) information. This requirement is amplified with something like roller blades, where vision of your own body is in the far periphery and extra-retinal, proprioceptive information should be of primary importance. We argue therefore that in steering a car, motorcycle, bicycle or roller blades the quality of VD information will vary and more or less emphasis may be placed on non-visual sources of information (GAD).

In principle some optic flow is always available, but this will also vary across environmental conditions. Quantifying flow is not straightforward since the optimal texture density to generate flow depends upon the speed of locomotion and the eye height in the world. In principal the higher the contrast between adjacent elements the better, but conventional tarmac is too fine grained to provide rich optic flow at speed, and a driver on a wide highway or racetrack may have considerably less flow information available than a driver on a narrow paved street. As night falls the quality of optic flow will also change for the pedestrian or cyclist and the erratic beam of a bicycle headlight does little to offset the loss of ambient light. So although modern highway designs have introduced devices to enhance ground-flow, such as retro-reflective cats-eyes and road markings, we have evolved to cope with

Figure 5. Left Panel - A bicyclist approaching a bend. Again the cyclist's head is offset from her locomotor axis supplying GAD information. *Right Panel* – The view from the bicycle with the point of fixation marked with a black disc. There is more optic flow available to the cyclist since it is not obscured by a helmet or the vehicle body, and peripherally the handlebars only supply a weak retinal reference (RRD) for the locomotor axis.

environments where the quality of flow may change and we still encounter these in everyday situations.

To explore the weighting attached to either retinal flow (RF), retinally referenced direction (RRD) or gaze angle direction (GAD) Wilkie & Wann (2003) created a virtual environment steering task where each information source could be manipulated. The participant's task was to steer a smooth, curved trajectory (speed: 18 mph) to a pair of target gates, so the retinal flow and VD information discussed in this chapter was available to them. During some of the trials the texture on the ground plane was rotated around the target at a rate that was equivalent to them being on a different trajectory. Hence on the RF-bias trials they received veridical RRD and GAD information, but the RF information suggested that they were either over-steering or under-steering. The participants displayed steering errors in the direction predicted by their use of RF as steering information. Wilkie & Wann also manipulated the availability of GAD information and the availability of RRD information such as a windscreen surround. Key findings were:

i) When Retinal flow (RF) was severely degraded, participants could steer accurately using gaze angle direction (GAD) information from somatosensory signals (head and eyes).

ii) Adding a strong retinally referenced direction (RRD) information source such as a windscreen or hood-mascot resulted in very accurate steering.
iii) If a rotational bias was added to RF in condition (ii), however, errors confirmed that RF was still being used for steering control.
iv) RF was not sufficient for accurate steering when GAD was absent, or RRD information was in conflict.
v) Introducing a similar bias into RF or RRD on interleaved trials allowed the estimation of the reliance that the participants placed on each source. As a first approximation RF, RRD and GAD appear to be weighted equivalently.

Wilkie & Wann (2002) took this paradigm one step further and estimated the change in reliance on RF, RRF and GAD as the ambient lighting in the scene changed from day to night. As might be expected RF is relied upon less, and VD (GAD and/or RRD) more, as the light levels fall. But surprisingly, even when the night-time illumination made the ground texture very difficult to perceive in a static scene, there was evidence that the low-contrast flow pattern was still influencing steering when other sources of information were available. In the context of actively negotiating curved paths at relatively high speed we concur with Warren et al. (2001) that the reported demise of flow as a primary control variable in locomotion was premature.

3 A MODEL USING FLOW AND VISUAL DIRECTION DRIVEN BY ACTIVE GAZE

The model proposed by Wilkie & Wann (2002, 2003), follows the lead of Warren et al. (2001) and Fajen & Warren (2003) by using a point attractor system. A point attractor, in simplified terms, is a control system that acts like a spring:

$$F = k(x - x_o) \qquad (5)$$

where F is the force generated by a spring with a current length of x, a natural rest length of x_o, and a stiffness defined by k. If $x > x_o$ then a restoring force is generated until $x = x_o$ and F drops to zero. An undamped spring such as this will respond very quickly but tend to overshoot and oscillate around its equilibrium position. By adding a second damping term b to the rate of response (speed: $\dot{x}$) the system will respond more slowly but reach its equilibrium position in a steady manner:

$$F = k(x - x_o) - b\dot{x} \qquad (6)$$

An example of a damped spring system such as this is a hydraulic door mechanism, that always returns the door to a closed position (an "attractor"), but does so without allowing the door to slam into the frame. Such systems can also be set up as repellors, such that the force F pushes the state away from a specified position. Mass-spring models such as this have been used to describe the behavior of neural systems such as agonist and antagonist muscle pairs (e.g. biceps and triceps). Fajen & Warren (2003) propose that such models can be used to describe the coupling between perception and action in steering. This is not proposing that there are any actual "springs" in the system, but that certain classes of information act to push the system towards a particular state or trajectory. Their model is quite complex and includes attractors (goals you wish to move to) and repellors (obstacles you wish to avoid), in simplified form this is equivalent to

$$\ddot{\theta} = -k_g\,(goal\ VD) + k_o\,(obstacle\ VD) - b\dot{\theta} \tag{7}$$

In this case the output of the system is not a force F, but a turning response specified by angular acceleration ($\ddot{\theta}$), where θ is the instantaneous heading and $\dot{\theta}$ is the rate of change of steering. In this case k_g and k_o tune the response strength of the goal and obstacle respectively. The system acts to cancel the offset angle for attractors (goals) while maximizing the passing angle for repellors (obstacles). For the Fajen & Warren model the locations of both goals (ψ_g) and obstacles (ψ_o) are referenced to instantaneous heading (θ). This system actually operates on the basis of the VD of the target or repellor, not flow, but flow may be required to recover instantaneous heading if there isn't a clear reference to the locomotor axis (e.g. Figures 3-5). To account for observed performance the VD angles ($\theta-\psi_g$ and $\theta-\psi_o$, respectively) are weighted by exponential functions that include goal distance (d_g) and obstacle distance (d_o). The resulting model is then equivalent to:

$$\ddot{\theta} = -k_g(\theta-\psi_g)(e^{-c1dg}+C_2) + k_o((\theta-\psi_o)(e^{-c3\,|\theta-\psi o|}))(e^{-c4dg}) - b\dot{\theta} \tag{8}$$

Although this model seems very complex, we should accept that neural systems can exhibit behavior that may appear complex when described mathematically. But there are limitations of the model in that it does require the recovery of heading (θ) and absolute distance of both the goal (d_g) and all obstacles (d_o). It is questionable as to whether we can access heading in an "instant", as opposed to after 0.5s of viewing (at which point a manoeuvre may have been completed). It is also debatable as to whether we can judge absolute distance with precision when targets are beyond 3-4m from the observer.

Our version of a point attractor steering model avoids these issues by operating on the rate of change of VD and/or the rotation component in retinal flow that arise when the observer fixates their intended target. We have already discussed that steering on the basis of VD requires the VD angle to be cancelled in a steady systematic manner (Eqs. 3, 4). We have also outlined that rotation in the retinal flow field indicates that the observer is not heading to the point of fixation (principle 1), so a goal is also to cancel the rotation component. The rotation components of VD and retinal flow are both functions of $V\sin\theta/D$, where V is the travel speed, d_f is the distance of the fixated target and θ the offset of the fixated target from the current direction of travel. So if inputs to the system are the rotation components of VD and flow, then there is no need to specify V, d_f or θ as these are reflected in the rotation rates. This may be represented as

$$\ddot{\theta} = k\,(\beta_1 RF + \beta_2 VD) - b\dot{\theta} \qquad (9)$$

where β_1 and β_2 are the respective weights or reliance placed on the two inputs. As discussed previously, VD may in turn be split into pure retinally referenced direction (RRD) when there is a visible reference for the target angle (Figure 3) and a non-visual gaze angle direction signal (GAD) when the observer must rely upon head, eye and body position information. A limitation for the model in Eq. 9 might be the extent to which an observer can accurately estimate the rotation component in retinal flow (RF) or even the rotation component of VD. Wilkie & Wann (2003) demonstrated that precise estimates are not required. The model will achieve a smooth curved trajectory even when the RF and VD inputs are crudely quantized into a 0–3 point integer scale[3]. We argue that the observer does not need to precisely estimate RF rotation, but just recognize between "on-course" (0) a small (1), medium (2) or major (3) steering error and whether it is an understeer (-ve) or oversteer (+ve) error. The retinal flow principles outlined in the previous section provide the basis for such judgments. The β weighting factors allow for adjustments in reliance on RF or VD as conditions change, such as the onset of night (Wilkie & Wann, 2002).

[3] Irrespective of what the rotation rate might be at the eye (deg/s) it will be encoded into different co-ordinate systems at the retina or by the head-eye GAD system. Irrespective of the units, we can scale the input down to integers of 3, 2, 1, 0, -1, -2, -3 and the system still steers effectively demonstrating that high sensitivity is not required by either the RRD or GAD system.

3.1 Application of the Model to Ecological Contexts

An important difference between the Fajen & Warren model, in Eq. 8 and Eq. 9, is that once the k and b parameters are set the trajectory that results from equation 8 is dictated by the location of the targets and repellors, not by the performer. The model we propose in Eq. 9 is an attractor to the *point of fixation.* This puts trajectory control in the eye of the beholder. The task for the driver is to fixate the future locations that he/she wishes to pass through at appropriate times during the forward trajectory. This provides the driver with the option of executing a safe lane-following strategy or taking a "racing-line" and cutting the corner. The skill is in identifying where the critical via-points lie on a path and when to direct attention to each in turn. In our model (Eq. 9) parameters k and b dictate the response speed and may reflect a combination of the characteristics of the vehicle and the performers input. A high k and low value of b, would equate to a driver who decides to snatch at the wheel, causing the car to pivot sharply towards the target, but at the risk of a loss of control and oscillation. The optimal values of k and b can be acquired through experience and will change as the performer adopts different modes of transport. For example a commuter bicycle has a very different steering geometry from a lightweight racing bicycle, a family saloon has different steering characteristics from a sports car. A driver may respond in a manner equivalent to a high k if they have failed to notice a turning (or are trying to impress a passenger), but the perceptual inputs to the system remain unchanged. So the selection of path fixations and response speed (k) allow the performer to tackle a bend in an aggressive or relaxed manner. Learning to take a racing-line in a car or on a bicycle[4] would be to learn the optimal path settings for characteristic bends.

How well does this model correlate with what we can observe, experimentally or anecdotally, about steering in natural contexts? The inclusion of both retinal flow (RF) and visual direction (VD) inputs is based on the findings of Wilkie & Wann (2002, 2003). We don't know of any evidence that rules out either of these inputs. The weighting factors (β) allow for environmental changes such as thick fog, where the reliance on global RF may be zero and the driver may just follow an illuminated feature. Gaze fixation as the mechanism for trajectory aiming would seem to conflict with

[4] The turning rate of a bicycle it dictated primarily by the lean of the rider with a small amount of handlebar adjustment. So on a bicycle, or on skis, $\ddot{\theta}$ results from the controlled shift in weight, rather than a steering wheel. The parameters k and b are still valid, the rate at which weight is shifted is critical to the stability of the bicycle, but there is the need for an additional control system that incorporates the maximum degree of lean that is possible at a given speed of travel.

the findings of Land & Lee (1994). In our model looking at the tangent point would cause you to steer towards the tangent point. The Land & Lee study, however, was conducted on a one-way road around a mountain (Arthur's Seat in Edinburgh). Hence there was no requirement for the driver to stay in the offside lane and because the inside verge of the road was steeply banked, it is difficult to differentiate between looking *at the apex* and looking *through the apex* to the most distant point on the path that was visible.

We have repeated a similar study in a simulation setting with open field bends, where the performer could look across the apex, and found that although participants looked to the tangent point for 64% of the time, they fixated their future path 32% of the time, as coded by naive observers (Wann & Swapp, 2001). Land & Tatler (2001) recently observed that racing drivers, on a closed circuit, did not fixate the tangent point but oriented to the bend with their head. This is generally consistent with our model, but suggests that they may be using VD as a stronger input than RF, which may in turn be due to the lower quality of flow that arises on a broad racing circuit with smooth tarmac and no central road markings. This is speculation until we have further data of eye-movements from natural contexts such as these. If drivers do habitually seek out points on their future path, we might suppose that they can identify these more effectively than their instantaneous heading, which we have suggested is not essential. Wilkie & Wann (2002b) did find that participants' ability to fixate a point on their future path was significantly more accurate than their ability to fixate their instantaneous heading. Finally, a control mechanism based on gaze fixation seems to be reflected in the advice given in advanced driver manuals (1992; 1998). These discuss orienting gaze on approaching a bend and also warn that if an unexpected obstacle is encountered, the motorcyclist/driver should not look at the obstacle, because they will hit it, but look to the side of the obstacle that they wish to pass. This seems very much in tune with the mechanisms we propose.

In summary there are multiple sources of information that can be used to control direction of motion when locomoting through the world. These can be broadly categorized as either retinal flow or visual direction information (both egocentric and exocentric), but all can be encoded as a rate of change of rotation, which directly informs the required degree of adjustment to the current course. This also allows multiple estimates from different information sources to be pooled providing robust steering estimates across varying visual conditions (Figure 6).

A series of articles and letters in Trends in Cognitive Sciences saw a heated debate over the roles of retinal flow and visual direction information in human locomotion (Harris & Rogers, 1999; Lappe et al., 1999a,b; Fajen & Warren, 2000; Wann & Land, 2000; Harris, 2001; Rushton & Salvucci, 2001; Wann & Land, 2001). This highlighted the variety of explanations that are available to describe the everyday activity of locomoting through the world.

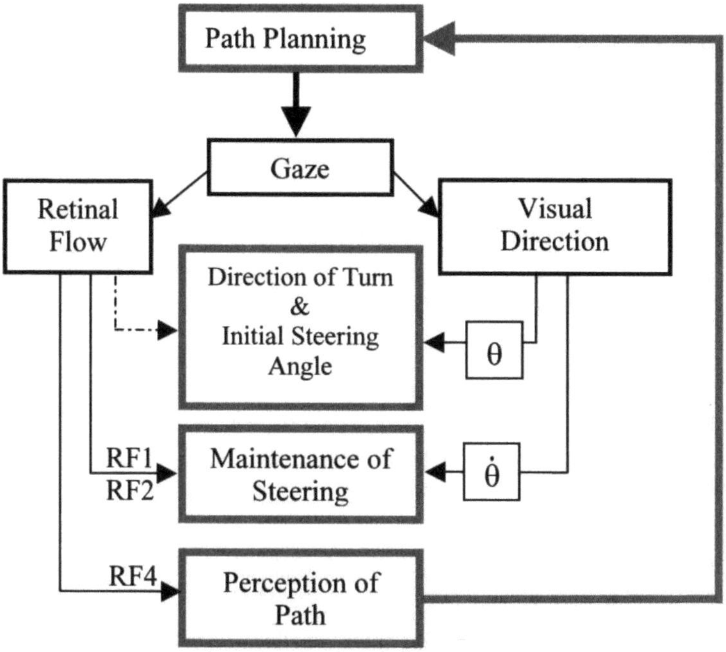

Figure 6. Potential inputs from retinal flow or visual direction for planning and controlling steering behaviour. RF1, RF2, RF4 indicate the use of the principles outlined earlier.

Significantly, there was considerable agreement over which information sources are used, even if there was not agreement as to the weighting attached or the actual method for using them. Many of the comments support the combined use of both retinal flow and visual direction information, and there was consensus that empirical evidence needs to be gathered to discover the true role if vision in steering.

The future agenda for investigating the control of locomotion is twofold:

i) to determine how we use gaze to sample from the world, and the way in which this simplifies the information received.
ii) to determine how the information that is made available to the visual system as a result of directed gaze can be used flexibly to control direction of motion.

ACKNOWLEDGEMENTS

This work was supported by a grant from the UK EPSRC

REFERENCES

Experienced Motorcycle Rider Course (1992). Motorcycle Safety Foundations Inc.

Road Craft: The Police Drivers Handbook (1998). HMSO, UK.

Banks, M. S., Ehrlich, S. M., Backus, B. T. & Crowell, J. A. (1996). Estimating Heading During Real and Simulated Eye Movements, *Vision Res., 36,* 431-443.

Crowell, J. A., Banks, M. S., Shenoy, K. V. & Andersen, R. A. (1998). Visual self-motion perception during head turns, *Nat. Neurosci., 1,* 732-737.

Fajen, B. R. & Warren, W. H. (2000). Go with the flow, *Trends Cogn. Sci., 4,* 319-324.

Fajen, B. R. & Warren, W. H. (2003). Behavioral dynamics of steering, obstacle avoidance, and route selection, *J. Exp. Psychol. Hum. Percept. Perform., 29(2),* 343-362.

Gibson, J. J. (1958). Visually controlled locomotion and visual orientation in animals, *Br. J. Psychol., 49,* 182-194.

Harris, J. M. (2001). The future of flow? *Trends Cogn. Sci., 5,* 7-8.

Harris, J. M. & Bonas, W. (2002). Optic flow and scene structure do not always contribute to the control of human walking, *Vision Res., 42,* 1619-1626.

Harris, J. M. & Rogers, B. J. (1999). Going against the flow. *Trends Cogn. Sci., 3,* 449-450.

Kim, N. G. & Turvey, M. T. (1999). Eye movements and a rule for perceiving direction of heading, *Ecol. Psychol., 11,* 233-248.

Land, M. F. (1998). The Visual Control of Steering. In: Harris, L. R. & Jenkins, M. (Eds.) *Vision and Action* (pp. 163-180). Cambridge University Press.

Land, M. F. & Lee, D. N. (1994). Where we look when we steer, *Nature, 369,* 742-744.

Land, M. F. & Tatler, B. W. (2001). Steering with the head: the visual strategy of a racing driver, *Curr. Biol., 11,* 1214-1220.

Lappe, M., Bremmer, F. & van den Berg, A. V. (1999). Going against the flow - Reply to Harris and Rogers, *Trends Cogn. Sci., 3,* 450.

Lappe, M., Bremmer, F. & van den Berg, A. V. (1999). Perception of self-motion from visual flow, *Trends Cogn. Sci., 3,* 329-450.

Lee, D. N. & Lishman, R. (1977). Visual control of locomotion, *Scand. J. Psychol., 18,* 224-230.

Li, L. & Warren, W. H. (2000). Perception of heading during rotation: sufficiency of dense motion parallax and reference objects, *Vision Res., 40,* 3873-3894.

Regan, D. & Beverley, K. I. (1982). How do we avoid confounding the direction we are looking and the direction we are moving? *Science, 215,* 194-196.

Rogers, B. J. & Dalton, C. (1999). The role of (i) perceived direction and (ii) optic flow in the control of locomotion and for estimating the point of impact, *Invest. Opthalmol. Vis. Sci. (Suppl), 40,* S764.

Royden, C. S., Crowell, J. A. & Banks, M. S. (1994). Estimating heading during eye movements, *Vision Res., 34,* 3197-3214.

Rushton, S. K., Harris, J. M., Lloyd, M. R. & Wann, J. P. (1998). Guidance of locomotion on foot uses perceived target location rather than optic flow, *Curr. Biol., 8,* 1191-1194.

Rushton, S. K. & Salvucci, D. D. (2001). An egocentric account of the visual guidance of locomotion, *Trends Cogn. Sci., 5,* 6-7.

Stone, L. S. & Perrone, J. A. (1997). Human heading estimation during visually simulated curvilinear motion, *Vision Res., 37,* 573-590.

van den Berg, A. V. (1993). Perception of heading, *Nature, 365,* 497-498.

van den Berg, A. V., Beintema, J. A. & Frens, M. A. (2001). Heading and path percepts from visual flow and eye pursuit signals, *Vision Res., 41,* 3467-3486.

Wann, J. P. & Land, M. (2000). Steering with or without the flow: is the retrieval of heading necessary? *Trends Cogn. Sci., 4,* 319-324.

Wann, J. P. & Land, M. (2001). Heading in the wrong direction, *Trends Cogn. Sci., 5,* 8-9.

Wann, J. P. & Swapp, D. K. (2000). Why you should look where you are going, *Nat. Neurosci., 3,* 647-648.

Wann, J. P. & Swapp, D. K. (2001). Where do we look when we steer and does it matter? *J. Vis., 1,* 185a.

Warren, W. H. & Hannon, D. J. (1988). Direction of self-motion is perceived from optical flow, *Nature, 336,* 162-163.

Warren, W. H. & Hannon, D. J. (1990). Eye movements and optical flow, *J. Opt. Soc. Am., 7,* 160-169.

Warren, W. H., Kay, B. A., Zosh, W. D., Duchon, A. P. & Sahuc, S. (2001). Optic flow is used to control human walking, *Nat. Neurosci., 4,* 213-216.

Wilkie, R. M. & Wann, J. P. (2002). Driving as night falls: the contribution of retinal flow and visual direction to the control of steering, *Curr. Biol., 12,* 2014-2017.

Wilkie, R. M. & Wann, J. P. (2002b). Looking to your future path: is heading off on a tangent? *J. Vision, 2(7),* 626a, http://journalofvision.org/2/7/626/, DOI 10.1167/2.7.626.

Wilkie, R. M. & Wann, J. P. (2003). Controlling steering and judging heading: retinal flow, visual direction and extra-retinal information, *J. Exp. Psychol. Hum. Percept. Perform.*, 29(2), 363-378.

Wood, R. M., Harvey, M. A., Young, C. E., Beedie, A. & Wilson, T. (2000). Weighting to go with the flow? *Curr. Biol., 10,* R545-R546.

19. Model-Based Control of Perception/Action

Jack M. Loomis and Andrew C. Beall

Department of Psychology
University of California, Santa Barbara, CA, USA

1 INTRODUCTION

The coupling of perception and action in humans and other species is one of the major questions in cognitive science and neuroscience. The empirical domain of perception/action includes reaching and grasping, skilled use of tools, visual control of locomotion (e.g., driving and flying), dance, and individual and team sports. Understanding of perception/action in the human requires research on perceptual and cognitive representations of 3-D space, internal representation of the body, motor control of the limbs and effectors, and internalization of vehicle dynamics. Like many others in the field, we believe that research on perception/action will advance more quickly in the future by giving greater emphasis to active control tasks over simple psychophysical judgments and by studying skilled actions of moderate complexity, like driving and sport activity, instead of the stripped down tasks of the past that have permitted experimental rigor at the cost of failing to engage many of the perceptual, cognitive, and motor processes involved in complex action.

2 THREE CONCEPTIONS OF PERCEPTION/ACTION

There are three prominent conceptions of how perception controls action. These are presented in Figure 1 in the context of visual control of action (see also Philbeck et al., 1997; Warren, 1998). The first of these, the "ecological" conception, originated with Gibson (Gibson, 1950, 1958, 1966, 1979) and has since been elaborated by many others (e.g., Flach & Warren, 1995; Lee, 1993; Turvey, 1977; Turvey & Carello, 1986; Warren & Wertheim, 1990; Warren, 1998). An important assumption is that the rich and multifaceted sensory

L.M. Vaina, S.A. Beardsley and S.K. Rushton (eds.), Optic Flow and Beyond, 421–441.

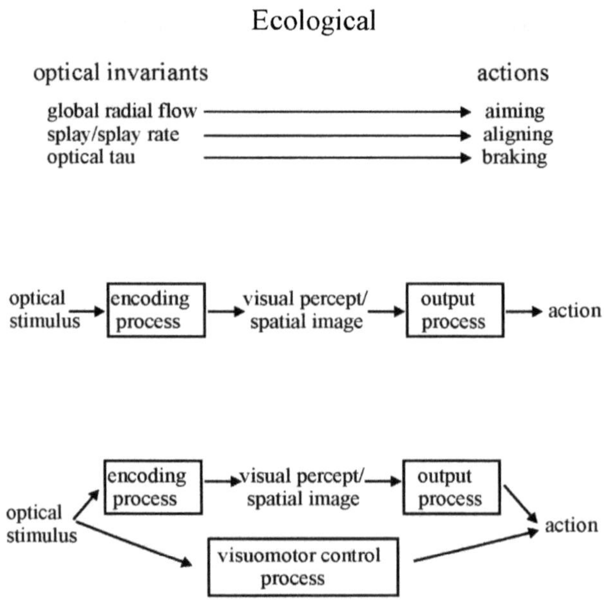

Figure 1. The three conceptions of how visual perception controls action.

stimulation of a moving actor is sufficient for understanding the control of action (i.e., additional information, like that stored in the actor's memory, need not be brought to bear). Moreover, it is assumed that very specific aspects of this changing sensory stimulation are tightly coupled to particular aspects of the desired action. Thus, in connection with the visual control of locomotion, aiming toward a point target might be controlled by global radial outflow (e.g., Warren, 1998; Warren et al., 1988), braking might be controlled by the temporal derivative of tau (Lee, 1976; Yilmaz & Warren, 1995), and aligning with a straight path might be controlled using splay (the angle between the perspective image of the path and environmental vertical) and its temporal derivative, splay rate (Beall & Loomis, 1997; Calvert, 1954; Riemersma, 1981). Specific optical variables that are sufficient for controlling particular actions are referred to as optical invariants. Strategies for utilizing these optical invariants are referred to as optic flow rules (e.g., Loomis & Beall, 1998) or laws of control (Warren, 1998). Some ecological researchers consider the mechanisms of perception and cognition and the internal representations resulting from them to be superfluous to functional explanations of action.

The second conception, referred to here as the "model-based" conception (Loomis & Beall, 1998; Warren, 1998), derives from the long tradition of research on space perception. In this conception, spatiotemporal stimulation arising from activity within the physical world is registered by the sense organs and processed by the central nervous system to yield an internal representation of the perceiver's environment. In the case of vision, the internal representation of a single target is the visual percept, and the visual representation of surrounding 3-D space is referred to as visual space. To generalize to all of the spatial senses (vision, audition, and touch), we use the corresponding terms of percept and perceptual space. These internal representations are much simpler in informational content than the spatiotemporal stimulation to which they correspond, for much of the stimulus information is lost in the perceptual processing. A great deal of research over the years has been directed toward establishing the correspondence between physical and perceptual space and toward understanding the sensory and perceptual processes underlying this representation. It is generally recognized that sensory stimulation, when available, is a primary determinant of perceptual space. Thus, identical patterns of sensory stimulation, whether arising from real objects or virtual objects (those simulated by a virtual reality system) ought to result in nearly identical percepts (e.g., Loomis, 1992). In addition, internal assumptions (such as smoothness of optic flow, smoothness of surfaces, and rigidity of moving objects) and expectancies based on past experience also serve as important determinants of perceptual space. Thus, when stimulus determinants are greatly reduced (as in dimly lit environments), perceptual space is still well defined even when discrepant with respect to physical space (e.g., Ooi et al., 2001; Philbeck et al., 1997).

In this conception, concurrent percepts play a causal role in the control of action (for a more extensive argument, see Philbeck et al., 1997). The actor plans his/her action with respect to perceptual space and observes the consequences of the action within it. Because of the close correspondence between physical space (or action space) and perceptual space, the actor is usually successful in carrying out the desired action. However, even when perceptual input is interrupted (e.g., by closing the eyes or blocking the ears), continued action towards some goal is still often successful, indicating that more abstract spatial images mediate such actions; in this case, these spatial images are the causes of behavior in the same way that percepts are causes of behavior when sensory input is available. Besides these internal representations of the physical world (percept and spatial image), the model-based conception depicted in Figure 1 postulates an internal representation in the output process that intervenes between the percept or spatial image and the action. More will be said about this later.

The third conception, termed “two visual systems”, has attained prominence in recent years because of the confluence of two trends: the increasing recognition of the functional modularity of mind (e.g., Fodor, 1983) and the deepening understanding of the functional architecture of the brain, especially those pathways dealing with vision (Milner & Goodale, 1995). Turvey (1977) was one of the first to express the view that visually-based action might be controlled by levels of the nervous system that are inaccessible to conscious awareness. More recently, evidence has been accumulating that conscious visual perception is sometimes dissociated from those processes controlling visually-based action. "Blindsight", the visually-based capacity for performing certain actions like pointing to and reaching for targets in the absence of visual experience but with visual input available, is perhaps the strongest evidence of dissociation (Weiskrantz, 1986). Milner and Goodale (1995) summarize much of the evidence from patients with brain lesions, including their own work with a brain damaged but sighted patient who can readily reach for visual objects that she cannot otherwise discriminate. Their research and other research like it (e.g., Bhalla & Proffitt, 1999) point to the existence of two distinct visual systems. The visuomotor control system acts on concurrent perceptual input and need not involve conscious awareness. The other system is the same as that discussed in connection with the model-based approach.

For our present purposes, we group visuomotor control with the ecological conception, for both operate solely on concurrent sensory input. Here, we will be most interested in the contrast between the idea of control of action by concurrent optic flow information and that of control by a 3-D representation of surrounding space that may be concurrent with sensory input (i.e., a percept) or not (i.e., a spatial image).

3 SIGNIFICANCE OF CHANGE BLINDNESS

Recent research on change blindness has shown that the impression of a perceptual space that is coherent over extended epochs and regions of surrounding space is illusory (Rensink, 2002); instead, the perceptual representation that is constructed concurrently with sensory input exhibits coherence only over very limited regions of space and time. However, to the extent that perceptual representations of surrounding space are involved in controlling action, change blindness implies that these representations must be limited in spatiotemporal scope. Even so, these perceptual representations must be sufficient, given the success of most perception/action. Thus, for an actor running over terrain, the actor's brain creates a representation of the ground surface "just in time", a representation that is usually sufficient to permit safe travel. The recent work of Land and Hayhoe (2001) on the patterns of eye fixations while performing mundane tasks is consonant with the idea that visual perception is highly selective in creating representations that are "just in time" and sufficient for the action being controlled.

4 ELABORATION OF MODEL-BASED CONTROL

The brief discussion above of model-based control focused on perception, the input component of the perception/action cycle. Here we elaborate on this input component as well as extend the idea of model-based control to the output side as well.

4.1 Input Side

4.1.1 Percept

Our starting point is the recognition that percepts (perceptual representations) of the surrounding environment mediate most interactions between the human actor and his/her environment. One of the profound ideas to come out of philosophical and scientific inquiry is the realization that the world we see, hear, and touch in everyday life is not the physical world itself but an amazingly functional representation of the physical world created by our senses and central nervous system (e.g., Lehar, 2002; Loomis, 1992; Russell, 1948; Smythies, 1994).

One convincing argument comes from the study of color vision. Research on color vision has made it abundantly clear that the tri-dimensionality of color is a consequence of visual processing rather than a property of spectrally varying light. The fact that color is part of the fabric of the surfaces in visual space is a strong clue that these surfaces are themselves representations of the corresponding physical surfaces. An interesting phenomenon that also makes this point is "apparent concomitant motion", the experience of seeing compelling motion in depth-reversed objects as the person moves his/her head side-to-side (see Gogel, 1990 for an explanation); the apparent movement of a concave mask of a human face is a familiar example but much more compelling are the 3-D paintings by the British artist Patrick Hughes that are extremely rich in visual perspective.

The idea that visual space is an internal representation can be extended to the other senses, as well as to the cognitive, social, and emotional realms. Accordingly, philosophers have referred to the world of everyday experience as the phenomenal world (e.g., Smythies, 1994). It includes all of perceptually external space, the phenomenal body (body image), and the province of subjective experience (Brain, 1951; Lehar, 2002; Loomis, 1992; Ramachandran & Blakeslee, 1998; Russell, 1948; Simmel, 1966; Smythies, 1994). The evidence for this view provided by perceptual science, neurology, and cognitive neuroscience is overwhelming. Because the phenomenal world is such a functional representation of the physical world, most laypeople and many scientists, even some perceptual scientists, fail to appreciate that contact with the physical world is indirect. Given that the phenomenal world can be so functional as to be self-concealing of its very nature, it is difficult to imagine that the phenomenal world is merely epiphenomenal. In a later section, we will present evidence that visual and auditory percepts in 3-D space are indeed causes of action.

4.1.2 Spatial Image

Even when visual, auditory, and haptic stimulation about the surrounding environment is temporally blocked, people continue to maintain some sort of internal 3-D representation of the environment that is of sufficient fidelity to support continued action for a short period of time (see section below on perceptually directed action). We refer to this representation as a spatial image. As the person moves through space, this representation is updated in much the way that the percept would be updated were sensory stimulation still available.

4.2 Output Side

The idea of model-based control of perception/action is not limited to the idea of an internal representation of the surrounding environment but applies to the output side as well. We begin by recognizing the complexity of motor control. Besides moving ourselves about under our own propulsion, we also perform many activities in which we maneuver self-propelled vehicles through space or in which we manipulate tools and other implements to achieve our goals. In many cases, feedback control is sufficient to perform the action (e.g., tracing a line with a pencil, steering a motorboat so as remain directly behind a leader moving with constant speed) but in many other cases, feedforward control is either essential (as when sensory feedback is temporarily lost) or is used to reduce tracking error (as when steering along a curving road) to a manageable level that allows for stable feedback control (McRuer et al., 1977) . Fluent and effective control of complex behavior requires that the human has internalized the dynamics of the body, vehicle, and/or tool (Godthelp, 1986; Godthelp & Käppler, 1988; Loomis, 1992; Loomis & Beall, 1998; McRuer et al., 1977; Miall et al., 1993; Wolpert & Ghahramani, 2000; Wolpert et al., 1995). It is obvious that the skill of performing gymnastic maneuvers, of manipulating a tool (e.g., using a scalpel in surgery, wielding a tennis racket in tennis play) or of controlling a vehicle (e.g., automobile, airplane, helicopter, vessel) comes only with extensive practice. In such cases, the skilled actor has created an internal model of the dynamics of the plant (body, vehicle, or tool). Once acquired, the flexible internal model can be modified by parametric adjustment to allow for transfer of the skill to other plants that are functionally similar (e.g., when transitioning from a light and maneuverable aircraft to a heavy and sluggish one).

5 LEVELS OF CONTROL

It is widely recognized that the control of locomotion, like the control of action more generally, is a hierarchy of control problems (Dickmanns, 1992; Lee & Lishman, 1977; McRuer et al., 1977); for each problem, a goal is set and actions are then taken to achieve that goal. With respect to controlling locomotion, McRuer et al. (1977), identified three levels of control: "navigation", "guidance", and "control". The uppermost level, "navigation", refers to the overall selection of a route within geographical space and will not be discussed further. The next lower level, "guidance" (McRuer et al., 1977) involves planning an optimal path to the goal based on current perceptual information and then attempting to follow that path. For steering a car along a curving road, McRuer et al. (1977) assume that guidance depends on perception of the road ahead and that control inputs to the vehicle are largely feedforward (see also Donges, 1978; Godthelp, 1986; Godthelp & Käppler, 1988; Land, 1998). Gibson and Crooks (1938) presented an intriguing analysis of guidance in connection with car driving employing the notion of the "field of safe travel", an internal representation of the open space in front of the driver that is momentarily suitable for forward progress (for a similar idea in the context of autonomous vehicles see Dickmanns, 1992). The lower level, referred to as "control" (McRuer et al., 1977) or "stabilization" (Donges, 1978), involves feedback control of the vehicle's motion relative to edges of the road. There are two obvious possibilities for this level of control corresponding to the model-based and ecological conceptions – either the actor is perceiving 3-D layout of the road and judging the location of the vehicle in relation to this 3-D representation or the actor is using optical invariants.

Although the two lower levels, guidance and control, were identified in connection with the control of locomotion, they obviously apply to many other forms of action (e.g. Patla, 1998). A basketball player intending to shoot a goal obviously uses feedforward control in planning how to approach the basket while using concurrent visual feedback to judge his/her movement with respect to the basket and with respect to other players. A dancer on stage has a plan of how and where to maneuver in space but uses visual information to monitor how well the executed action compares with what is intended. Finally, a person using heavy machinery to grab a large object and place it on top of another first plans the action and then uses visual feedback to monitor its execution.

6 PERCEPTUALLY-DIRECTED ACTION

In many circumstances, perceptual information is continuously available to provide closed-loop control of action. Contrasting with this is action that is carried out after the feedback loop with the surrounding environment has been opened (e.g., by occluding vision). An example is perceiving a target's location from a fixed vantage point (visual or auditory), and then attempting to walk to the target without further perceptual information about its location during the traverse. When vision provides the initial information, this type of open-loop behavior has been referred to as visually-directed action (Foley & Held, 1972). Perceptually directed action is the general term involving any of the spatial senses (Loomis et al., 1998).

Thomson (1980, 1983) was the first to do systematic research on visually directed action. He showed that, on average, adults walk blindly with little systematic error to targets up to 20m away. Since then, a number of studies have confirmed this basic result for vision using blind walking and related tasks when visual cues to distance are plentiful (e.g. Amorim et al., 1997; Corlett, 1986; Elliot, 1987; Elliot et al., 1990; Farrell & Thomson, 1999; Fukusima et al., 1997; Loomis et al., 1992, 1998; Rieser et al., 1990; Steenhuis & Goodale, 1988). In addition, several studies have shown that when distance information is inadequate for veridical perception, actors performing the same tasks exhibit systematic error consistent with visual distance errors measured in other ways (Ooi et al., 2001; Philbeck & Loomis, 1997; Philbeck et al., 1997). Other research has demonstrated perceptual directed action with auditory targets (e.g., Ashmead et al., 1995; Loomis et al., 1998) and with haptic targets (e.g., Hollins & Kelley, 1988).

A notable feature of perceptually directed action is its flexibility. It might be thought that a person performing perceptually directed action preprograms the motor response prior to moving and, once underway, simply executes the preprogrammed response ballistically; this is what Miall et al. (1993) refer to as control by an inverse dynamic model. Thomson (1983) showed ballistic execution of a motor program could not explain some types of perceptually directed action. Actors were visually presented with a target on the ground some distance ahead. While walking without vision toward the target, actors were cued at some point to stop and then throw a beanbag the rest of the way.

Accuracy was high even though they were unaware at which point they would be cued to throw.

Fukusima et al. (1997) provided similar evidence. On each trial of the experiment, the actor viewed a target and then walked without vision along a straight path in a direction oblique to the target. On command from an experimenter, the actor then turned and walked in the direction of the target. Because the actor did not know when to turn, the accurate terminal walking direction indicates that the actor could not have been preprogramming the response. Still other evidence comes from an experiment by Philbeck et al. (1997). Here actors viewed a target, after which they walked blindly to the target along one of three paths. Because they were not cued as to which path to take until after vision was blocked, the excellent updating performance indicates that actors were not preprogramming the response (Figure 2). As Philbeck et al. (1997) noted, the mode of locomotion toward a goal is completely flexible; once the goal has been established a person can switch from walking to sidestepping to crawling on the knees. In a similar vein, the first author has informally demonstrated the ability to walk blindly to a previewed target while another person pushes against him to impede direct access to the target. These perturbations result in surprisingly little error. This is further evidence of the flexible nature of perceptually directed action.

An explanation of perceptually directed action consistent with the above results relies on several internal representations. First, visual, auditory, or haptic input specifies the perceived target location within the current perceptual model. Accompanying this percept is the spatial image, mentioned earlier. This spatial image of the target location is initially coincident with the percept. It continues to exist even after cessation of perceptual input causes the percept to disappear (as when closing the eyes). Depending on the circumstances, the percept may or may not be veridical (see the results for the Light on and Lights off conditions in Figure 3). Because the spatial image is coincident with the percept, error in the percept is carried over to the spatial image. When the actor begins moving, sensed changes in position and orientation (path integration) result in spatial updating of the spatial image. Clear evidence for continuous updating of the spatial image was provided by an experiment by Loomis et al. (1992) in which actors walked without vision and attempted to continually point toward a target that had been viewed prior to walking; pointing direction was accurate during the entire traverse. The fact that errors in perceiving target locations, as measured using verbal report and other non-motor perceptual judgments, are also reflected in perceptually directed action (vision-based reaching: Foley, 1985; vision-based locomotion: Loomis & Knapp, in press; Philbeck & Loomis, 1997; audition-based locomotion: Loomis et al., 1998) is evidence that perceptual space (or more generally, the phenomenal world) is not an epiphenomenon but indeed a cause of action. Incidentally, performance of perceptually directed action is not

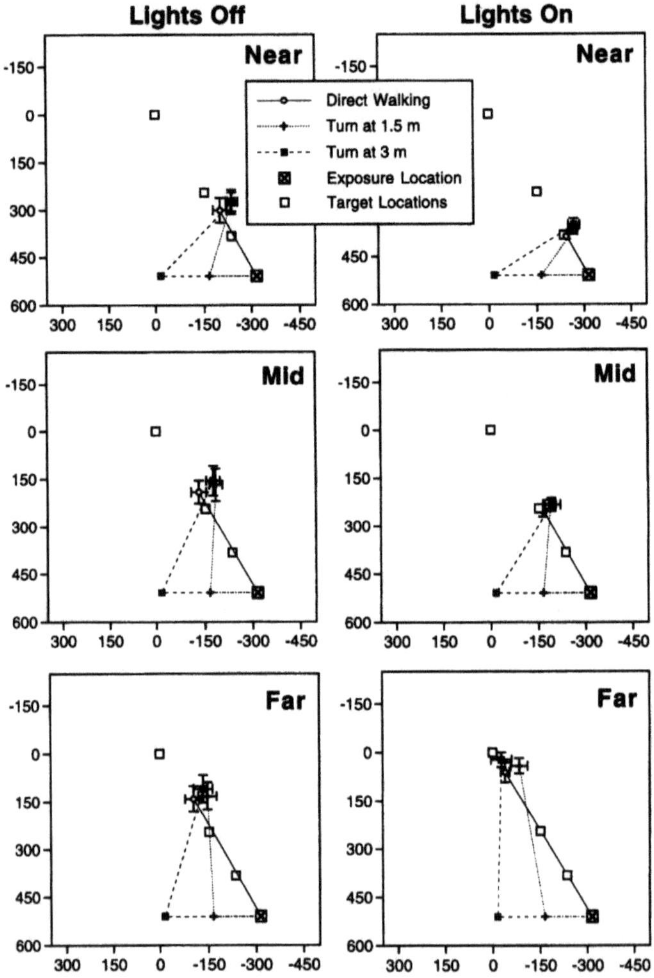

Figure 2. Results of an experiment on perceptually directed action. Glowing rectangular targets of constant angular size were viewed in darkness or with the room lights on. After the observer viewed the target on a given trial, he/she closed the eyes and was then informed whether to walk directly to the target or along one of two indirect paths. For the latter, the initial segment was aligned with a wall, which the observer could feel while walking. On each trial, the target was placed 1.5, 3.1, or 6.0 m from the origin (lower-right corner of each panel). The data are the centroids of the terminal locations of six walking trajectories of each of eight observers in response to each target and condition along each of the three paths. The convergence of the walking paths in each panel indicates that observers were directing their walking behavior to the same points in space without knowing in advance which path they would be asked to walk. Perception was more accurate with lights on, as indicated by the fact that the centroids in this condition were closer to the targets. [Fig. 5 from Philbeck, Loomis, & Beall, 1997; used by permission]

easily interpretable within the ecological conception. In vision-based tasks, for example, once perceptual input about the target's location is removed, there is no stimulus support for the ensuing action.

Because we rarely close our eyes or block our hearing while acting in space, perceptually directed action as it has been typically studied may seem largely irrelevant to the understanding of perception/action. We note, however, that perceptually directed action is more common than it first appears. When one is moving toward a visually specified goal, its temporary occlusion by intervening objects poses no problem, for the goal is spatially updated in the meantime. Similarly, team sports and dancing with a partner involve spatial updating of objects and other people (who are frequently in motion!) when they pass out of the field of view. Although certainly keeping targets of interest in sight ought to result in more accurate interaction with them than spatial updating allows, the importance of spatial updating is unmistakable in providing a robustness and flexibility to action that would otherwise not exist.

Besides these examples of perceptually directed action in which the actors move through space under their own propulsion, there have been a few studies on car steering in which the visual feedback loop has been opened, forcing the actors to continue steering without visual feedback (Hildreth et al., 2000; Godthelp, 1986; Godthelp et al., 1984; Land, 1998; Wallis et al., 2002; see also Larish & Andersen, 1995, in connection with flying). The research shows that drivers can continue steering for very brief periods following the cutoff of vision, but performance quickly degrades. In open-loop lane changing in particular, actors make the correct initial turn toward the next lane but after arriving in the lane, fail to make the opposite turn to align the vehicle with the turn (Wallis et al., 2002). This is a clear case in which perceptually directed action fails rather drastically. However, lane changing at highway speeds, which involves small control inputs and subtle inertial cues about the car's motion, may not be a particularly good candidate for open-loop control, and there may be better candidates. For example, it is possible that an expert driver who has formed a very good internal model of the dynamics of his/her car and who has strong inertial cues available, might be able to view a layout of several markings in a parking lot and, from a standing start, travel over them without vision. Performance would surely fall short of that possible with continuous vision, but if the person could do the task at all, it would support the idea of model-based control in connection with such a complex behavior.

The examples of perceptually directed action discussed above indicate that an actor, after receiving preview information about the environment, can temporarily control his/her action with respect to the environment without concurrent information specifying errors in position and orientation relative to the environment. Without this concurrent information, the actor has no feedback from the environment. To the extent that the action is still reliably related to the environment, the actor is exercising feedforward control with respect to the environment. Is there any sense in which the actor can be said to be exercising feedback control? The answer is yes, for the person is updating an estimate of his/her position and orientation with respect to a spatial image of the environment. The person modifies commands to the musculature in order for the internally represented trajectory to match the desired trajectory through the represented environment.

Systematic error in the person's observed trajectory can come about as the result of systematic error in the percept and in the updating of the spatial image (Loomis et al., 2002). We have already seen the effects of systematic error in perception (Figure 3); actors consistently traveled to the wrong location in the Lights off condition in the study by Philbeck et al. (1997), regardless of the path traveled. One way that systematic error in spatial updating might come about is through error in updating the actor's estimate of position and orientation. In the absence of direct sensory information about position and orientation (e.g., a position fix), the actor must use path integration, based on information about velocity and/or acceleration of one's displacements and turns. Lacking allothetic information from the environment about velocity and acceleration, the actor needs to rely on idiothetic information (e.g., inertial cues, proprioceptive cues, and efference copy of the commands sent to the musculature). The accuracy with which people perform perceptually directed action using walking indicates that path integration is quite accurate with walking, at least over the short term (Loomis et al., 1999).

Would it be possible to perform perceptually directed action while driving a car without vision, as in the above thought experiment involving targets on the ground? Here, proprioceptive cues would be non-existent, leaving only inertial cues and efference copy. Taking this even further, what if the person had to perform the task while driving a car simulator without a motion base? Here, after initial preview of the scene, the actor would have only efference copy of the control inputs to the simulator as the basis for keeping track of his/her trajectory. Although control might at first seem impossible in this task, an actor who has internalized the vehicle dynamics through extensive practice might nevertheless be able to perform moderately well. In this case, we would say that the actor is using feedforward control with respect to the external world and feedback control with respect to the internal model (updated spatial image).

7 A CLOSER LOOK AT OPTIC FLOW

We now consider the case of feedback control of perception/action, with a focus on optic flow. The prototypical optic flow field of a translating observer is that depicted in the well known image of a flying bird (Gibson, 1966). For pure translation, the optic flow diverges from the point in the optic array (focus of expansion), which corresponds to the direction of travel and converges on the opposite pole (e.g., Gibson et al., 1955; Nakayama & Loomis, 1974). This prototypical flow field, corresponding to point features, is the stimulus used in the many studies of heading perception (e.g., Crowell & Banks, 1993; Warren et al., 1988). Aside from the important issue of whether heading perception is actually involved in directional control (aiming toward a point, moving into and maintaining alignment with a straight path, and steering along a curving path), (Beall & Loomis, 1996; Rushton et al., 1998; Wann & Land, 2000; Warren et al., 2001), a more basic issue is whether the optic flow used in directional control need be at all like the prototypical flow field.

Beall and Loomis (1996) observed that in the case of driving along a straight road with only solid lane markers visible, maintaining alignment with the road in the presence of crosswinds does not require sensing of the longitudinal component of motion (parallel to the road); the splay of the lane markers is sufficient for lateral control and information about the longitudinal component (by providing ground texture) does not improve steering performance. Gordon (1966; see also Biggs, 1966; Donges, 1978; and Wann & Land, 2000) made a similar point in connection with steering a curving path of constant curvature in which solid lane markers are present (Figure 3a). Once one is moving about such a curve, it is sufficient to provide steering inputs to keep the perspective view of the road constant. If the perspective view of the road begins to move to the left and to distort in shape, one can compensate with a steering input that causes the perspective view to once again remain constant. As with the case of steering a straight path, it is quite possible that adding information about the longitudinal component of vehicle motion (e.g., by adding ground texture) would fail to improve steering performance (Figure 3b). Although these cases of steering are special cases where sensing one's longitudinal speed is irrelevant to the task and other cases of steering surely do depend upon sensing the full motion state of the vehicle from optic flow, they remind us that no single optic flow analysis, based on the prototypical flow field, is suited for understanding the control of locomotion and other forms of perception/action.

A separate issue concerning closed-loop control of action using vision is whether continuous optic flow or intermittent sampling of the optic array (the perspective view of visual space) is important. As an ideal, optic flow corresponds to continuous change in the angular positions of visible features (Gibson et al., 1955; Nakayama & Loomis, 1994). In practice, however, experimenters use a rapid succession of static images (e.g. on a CRT) that is visually indiscriminable from continuous flow or very nearly so. An important question is whether perceptually smooth optic flow is required in most tasks

Figure 3. a) A driver's view of a curving road with constant curvature at slightly different times while the vehicle is moving laterally within the lane. The change in perspective, even without flow signaling longitudinal motion of the vehicle, is sufficient to steer. b) Adding ground texture to the information of panel (a) adds information about the longitudinal motion of the vehicle but quite possibly does not contribute to steering performance.

of visually controlled perception/action. The answer is probably yes in some cases (e.g., postural control; see Bardy et al., 1996) but decidedly no in many others. For directional control of a vehicle (aiming toward a point, maintaining alignment with a straight path, steering a curving path), it is clear that intermittent view of the environment ahead will often be as good as visually smooth flow. However, the intermittency required for accurate control of steering is surely related to the variation in road curvature per unit time, the vehicle dynamics, and other factors (Godthelp et al., 1984). Thus, we conclude that "optic change" in the sampled images of the optic array, rather than continuous optic flow, is often sufficient for precise visual control of action. To keep the terminology simple, however, we will generally use optic flow to refer to both continuous flow and optic change.

8 OPTIC FLOW AND PERCEIVED FLOW

In many situations of visually controlled action, the optic flow produced by locomotion results in a visual streaming or perceived flow of the environment. Indeed, this perceived flow is surely considered by some researchers to be identical with the optic flow that causes it. Yet, it is important to keep the two separate, for one is the stimulus and the other is perceptual. Indeed, Gogel (1990) has provided abundant evidence that perceived depth is a fundamentally important determinant of how optic flow is perceived.

Because optic flow and its discrete version, optic change, often have perceptual concomitants, the question arises as to whether perceived flow and changes in the perceived visual scene are instead the actual basis for visually controlled perception/action. This is a broad question about which little is known. However, one study addressing this issue (Beusmans, 1998) showed that changes in how a visual scene was perceived in terms of depth did influence the perception of heading. This is a hint that optic flow, when it is the basis for heading perception, might act through perceived flow.

We have recently discovered that a variety of active control tasks of various spatial behaviors can be effectively accomplished without any optic flow whatsoever. For this research, we have exploited the random dot cinematogram (RDCs) with single-frame lifetimes (Julesz, 1971); this is a sequence of random-dot stereograms portraying a changing 3-D stimulus in which the base pair of random dots changes every frame. Julesz (1971) first demonstrated that observers are readily able to perceive a stereo-defined square moving laterally across a visual display despite the very strong perception of scintillation. More recently, Gray and Regan (1996, 1998) have shown that observers are able to judge time to contact and the direction of motion of single moving targets presented in these types of displays.

In our work, the second author has developed a technique for combining RDCs with single-frame lifetimes with our immersive virtual reality system that we have been using to study complex behavior (Loomis et al., 1999). With this technique, the second author can switch from normal binocular display of a richly textured virtual environment to a scintillating RDC display of the same environment (without textures). At a 60 Hz graphics update rate in each eye, he can cast up to 1000 random dots onto the surfaces of the virtual environment within the observer's field of view. These dots are spread over the field of view with constant angular density. With the single-frame lifetimes of each video frame, each eye simply is exposed to a scintillating field of uniform dot density. Thus, there is no optic flow signal correlated with the spatial structure of the virtual environment. Amazingly, when viewing with both eyes, observers are able to perceive complex environments almost immediately. Even more amazing is the ease with which they can perform complex spatial tasks, such as walking through a cluttered field of obstacles, walking to intercept a rapidly moving object, and steering a simulated vehicle along a curving road. Figure 4a is a stereogram (for cross-fusion) depicting a path represented by a raised centerline. Figure 4b is the same path rendered as a random-dot stereogram.

Research underway is exploring the extent to which performance of visually controlled actions carried out with the RDC approaches that of the same actions using the traditional optic flow stimulus. Because visual

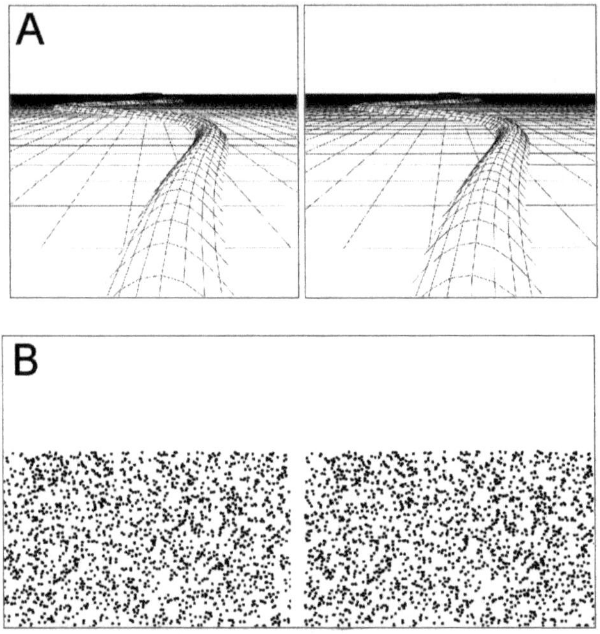

Figure 4. a) Cross-fusion stereogram of a path represented by a raised centerline. b) Random-dot stereogram of the same view. A scintillating RDC consists of many frames like this, each with a different base pair of random dots (Julesz, 1971). Surprisingly, people can perform a wide variety of visually controlled actions with scintillating RDCs even though there is no optic flow relating to the task.

perception of RDC stimuli is impaired relative to vision of normal optic flow stimuli (Gray & Regan, 1998), performance of complex actions is somewhat poorer than it is with normal stimulation. For example, in a steering task using the stimuli of Figure 4, the root-mean-squared error showing how closely the simulated vehicle tracks the path is approximately 75% greater with the scintillating RDC (Loomis & Beall, 2003). The important question is whether, when monocular and stereoscopic vision are matched for these lower level processing differences, performance of these complex actions is also matched (see Gray & Regan, 1998 for a treatment of the same issue in connection with the processing of single moving targets).

Our work so far indicates that when adjusting for these processing differences, performance of complex action is even more closely matched (Loomis & Beall, 2003). If further research shows a near functional equivalence between optic flow and this binocularly-mediated "flow", for example, with respect to eye fixations and pursuit eye movements during the performance of complex action, this would support the hypothesis of a final common pathway of perceived flow through which both forms of flow exert their influence in controlling action. This would implicate internal representation in the control of perception/action, even in this case of feedback control involving optic flow. However, we note that conventional optic flow, even if acting through perceived flow, surely will allow for better performance of certain complex actions (e.g., catching a baseball) than binocularly-mediated flow when those actions require higher spatial and temporal bandwidths of visual processing than is afforded by pure binocular vision.

ACKNOWLEDGEMENTS

The authors' ideas and empirical results presented in this chapter are based on research supported by Office of Naval Research grant N00014-95-1-0573 and Air Force Office of Scientific Research grant F49620-02-1-0145. The authors thank Willard Larkin for comments on an earlier version of the manuscript.

REFERENCES

Amorim, M. A., Glasauer, S., Corpinot, K., & Berthoz, A. (1997). Updating an object's orientation and location during nonvisual navigation: A comparison between two processing modes, *Percept. Psychophys.*, *59*, 404-418.

Ashmead, D. H., DeFord, L. D., & Northington, A. (1995). Contribution of listeners' approaching motion to auditory distance perception, *J. Exp. Psychol. Hum. Percept. Perform.*, *21*, 239-256.

Bardy, B. G., Warren, W. H., & Kay, B. A. (1996). Motion parallax is used to control postural sway during walking, *Exp. Brain Res.*, *111*, 271-282.

Beall, A. C. & Loomis, J. M. (1996). Visual control of steering without course information, *Perception*, *25*, 481-494.

Beall, A. C. & Loomis, J. M. (1997). Optic flow and visual analysis of the base-to-final turn, *Int. J. Aviat. Psychol.*, *7*, 201-223.

Beusmans, J. M. H. (1998). Perceived object shape affects the perceived direction of self-movement, *Perception*, *27*, 1079-1085.

Bhalla, M. & Proffitt, D. (1999). Visual-motor recalibration in geographical slant perception, *J. Exp. Psychol. Hum. Percept. Perform.*, *25*, 1076-1096.

Biggs, N. L. (1966). Directional guidance of motor vehicles: A preliminary survey and analysis, *Ergonomics*, *9*, 193-202.

Brain, W. R. (1951). *Mind, Perception, and Science*. London: Blackwell.

Calvert, E. S. (1954). Visual judgments in motion, *J. Inst. Nav.*, *7*, 233-251.

Corlett, J. T. (1986). The effect of environmental cues on locomotor distance estimation by children and adults, *Hum.. Movement Sci.*, *5*, 235-248.

Crowell, J. A. & Banks, M. S. (1993). Perceiving heading with different retinal regions and types of optic flow, *Percept. Psychophys.*, *53*, 325-337.

Dickmanns, E. D. (1992). A general dynamic vision architecture for UGV and UAV. *J. App. Intell.*, *2*, 251-270.

Donges, E. (1978). A two-level model of driver steering behavior, *Hum. Factors*, *20*, 691-707.

Elliott, D. (1987). The influence of walking speed and prior practice on locomotor distance estimation, *J. Mot. Behav.*, *19*, 476-485.

Elliott, D., Jones, R., & Gray, S. (1990). Short-term memory for spatial location in goal-directed locomotion, *Bull. Psychon. Soc.*, *8*, 158-160.

Farrell, M. J. & Thomson, J. A. (1999). On-line updating of spatial information during locomotion without vision, *J. Mot. Behav.*, *3*, 39-53.

Flach, J. M., & Warren, R. (1995). Active Psychophysics: The Relation Between Mind and What Matters. In: J. M. Flach, P. A. Hancock, J. Caird, & K. J. Vicente (Eds.) *Global Perspectives on the Ecology of Human-Machine Systems, Vol. 1. Resources for Ecological Psychology* (pp. 189-209). Hillsdale, NJ: Lawrence Erlbaum Associates.

Fodor, J. A. (1983). *The Modularity of Mind: An Essay on Faculty Psychology*. Cambridge, MA: MIT Press.

Foley, J. M. (1985). Binocular distance perception: egocentric distance tasks, *J. Exp. Psychol. Hum. Percept. Perform.*, *11*, 133-149.

Foley, J. M. & Held, R. (1972). Visually directed pointing as a function of target distance, direction, and available cues, *Percept. Psychophys.*, *12*, 263-268.

Fukusima, S. S., Loomis, J. M., & Da Silva, J. A. (1997). Visual perception of egocentric distance as assessed by triangulation, *J. Exp. Psychol. Hum. Percept. Perform.*, *23*, 86-100.

Gibson, J. J. (1950). *The Perception of the Visual World.* Boston: Houghton-Mifflin.

Gibson, J. J. (1958). Visually controlled locomotion and visual orientation in animals, *Br. J. Psychol.*, *49*, 182-194.

Gibson, J. J. (1966). *The Senses Considered as Perceptual Systems*. Boston: Houghton-Mifflin.

Gibson, J. J. (1979). *The Ecological Approach to Visual Perception.* Boston: Houghton-Mifflin.

Gibson, J. J. & Crooks, L. E. (1938). A theoretical field-analysis of automobile driving, *Am. J. Psychol.*, *51*, 453-471.

Gibson, J. J., Olum, P., & Rosenblatt, F. (1955). Parallax and perspective during aircraft landings, *Am. J. Psychol.*, *68*, 372-385.

Godthelp, H. (1986). Vehicle control during curve driving, *Hum. Factors*, *28*, 211-221.

Godthelp, H. & Käppler, W. D. (1988). Effects of vehicle handling characteristics on driving strategy, *Hum. Factors*, *30*, 219-229.

Godthelp, H., Milgram, P., & Blaauw, G. J. (1984). The development of a time-based measure to describe driving strategy, *Hum. Factors*, *26*, 257-268.

Gogel, W. C. (1990). A theory of phenomenal geometry and its applications, *Percept. Psychophys.*, *48*, 105-123.

Gordon, D. A. (1966). Perceptual basis of vehicular guidance, *Public Roads*, *34*, 53-68.

Gray, R. & Regan, D. (1996). Cyclopean motion perception produced by oscillations of size, disparity and location, *Vision Res.*, *36*, 655-665.

Gray, R. & Regan, D. (1998). Accuracy of estimating time to collision using binocular and monocular information, *Vision Res.*, *38*, 499-512.

Hildreth, E. C., Beusmans, J. M. H., Boer, E. R., & Royden, C. S. (2000). From vision to action: Experiments and models of steering control during driving, *J. Exp. Psychol. Hum. Percept. Perform.*, *26*, 1106-1132.

Hollins, M. & Kelley, E. K. (1988). Spatial updating in blind and sighted people, *Percept. Psychophys.*, *43*, 380-388.

Julesz, B. (1971). *Foundations of Cyclopean Perception*. Chicago: University of Chicago Press.

Land, M. F. (1998). The Visual Control of Steering. In: L. R. Harris & M. Jenkin (Eds.), *Vision and Action* (pp. 163-180). Cambridge: Cambridge University Press.

Land, M. F., & Hayhoe, M. (2001). In what ways do eye movements contribute to everyday activities? *Vision Res.*, *41*, 3559-3565.

Larish, J. F. & Andersen, G. J. (1995). Active control in interrupted dynamic spatial orientation: the detection of orientation change. *Percept. Psychophys.*, *57,* 533-545.

Lee, D. N. (1976). A theory of visual control of braking based on information about time-to-collision, *Perception*, *5*, 437-459.

Lee, D. N. (1993). Body-Environment Coupling. In: U. Neisser (Ed.), *The Perceived Self: Ecological and Interpersonal Sources of Knowledge* (pp. 43-67). New York: Cambridge University Press.

Lee, D. N. & Lishman, R. (1977). Visual control of locomotion, *Scand. J. Psychol.*, *18*, 224-230.

Lehar, S. (2002). *The World in Your Head: A Gestalt View of the Mechanism of Conscious Experience.* Mahwah, N.J.: Lawrence Erlbaum Associates.

Loomis, J. M. (1992). Distal attribution and presence, *Presence*, *1*, 113-119.

Loomis, J. M. & Beall, A. C. (1998). Visually-controlled locomotion: Its dependence on optic flow, 3-D space perception, and cognition, *Ecol. Psychol.*, *10*, 271-285.

Loomis, J. M. & Beall, A. C. (2003). Visual control of locomotion without optic flow. *Third annual meeting of Vision Sciences Society, Sarasota, FL*, May 9-14, 2003.

Loomis, J. M., Blascovich, J.J., & Beall, A. C. (1999). Immersive virtual environment technology as a basic research tool in psychology, *Behav. Res. Meth. Instrum. Comput.*, *31*, 557-564.

Loomis, J. M., Da Silva, J. A., Fujita, N., & Fukusima, S. S. (1992). Visual space perception and visually directed action, *J. Exp. Psychol. Hum. Percept. Perform.*, *18*, 906-921.

Loomis, J. M., Klatzky, R. L., Golledge, R. G., & Philbeck, J. W. (1999). Human Navigation by Path Integration. In: R. G. Golledge (Ed.), *Wayfinding: Cognitive Mapping and Other Spatial Processes* (pp. 125-151). Baltimore: Johns Hopkins.

Loomis, J. M., Klatzky, R. L., Philbeck, J. W., & Golledge, R. G. (1998). Assessing auditory distance perception using perceptually directed action, *Percept. Psychophys.*, *60*, 966-980.

Loomis, J. M. & Knapp, J. M. (in press). Visual Perception of Egocentric Distance in Real and Virtual Environments. In: L. J. Hettinger and M. W. Haas (Eds.), *Virtual and Adaptive Environments.* Mahwah N.J.: Erlbaum.

Loomis, J. M., Lippa, Y., Klatzky, R. L., & Golledge, R. G. (2002). Spatial updating of locations specified by 3-D sound and spatial language, *J. Exp. Psychol. Learn. Mem. Cogn.*, *28*, 335-345.

McRuer, D. T., Allan, R. W., Weir, D. H., & Klein, R. H. (1977). New results in driver steering control models, *Hum. Factors*, *19*, 381-397.

Miall, R. C., Weir, D. J., Wolpert, D. M., & Stein, J. F. (1993). Is the cerebellum a Smith Predictor? *J. Mot. Behav.*, *25*, 203-216.

Milner, A. D. & Goodale, M. A. (1995). *The Visual Brain in Action.* New York: Oxford University Press.

Nakayama, K. & Loomis, J. M. (1974). Optical velocity patterns, velocity sensitive neurons, and space perception: a hypothesis, *Percept.*, *3*, 63-80.

Ooi, T.L., Wu, B., & He, Z. J. (2001). Distance determined by the angular declination below the horizon, *Nature*, *414*, 197-200.

Patla, A. E. (1998). How is human gait controlled by vision? *Ecol. Psychol.*, *10*, 287-302.

Philbeck, J. W. & Loomis, J. M. (1997). Comparison of two indicators of visually perceived egocentric distance under full-cue and reduced-cue conditions, *J. Exp. Psychol. Hum. Percept. Perform.*, *23*, 72-85.

Philbeck, J. W., Loomis, J. M., & Beall, A. C. (1997). Visually perceived location is an invariant in the control of action, *Percept. Psychophys.*, *59*, 601-612.

Ramachandran, V. S. & Blakeslee, S. (1998). *Phantoms in the Brain: Probing the Mysteries of the Human Mind.* New York: William Morrow.

Rensink, R. A. (2002). Change detection, *Ann. Rev. Psychol.*, *53*, 245-277.

Riemersma, J. B. (1981). Visual control during straight road driving, *Acta Psychologica*, *48*, 215-225.

Rieser, J. J., Ashmead, D. H., Talor, C. R., & Youngquist, G. A. (1990). Visual perception and the guidance of locomotion without vision to previously seen targets, *Perception*, *19*, 675-689.

Rushton, S. K., Harris, J. M., Lloyd, M. R. & Wann, J. P. (1998). Guidance locomotion on foot uses perceived target location rather than optic flow, *Curr. Biol.*, *21*, 1191-1194.

Russell, B. (1948). *Human Knowledge: Its Scope and Limits.* New York: Simon and Schuster.

Simmel, M. L. (1966). Developmental aspects of the body scheme, *Child Develop.*, *37*, 83-96.

Smythies, J. R. (1994). *The Walls of Plato's Cave.* Aldershot, England: Avebury.

Steenhuis, R. E. & Goodale, M. A. (1988). The effects of time and distance on accuracy of target-directed locomotion: does an accurate short-term memory for spatial location exist? *J. Mot. Behav.*, *20*, 399-415

Thomson, J. A. (1980). How do we use visual information to control locomotion? *Trends Neurosci.*, *3*, 247-250.

Thomson, J. A. (1983). Is continuous visual monitoring necessary in visually guided locomotion? *J. Exp. Psychol. Hum. Percept. Perform.*, *9*, 427-443.

Turvey, M. T. (1977). Preliminaries to a Theory of Action with Reference to Vision. In: R. J. Shaw & J. Bransford (Eds.), *Perceiving, Acting, and Knowing: Toward an Ecological Psychology*. Hillsdale, N.J.: Erlbaum.

Turvey and Carello, C. (1986). The ecological approach to perceiving-acting: a pictorial essay, *Acta Psychologica*, *63*, 133-155.

Wallis, G.M., Chatziastros, A., & Bülthoff, H. H. (2002). An unexpected role for visual feedback in vehicle steering control, *Curr. Biol.*, *12*, 295-299.

Wann, J. & Land, M. (2000). Steering with or without the flow: is the retrieval of heading necessary? *Trends Cogn. Sci.*, *4*, 319-324.

Warren, R. & Wertheim, A. H. (1990). *Perception & Control of Self-Motion*. Hillsdale, NJ: Lawrence Erlbaum.

Warren, W. H. (1998). Visually controlled locomotion: 40 years later, *Ecol. Psychol.*, *10*, 177-219.

Warren, W. H., Kay, B. A., Zosh, W. D., Duchon, A. P., & Sahuc, S. (2001). Optic flow is used to control human walking, *Nat. Neurosci.*, *4*, 213-216.

Warren, W. H., Morris, M. W., & Kalish, M. (1988). Perception of translational heading from optical flow, *J. Exp. Psychol. Hum. Percept. Perform.*, *17*, 28-43.

Weiskrantz, L. (1986). *Blindsight: A Case Study and Implications*. Oxford: Oxford University Press.

Wolpert, D. M. & Ghahramani, Z. (2000). Computational principles of movement neuroscience, *Nat. Neurosci. Supp.*, *3*, 1212-1217.

Wolpert, D. M., Ghahramani, Z., & Jordan, M. I. (1995). An internal model for sensorimotor integration, *Science*, *269*, 1880-1882.

Yilmaz, E. H. & Warren, W. H. (1995). Visual control of braking: a test of the tau-dot hypothesis, *J. Exp. Psychol. Hum. Percept. Perform.*, *21*, 996-1014.

20. A Neural Model for Biological Movement Recognition: A Neurophysiologically Plausible Theory

Martin A. Giese[1,2]

[1]ARL, Department of Cognitive Neurology,
University Clinic Tübingen, Germany

[2]Center for Biological and Computational Learning,
M.I.T., Cambridge, USA

1 INTRODUCTION

Complex *biological movements,* like actions, body movements, or facial expressions, are frequently occurring important visual stimuli. Psychophysical experiments show that humans can process them with amazing robustness and accuracy and individual neurons and cortical areas involved in the processing of such motion stimuli have been identified in electrophysiological and fMRI experiments. However, the understanding of the underlying neural mechanisms is still lacking. A neural model is presented that accounts for a number of key experimental results by exploiting simple physiologically plausible mechanisms. The model is based on the central assumption that biological movements are encoded by learning. Learned movement patterns can be encoded by neurons in the ventral and the dorsal pathway that are selective for sequences of body configurations and complex optic flow patterns. From the model several predictions can be derived that motivate new experiments.

The first section of this chapter reviews experimental results on the neural basis of motion recognition. Subsequently, the model is presented, and some justifications of the underlying assumptions are given. The following section presents a number of simulation results that demonstrate the quantitative consistency of the model with existing experimental data, and that illustrate how experimentally testable predictions can be derived. Some key predictions and different implications of the model are discussed in the concluding section.

L.M. Vaina, S.A. Beardsley and S.K. Rushton (eds.), Optic Flow and Beyond, 443–470.

2 BIOLOGICAL MOVEMENT RECOGNITION

The amazing sensitivity of human vision for biological movements has been demonstrated by G. Johansson (1973) in a well-known psychophysical experiment. He generated complex motion stimuli consisting of a small number of moving illuminated dots by fixing light bulbs on the joints of actors that performed different actions, e.g. walking or dancing. Based on these strongly impoverished stimuli subjects were effortlessly able to recognize the actions. Subsequent studies (e.g., Kozlowski & Cutting, 1977; Cutting & Kozlowski, 1977) have shown that even subtle details, such as the gender or the identity of the walker could be determined from such "point light stimuli".

Our knowledge about the neural mechanisms of biological movement recognition is quite limited. In electrophysiological experiments neurons in different parts of the *superior temporal sulcus* (STS) have been found that respond selectively for full-body and hand movements (e.g., Perrett et al. 1985; Oram & Perrett, 1996). Some of these neurons are view-dependent, i.e. they respond only when the action is presented from a particular viewing direction. Functional imaging studies suggest the existence of similar neural structures in humans (Decety & Grèzes, 1999). Selective activation of areas in the STS, and in the *occipital and the fusiform face area* during observation of biological motion has been reported in PET and fMRI studies using point light displays and natural stimuli (e.g., Bonda et al., 1996; Allison et al., 2000; Vaina et al. 2001; Grossman & Blake, 2002).

Another class of action-selective neurons, "mirror neurons" in area F5 in the premotor cortex of monkeys (Rizzolatti et al., 2001), are also activated during motor planning. An analog of this area has been found in the *inferior frontal gyrus* of humans. In addition, biological movements activate other non-visual areas, such as the *amygdala* and the *cerebellum* (e.g. Bonda et al. 1996; Vaina et al. 2001).

These experimental results provide some insights regarding which neural structures are involved in the recognition of biological movements. However, they do not provide precise information about the underlying neural mechanisms and circuits. Theoretical models help to develop well-defined testable hypotheses about such mechanisms, and they provide a systematic way to evaluate their quantitative consistency with existing experimental data. In the following section a neural model is presented that formulates one set of neural mechanisms that are physiologically plausible. The architecture and the components of the model are consistent with the basic anatomical and physiological facts that are known about monkey visual cortex

3 MODEL

The constraint of biological plausibility makes many popular technical solutions of the motion recognition problem rather unlikely as explanations for movement recognition in visual cortex (for review see Aggarwal & Cai, 1999; Gavrila, 1999). Many technical algorithms, e.g., the fitting of complex kinematic models or optimization on graphs, have no obvious implementation in terms of physiologically plausible neural mechanisms.

It is shown in the following model that much of the existing data can be accounted for without assuming sophisticated computational algorithms. Quite robust recognition can be realized by exploiting the following three neural principles:

a) Analysis of visual stimuli occurs in two pathways that are specialized for the processing of form and motion information.
b) Each pathway is organized in terms of a hierarchy of neural feature detectors with an increase of the complexity of the extracted features and of the invariance of the detectors against translation and scaling along the hierarchy.
c) Complex movements are represented in terms of learned prototypical example patterns. Such patterns are encoded by model neurons that are selective for "snapshots" representing body shapes, and for the complex instantaneous optic flow patterns that are characteristic for biological movements.

The first two principles match well-established neuroanatomical facts; the third principle is a central assumption of the model.

The basic architecture of the model is illustrated in Figure 1. The different parts of the model try to approximate neural structures that are involved in visual movement recognition. Consistent with the known functional division of the visual cortex into a ventral and a dorsal pathway (e.g. Ungerleider & Mishkin, 1982; Felleman & van Essen, 1991; Goodale & Milner, 1992) the model includes two parallel processing streams that are specialized for the analysis of form and optic flow information. In the monkey and human cortex the two processing streams are connected at multiple levels (e.g. Felleman & van Essen, 1991; Kourtzi & Kanwisher, 2000). Particularly strong connections exist between the *inferotemporal cortex* and the STS in monkeys (Morel & Bullier, 1990; Saleem et al., 2000). For evaluating separately the contributions of form and motion information the model is based on the idealizing assumption that the form and the motion pathway are not coupled. The model can be extended by the addition of such couplings. A version of

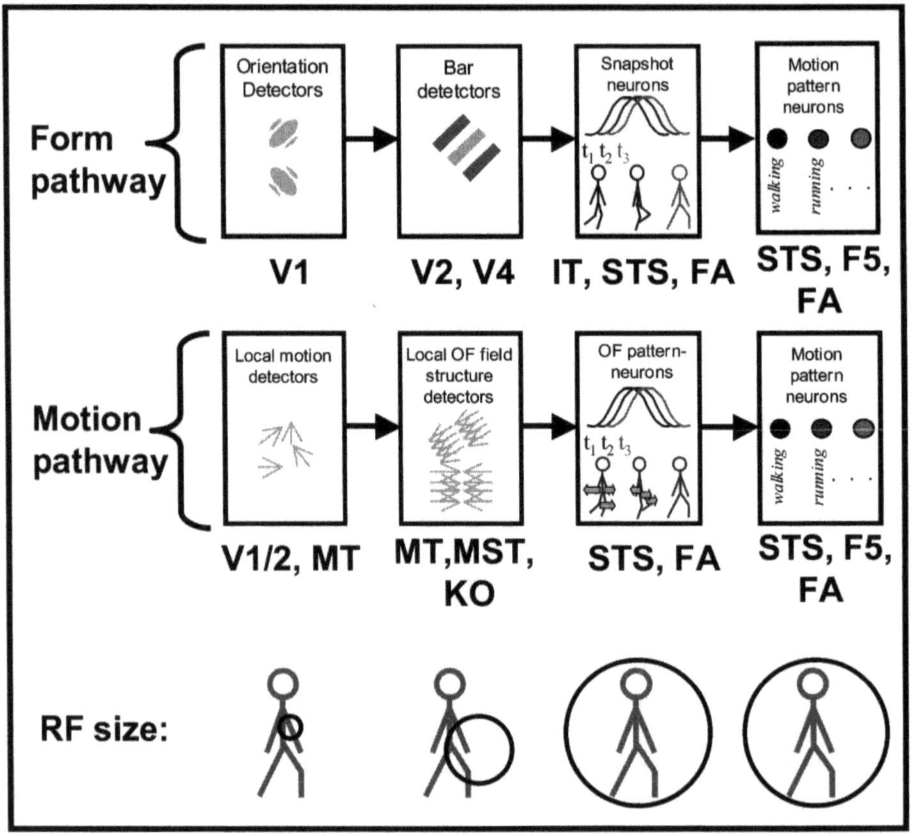

Figure 1. Overview of the model with two pathways for the processing form and optic flow information. Abbreviations indicate potential corresponding areas in monkey and human cortex. The lower panel shows the approximate sizes of the receptive fields compared to a walker stimulus.

the model with a fusion of the two pathways at a level that coarsely corresponds to the STS has been presented in Giese & Vaina (2001).

Each pathway consists of a hierarchy of neural feature detectors. Their properties are, where possible, coarsely matched with known properties of neurons in monkey cortex. Both pathways of the model include a neural mechanism that makes recognition selective for the temporal order of the stimulus sequence. The following sections describe the properties of the neural detectors in both pathways and simple a biologically plausible neural mechanism for sequence selectivity.

3.1 Form Pathway

The form pathway analyses biological movements by recognizing sequences of "snapshots" of body shapes. The recognition of body shapes might be accomplished with neural principles similar to the recognition of stationary objects. The form pathway of the model subsumes a model for the recognition of stationary form. A variety of neurophysiologically plausible models for the recognition of stationary objects have been proposed (e.g., Perrett & Oram, 1989; Riesenhuber & Poggio, 1999; Rolls & Milward, 2000). Deviating from these earlier models, the recognition of biological movements must integrate form information over time. This integration is accomplished by a neural mechanism for sequence selectivity that is described below.

The form pathway has four hierarchy levels that consist of neural detectors for form features of increasing complexity (Figure 1). The sizes of the receptive fields of the detectors increase along the hierarchy (cf. Table 1), as does the invariance of their responses against translation and scaling of the stimuli. This assumption is consistent with the tuning properties of neurons in visual cortex. Some specific details about the neural detectors discussed in the model are listed in Table 1. See Giese & Poggio (2003) for further details.

The first level of the form pathway consists of *local orientation detectors* that are modeled by Gabor filters. Similar tuning properties have been reported for simple cells in primary visual cortex (area V1 in monkeys) (Hubel & Wiesel, 1962; Jones & Palmer, 1977). We implemented detectors with eight different preferred orientations and two spatial scales that differ by a factor 2. Receptive field sizes are in a range that is consistent with V1 neurons in monkey cortex (e.g., Dow et al. 1981).

The next hierarchy level contains position- and scale- *invariant bar detectors.* These neurons extract local orientation information, but irrespective of the exact spatial position and scale of oriented contours. Similar behavior has been reported for complex cells in the primary visual cortex (Hubel & Wiesel, 1962). Invariance against position changes has also been reported for neurons in areas V2 and V4 (Hegdé & van Essen, 2000; Gallant et al. 1996). The receptive fields of the invariant bar detectors are bigger than those of the orientation detectors on the previous hierarchy level. Their size is in the range that has been reported for neurons in area V4 (e.g., Pinon et al., 1998). Many neurons in areas V2 and V4 are selective for more complex form features, like corners or junctions (e.g., Pasupathy & Connor, 1999; Hegdé & van Essen, 2000). Corresponding detectors for more complex features can be easily added on the medium levels of the hierarchy (Riesenhuber & Poggio, 1999). They make recognition more selective, but they were not necessary to account for the results discussed in this chapter.

Table 1. Overview of the properties of the neural detectors, and potential corresponding areas in the monkey and human cortex.

Neuron Type	Features	Cortical Areas	# of neurons	RF Size [deg]
FORM PATHWAY				
Orientation detectors	Local orientation (2 spatial scales)	V1, V2	1010	0.6/ 1.2
Invariant bar detectors	Oriented bars (pos./scale invariant)	V1, V2, V4	128	4
Snapshot neurons	Body postures	IT, STS, FA	63...840	> 8
Motion pattern neurons	Movement patterns	STS, FA, F5	3...40	> 8
MOTION PATHWAY				
Local motion detector	Local motion energy	V1,V2, MT	1147	0.9
Local OF pattern detector	Translation Opponent motion	MT,MST(l) MST(l),KO	72 100	3.5
Complex OF pattern det.	Instantaneous OF patterns	STS, FA	63...840	>8
Motion pattern neurons	Movement patterns	STS, FA, F5	3...40	> 8

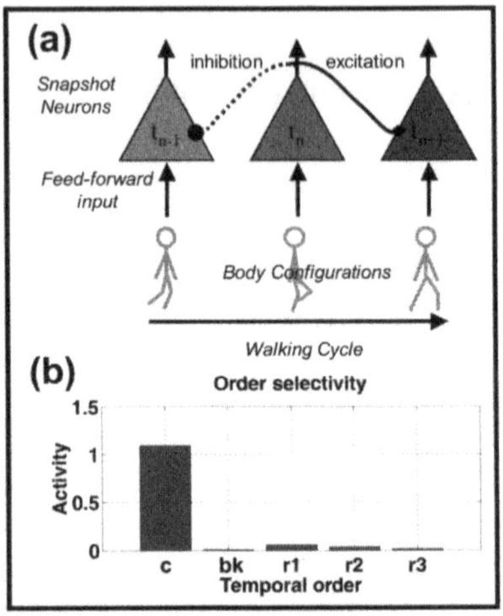

Figure 2. a) Simple mechanism for sequence selectivity based on asymmetric lateral connections. b) Activity of the snapshot neurons for correct (c), reverse (bk), and random (r1, ..., r3) temporal orders of frames.

A simple neural implementation of position and scale invariance is the pooling of the responses of neural detectors with the same preferred orientation, but different receptive field positions and spatial scales (e.g., Fukushima, 1980; Oram & Perrett, 1993). Recent theoretical studies show it is critical that this pooling be realized using a maximum-like operation, as opposed to linear summation, to achieve high degrees of feature selectivity together with invariance (e.g., Riesenhuber & Poggio, 1999; Mel & Fieser, 2000). Indeed, subpopulations of complex cells in the visual cortex of cats [Lampl, I., Riesenhuber, M., Poggio, T. & Ferster, D. (2001) *Society of Neuroscience Abstracts* 30, 619] and neurons in area V4 of macaques (Gawne & Martin, 2002) show behavior that is compatible with a maximum computation. Maximum computation can be realized with several physiologically plausible neural circuits (Yu et al., 2002).

The next hierarchy level of the form pathway contains detectors that are selective for complex shapes. These detectors are selective for body configurations that correspond to "snapshots" from a biological movement sequence. These *snapshot neurons* are similar to the view-tuned neurons that have been described in area IT of monkeys (Logothetis & Sheinberg, 1996;

Tanaka, 1996). Such neurons respond selectively for views of natural and artificial objects. Also it has been shown that they can learn individual views of synthetic artificial objects (Logothetis & Sheinberg, 1996).

The proposed theory postulates the existence of similar shape-selective neurons that respond selectively for human body configurations. The receptive fields of the snapshot neurons are large (about 8 deg). This is a regime that has been observed in (anterior) area IT of monkeys (Op de Beeck & Vogels, 2000). Neurons that are selective for body configurations have been found in the STS of monkeys (e.g., Oram & Perrett, 1996). In fMRI studies[1] selective activation for human bodies has been found in the STS, the *extrastriate body part area* (EBA), (Downing et al., 2001), and the *occipital and fusiform face area* (abbreviated by FA in Figure 1), (Grossman & Blake, 2002).

Consistent with previous quantitative models for IT neurons (e.g., Riesenhuber & Poggio, 1999) the *snapshot neurons* are modeled by gaussian radial basis functions. Their input signals are given by the responses of the invariant bar detectors[2]. The centers of these basis functions were trained with key frames from training movement sequences. In our model each training pattern is represented by 21 snapshot neurons. This number is not critical for the performance of the model. The snapshot neurons are laterally connected and form a recurrent neural network that responds maximally only if the snapshots arise with the right temporal order – the network dynamics are discussed below.

The highest hierarchy level of the model is formed by *motion pattern neurons* that temporally smooth and summate the activity of all snapshot neurons that represent the same movement pattern. Temporal smoothing is modeled by a leaky integrator. The dynamics of the membrane potential $z(t)$ of a motion pattern neuron that sums the input signals $y_n(t)$ is given by the differential equation:

$$\tau_p \dot{z}(t) = -z(t) + \sum_n y_n(t) \tag{1}$$

The time constant τ_p was *150* ms and each motion pattern neuron encoded a complete motion sequence or action, e.g. “walking”, “boxing”. To keep the model simple, each training pattern was encoded by a single motion pattern neuron. Such “grandmother cells” are an over-simplification, and it seems likely that in the cortex motion patterns are encoded rather by some form of a neural population code, as shown before for the neural encoding of faces (e.g., Young & Yamane, 1992). Neurons encoding biological movement patterns have been found in different parts of the STS of monkeys (e.g., Perrett et al., 1985; Oram & Perrett, 1996). Functional imaging studies suggest that in humans similar neurons might exist in the STS (e.g., Bonda et al., 1996;

Vaina et al., 2001), and in the *fusiform* and the *occipital face area* (Grossman & Blake, 2002). Another possible site for neurons encoding whole actions is the premotor cortex (area F5) in monkeys, and homologous areas in humans (Rizzolatti et al., 2001).

3.2 Motion Pathway

The motion pathway of the model consists of neural detectors that are specialized for the analysis of optic flow information. Like the form pathway, the motion pathway has a hierarchical structure. Consistent with neurophysiological data from the dorsal visual pathway (e.g., Saito, 1993) the receptive field sizes of the detectors and the complexity of the extracted features increase along the hierarchy.

The first level of the motion pathway consists of *local motion energy detectors*. Many neurophysiologically plausible models for local motion energy extraction have been proposed in the literature (e.g., Sereno, 1993; Smith & Snowden, 1994; Simoncelli & Heeger, 1998). Local motion extraction in the cortex is accomplished by direction-selective neurons in the primary visual cortex, in particular areas V1 and V3 (see Smith & Snowden, 1994, for a review), and component motion-selective neurons in area MT (e.g., Rodman & Albright, 1989) .

To accelerate the simulations no detailed neural model for local motion energy extraction was implemented. Since our stimuli were generated using an articulated walker model the optic flow generated by the articulating figure could be directly computed from the stimulus geometry. For point light stimuli optic flow fields were derived by matching each point with the closest point in the subsequent frame. The obtained optic flow fields were sampled in space, and equivalent responses of local motion detectors were derived by counting the number of optic flow vectors within a limited spatial receptive field, and with a limited regime of motion directions and speeds. The spatial receptive fields had a size that is in the range of direction-selective neurons in the primary visual cortex and of foveal neurons in area MT of monkeys (e.g., Dow et al., 1981; Albright & Desimone, 1987).

This implementation provides a course approximation of motion energy detectors if it can be assumed that the visual system can estimate the local motion vectors of stick figures and moving dots with reasonable accuracy. Though this assumption seems to be consistent with perceptual experience it is non-trivial. We are presently extending the model by including biologically more realistic mechanisms for local motion estimation that work for real video sequences. Since recognition works even on stimuli with strongly degraded local motion information (see below), it seems that the exact mechanism of local motion estimation is not critical for the function of the model.

The second level of the motion pathway consists of neural detectors with medium-sized receptive fields that analyze the local structure of the optic flow patterns generated by biological movements. Two types of such *local optic flow detectors* are included in the model.

One class is selective for *translation flow* with four different directions and slow and fast speeds. Their responses are computed by summing the responses of local motion detectors with similar direction and speed selectivity. Neurons with similar properties are found in area MT of monkeys. Consistent with MT neurons, these detectors have low or bandpass tuning with respect to speed (cf. Rodman & Albright, 1987), and direction tuning curves with a width of about 90 deg (cf. Lagae et al., 1993). Their receptive field size is in the range that has been reported for neurons in area MT of monkeys (e.g., Albright & Desimone, 1987).

The second class of local optic flow detectors is selective for *motion edges* (horizontal and vertical). Their output signals are computed by combining the responses of two adjacent sub-fields with opposite direction preference in a multiplicative way. Neurons with such opponent motion selectivity have been found in multiple areas in the dorsal processing stream including area MT (e.g., Allman et al., 1985; Xiao et al. 1995; Born, 2000), area MSTd (e.g., Tanaka et al., 1989; Saito, 1993), and area MSTl (e.g. Eifuku & Wurtz, 1998). They are likely also present in the *kinetic occipital area* (KO) in humans that has been reported in fMRI experiments (e.g. Orban et al., 1995). The model postulates that the responses of the motion edge detectors show (partial) position invariance, like the invariant bar detectors. This position invariance is modeled by pooling the signals from position-specific motion edge detectors using a maximum operation. The receptive field size of the motion edge detectors is in the range of neurons in areas MT and MSTl in the macaque (e.g., Albright & Desimone, 1987; Eifuku & Wurtz, 1998).

The *optic flow pattern neurons* on the third level of the motion pathway are equivalent to the snapshot neurons in the form pathway. Their existence is a postulate in the proposed theory. They are selective for complex optic flow patterns that are characteristic for individual moments of biological movement patterns. An example would be the characteristic opponent movements of the opposite arms and legs during walking (see Figure 1). The receptive fields of these neurons are large (about 8 deg in the model), and they integrate information over the whole biological movement stimulus. Like the snapshot neurons the motion pattern neurons are modeled by radial basis functions that receive inputs from the previous level[2]. Their centers are trained with optic flow patterns from example movement sequences (21 centers per stored pattern were implemented). Like the snapshot neurons, the motion pattern neurons are laterally connected resulting in sequence-selective responses (see the following section). Real neurons with similar properties might be found in the STS or the *fusiform and occipital face area* (e.g., Bonda et al., 1996; Vaina et al. 2001; Grossman & Blake, 2002). Their receptive fields likely exceed the size assumed for the model, which was limited by the size of the simulated visual field (8 deg). For example, in area STP of monkeys receptive fields larger than 100 deg have been reported (e.g., Bruce et al., 1981).

The output signals of the optic flow pattern neurons are summed and temporally smoothed by the *motion pattern neurons* that form the highest level of the motion pathway. Alternatively a single set of motion pattern neurons can be implemented that integrates the information from both pathways (Giese & Vaina, 2001). Equivalent neurons in the cortex might be located in the STS (e.g., Oram & Perrett, 1996; Bonda et al., 1996; Vaina et al., 2001), the *fusiform* and the *occipital face area* (Grossman & Blake, 2002). Another possible site is area F5 in the premotor cortex (Rizzolatti et al., 2001). Likely such neurons have larger receptive fields than assumed in the model.

3.3 Sequence Selectivity

The model components described so far are suitable for recognizing stationary shapes and instantaneous optic flow patterns. They are, however, not sufficient to account for the recognition of biological movements since movement recognition is order-selective: Presentation of the frames of a movie in random temporal order typically does not lead to perception of a biological movement even though the temporally scrambled movie contains exactly the same "snapshots". In the model, sequence selectivity is implemented at the level of the snapshot neurons and the optic flow pattern neurons. The activity of these neurons is strongly dependent on the temporal order with which the individual snapshots (or optic flow patterns) arise in the stimulus.

Sequence selectivity can be accounted for by a variety of biologically plausible neural mechanisms, including the classical Reichardt detector (Hassenstein & Reichardt, 1956) and mechanisms based on dendritic delays. Many of these mechanisms have been discussed in the context of direction selectivity in the visual cortex. The mechanism in our model is based on asymmetric lateral connections between the snapshot and optic flow pattern neurons (cf. Mineiro & Zipser, 1998). Directional asymmetries between (feed-forward and lateral) inhibition and excitation have been discussed previously as a basis for direction selectivity in the primary visual cortex (Livingstone, 1998; Murty & Humphrey, 1999).

Let $u_n(t)$ be the activity of a snapshot neuron and $f(u)$ a sigmoidal nonlinear threshold function. The dynamics of the snapshot neurons is given by the differential equations

$$\begin{aligned} \tau \dot{u}_n(t) + u_n(t) &= \sum_k w(n-k)\, y_k(t) + s_n(t) \\ y_n(t) &= f(u_n(t)) \end{aligned} \tag{2}$$

where τ is a time constant and $s_n(t)$ is the output signal of the gaussian radial basis functions that have been trained with the snapshots. $w(k)$ is an asymmetric lateral interaction kernel. Through these lateral connections the activated snapshot neurons pre-excite neurons encoding temporally subsequent configurations and inhibit the neurons that encode other configurations. The principle and the network topology are illustrated schematically in Figure 2a.

For and adequate profile and strength of the lateral connections significant activity arises only when the individual snapshot neurons are activated in the right temporal order. In this case the network stabilizes a propagating activation pulse with high amplitude since its feed-forward input and the recurrent activation interact in a synergistic way. If the body configurations in the stimulus arise in the wrong temporal order the feed-forward and the recurrent feedback signals in the snapshot neurons compete, leading to a suppression of the activation. This is illustrated in Figure 2b, which shows the activation levels for correct, reverse and random temporal order. More details about this sequence selectivity mechanism are discussed in Giese & Poggio (2003), and a mathematical analysis of the network dynamics can be found in Xie & Giese (2002).

The proposed recurrent mechanism seems interesting for movement recognition for several reasons: (a) Simulations have shown that the necessary lateral connectivity can be learned robustly with a simple biologically plausible time-dependent Hebbian learning rule within 30 to 40 stimulus repetitions. This learning rate is consistent with psychophysical results (Giese et al., 2002). (b) Experimentally observed memory and delay effects (e.g., Miyashita & Chang, 1991) provide evidence for substantial effective recurrent connectivity in area IT, which might be a possible site for the snapshot neurons. Sequence selectivity and short-term memory for stationary images might thus be mediated by similar recurrent neural mechanisms (Stryker, 1991). (c) The proposed recurrent mechanism leads to short response latencies, less than 200 ms. This is consistent with electrophysiological data from neurons in the STS (Oram & Perrett, 1996) and with psychophysical observations (Johansson, 1976).

4 SIMULATION RESULTS

In the following section a number of example simulations are shown that illustrate the model's ability to reproduce multiple experimentally established properties of the recognition of biological movements within a quantitatively consistent framework. These properties are: (a) High selectivity, (b) gradual invariance against translation and scaling, and (c) robustness. In addition, the model can be used to predict results from functional imaging experiments. More simulation results are discussed in Giese & Poggio (2003).

Several of these simulations are predictions from the model since neither its parameters, nor its structure was specifically designed for approximating these results. Others results follow in a straightforward manner from the model assumptions. The model accounts for all presented results with a single set of parameters. This quantitative consistency with an extended body of experimental results provides strong constraints for the modeling and its underlying concepts.

The model was tested with biological movement stimuli that were generated by tracking the movements of human actors in more than 100 video sequences. The trajectories of the joint positions were smoothed, normalized, and used to animate point light stimuli and stick figures that had approximately the same outline as the original actor. From these animations pixel maps and optic flow fields were computed that served as inputs for the two pathways of the model.

4.1 Selectivity

The recognition of biological movements has been shown to be highly selective. However, in terms of coding efficiency it is important that the neural representations of biological movements are also capable of generalizing over multiple realizations of the same basic movement by the same, or even different actors.

As a first test of selectivity, the model was trained with 40 natural actions, including locomotion, physical exercises, dancing movements, and techniques from martial arts. The insets in Figure 3 show gray level plots of the activities of the motion pattern neurons of the two pathways. All motion pattern neurons become significantly activated only for the training pattern. This shows that either pathway can reliably distinguish between natural action categories. In an additional set of simulations each pathway was shown to generalize over multiple repetitions of the same action by the same or different actors (Giese & Poggio, 2003).

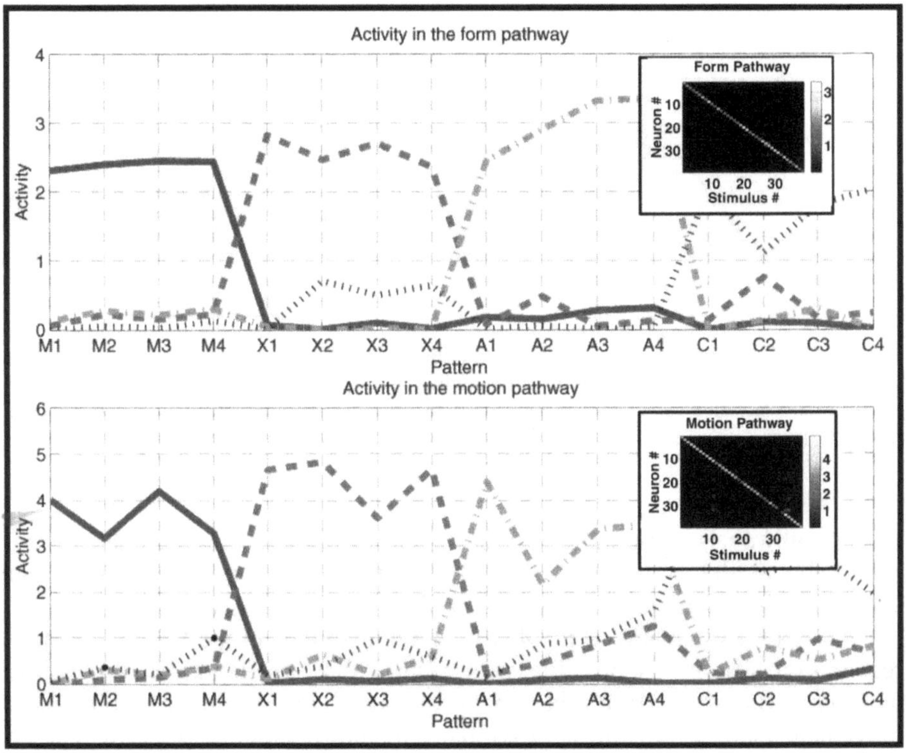

Figure 3. Activity of motion pattern neurons of the two pathways after training with the gaits of four actors (M, X, A and C). Different line styles indicate different motion pattern neurons. The insets show the (grayscale-coded) activations of 40 motion pattern neurons trained and tested with 40 different natural action categories for the form and the motion pathway. [Adapted from Giese & Poggio (2003) with permission of Nature Publishing Group]

A more striking experimental demonstration of the selectivity of biological movement perception is the recognition of identity by gait. It has been psychophysically shown that such recognition is even possible from point light displays (Cutting & Kozlowski, 1977). Figure 3 shows that the model reproduces this astonishing selectivity in spite of the absence of any sophisticated computational mechanism. After training with examples of the locomotion pattern "walking" from four different actors (M, X, A and C; two males and two females) the model was tested with multiple repetitions of walks of the same actors. Only the patterns M1, ..., C1 were used for training. The different line styles correspond to the activities of the motion pattern neurons trained with the different actors. The motion pattern neurons in both pathways show excellent discrimination between the gaits of different actors, but generalization over the multiple repetitions of the gait of the learned actor.

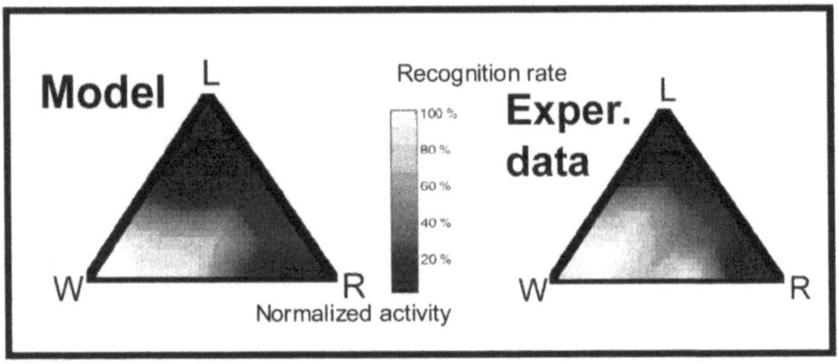

Figure 4. Left – Normalized activity of the motion pattern neurons trained with "walking" for morphs between the prototypes Walking, Running, and Limping. *Right* – Probabilities that a morph was classified as "walking" in a psychophysical experiment with seven subjects (adapted from Giese & Lappe, 2002). Pixel positions signify the weights of the three prototypes.

Either pathway is computationally sufficient to achieve this high selectivity. When the model is tested with point light stimuli one obtains a similar selectivity, but more noisy responses for the motion pathway.

A more systematic test of the generalization properties of motion recognition systems is possible with stimuli that are generated by *motion morphing*. Based on computing spatio-temporal correspondences, biological movement patterns can be approximated by linear combinations of learned prototypical example trajectories (Giese & Poggio, 2000). In this way, "slow running" can be approximated by a linear combination of "walking" and "running":

$$\text{“slow running”} = \alpha_1 * \text{“walking”} + \alpha_2 * \text{“running”} \qquad (3)$$

The weights α_1 and α_2 determine the contributions of the prototypes to the morph. At the same time, they parameterize a whole class of biological movement patterns that interpolate smoothly between walking and running. The movement pattern "slow running" can be mapped onto a point in a two-dimensional Euclidian movement pattern space with 'coordinates' $[\alpha_1, \alpha_2]$. The same argument applies to linear combinations of more than two prototypes.

We have conducted a psychophysical experiment during which subjects had to categorize motion morphs between the different prototypical locomotion patterns (Giese & Lappe, 2002). The right panel in Figure 4 shows the grayscale-coded plot of the measured categorization probability of "walking", for morphs between the three prototypical patterns "walking", "running", and "limping". The grayscale of the pixels encodes the categorization frequencies averaged over seven subjects. The position of each pixel encodes the position of the morph in the three dimensional motion pattern space. The weights of the three prototypes **W**alking, **R**unning, and **L**imping change inversely with the distance of the pixel from the corners of the triangle. The corners correspond to the pure prototypes. The midpoint of the triangle corresponds to a morph with the weight combination $\alpha_1 = \alpha_2 = \alpha_3 = 1/3$.

The model was tested with exactly the same stimuli after training with the prototypes. The left panel in Figure 4 shows the normalized activity of the motion pattern neuron (motion pathway) that encodes "walking" using the same type of representation. The model reproduces the gradual variation of categorization probability with the weights of the prototypes, and also the approximate relative sizes of the "generalization fields". The generalization field for "walking" is the region in pattern space for which the stimulus is classified walking, i.e., for which the "walking" pattern neuron is strongly activated. The generalization field for "walking" is significantly larger than those for running and limping (relative sizes, *0.39 / 0.27 / 0.29* for the psychophysical data, and *0.54 / 0.23 / 0.16* for the neural activities in the model[3]). A possible explanation for this asymmetry is that walking, by absence of extreme movements, is located more centrally in the movement pattern space than "running' and "limping". Similar results are obtained for the form pathway.

4.2 Invariance

Further simulations tested the invariance properties of the recognition model with respect to changes of the stimulus position, rotation, scaling and speed. Biological movement recognition is known to be view-dependent. This has been demonstrated in many psychophysical experiments (e.g., Sumi, 1984; Bülthoff et al., 1998). Consistent with these psychophysical results, neural responses in area TPO in the STS have been found to be dependent on rotations of the walker in depth (Perrett et al., 1985). Since the model is based on the assumption that movement patterns are encoded in terms of learned two-dimensional views and optic flow patterns it reproduces this view-dependence quantitatively (Giese & Poggio, 2003).

The model also quantitatively reproduces the position invariance of biological movement recognition. This is illustrated in Figure 5a. Here the model was trained with locomotion patterns ("walking", "running", and "limping") presented at a fixed position in the visual field. It was then tested by presenting a "walking" stimulus with different displacements along the horizontal axis relative to the training pattern. Responses of the "walking" motion pattern neuron (motion pathway) decay substantially for translations

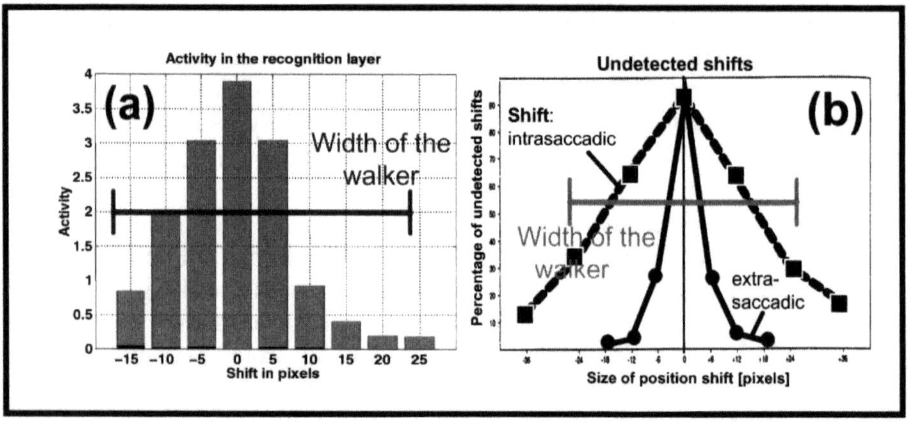

Figure 5. a) Position invariance of the model when the test view is displaced against the training view. Activity decays strongly for shifts that exceed approximately 0.5 deg. The walker had a width of about 40 pixels. b) Experimental data on the detection of shifts of a walker during saccades (dashed curve). Shifts that exceed approximately 0.5 deg are detected. Extrasaccadic shifts, cf. solid curve, are detected more easily. The walker in the experiment had a width of about 50 pixels. [Adapted from Verfaillie et al. (1993); used by permission of the American Psychological Association].

that exceed approximately 0.5 deg (about one quarter of the width of the walker). The gray bars indicate the activity of the motion pattern neuron that was trained with "walking". The black bars indicate the maximum of the activities of the other motion pattern neurons.

About the same amount of position invariance has been found in a psychophysical experiment. Subjects had to detect position changes of a point light walker that was translated during saccades (Verfaillie et al., 1993). Figure 5b shows the frequency of undetected shifts as function of the displacement size. When the displacement of the walker exceeded 0.5 deg the shifts became detectable. This is consistent with the interpretation that such position shifts should remain undetected when they are within the range of the position invariance of the underlying neural representation. Extrasaccadic shifts provide strong motion cues and are thus detected much more easily. Additional simulations show that the invariance regime of the model for scale changes is about 1.3 octaves, and the one for speed changes about two octaves. These values are consistent with electrophysiological and psychophysical data.

4.3 Robustness

Biological movement recognition is highly robust against occlusion and clutter. This is impressively demonstrated by the recognition of point light displays. It is sometimes assumed that this robustness is achieved by sophisticated computational mechanisms that reconstruct the missing information, e.g. the links between the dots, or the kinematic structure of the walker. However, it seems difficult to imagine simple physiologically plausible mechanisms that could realize such complex computations. To account for the fast recognition of point light displays in less than *300* ms such neural mechanisms would have to operate extremely quickly, not leaving much time for closed-loop optimization or complex search processes. The following simulations show that several robustness results on the recognition of point light walkers can be reproduced using simple neural mechanisms, without complex computational operations. The simulations also yield predictions about the involvement of the form and the motion pathway in the recognition of point light displays.

The most astonishing fact about Johansson's experiment is that naive observers can recognize point light stimuli immediately, without any perceptual experience. The challenge for a learning-based model is thus to account for the generalization to point light walkers after training with normal biological movement stimuli. The proposed model shows such generalization. Figure 6 shows the sum activities of the motion pattern neurons of the form and the motion pathway after training with normal full-body stimuli. The gray filled bars indicate the sums of the activities for the motion pattern neurons (labeled "M") in the motion pathway. On top of these bars the sum of the activities of the motion pattern neurons of the form pathway are plotted as black open bars (labeled "F"). The activity values induced by a normal "walking" stimulus (W) are normalized to one. Very similar sum activities arise for the normal full-body "running" stimulus (R). For testing with a "walking" point light stimulus (Wj) an interesting dissociation occurs: Significant but lower activity arises in the motion pathway, whereas the form pathway remains silent.

A more precise analysis of the activities on the different hierarchy levels shows that the reason why the form pathway does not respond is not that point light stimuli do not activate the local orientation detectors and the invariant bar detectors. Simulations with optimization of the detector properties show that it is very difficult to obtain transfer from normal to point light stimuli in the form pathway. The reason is that the form information specified by the point light stimuli deviates to strongly from the form features of the full body stimuli. This is different in the motion pathway: The optic flow field induced

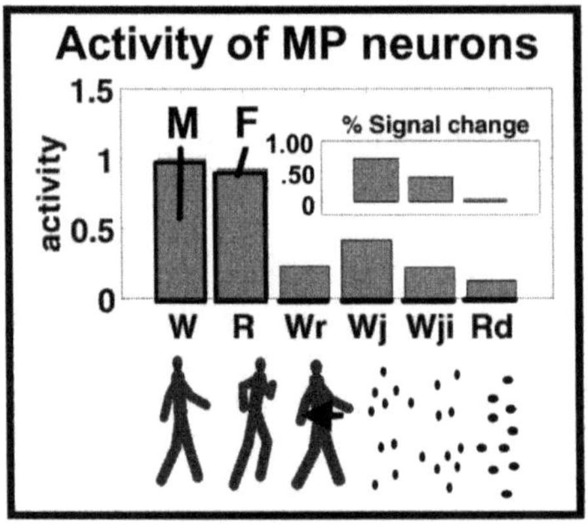

Figure 6. Responses of the motion pattern neurons of both pathways (F: form and M: motion). Gray bars indicated the sum activity of the motion pattern (MP) neurons in the motion pathway (STS), and the open black bars indicate the activities of the MP neurons in the form pathway (IT) for the following stimuli: W: walking, R: running, Wr: reverse walking (full-body), Wj: point light walker, Wji: rotated point light walker, Rd: scrambled point light walker. The inset shows the BOLD signal changes in an fMRI experiment by Grossman & Blake (2000) using the same stimuli.

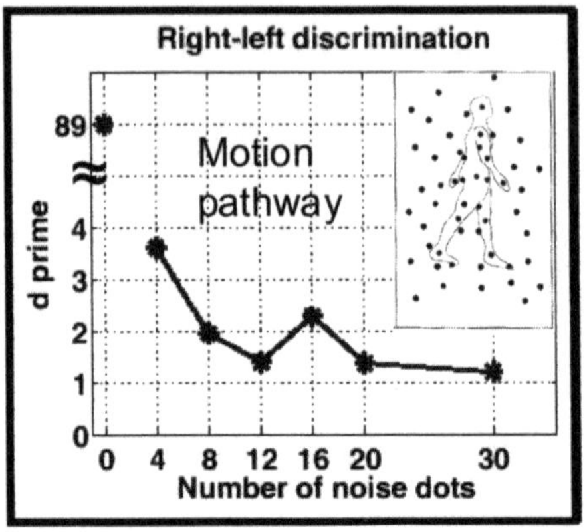

Figure 7. Values of the discrimination parameter d' computed from the responses of a motion pattern neuron trained with a normal walker stimulus walking rightwards, and tested with point light walkers walking rightwards and leftwards. If the measure $d' = (\mu_{right} - \mu_{left})([\sigma_{right}^2 + \sigma_{left}^2]/2)^{-1/2}$ exceeds one a reliable right / left discrimination is possible based on the neural signal. (μ and σ^2 are the mean and the variances of the neural activities for walking right / left. Data was obtained from 20 repeated simulations with randomly moving noise dots.)

by a point light walker is a sparsely sampled version of the optic flow field of the normal walker. By the maximum operation in the motion pathway the local optic flow pattern detectors respond well for sparsely sampled optic flow fields with similar local structure. This explains why transfer from normal to point light stimuli occurs in the motion pathway.

This result predicts that point light stimuli *should not* be specifically processed in areas that are predominantly selective for form. This prediction has been confirmed by a recent fMRI experiment during which the activity induced by point light walkers was not significantly higher than the activity induced by scrambled point light walkers in the predominately form-selective areas like the *lateral occipital complex* (LOC) and the *extrastriate body part area* EBA (Grossman & Blake, 2002).

The model predicts robustness even for further degradations of point light stimuli. Detailed simulations (Casile & Giese, submitted) show that the motion pathway of the model, consistent with psychophysical observations, successfully recognizes point light stimuli with limited dot lifetimes, even if the dots are displaced on the skeleton of the walker in every frame (Beintema & Lappe, 2002). If point light walkers are degraded by removing the dots on individual joints the model predicts, consistent with psychophysical results, that elbows and feet are most important for recognition (Giese & Poggio, 2003).

Masking experiments provide a further demonstration of robustness (e.g., Cutting et al., 1988). A substantial number of moving noise dots can be added to point light walkers without abolishing recognition. Such robustness is even observed when the background dots are created by "scrambling" point light walkers. In this case, each masking dot has the same movement as one dot of the point light walker, but a different average position.

High robustness against masking is also predicted by the model, as illustrated in Figure 7. The model was trained with normal walker stimuli walking rightwards, and tested with point light stimuli walking to the right and to the left. Different numbers of moving background dots were added to the stimulus. The figure shows a discriminability parameter d' computed from the activity levels of a motion pattern neuron that were trained with a normal walker stimulus walking rightwards[4]. Even if the number of noise dots exceeds the number of dots of the point light walker (ten) by a factor of almost three, values of $d'>1$ are obtained. This implies that, based on this neural response, a reliable discrimination of right vs. left walking can be accomplished in the majority of the trials. The proposed neural mechanism thus shows high robustness against masking. In the brain, it is likely that even higher robustness is achieved by attentional selection of the informative area in the display (Cavanagh et al., 2001).

4.4 Predictions of Imaging Results

Because the levels of the model correspond, at least coarsely, to neural structures in the visual cortex it can be used to predict results from functional imaging studies. For this purpose the sum activities of the different postulated model "areas" are computed for stimulus classes that have been used in functional imaging experiments. The assignment of "areas" to computational functions in the model must be interpreted with caution. In particular, for the higher hierarchy levels of the model there is not much neurophysiological data is available. In addition, the homology between higher areas in monkey and human visual cortex is still a topic of ongoing research. However, neural modeling seems to be beneficial in this situation for the following reasons: (1) it delineates well-defined computational steps, and (2) it provides a tool for deriving distinctive predictions that can be falsified in fMRI experiments. A few examples for predictions are discussed below.

Figure 6 shows the sum activity of the motion pattern neurons for a model with separate motion and form pathways. One prediction is that the activity in biological motion-selective areas (like the STS) should be strongly reduced when stimuli presented in reverse order are compared with stimuli presented in normal temporal order (W vs. Wr). This prediction could be readily tested in fMRI adaptation experiments (e.g., Grill-Spector & Mallach, 2001) that allow distinguishing of the activity of neural ensembles that encode forward and reverse walking. The prediction seems consistent with results from psychophysical priming experiments (Verfaillie & Daems, 2002).

Another prediction, which also illustrates view-dependence, arises when the model, previously trained with normal walker stimuli, is tested with point light stimuli (Wj), point light stimuli rotated by 180 deg (Wji), and scrambled point light stimuli (Rd). The same stimuli have been used in a recent fMRI experiment (Grossman & Blake, 2001). In this experiment, first the STS was localized and then the activity in the STS induced by the three stimuli was quantified. The inset in Figure 6 shows the measured percentages of change of the BOLD signal. The predicted variation of the sum activities of the motion pattern neurons in the model matches qualitatively the changes in the BOLD signal. This prediction is highly nontrivial, given that the model was not even specifically designed to recognize point light displays.

5 DISCUSSION

The recognition of biological movements is an important perceptual function, for communication and recognition at a distance, but potentially also in the context of imitation learning of complex motor actions (Rizzolatti et al., 2001). The neural model presented in this chapter achieves recognition of biological movements with well-established biologically plausible neural mechanisms. The key assumption of the model was that biological movement patterns are encoded in terms of learned prototypes. Prototype-based representations have been a powerful theoretical concept in the domain of stationary object recognition (e.g., Poggio, 1990; Tarr & Bülthoff, 1998). One major conceptual contribution of this work is the transfer of this concept to the domain of motion recognition at the level of neural mechanisms. (For a related discussion in the domain of psychophysics see also Verfaillie, 2000.)

As second major result it was shown that, assuming neurophysiologically plausible neural detectors, either visual pathway seems to be computationally sufficient to achieve highly selective motion recognition for normal stimuli. However, our results suggest that the recognition of point light stimuli might be mainly based on motion information. This prediction is consistent with the results of a recent fMRI experiment (Grossman & Blake, 2002).

The restriction to physiologically plausible mechanisms sets the presented model apart from most computational approaches to movement recognition in computer vision. Many algorithms exploit concepts, like articulated body models or graph structures that have no evident neural implementation (e.g., Aggarwal & Cai, 1999; Gavrila, 1999; Song et al. 2001). Our simulations show that several basic results on the recognition of complex movements can be explained without such complex computational principles. In the domain of computer vision neural networks have been used for motion recognition before (e.g., Goddard, 1992; Rosenblum et al., 1996). Such systems exploit principles similar to those in the proposed model. However, most technical solutions do not take the specific constraints of the visual cortex into account. This makes it difficult to compare them directly, or even quantitatively, with experimental data.

The model makes a number of predictions that can be experimentally tested. One major prediction is that arbitrary complex movement patterns can be learned, as long as they provide adequate stimulation for the low- and mid-level detectors. This prediction implies that the recognition of biological movements is not "special" in any sense, except for the fact that movements like walking are ecologically highly relevant and frequently occurring. This prediction can be tested directly by verifying whether artificial movement patterns can be learned, and whether the learned representations show the same invariance properties as representations of natural movements. We have conducted psychophysical experiments that confirm this prediction: artificial movements can be learned very quickly and the learned representations are view-dependent (Giese et al., 2002).

Further predictions follow from the postulated hierarchical structure of the motion pathway. Simulations show that the opponent motion detectors might be crucial for the recognition of biological movements. Their elimination abolishes recognition, whereas this is not the case when the detectors for translation flow are removed. This seems to be consistent with the observation in fMRI experiments that the opponent motion sensitive area KO becomes active during biological motion recognition (Vaina et al., 2001). Patient studies and direct recordings in opponent motion-sensitive neurons in presence of biological movement stimuli might help verify this prediction.

Another set of predictions follows from the assumed representation of biological movements in terms of sequences of snapshots and optic flow patterns. Responses of neurons encoding snapshots (or instantaneous optic flow patterns) should reflect this sequence selectivity. A test of this prediction seems possible in electrophysiological experiments: The activity of neurons that are selective for body postures (e.g., in area STP) should be higher when the corresponding snapshot is embedded in a temporal sequence presenting frames from a biological movement in right temporal order, than for sequences with a different order of the frames (cf. Figure 2b).

The association of sequential information over time seems, at first look, relevant only for the recognition of image sequences. Interestingly, the same principle has been previously discussed for the learning of view invariance (Wallis & Rolls, 1997). This implies that sequence recognition and the learning of view-invariance might share similar neural mechanisms.

Of course, the proposed model provides only an extremely simplified framework that will have to be modified and refined substantially based on further experimental data. One major limitation of the model is the lack of top-down influences and task-dependent modulation. Without doubt there are strong influences of attention and of the task on the perception of biological movements and the related neural activity (e.g., Cavanagh et al., 2001; Vaina et al. 2001). A second over-simplification is the strict separation of the two pathways in the model. Future work will have to address these limitations by appropriate extensions of the model.

Given these limitations, it seems intriguing that the proposed skeleton feed-forward architecture is already sufficient to account for a variety of the known experimental results. This result implies in particular that no sophisticated top-down mechanisms are required to accomplish basic tasks in biological movement recognition. Experimental evidence suggests that such factors are important for more complex tasks (e.g., Cavanagh et al., 2001).

The type of the theory proposed in this chapter differs from other applications of theoretical modeling in vision science. Instead of fitting a relatively limited amount of experimental results using a model with a minimum number of free parameters the proposed theory tries to integrate a whole body of experimental results obtained with multiple experimental methods. This was achieved at the expense of less accurate quantitative fitting of individual experiments. It is important to recognize that, in spite of such 'inaccurate quantitative fitting' the proposed theory is highly constrained – by the requirements of coarse quantitative consistency with multiple key experimental results and neurophysiological plausibility.

Given the still quite limited amount of neurophysiological data on motion recognition, this approach seems more suitable than exact quantitative fitting of individual experiments for several reasons: (a) Given the strong modulation of motion recognition by top-down influences exact quantitative fitting seems questionable, unless such factors are exactly controlled in the experiments (which is not the case for most existing data). (b) All known computational approaches for the recognition of complex movements are based on mathematical descriptions with a large effective number of unknown parameters (structure parameters of kinematic models, transition probabilities in predictive Bayesian networks, etc.). This implies that it is difficult to falsify such computational models by fitting small data sets. The proposed model makes specific assumptions about neural mechanisms that might be wrong, but which can be falsified experimentally. (c) By combining the constraints obtained from different experiments and using different experimental methods it seems possible to rule out possible mechanisms of movement recognition. This seems to be much more difficult with models that are based on concepts that are not accessible experimentally, even if they fit individual data sets very accurately. (d) The proposed theory points to relationships between different experimental results that are not obvious.

It seems likely that quantitative modeling of individual data sets will become more important once constraining neurophysiological data is available. Quantitative modeling of decisive neurophysiological data sets might finally be essential to unravel the true neural mechanisms of biological movement recognition.

ACKNOWLEDGMENTS

I am grateful to L. Vaina for providing helpful comments, and for the invitation to this chapter. Special thanks to T. Poggio for supporting this work and many insightful remarks. Thanks also to R. Blake, I. Bülthoff, S. Edelman, C. Koch, Z. Kourtzi, D. Perrett, M. Riesenhuber, P. Sinha, T. Sejnowski, and I. Thornton for reading previous versions of this manuscript. A. Benali, Z. Kourtzi, and C. Curio helped with the data acquisition. M. Giese was supported by the Deutsche Forschungsgemeinschaft, Honda R&D Americas Inc., and the Deusche Volkswagen Stiftung. Additional support was provided by the Max Planck Institute for Biological Cybernetics in Tübingen. CBCL was sponsored by the National Science Foundation, Central Research Institute of Electric Power Industry, Eastman Kodak Company, DaimlerChrysler AG, Compaq, Komatsu, Ltd., NEC Fund, Nippon Telegraph & Telephone, Siemens Corporate Research, Inc.

NOTES

1. The activity in the extrastriate body part area (EBA) seems not to show sequence selectivity (Downing et al., 2001; Grossman & Blake, 2002). Sequence selective activation for human bodies has been found in the *fusiform* and *parietal face* area (Grossman & Blake, 2002). It seems likely that face movements can be decoded with the same neural mechanisms even though this was not tested with our model (see Rosenblum et al., 1996, for a related technical implementation).

2. Only a subset of the orientation detectors from the middle level of the form pathway are connected with the snapshot neurons. This set of detectors was selected by the requirement that their responses should vary significantly over the training patterns. This "feature selection" excludes orientation features that do not carry relevant information about biological movement patterns and makes the recognition more robust. Such a selection could follow from many physiologically plausible competitive learning rules. The same

selection rule was applied for the afferents of the optic flow pattern neurons in the motion pathway.

3. The size estimate from the psychophysical data was the average classification probability of the pattern over all tested morphs. The size estimate from the model was obtained by normalizing the activities of the motion pattern neuron to the interval [0, 1] and computing the average activity over all tested morphs.

4. The model does not contain internal stochastic noise. Noise is however introduced by the random masking stimuli leading to a distribution of hits and false alarms. The coefficient *d'* is a coarse measure for the discrimination performance of the model. If the underlying distributions were gaussian this parameter would coincide with the discriminability measure *d'* from signal detection theory.

REFERENCES

Aggarwal, J. K., & Cai, Q. (1999). Human motion analysis: a review, *Comput. Vis. Im. Underst., 73,* 428-440.

Albright, T. D., & Desimone, R. (1987). Local precision in the middle temporal visual area MT of the macaque, *Exp. Brain Res. 65,* 582-592.

Allison, T., Puce, A., & McCarthy, G. (2000). Social perception from visual cues: role of the STS region, *Trends Cogn. Sci., 4,* 267-278.

Allman, J., Miezin, F., & Mc Guinness, E. (1985). Direction- and velocity-specific responses from beyond the classical receptive field in the middle temporal visual area (MT), *Perception, 14,* 105-126.

Beintema, J. A., & Lappe M. (2002). Perception of biological motion without local image motion, *Poc. Nat.l. Acad.. Sci. U S A*, *99,* 5661-5663.

Bonda, E., Petrides, M., Ostry, D., & Evans, A. (1996). Specific involvement of human parietal systems and the amygdala in the perception of biological motion, *J. Neurosci.,16,* 3737-2744.

Born, R. T. (2000) Center-surround interactions in the middle temporal visual area of the owl monkey, *J. Neurophysiol., 84,* 2658-2669.

Bülthoff, I. Bülthoff, H. H. & Sinha, P. (1998). Top-down influences on stereoscopic depth-perception, *Nat. Neurosci.*, *1,* 254-257.

Cavanagh, P., Labianca, A. P., & Thornton, I. M. (2001). Attention-based visual routines: sprites, *Cognition, 80,* 47-60.

Cutting, J., & Kozlowski, L.T. (1977). Recognizing friends by their walk: Gait perception without familiarity cues, *Bull. Psychon. Soc., 9,* 353-356.

Cutting, J. E., Moore, C., & Morrison, R. (1988). Masking the motions of human gait, *Percept. Psychophys., 44,* 339-347.

Decety, J., & Grèzes, J. (1999). Neural mechanisms subserving the perception of human actions, *Trends Cogn. Sci., 3,* 172-178.

Dow, B. M., Snyder, A. Z., Vautin, R. G., & Bauer, R. (1981). Magnification factor and receptive field size in foveal striate cortex of the monkey, *Exp. Brain Res., 44,* 213-228.

Downing, P., Jiang, Y., Shuman, M. & Kanwisher, N. (2001). A cortical area selective for visual processing of the human body, *Science, 293,* 2470-2473.

Eifuku, S., & Wurtz, R. H. (1998). Response to motion in extrastiate area MSTl: center-surround interactions, *J. Neurophysiol., 80,* 282-296.

Felleman D. J., & van Essen, D. C. (1991). Distributed hierarchical processing in the primate visual cortex., *Cereb. Cortex, 1,* 1-49.

Fukushima, K. (1980). Neocognitron. A self-organizing neural network model for a mechanism of pattern recognition unaffected by shift in position, *Biol. Cybern., 36,* 193-202.

Gavrila, D. M. (1999). The visual analysis of human movement: a survey, *Comput. Vis. Image Underst., 73,* 82-98.

Gawne, T. J., & Martin, J. (2002). Response of primate visual cortical V4 neurons to two simultaneously presented stimuli, *J. Neurophysiol., 88,* 1128-1135.

Giese, M. A., Jastorff, J., & Kourtzi, Z. (2002). Learning of the discrimination of artificial complex biological motion, *Perception, 31 (Suppl.),* 117.

Giese, M. A., & Lappe, M. (2002). Measuring generalization fields for the recognition of biological motion, *Vision Res., 42,* 1847-1858.

Giese, M. A., & Poggio, T. (2000). Morphable models for the analysis and synthesis of complex motion patterns, *Int. J. Comput. Vis., 38,* 59-73.

Giese, M. A., & Poggio, T. (2003). Neural mechanisms for the recognition of biological movements, *Nat. Rev. Neurosci., 4,* 179-192.

Giese, M. A., & Vaina, L. M. (2001). Pathways in the analysis of biological motion: computational model and fMRI results, *Perception, 30 (Suppl.),* 116.

Goddard, N. H. (1992). *The Perception of Articulated Motion: Recognizing Moving Light Displays.* Thesis: Dept. of Computer Science, Univ. of Rochester.

Goodale, M. A., & Milner, A. D. (1992). Separate visual pathways for perception and action, *Trends Neurosci., 15,* 97–112.

Grill-Spector, K., & Malach, R. (2001). fMR-Adaptation: a tool for studying the functional properties of human cortical neurons, *Acta Psychol. (Amst.), 107,* 293-321.

Grossman, E. D., & Blake, R. (2001). Brain activity evoked by inverted and imagined biological motion, *Vision Res., 41,* 1475-1482.

Grossman, E. D., & Blake, R. (2002). Brain areas active during visual perception of biological motion, *Neuron, 35,* 1167-1175.

Hassenstein, B., & Reichardt, W. E. (1956). Systemtheoretische analyse der zeit-, reihenfolgen- und vorzeichenauswertung bei der bewegungsperzeption des rüsselkäfers chlorophanus, *Z. Naturforsch., 11b,* 513-524.

Hegdé J., & van Essen, D. C. (2000). Selectivity for complex shapes in primate visual area V2, *J. Neurosci., 20, RC61,* 1-6.

Hubel, D. H., & Wiesel, T. N. (1962). Receptive fields, binocular interaction and functional architecture in the cat's visual cortex, *J. Physiol. (Lond..), 160,* 106-154.

Johansson, G. (1973). Visual perception of biological motion and a model for its analysis, *Percept. Psychophys.*, *14,* 201-211.

Johansson, G. (1976). Spatio-temporal differentiation and integration in visual motion perception. An experimental and theoretical analysis of calculus-like functions in visual data processing, *Psych. Res.*,379-393.

Jones, J. P., & Palmer, L. A. (1987). An evaluation of the two-dimensional Gabor filter model of simple receptive fields in cat striate cortex, *J. Neurophysiol.*, *58,* 1233-1258.

Kourtzi, Z., & Kanwisher, N. (2000). Activation in human MT/MST by static images with implied motion, *J. Cogn. Neurosci.*, *2,* 48-55.

Kozlowski, L. T., & Cutting, J. E. (1977). Recognizing the sex of a walker from a dynamic point light display, *Percept. Psychophys.*, *21,* 575-580.

Lagae, L., Raiguel, S., & Orban, G. A. (1993). Speed and direction selectivity of macaque middle temporal neurons, *J. Neurophysiol.*, *69,* 19-39.

Livingstone, M. S. (1998). Mechanisms of direction selectivity in macaque V1, *Neuron, 20,* 509-526.

Logothetis, N. K., & Sheinberg, D. L. (1996). Visual object vision, *Ann. Rev. Neurosci.*, *19,* 577-621.

Mel, B, & Fieser, J. (2000). Minimizing binding errors using learned conjunctive features, *Neural Comput.*, *9,* 779-796.

Mineiro, P., & Zipser, D. (1998). Analysis of direction selectivity arising from recurrent cortical interactions, *Neural Comput.*, *10,* 353-371.

Miyashita, Y. & Chang, H. S. (1988). Neural correlate of pictorial short-term memory in the primate temporal cortex, *Nature, 331,* 68-70.

Morel, A., & Bullier, J. (1990). Anatomical segregation of two cortical visual pathways in the macaque monkey, *Vis. Neurosci.*, *4,* 555-578.

Murthy, A., & Humphrey, A. L. (1999). Inhibitory contributions to spatio-temporal receptive-field structure and direction selectivity in simple cells of cat area 17, *J. Neurophysiol.*, *81,* 1212-1224.

Op De Beeck, H., & Vogels, R. (2000). Spatial sensitivity of macaque inferior temporal neurons, *J. Comp. Neurol.*, *30,* 505-518.

Oram, M. W., & Perrett, D. I. (1996). Integration of form and motion in the anterior temporal polysensory area (STPa) of the macaque monkey, *J. Neurophysiol.*, *76,* 109-129.

Orban, G., Dupont, P., De Bruyn, B., Vogels, R., Vandenberghe, R., Mortelmans, L. (1995). A motion area in human visual cortex, *Proc. Nat. Acad. Sci. USA, 92,* 993-997.

Pasupathy, A. Connor, C. E. (1999). Response to contour features in macaque area V4, *J. Neurophysiol.*, *82,* 2490-2502.

Perrett, D. I., Smith, P. A., Mistlin, A. J., Chitty, A. J., Head, A. S., Potter, D. D., Broennimann, R., Milner, A. D., & Jeeves, M. A. (1985). Visual Analysis of body movements by neurones in the temporal cortex in the macaque monkey: a preliminary report, *Behav. Brain Res.*, *16,* 153-170.

Perrett, D. I., & Oram, M. W. (1993). Neurophysiology of shape processing, *Im.. Vis. Comput.*, *11,* 317-333.

Pinon, M. C., Gattass, R., & Sousa, A. P. (1998). Area V4 in Cebus monkey: extent and visuotopic organization, *Cereb. Cortex, 8,* 685-701.

Poggio, T. (1990). A Theory of How the Brain Might Work. In: *Proceedings of Cold Spring Harbor Symposia on Quantitative Biology 4* .(pp. 899-910). Cold Spring Harbor: Cold Spring Harbor Laboratory Press.

Riesenhuber, M., & Poggio, T. (1999). Hierarchical models of object recognition, *Nat. Neurosci., 2,* 1019-1025.

Rizzolatti, G., Fogassi L., & Gallese, V. (2001). Neurophysiological mechanisms underlying the understanding and imitation of action, *Nat. Rev. Neurosci., 2,* 661-670.

Rodman, H. R., & Albright, T. D. (1987). Coding of visual stimulus velocity in area MT of the macaque, *Vision Res., 27,* 2035-2048.

Rolls, E. T., & Milward, T. (2000). A model of invariant object recognition in the visual system: learning rules, activation functions, lateral inhibition, and information-based performance measures, *Neural Comput., 12,* 2547-2572.

Rosenblum, M., Yacoob, Y, & Davis, L. (1996). Human emotion recognition from motion using a radial basis function network architecture, *IEEE Trans. Neural. Netw., 7,* 1121-1138.

Saito, H. (1993). Hierarchical Neural Analysis of Optical Flow in the Macaque Visual Pathway. In: T. Ono, L. R. Squire, M. E. Raichle, D. I. Perrett, & M. Fukuda (Eds.), *Brain Mechanisms of Perception and Memory* (pp. 121-140) Oxford: Oxford University Press.

Saleem, K. S., Suzuki, W., Tanaka, K., & Hashikawa, T. (2000). Connections between anterior inferotemporal cortex and superior temporal sulcus regions in the macaque monkey, *J. Neurosci., 20,* 5083-5101.

Sereno, M. E. (1993). *Neural Computation of Pattern Motion.* Cambridge: MIT Press.

Simoncelli, E. P., & Heeger, D. J. (1998). A model of neuronal responses in visual area MT, *Vision Res., 38,* 743-761.

Smith, A. T., & Snowden, R. J. (1994). *Visual Detection of Motion.* London: Academic Press, 1994.

Song, Y, Goncalves, L., Di Bernardo, E., & Perona, P. (2001). Monocular perception of biological motion in Johansson displays, *Comput. Vis. Im. Underst., 81,* 303-327.

Stryker, M. P. (1991). Neurobiology. Temporal associations, *Nature, 354,* 108-109.

Sumi, S. (1984). Upside-down presentation of the Johansson moving light-spot pattern, *Perception, 13,* 283-302.

Tanaka, K. (1996). Inferotemporal cortex and object vision, *Ann. Rev. Neurosci., 19,* 109-139.

Tanaka, K., Fukuda, Y., & Saito, H. (1989). Analysis of motion of the visual field by direction, expansion / contraction, and rotation cells clustered in the dorsal part of the medial superior temporal area of the macaque monkey, *J. Neurophysiol., 62,* 626-641.

Tarr M. J., & Bülthoff, H. H. (1998). Image-based object recognition in man, monkey and machine, *Cognition, 67,* 1-20.

Ungerleider, L. G., & Mishkin, M. (1982). Two Cortical Visual Systems. In: D. J. Ingle, M. A. Goodale, & R. J. W. Mansfield (Eds.), *Analysis of Visual Behavior* (pp. 549-586) Cambridge: MIT Press.

Vaina, L. M., Solomon, J., Chowdhury, S., Sinha, P., & Belliveau, J. W. (2001). Functional neuroanatomy of biological motion perception in humans, *Proc. Natl. Acad. Sci. USA, 98,* 11656-11661.

Verfaillie, K. De Troy, A., & van Rensbergen, J. (1994). Transsaccadic integration of biological motion, *J. Exp. Psychol. Learn. Mem. Cogn., 20,* 649-670.

Verfaillie, K. (2000). Perceiving human locomotion: priming effects in direction discrimination, *Brain Cogn., 44,* 192-213.

Verfaille, K., & Daems, A. (2002). Representing and anticipating actions in vision, *Vis. Cogn., 9,* 217-232.

Wallis, G., & Rolls, E. (1997). A model of invariant object recognition in the visual system, *Prog. Neurobiol., 51,* 167-194.

Xiao, D, K., Raiguel, S., Marcar, V., Koenderink, J., & Orban, G. A. (1995). Spatial heterogeneity of inhibitory surrounds in the middle temporal visual area, *Proc. Natl. Acad. Sci. USA, 92,* 11303-11306.

Xie, X., & Giese, M. A. (2002). Nonlinear dynamics of direction-selective nonlinear neural media, *Phys. Rev. E. Stat. Nonlin. Soft Matter Phys., 65,* 051904.

Young, M. P., Yamane, S. (1992). Sparse population coding of faces in the inferotemporal cortex, *Science, 256,* 1327-1331.

Yu, A. J., Giese, M. A., & Poggio, T. (2002). Biophysiologically plausible implementations of the maximum operation, *Neural Comput., 14,* 2857-2881.

21. Controlling Bipedal Movement Using Optic Flow

M. Anthony Lewis

Iguana Robotics, Inc.
Urbana, IL

1 INTRODUCTION

In the 1950's Gibson pointed out the importance of the flow of the 'optic array' (i.e. optic flow), a visual cue arising from relative movement of the environment, in the control of human locomotion (Gibson 1958). Relative motion of the environment can reveal environment structure. His pioneering ideas have influenced generations of experimentalists in the brain and behavioral sciences.

Visual information is extremely important for human locomotion. Locomotion is a complex behavior that requires close coordination with the environment. This coordination can sometimes be accomplished by mechanical coupling of the musculoskeletal system to the substrate of locomotion alone. However, large, fast moving animals in cluttered environments adopt a distal sensing strategy to activate, in an anticipatory manner, motor synergies appropriate to maintain this coordination.

Humans use various visual cues for navigation during locomotion including motion-based cues. This information is used to give a sense of the relationship of the viewer to the environment. For example, kinesthetic information derived from vision provides powerful cues to self-motion (Lee & Young, 1986). Motion information is apparently used for complex tasks such as stepping over obstacles. Monocular subjects can both judge distance to, and step over obstacles. However, these subjects exhibit greater variability in performance. See (Patla, 1997) for a review of the use of vision during locomotion.

Robots provide an interesting platform for theoretical investigation into visually guided locomotion. As an artificial entity, much as in a computer simulation, variables and algorithms can be carefully controlled. However, a robot, unlike a computer simulation, can interact with the world. Robots that

L.M. Vaina, S.A. Beardsley and S.K. Rushton (eds.), Optic Flow and Beyond, 471–485.

seek to simulate biological systems with the goal of gaining a better understanding of biological systems or to achieve purely engineering objectives are collectively referred to as biorobots or biomorphic robots.

In biomorphic robot investigations, optic flow has been generally considered as part of a feedback control loop and modeled based on insect behavior. Insect motion processing pathways are relatively well understood. A number of robotic implementations use optic flow as their primary input for visual navigation. For a sampling of that work, see (Franceschini et al., 1992; Sobey, 1996; Weber et al., 1996; Srinivasan & Venkatesh, 1997; Coombs et al., 1998; Lewis, 1998; Lewis & Nelson 1998; Harrison, 2000).

However, there has been little or no work using optic flow to control legged locomotion, although there has been interest in other visual techniques for the control of walking machines (e.g., visual servoing, color, stereopsis, and terrain reconstruction using laser ranging). Given the strong belief in the biological community that optic flow is a key component of visuomotor coordination we decided to investigate how optic flow might be used to control a legged robot.

In previous work we have shown how a robot can learn to step over obstacles using a reinforcement paradigm (Lewis & Simó 1999, 2001). The control architecture featured stereo input, and used sensory expectation, i.e., building up an expectation of the surface in front of the robot, for the detection of surface anomalies or obstacles and the association. These novelties were then associated with motor behavior, for example, the control of foot placement and triggering when to step over the obstacle. A key observation of that work was that what structural form of the environment constitutes an "obstacle" is dependent on the capabilities of the robot. The robot learned to perceive only those "obstacles" that could potentially destabilize it. An overview of the architecture is shown in Figure 1.

In this chapter, we focus on replacing the front-end of this model with a new visual cue: optic flow. At the outset, it is not clear whether optic flow is stable enough to allow detection of fine surface features. Our experiments here are designed to evaluate if this is possible.

The hypothesis which we will test is: (1) Optic flow can be used to detect potentially destabilizing environmental features during locomotion in a biped, even given significant up and down movement and jarring of the robot during locomotion; (2) Reliable detection is only possible if a prediction of the expected optic flow field is made at each instance; (3) This prediction can be accomplished in a distributed, biologically plausible framework.

The architecture of the model uses neural elements and is motivated by the structure of the vertebrate nervous system.

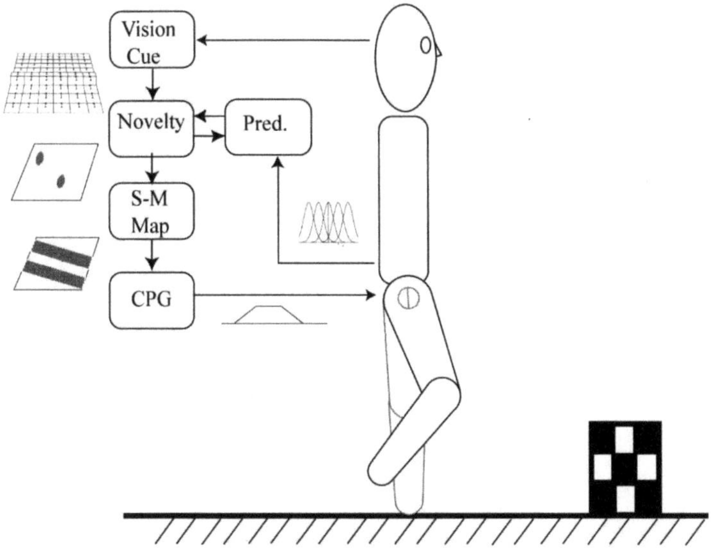

Figure 1. Overall control system for a robot that learns to step over obstacles. Visual cues, i.e. stereopsis or optic flow 'events' are compared to a prediction of these events. The prediction is driven by the phase of the step cycle. Any resulting difference is considered 'novel' stimuli. These novel stimuli are weighted via the "modulation weight fields" and used to adjust stride length before encountering an obstacle, as well as triggering stepping over the obstacle. In this chapter, we are concerned with the use of optic flow as the source of the visual cue. The behavior we use to assess performance is stopping.

2 MATERIALS AND METHODS

2.1 The Robot Mechanism

A 20-cm tall tethered biped is used in the following experiments. See Figure 2. The tether allows forward/backward and up/down translation of the body. The hip's rotation is held fixed. Two miniature cameras give the robot a view from its feet to about 40 cm in front of the robot. The robot itself uses four hobby type servos to actuate the limbs (Futaba 3002 for hips and 5203 for knees). The servos are controlled by a custom board (ServoX24 Board, Digital Designs and Systems, Cambridge, MA). Joint level commands are streamed over a RS-232 port from a Win 2000 Laptop (550MHz Dell). Visual computation is performed on a Linux workstation (1.7 GHz Dell).

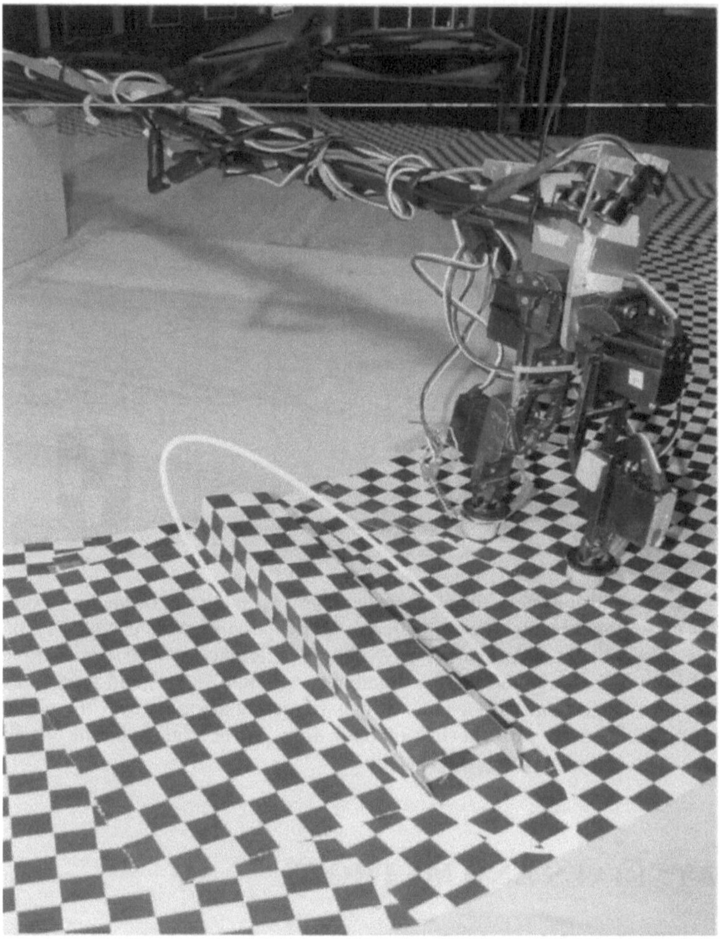

Figure 2. The walking biped mechanism and camouflaged obstacle. The obstacle has the same texture pattern as the track. A white oval is placed on the center of the obstacle to help the reader localize it.

2.2 The Locomotory Pattern Generator

The locomotory pattern is generated using a Central Pattern Generator (CPG) network. The network also generates a distributed representation of the phase of the left and right halves of the CPG network. In general, this phase information is used to coordinate visual processing with the step cycle. In the particular case presented here, this phase information and joint information is used to predict the expected optic flow at 18 horizontal band regions in the incoming image (see Figure 5). Balance is not a concern here as the tether provides a great deal of stability to the biped. The walking speed of the robot was about 2 cm/sec.

2.3 Using Optic Flow Triggered Stopping

Visual data caries a significant amount of information about the relationship of the animal or robot to the environment. The computation applied to an array of intensity values, as would be measured with a camera or in the retina, may provide misleading information. In particular, if a surface is smooth, such as a blank wall, there may be very little edge information. In addition, there are fundamental problems in computing optic flow from digitized information. A typical camera system digitizes information to only 256 gray levels. The computation for a gradient involves taking the difference between adjacent pixel levels. Any digitization noise is greatly amplified by this differencing operation. To address this problem, a spatial smoothing stage is applied before the gradient is computed, or a kernel can be designed that computes both operations at once. In this particular work, smoothing and differentiation occurred sequentially.

Normal flow is computed using an algorithm previously described. (Lewis, 1999). The components of flow are computed as:

$$\frac{dx}{dt} \propto \int \frac{E_t(\nabla E \cdot [1 \quad 0]^T)}{\|\nabla E\|^2} \tag{1}$$

and

$$\frac{dy}{dt} \propto \int \frac{E_t(\nabla E \cdot [0 \quad 1]^T)}{\|\nabla E\|^2} \tag{2}$$

where E_t is the temporal derivative and ∇E is the spatial gradient. The integral indicates averaging over a small patch of the image.

We have shown that this algorithm will give a good estimate of the true optic flow if the statistics of spatial gradient orientations are balanced in their statistical distribution, e.g. with the assumption of a flat distribution for edge orientation (Lewis, 1999). The results should also hold here where the ratio of horizontal and vertical edges is about 1:1.

Information is selectively filtered based on an optic flow "event" representation. For an optic flow estimate to be forwarded to further stages of

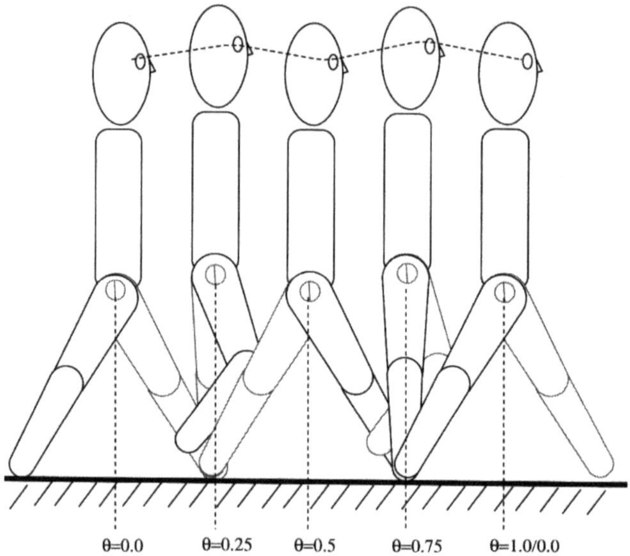

Figure 3. As a biped moves it generates considerable up and down movement, which confounds the interpretation of the optic flow field. In addition, there is jarring which can make the interpretation of the optic flow field unreliable.

processing, we check that E_t and ∇E are of sufficient magnitude to give a reliable estimate of optic flow. Thus, if sparse edges are available, few 'events' will be sent to latter stages of processing. The lack of an event does not imply that there is or is not an environmental surface present. The lack of an event only means that there is insufficient evidence to decide one way or another. This idea of detecting "events" is basic to communication between "Neuromorphic" chips (i.e., chips based on biological processing, some of which can compute 1- and 2-D flow fields at high rates). The interested reader is referred to (Lazzaro & Wawrzynek, 1995; Lewis, 1999) for more detail.

2.3.1 Optic Flow During Locomotion

The central problem of using optic flow to guide a *walking* robot during locomotion is that the up and down (and sideways) movement of the robot generates an optic flow field that varies significantly during a gait cycle, even when stepping over flat surfaces (see Figure 3). The robot used here captures the key features of this variable movement, except sideways movement due to the tether constraint.

Because the movement is periodic, we hypothesize that at different points in the step cycle, there is a different likelihood of a given optic flow vector occurring at a point in the image. If we can *predict* the most likely heading and its magnitude, we can then detect anomalous conditions in the flow field. We describe our approach to this below.

2.3.2 Generating Optic Flow Expectations

In previous work (Lewis & Simó, 2001) we describe a walking robot that learns to step over obstacles. A key component of this work was the generation of "sensory expectations". Here, we rely on the same computation to generate expectancies of optic flow.

Referring to Figure 1, visual cues are processed as described in Section 2.3 above. Simultaneously, a prediction is made of the expected optic flow vector, denoted by the "Pred" box. The predictor has 20 subsections, activated sequentially during different phases of movement.

The phase of movement, θ, is recoded into a distributed representation as a vector of twenty values using a family of Gaussian-like functions. The *ith* element is given as:

$$\Theta_i = \exp(-a \cdot (\theta - \theta_i)^2) \tag{3}$$

where Θ is a vector of 20 elements representing the phase angle, θ_i are evenly spaced centers of each kernel function and a is chosen to allow some overlap between adjacent coding functions. For example, over the interval [0.0, 0.05], the first subsection of the predictor is largely activated, at [0.05, 0.1] the second subsection is largely activated and so on. Each subsection specializes in predicting the optic flow at over a particular interval of phase. This architecture effectively creates 20 separate specialized predictive circuits. This recoding is represented as the graph of a family of "bump" functions in Figure 1.

The next stage of processing is denoted by the "Novelty" box in Figure 1. Here, the comparison between the actual versus the predicted optic flow vector is made. A comparison is only made if an optic flow event is generated.

The "Gain" stage of processing contains an adaptive sensitivity function. This adaptive sensitivity seeks to produce a certain probability of detecting events based on assumptions about the statistics of anomalies in the environment. In the work here, we assume 5% as the expected rate of anomalies per optic flow region per step. Thus, sensitivity is dynamically adjusted to maintain the expected rate of anomalies

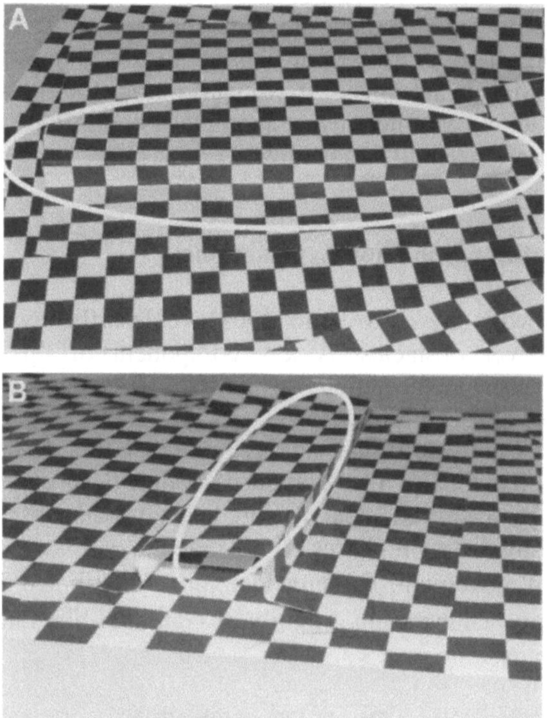

Figure 4. Typical obstacle used in the experiments. a) Obstacle from the robot view is almost 'invisible'. A white eclipse is placed on the obstacle to help the reader find it. b) The same obstacle, side view.

2.3.3 Criteria for 'Stopping' the Robot

Based on the 5% threshold for each firing, and given that in our model about 18 cells have the potential for detecting an obstacle, we set the threshold for obstacle detection at 140% of the expected spontaneous firing (in practice, this threshold can be found optimally for a given application). The detection is based on a running average of "novelty pixels" with a time constant of about 1 step cycle.

2.3.4 Obstacles

Figure 4a shows an obstacle from the robot's view. As can be seen, the obstacle is well camouflaged. In Figure 4b we see the same obstacle from the

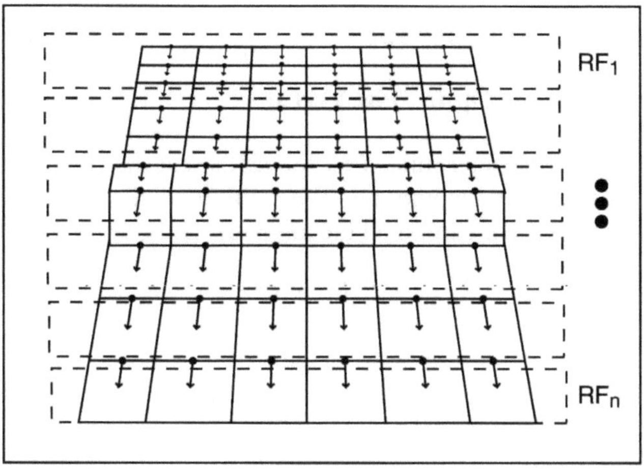

Figure 5. Obstacle induced optic flow field. The checkerboard pattern shown in Figure 4 induces an optic flow field as shown above. Flow vectors meeting minimum threshold criteria are averaged to create a value for a given receptive field. Horizontally orientated receptive fields are shown as dashed line boxes. The value of the center receptive field is subtracted from all receptive fields. This emulates fixation in the walking robot.

side. Because of the camouflage, the robot must use optic flow or a similar cue to detect the obstacle.

The obstacles were constructed from paper printed with a special pattern as shown in Figure 4. The paper was then folded appropriately to form obstacles of the needed heights. In Figure 5 we see an illustration of the optic flow field generated by the checkerboard pattern and some of the 18 bands that detect optic flow events.

2.3.5 Detecting Obstacles

The strategy used to detect obstacles needed to work in the case where prediction is used and in the control case where prediction is not used. We conditioned the data as follows. Animals normally fixate on aspects of the environment using eye movements. In our robot, the eyes are fixed to the body. To emulate fixation, we used a center patch in the image as a reference point. We then subtracted the velocity of this reference point from all of the other optic flow velocities in the image. In this way, we produced an approximation to a stable, fixed gaze in the environment. Under this strategy, we detected a sheering in the optic flow field.

2.4 Experimental Procedure

2.4.1 Statistics of the Optic Flow Vector

We first wanted to determine what the statistical distribution of the optic flow vector was during locomotion. We computed the optic flow field as described above. We then computed the angle of the flow vectors

$$\theta = a\tan 2(dy/dt, dx/dt) \tag{4}$$

and the magnitude

$$mag = \sqrt{(dx/dt)^2 + (dy/dt)^2} \tag{5}$$

We also collected data on the frequency at which each histogram value was chosen. We then computed magnitude and frequency histograms of this data. This information gives us some quantitative idea of the statistical variation of the flow field during normal locomotion.

2.4.2 Does the Use of Prediction Improve Obstacle Detection Performance?

To test the hypothesis that prediction of optic flow is necessary for effective detection of obstacles, we performed the following experiment. We tested the case of Prediction versus No-Prediction in detecting obstacles. The No-Prediction cases were the control group. We used five different height obstacles: 1, 2, 3, 4, and 5 cm. All obstacles were 3 cm wide and about 15 cm across.

For each obstacle, and within each group, 10 trials were performed that resulted in either a collision with the obstacle or the robot stopping short of the obstacle. Across five obstacles, and the prediction and the control cases, we performed a total of 100 experiments (5x2x10).

We also recorded false positives, or the number of times the robot stopped before encountering an obstacle. This number was very low in both groups.

3 RESULTS

3.1 Statistics of Optic Flow

Figure 6a shows the distribution of optic flow vector magnitudes versus direction (0 to 360 degrees). As can be seen, there is a significant probability

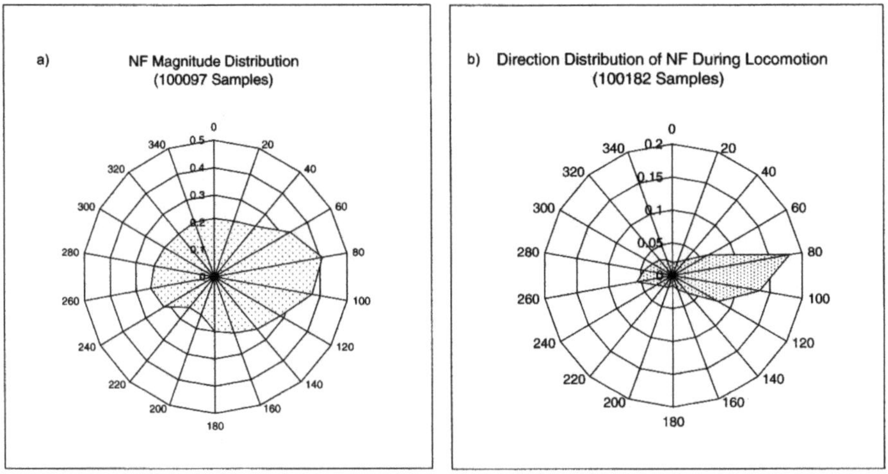

Figure 6. Statistical distribution of optic flow vectors during normal locomotion. In both figures, 90° indicates straight ahead. a) The average magnitude of optic flow versus vector angle. b) The probability of choosing any one vector at a given time.

of an optic flow vector taking on *any* possible direction. In Figure 6b we see however that the number of times vectors between 80 and 100 degree (in the direction of movement of the robot) are chosen is far higher than in other directions. However, there is a better than 50% probability that the robot has an optic flow vector not pointed in the major direction.

3.2 Use of Prediction

Figure 7 shows the results of the experiment comparing prediction versus control in the task of detecting anomalous surface features. As can be seen, the robot can detect obstacles reliably when the obstacle is about 4 cm tall in the control case. However, in the case where prediction is used, the results are much better. The obstacle is detected reliably even when the obstacle is 1 cm tall. These results indicate that the use of prediction significantly improves the detection of fine obstacles.

We also computed how many times the biped stopped when no obstacle was present. When prediction was used, it stopped seven times out of a total of 107 trials. In the control group, it stopped 2 times in 102 trials. Future experimentation will allow us to determine if this effect is significant, and the cause of this slight difference.

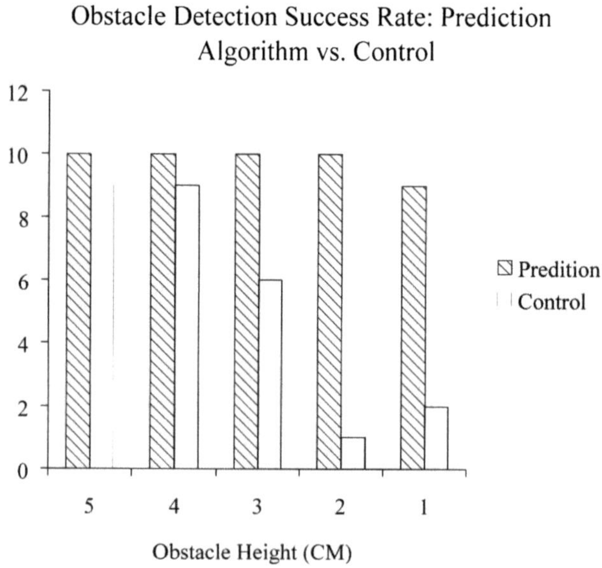

Figure 7. Obstacle detection success rate.

4 DISCUSSION

It is clear that motion information provides a valuable cue that can control a wide variety of behaviors, from self-stabilization versus the environment, and relevant here, to understanding when an obstacle might lie in a path. To identify that optic flow is a cue and to determine the mechanisms underlying how that cue is used is a difficult quest. Only the simplest control systems can be reduced to control laws for motion.

Patla has noted a qualitative difference in the use of vision for feedback control tasks, such as optomotor response, and for tasks involving highly non-linear mapping between sensory input and motor output and in tasks such as stepping over obstacles (Patla, 1997). Thus, we are forced to create more elaborate models to explain our observations. Biomorphic robots may provide a valuable modeling aid as they allow the simultaneous acquisition of real data from the environment as well as action on the environment. They may add a new dimension to modeling.

In the model presented here we found that optic flow is a viable cue for controlling locomotion. However, the raw optic flow signal is confounded by the inherent non-smooth movement of walkers. The use of prediction can effectively subtract out the consequences of self-motion.

The nervous systems of animals certainly use prediction as an essential adjunct of perception. An animal's sensory system is presented with volumes of sensory data. Some of that data results from changes in the surrounding environment. Yet significant sensory stimuli result from voluntary movement of the animal. Examples include: optic flow, tactile, and proprioceptive information from joints and muscles. The animal must disambiguate self-generated stimuli from environment-generated stimuli. This ability is ubiquitous in animals, from electrosense in fish (Bastian, 1998) to the prediction of tactile self-stimulation in humans (Blakemore et al., 1998, 1999). It seems likely that humans would rely on prediction to subtract out the optic flow consequences of self-movement. This is supported by the results presented here which clearly illustrate the advantages of using prediction.The model presented in this chapter draws upon this inspiration to produce a system level model of how optic flow might be incorporated into more complex motor tasks.

5 CONCLUSIONS

Our results lend compelling evidence to the idea that (1) it is possible to predict the optic flow based on the phase of gait and kinematic configuration of the robot. (2) This prediction is necessary (at least in this robot) for the reliable detection of surface features.

In previous work, we showed how a robot could learn to adjust its foot placement and timing to step over obstacles using stereopsis as a cue (Lewis & Simó, 2001). It is likely that optic flow can serve a similar function.

The results presented here are surprising. Optic flow signals can be relatively noisy. The detection of small amounts of changes in the optic flow field, especially with a walking robot, seems a difficult problem. Further work needs to be done to fully characterize the performance of this methodology. For example, what is the smallest object that can be reliably detected? Can the size of the obstacle be deduced from the optic flow signal? Finally, in our system, optic flow and stereopsis give redundant information about the environment. Under which conditions is a visual cue dominant in the task of avoiding obstacles and determining the transverability of the path immediately in front of the robot? What is the relationship between walking speed and visuomotor performance?

ACKNOWLEDGMENTS

The author wishes to thank Yoichiro Endo of the Georgia Institute of Technology for many interesting discussions related to this work, and for the

Linux vision setup and the vision processing code. The author thanks Richard Reeve and others at the 2001 Telluride Neuromorphic Engineering workshop for useful comments related to this work. This work was supported by the Office of Naval Research (ONR N0014-02-C0110 and N00014-99-0984).

REFERENCES

Bastian, J. (1998). Modulation of calcium-dependent postsynaptic depression contributes to an adaptive sensory filter, *J. Neurophysiol., 80(6),* 3352-3355.

Blakemore, S. J., Goodbody, S. J., & Wolpert, D. M. (1998). Predicting the consequences of our own actions: the role of sensorimotor context estimation, *J. Neurosci., 18(18),* 7511-8.

Blakemore, S. J., Wolpert, D.M., & Frith C. D. (1999). The cerebellum contributes to somatosensory cortical activity during self-produced tactile stimulation, *Neuroimage 10,* 448-459.

Coombs, D., Herman, M., Hong, T. H., & Nashman, M. (1998). Real-time obstacle avoidance using central flow divergence, and peripheral flow, *Trans. Rob. Automat., 14(1),* 49-59.

Franceschini, N., Pichon, J. M., & Blanes C. (1992). From insect vision to robot vision, *Phil. Trans. R. Soc. Lond. B, 337,* 283-294.

Gibson, J. J. (1958). Visually controlled locomotion and visual orientation in animals, *Br. J. Psychol., 49,* 182-194.

Harrison, R. R. (2000). An analog VLSI motion sensor based on the fly visual system, *CNS, Pasadena, California Institute of Technology.*

Lazzaro, J. & Wawrzynek, J. (1995). A multi-sender asynchronous extension to the AER protocol, *Proceedings of 1995 Conference on Advanced Research in VLSI.*

Lee, D. N. & Young, D. S. (1986). Gearing action to the environment, *Exp. Brain Res., 15,* 217-230.

Lewis, M. A. (1998). Visual navigation in a robot using zig-zag behavior, *Advances in Neural Information Processing Systems (NIPS*10), 10,* 822-828.

Lewis, M. A. (1999). Egomotion computation using template matching and inner product of optic flow filter, *Conference on Information Sciences and Systems, Baltimore, MD.*

Lewis, M. A. & Nelson, M. E. (1998). Look before you leap: peering behavior for depth perception, *Simulation of Adaptive Behavior, Zurich.*

Lewis, M. A. & Simó, L. S. (1999). Elegant stepping: a model of visually triggered gait adaptation, *Connect. Sci., 11(3/4),* 331-344.

Lewis, M. A. & Simó, L. S. (2001). Certain principles of biomorphic robots, *Auton. Rob., 11(3),* 221-226.

Patla, A. E. (1997). Understanding the roles of vision in the control of human locomotion, *Gait Post., 5,* 54-69.

Sobey, P. J. (1996). Active navigation with a monocular robot, *Biol. Cybern., 71,* 443-440.

Srinivasan, M. V. & Venkatesh, S. (1997). *From Living Eyes to Seeing Machines.* UK: Oxford University Press.

Weber, K., Venkatesh, S. & Srinivasan, M. V. (1996). Insect inspired behaviors for the autonomous control of mobile robots, *Proceedings of the 13th International Conference on Pattern Recognition.*

GLOSSARY

[This Glossary has been compiled by the editors from definitions submitted by the authors.]

2-D: Two-dimensional.

3-D: Three-dimensional.

Agent: Any autonomous self-propelled entity, including humans, animals, robots, and autonomous vehicles.

Allothetic Information: Information about translations and rotations of the body provided by external cues (e.g., visual and auditory stimulation).

Angle of Gaze (or Gain Field): A neuronal response of a visual cell modulated by the position of the eye in the orbit.

Angular Velocity: The rate of change of the angular position of an environmental element in the visual plane with respect to time.

Anterior-Posterior (A-P) Axis: The fore-aft axis of the trunk, running from back to front.

Apparent Concomitant Motion: The perceived motion of a stationary physical stimulus that is concomitant with translations of the observer's head.

Attractor: A region of state space toward which the trajectories of a *dynamical system* converge over time.

Bearing: A direction defined with respect to the A-P axis of the body.

Behavioral Dynamics: The time-evolution of an observable behavioral variable(s), such as the heading direction. It can be formalized as a *dynamical system.*

Bi-Circular Flow: Circular flow in which the motion vectors in one hemifield are directed clockwise and vectors in the other hemifield are directed anticlockwise (as introduced by van den Berg *et al.*, 2001).

L.M. Vaina, S.A. Beardsley and S.K. Rushton (eds.), Optic Flow and Beyond, 487–497.

Bifurcation: A change in the number or type of *attractors* and *repellors* defined by a *dynamical system* as the parameters of the system are varied.

Binocular Disparity: The slight differences in the two retinal images resulting from the differing viewpoint of the two eyes. It is important to distinguish between *absolute* disparity, which refers to the differences in the positions of the two retinal images of a given object, and *relative* disparity, which refers to the differences in the absolute disparities of different objects. Absolute disparity is the adequate stimulus for vergence eye movements whereas relative disparity is the adequate stimulus for stereopsis (depth perception).

Biological Motion: The complex pattern of movements exhibited by living animals and humans. In psychophysical tasks *point light stimuli* are typically used to characterize the perception of biological motion.

BOLD fMRI Signal: Measures the time-varying spatial maps of oxygen concentration in the brain.

Circular Flow: Those components of the motion vectors that are tangent to circular contours centered on a specific axis of translation

Coherence: In a random dot motion pattern refers to the percentage of dots moving in the indicated direction, while the rest move in random directions.

Complex Cells: A class of neurons in the primary visual cortex whose responses are partially invariant to the exact position of the stimulus in their receptive field.

Control Law: A mapping between informational variables and the control variables of the action system.

Curl: A scalar that describes the rate of rotation at a point in the flow field.

Differential Motion: The pattern of relative motion between environmental elements at different depths that is produced by translation of the eye. See *motion parallax*.

Direction Tuning: A property of neurons that respond most strongly to motion in a particular direction within their receptive field and whose response decreases as the motion deviates from this preferred direction.

Discriminability Parameter: A statistical parameter from Signal Detection Theory that characterizes the certainty of perceptual decisions.

Disparity Vergence Mechanism: A mechanism that generates vergence eye movements in response to binocular disparity and thereby helps to maintain binocular alignment on the object(s) of interest.

Divergence: A scalar that describes the rate of expansion at a point in the flow field. Note that the divergence is zero at the focus of expansion.

Dot Lifetime: The length of time or number of frames that a dot exists before it is removed from the display to reappear at a different location.

Dynamical System: A system of differential equations that characterizes the pattern of change in some variable(s). The solution to the set of equations describes the trajectory of the variable(s) in state space.

Efficient Visual Search: Occurs in a search task where the time to find the target remains constant or nearly constant as the number of distractors increases. This is also known as parallel search.

Extra-Retinal Signal: Used in the literature to refer to information about gaze/target direction that is not provided in the retinal image, but arises from other sources such as head-eye motion signals.

Face Area: A region in visual cortex that is selectively activated by face stimuli. Such an area has, for example, been reported in the *fusiform gyrus.*

Flow Field: The *velocity field* description of optic or retinal flow.

Flow General Neurons (FLO-G): Neurons able to distinguish a particular type of optic flow (i.e., radial vs. rotational motion).

Flow Lines: Lines through the flow field to which the velocity vectors are tangent at every point.

Flow Particular Neurons (FLO-P): Neurons able to distinguish a particular direction of optic flow (i.e., expansion vs. contraction).

Focus of Expansion (FOE): The fixed point in a radial pattern of optic flow from which velocity vectors diverge. It is produced by translation of the eye and specifies the direction of *heading*. Other types of fixed points may also occur in the retinal flow, e.g., a rotational center produced by rotation of the eye, or a spiral center produced by translation and rotation of the eye to track a point on a surface.

Functional Magnetic Resonance Imaging (fMRI) of the Brain: Magnetic resonance (MR) imaging uses radio waves and a strong magnetic field to provide clear and detailed pictures of internal organs and tissues. Functional magnetic resonance imaging (fMRI) uses MR signals to measure the quick, small metabolic changes that take place in an active part of the brain. fMRI is based on the increase in blood flow to the local vasculature that accompanies neural activity in the brain.

Gabor Filter: Consists of a sinusoidal function that is windowed by a Gaussian function. Such filters are localized in space as well as in the spatial frequency domain and are a well-established model for V1 neurons.

Gain Field: The modulation by the eye position of a cell's response to a visual stimulus. It describes the modulation of a retinal receptive field by a head-centric eye signal.

Gain Field Model: A heading scheme proposed by Beintema and van den Berg (1998) that uses eye velocity gain fields to compensate for the effect of eye rotation on the retinal flow.

Gaussian Radial Basis functions (RBF): Computational units used to model neurons tuned for complex stimuli. Its activation is described by a multi-dimensional Gaussian function.

Gaze Stabilization: The act of keeping a point in the environment centered on the eye's fovea by means of eye and/or head movement.

Heading: Defined as the instantaneous direction of travel. On a straight path it corresponds to the current destination, whereas on a curved path it is tangent to the path and continuously changing. "Absolute heading" refers to the direction of travel in 3D space, "relative heading" to the angle of the current heading with respect to an environmental object, and "nominal heading" to the categorical direction of travel to the left or right of an object or the fixation point.

Horizontal Connections: Connections between neurons at the same processing level of the visual system (i.e., within the same cortical visual area).

Idiothetic Information: Information about translations and rotations of the body provided by internal sensorimotor cues (e.g., vestibular and proprioceptive stimulation and efferent commands to the musculature).

Inefficient Visual Search: Occurs in a search task in which the time to find the target increases significantly with the number of distractors. This is also known as serial search.

Inferotemporal Cortex: A region in the temporal cortex that is involved in the recognition of objects and visual memory.

Instantaneous Heading: The current direction of travel. For a straight trajectory it will be the current destination, whereas for a curved trajectory it will be the tangent to the curve and constantly changing.

Invariance: Signifies the capability of the visual system to abstract visual recognition from certain properties of stimulus, like the size or the position of an object in the visual field.

Jitter Motion: Corresponds to dot motion along a given motion vector in which the motion direction switches randomly between two opposing directions.

Kinematic Models: Models of a human consist of multiple links (bones) that are connected by joints. Contrasted with dynamic models, kinematic models neglect the influences of masses and inertia.

Lamellar: Refers to a flow field with parallel flow lines.

Lateral Inhibition: Corresponds to the inhibition between neurons at the same processing level of the visual system (i.e., within the same cortical visual area).

Leaky Integrator Model: A strongly simplified model for the dynamics of the membrane potential of a neuron. The dynamics is given by a linear differential equation.

Locomotor: A generic means of transport, as opposed to the specification of, on foot, by bike or car.

Locomotor Axis: The axis of the body that corresponds to the instantaneous heading direction. It corresponds to the A-P axis of the body when walking forward, but can diverge from the A-P axis when walking sideways ("crabbing").

Maximum-Like Operation: Operation that results in an output signal that approximates the maximum of multiple input signals. Maximum computation can be approximated with physiologically plausible neural circuits.

Medial Superior Temporal Area (MST): An area in primate visual cortex whose neurons are selective for radial, circular, and/or planar fields of motion.

Medio-Lateral (M-L) Axis: The left-right axis of the trunk, running from shoulder to shoulder.

Middle Temporal Area (MT): An area in primate visual cortex whose neurons are selective for the direction and speed of visual motion.

Mirror Neurons: Neurons located in the prefrontal cortex of monkeys that respond when the animal prepares the execution of a certain motor action, or when it observes another individual during execution of the same action. Mirror neurons have been postulated to play an important role for imitation learning.

Motion Energy: The phase invariant motion information contained within a local region of the visual field.

Motion Morphing: An algorithm that continuously interpolates between different movement patterns, e.g., walking and running. Good morphing algorithms result in interpolations that are very similar to natural movements.

Motion-Opponency: Refers to the property of specific neurons that are velocity-selective in one part of their receptive field, and exhibit an antagonistic velocity-selectivity in their counter part.

Motion-Opponent Operators: Operators with two adjacent responsive regions: an excitatory region, which has a positive response to a given preferred stimulus and an inhibitory region, which has a negative response to the same stimulus.

Motion Parallax: As described by Helmholtz (1866) is the change in visual direction of a single environmental element with a displacement of the eye. The term is often used to refer to the relative motion between two elements at different depths in the same visual direction that is produced by translation of the eye. Such relative motion is not produced by rotation of the eye.

Motion Templates: Pattern detectors for image motion that evaluate the evidence for a global match by summing local motion sensor outputs (Perrone 1992).

Ocular Following Responses (OFR): The initial output of an ultra-rapid visual reflex that helps to maintain binocular alignment on the object(s) of interest by generating version eye movements in response to object motion in the plane of fixation, such as that which occurs when the moving observer looks off to one side.

Optic Flow: As described by Gibson (1958) is the pattern of change in visual directions of environmental elements at a moving point of observation. It can be represented as a pattern of motion on a fixed projection surface at the observation point (e.g., the retina of a moving observer).

Optokinetic Nystagmus (OKN): The pattern of eye movements elicited when an observer is placed inside a rotating drum. There are two components to the response: a slow (or smooth) component that tracks the drum's motion and a quick (or saccadic) component that resets the eyes when they become eccentric. The smooth tracking component in turn has two components with different dynamics: an early component with brisk dynamics that dominates the initial responses (OKNe) and a delayed component with sluggish dynamics that dominates the later responses (OKNd).

Otolith Organs: Vestibular end-organs that are embedded in the base of the skull and sense linear accelerations of the head.

Parallel Flow: The pattern of optic flow produced by lateral eye movements (e.g., smooth pursuit), where the images of all objects move in the same direction (i.e., parallel motion vectors).

Path: The trajectory of locomotion in space. A path that is currently linear, circular, or a higher-order curve might be extrapolated in space, but the shape of a path may also change over time.

Perceptually Directed Action: An action, such as reaching or walking, that is directed toward a target location that was initially perceived using vision, hearing, or touch but is no longer perceptually accessible during the action.

Point Light Stimuli: Introduced by Johansson (1973). They consist of a small number of dots that move like the joints of a human or an animal performing different actions, like walking.

Population Code: Refers to the representation of information (e.g., visual motion) across a large number of neurons. Within the population individual neurons typically have different but overlapping selectivities (e.g., different preferred directions) such that a subset of the neurons generally responds to a given stimulus.

Population Model: A neural network that encodes heading in a distributed way by summing over a population of optic flow neurons, each neuron receiving local motion inputs with synaptic strengths set according to a minimalization scheme (e.g., Lappe & Rauschecker 1993).

Prediction Motion (PM) Task: An experimental paradigm often used to study time-to-collision (TTC) judgments. A stimulus on collision course with an observer is displayed for some time and then occluded from view. Observers are required to indicate, typically be means of a button press or a catching movement, when the object would collide with them had it continued its motion.

Presynaptic Inhibition: The local inhibition of a specific input to a neuron by an inhibitory connection on the axon of the incoming neural fiber immediately preceding the synapse to the next neuron.

Proprioception: The information arising from muscle and joint receptors that gives rise the to sensation of limb position.

Radial Flow: The pattern of optic flow produced by the organism's forward or backward movement through the environment whose local vector components lie perpendicular to circles centered on a specific axis of translation (typically referred to as the focus of expansion).

Radial-Flow Vergence: The initial transient output of an ultra-rapid visual reflex that helps to maintain binocular alignment on objects that lie ahead by generating vergence eye movements in response to the expanding pattern of optic flow experienced by the moving observer who looks in the direction of heading.

Random Dot Cinematogram (RDC). A rapid sequence of random-dot stereograms that can be used to depict rigid and non-rigid motion of a two-dimensional or three-dimensional stimulus. Also commonly referred to as random dot kinematograms (RDKs) and dynamic random dot displays.

Rate of Expansion: The fraction by which the size of an object grows per unit of time.

Receptive Field: The region of the visual image to which a given neuron responds. For an optic flow responsive neuron it is the collection of image motions that influences the neuron' firing rate.

Reichardt Detector: A simplified model of local motion detection. The basic Reichardt detector multiplies the output signals of two receptors with different positions in the visual field, after delaying the output of one of them. The output of the detector is direction-selective. Modified versions of the Reichardt detector have been used to model direction-selective neurons in primary visual cortex.

Repellor: A region of state space from which the trajectories of a *dynamical system* diverge over time.

Retinal Flow: The motion of visible features of the environment as projected onto the retinal. As the orientation of the motion will change with the observers gaze, retinal flow may differ from optic flow in its classic sense.

Rotational Component of Retinal Flow: The motion field produced by rotation of the eye with respect to the environment. It has a *solenoidal* pattern, without sources or sinks. Pitch or yaw of the eye creates parallel or *lamellar* flow, whereas roll about the line of sight creates rotary or circular flow.

Rotation Problem: Refers to the problem of how the visual system recovers heading during simultaneous translation and rotation of the eye based on the retinal flow and/or extra-retinal signals.

Rotational Vestibuloöcular Reflex (RVOR): A reflex that uses inputs from the semicircular canals to help stabilize gaze by generating eye movements that compensate for rotations of the head.

Saccadic Eye Movements: Rapid conjugate eye movements that are used to transfer binocular gaze to objects of interest, thereby, bringing their retinal images into the foveas.

Semicircular Canals: Vestibular end-organs that are embedded in the base of the skull and sense angular accelerations of the head.

Signal Detection Theory (SDT): A statistical theory about the formation of perceptual decisions based on noisy signals.

Smooth Pursuit System: A voluntary visual tracking mechanism that is used to track moving objects with the eyes and can perform even when those objects move across a textured background.

Solenoidal: Refers to a flow field without sources (foci of expansion) or sinks (foci of contraction).

Somatosensory: A general term that refers to information about limb and body motion from the various body senses, including proprioceptive, cutaneous, haptic, and vestibular information.

Speed-Tuning: The response variation observed in motion responsive neurons when stimulated with flow fields of different speeds.

Splay: For a straight line on the ground plane, splay is the angle in the perspective view between the image of the line and the environmental vertical.

Splay rate: The temporal derivative of splay.

Tau: The inverse of the relative rate of change in a variable. For example, the ratio between the current visual angle of a surface and its rate of change specifies the first-order time-to-contact with that surface, under certain conditions.

Template Cells: Neurons that respond most strongly to a given pattern of image velocities within their receptive fields. Their responses decrease with increasing deviation of the input velocity pattern from the preferred pattern.

Time-to-Contact (TTC) or Tau Hypothesis: The ecological hypothesis that humans base their timing of collision events on the relative rate of optical expansion. Thus, they estimate contact time without making time or distance estimates.

Translational Component of Retinal Flow: The motion field produced by translation of the eye with respect to the environment. It has a radial pattern of motion with a *focus of expansion* in the instantaneous heading direction.

Translational Vestibuloöcular Reflex (TVO): A reflex that uses inputs from the otolith organs to help stabilize gaze by generating eye movements that compensate for translations of the head.

Uni-Circular Flow: Circular flow in which all motion vectors are directed either clockwise or anti-clockwise.

Velocity Field: An instantaneous representation of the optic or retinal flow. A two-dimensional position field represents the visual directions (azimuth and elevation) of environmental elements, with the optical motion at each location represented by a velocity vector. Each vector has a magnitude that corresponds to the optical speed of motion, and a direction that corresponds to the direction of motion.

Vergence: The disconjugate component of eye movements that shifts binocular fixation to new depth planes and is computed from the difference in the movements of the two eyes.

Version: The conjugate component of eye movements that shifts gaze within the plane of fixation and is computed by averaging the movements of the two eyes.

View Dependence: Refers to recognition performance that varies with the two-dimensional view of the recognized object. View dependence has been reported for the recognition of static objects, and for biological motion. The responses of some neurons in the *inferotemporal cortex* and the superior temporal sulcus have been shown to be view-dependent.

View-Tuned Neurons: Neurons in the *inferotemporal cortex* that encode individual views of three-dimensional objects.

Visually Directed Action: Action, such as reaching or walking, that is directed toward a target location that was initially seen but is no longer in sight during the action.

Visual Search: A task in which observers are required to locate a visual target within a field of non-target items referred to as distractors.

Index

L.M. Vaina, S.A. Beardsley and S.K. Rushton (eds.), Optic Flow and Beyond, 499–507.

243. G. Debrock and M. Hulswit (eds.), *Living Doubt.* Essays concerning the epistemology of Charles Sanders Peirce. 1994 ISBN 0-7923-2898-1
244. J. Srzednicki, *To Know or Not to Know.* Beyond Realism and Anti-Realism. 1994 ISBN 0-7923-2909-0
245. R. Egidi (ed.), *Wittgenstein: Mind and Language.* 1995 ISBN 0-7923-3171-0
246. A. Hyslop, *Other Minds.* 1995 ISBN 0-7923-3245-8
247. L. Pólos and M. Masuch (eds.), *Applied Logic: How, What and Why.* Logical Approaches to Natural Language. 1995 ISBN 0-7923-3432-9
248. M. Krynicki, M. Mostowski and L.M. Szczerba (eds.), *Quantifiers: Logics, Models and Computation.* Volume One: Surveys. 1995 ISBN 0-7923-3448-5
249. M. Krynicki, M. Mostowski and L.M. Szczerba (eds.), *Quantifiers: Logics, Models and Computation.* Volume Two: Contributions. 1995 ISBN 0-7923-3449-3
Set ISBN (Vols 248 + 249) 0-7923-3450-7
250. R.A. Watson, *Representational Ideas from Plato to Patricia Churchland.* 1995 ISBN 0-7923-3453-1
251. J. Hintikka (ed.), *From Dedekind to Gödel.* Essays on the Development of the Foundations of Mathematics. 1995 ISBN 0-7923-3484-1
252. A. Wiśniewski, *The Posing of Questions.* Logical Foundations of Erotetic Inferences. 1995 ISBN 0-7923-3637-2
253. J. Peregrin, *Doing Worlds with Words.* Formal Semantics without Formal Metaphysics. 1995 ISBN 0-7923-3742-5
254. I.A. Kieseppä, *Truthlikeness for Multidimensional, Quantitative Cognitive Problems.* 1996 ISBN 0-7923-4005-1
255. P. Hugly and C. Sayward: *Intensionality and Truth.* An Essay on the Philosophy of A.N. Prior. 1996 ISBN 0-7923-4119-8
256. L. Hankinson Nelson and J. Nelson (eds.): *Feminism, Science, and the Philosophy of Science.* 1997 ISBN 0-7923-4162-7
257. P.I. Bystrov and V.N. Sadovsky (eds.): *Philosophical Logic and Logical Philosophy.* Essays in Honour of Vladimir A. Smirnov. 1996 ISBN 0-7923-4270-4
258. Å.E. Andersson and N-E. Sahlin (eds.): *The Complexity of Creativity.* 1996 ISBN 0-7923-4346-8
259. M.L. Dalla Chiara, K. Doets, D. Mundici and J. van Benthem (eds.): *Logic and Scientific Methods.* Volume One of the Tenth International Congress of Logic, Methodology and Philosophy of Science, Florence, August 1995. 1997 ISBN 0-7923-4383-2
260. M.L. Dalla Chiara, K. Doets, D. Mundici and J. van Benthem (eds.): *Structures and Norms in Science.* Volume Two of the Tenth International Congress of Logic, Methodology and Philosophy of Science, Florence, August 1995. 1997 ISBN 0-7923-4384-0
Set ISBN (Vols 259 + 260) 0-7923-4385-9
261. A. Chakrabarti: *Denying Existence.* The Logic, Epistemology and Pragmatics of Negative Existentials and Fictional Discourse. 1997 ISBN 0-7923-4388-3
262. A. Biletzki: *Talking Wolves.* Thomas Hobbes on the Language of Politics and the Politics of Language. 1997 ISBN 0-7923-4425-1
263. D. Nute (ed.): *Defeasible Deontic Logic.* 1997 ISBN 0-7923-4630-0
264. U. Meixner: *Axiomatic Formal Ontology.* 1997 ISBN 0-7923-4747-X
265. I. Brinck: *The Indexical 'I'.* The First Person in Thought and Language. 1997 ISBN 0-7923-4741-2
266. G. Hölmström-Hintikka and R. Tuomela (eds.): *Contemporary Action Theory.* Volume 1: Individual Action. 1997 ISBN 0-7923-4753-6; Set: 0-7923-4754-4

267. G. Hölmström-Hintikka and R. Tuomela (eds.): *Contemporary Action Theory*. Volume 2: Social Action. 1997 ISBN 0-7923-4752-8; Set: 0-7923-4754-4
268. B.-C. Park: *Phenomenological Aspects of Wittgenstein's Philosophy*. 1998 ISBN 0-7923-4813-3
269. J. Paśniczek: *The Logic of Intentional Objects*. A Meinongian Version of Classical Logic. 1998 Hb ISBN 0-7923-4880-X; Pb ISBN 0-7923-5578-4
270. P.W. Humphreys and J.H. Fetzer (eds.): *The New Theory of Reference*. Kripke, Marcus, and Its Origins. 1998 ISBN 0-7923-4898-2
271. K. Szaniawski, A. Chmielewski and J. Woleński (eds.): *On Science, Inference, Information and Decision Making*. Selected Essays in the Philosophy of Science. 1998 ISBN 0-7923-4922-9
272. G.H. von Wright: *In the Shadow of Descartes*. Essays in the Philosophy of Mind. 1998 ISBN 0-7923-4992-X
273. K. Kijania-Placek and J. Woleński (eds.): *The Lvov–Warsaw School and Contemporary Philosophy*. 1998 ISBN 0-7923-5105-3
274. D. Dedrick: *Naming the Rainbow*. Colour Language, Colour Science, and Culture. 1998 ISBN 0-7923-5239-4
275. L. Albertazzi (ed.): *Shapes of Forms*. From Gestalt Psychology and Phenomenology to Ontology and Mathematics. 1999 ISBN 0-7923-5246-7
276. P. Fletcher: *Truth, Proof and Infinity*. A Theory of Constructions and Constructive Reasoning. 1998 ISBN 0-7923-5262-9
277. M. Fitting and R.L. Mendelsohn (eds.): *First-Order Modal Logic*. 1998 Hb ISBN 0-7923-5334-X; Pb ISBN 0-7923-5335-8
278. J.N. Mohanty: *Logic, Truth and the Modalities from a Phenomenological Perspective*. 1999 ISBN 0-7923-5550-4
279. T. Placek: *Mathematical Intiutionism and Intersubjectivity*. A Critical Exposition of Arguments for Intuitionism. 1999 ISBN 0-7923-5630-6
280. A. Cantini, E. Casari and P. Minari (eds.): *Logic and Foundations of Mathematics*. 1999 ISBN 0-7923-5659-4 set ISBN 0-7923-5867-8
281. M.L. Dalla Chiara, R. Giuntini and F. Laudisa (eds.): *Language, Quantum, Music*. 1999 ISBN 0-7923-5727-2; set ISBN 0-7923-5867-8
282. R. Egidi (ed.): *In Search of a New Humanism*. The Philosophy of Georg Hendrik von Wright. 1999 ISBN 0-7923-5810-4
283. F. Vollmer: *Agent Causality*. 1999 ISBN 0-7923-5848-1
284. J. Peregrin (ed.): *Truth and Its Nature (if Any)*. 1999 ISBN 0-7923-5865-1
285. M. De Caro (ed.): *Interpretations and Causes*. New Perspectives on Donald Davidson's Philosophy. 1999 ISBN 0-7923-5869-4
286. R. Murawski: *Recursive Functions and Metamathematics*. Problems of Completeness and Decidability, Gödel's Theorems. 1999 ISBN 0-7923-5904-6
287. T.A.F. Kuipers: *From Instrumentalism to Constructive Realism*. On Some Relations between Confirmation, Empirical Progress, and Truth Approximation. 2000 ISBN 0-7923-6086-9
288. G. Holmström-Hintikka (ed.): *Medieval Philosophy and Modern Times*. 2000 ISBN 0-7923-6102-4
289. E. Grosholz and H. Breger (eds.): *The Growth of Mathematical Knowledge*. 2000 ISBN 0-7923-6151-2

290. G. Sommaruga: *History and Philosophy of Constructive Type Theory*. 2000 ISBN 0-7923-6180-6
291. J. Gasser (ed.): *A Boole Anthology*. Recent and Classical Studies in the Logic of George Boole. 2000 ISBN 0-7923-6380-9
292. V.F. Hendricks, S.A. Pedersen and K.F. Jørgensen (eds.): *Proof Theory*. History and Philosophical Significance. 2000 ISBN 0-7923-6544-5
293. W.L. Craig: *The Tensed Theory of Time*. A Critical Examination. 2000 ISBN 0-7923-6634-4
294. W.L. Craig: *The Tenseless Theory of Time*. A Critical Examination. 2000 ISBN 0-7923-6635-2
295. L. Albertazzi (ed.): *The Dawn of Cognitive Science*. Early European Contributors. 2001 ISBN 0-7923-6799-5
296. G. Forrai: *Reference, Truth and Conceptual Schemes*. A Defense of Internal Realism. 2001 ISBN 0-7923-6885-1
297. V.F. Hendricks, S.A. Pedersen and K.F. Jørgensen (eds.): *Probability Theory*. Philosophy, Recent History and Relations to Science. 2001 ISBN 0-7923-6952-1
298. M. Esfeld: *Holism in Philosophy of Mind and Philosophy of Physics*. 2001 ISBN 0-7923-7003-1
299. E.C. Steinhart: *The Logic of Metaphor*. Analogous Parts of Possible Worlds. 2001 ISBN 0-7923-7004-X
300. To be published.
301. T.A.F. Kuipers: *Structures in Science Heuristic Patterns Based on Cognitive Structures*. An Advanced Textbook in Neo-Classical Philosophy of Science. 2001 ISBN 0-7923-7117-8
302. G. Hon and S.S. Rakover (eds.): *Explanation*. Theoretical Approaches and Applications. 2001 ISBN 1-4020-0017-0
303. G. Holmström-Hintikka, S. Lindström and R. Sliwinski (eds.): *Collected Papers of Stig Kanger with Essays on his Life and Work*. Vol. I. 2001 ISBN 1-4020-0021-9; Pb ISBN 1-4020-0022-7
304. G. Holmström-Hintikka, S. Lindström and R. Sliwinski (eds.): *Collected Papers of Stig Kanger with Essays on his Life and Work*. Vol. II. 2001 ISBN 1-4020-0111-8; Pb ISBN 1-4020-0112-6
305. C.A. Anderson and M. Zelëny (eds.): *Logic, Meaning and Computation*. Essays in Memory of Alonzo Church. 2001 ISBN 1-4020-0141-X
306. P. Schuster, U. Berger and H. Osswald (eds.): *Reuniting the Antipodes – Constructive and Nonstandard Views of the Continuum*. 2001 ISBN 1-4020-0152-5
307. S.D. Zwart: *Refined Verisimilitude*. 2001 ISBN 1-4020-0268-8
308. A.-S. Maurin: *If Tropes*. 2002 ISBN 1-4020-0656-X
309. H. Eilstein (ed.): *A Collection of Polish Works on Philosophical Problems of Time and Spacetime*. 2002 ISBN 1-4020-0670-5
310. Y. Gauthier: *Internal Logic*. Foundations of Mathematics from Kronecker to Hilbert. 2002 ISBN 1-4020-0689-6
311. E. Ruttkamp: *A Model-Theoretic Realist Interpretation of Science*. 2002 ISBN 1-4020-0729-9
312. V. Rantala: *Explanatory Translation*. Beyond the Kuhnian Model of Conceptual Change. 2002 ISBN 1-4020-0827-9
313. L. Decock: *Trading Ontology for Ideology*. 2002 ISBN 1-4020-0865-1

314. O. Ezra: *The Withdrawal of Rights*. Rights from a Different Perspective. 2002
ISBN 1-4020-0886-4
315. P. Gärdenfors, J. Woleński and K. Kijania-Placek: *In the Scope of Logic, Methodology and Philosophy of Science*. Volume One of the 11th International Congress of Logic, Methodology and Philosophy of Science, Cracow, August 1999. 2002
ISBN 1-4020-0929-1; Pb 1-4020-0931-3
316. P. Gärdenfors, J. Woleński and K. Kijania-Placek: *In the Scope of Logic, Methodology and Philosophy of Science*. Volume Two of the 11th International Congress of Logic, Methodology and Philosophy of Science, Cracow, August 1999. 2002
ISBN 1-4020-0930-5; Pb 1-4020-0931-3
317. M.A. Changizi: *The Brain from 25,000 Feet*. High Level Explorations of Brain Complexity, Perception, Induction and Vagueness. 2003 ISBN 1-4020-1176-8
318. D.O. Dahlstrom (ed.): *Husserl's* Logical Investigations. 2003 ISBN 1-4020-1325-6
319. A. Biletzki: *(Over)Interpreting Wittgenstein*. 2003
ISBN Hb 1-4020-1326-4; Pb 1-4020-1327-2
320. A. Rojszczak, J. Cachro and G. Kurczewski (eds.): *Philosophical Dimensions of Logic and Science*. Selected Contributed Papers from the 11th International Congress of Logic, Methodology, and Philosophy of Science, Kraków, 1999. 2003 ISBN 1-4020-1645-X
321. M. Sintonen, P. Ylikoski and K. Miller (eds.): *Realism in Action*. Essays in the Philosophy of the Social Sciences. 2003 ISBN 1-4020-1667-0
322. V.F. Hendricks, K.F. Jørgensen and S.A. Pedersen (eds.): *Knowledge Contributors*. 2003
ISBN Hb 1-4020-1747-2; Pb 1-4020-1748-0
323. J. Hintikka, T. Czarnecki, K. Kijania-Placek, T. Placek and A. Rojszczak † (eds.) *Philosophy and Logic In Search of the Polish Tradition*. Essays in Honour of Jan Woleński on the Occasion of his 60th Birthday. 2003 ISBN 1-4020-1721-9
324. L.M. Vaina, S.A. Beardsley and S.K. Rushton (eds.) *Optic Flow and Beyond*. 2004
ISBN 1-4020-2091-0

Previous volumes are still available.

KLUWER ACADEMIC PUBLISHERS – DORDRECHT / BOSTON / LONDON

MIX
Papier aus verantwortungsvollen Quellen
Paper from responsible sources
FSC® C105338

If you have any concerns about our products,
you can contact us on
ProductSafety@springernature.com

In case Publisher is established outside the EU,
the EU authorized representative is:
Springer Nature Customer Service Center GmbH
Europaplatz 3, 69115 Heidelberg, Germany

Printed by Libri Plureos GmbH
in Hamburg, Germany